European ...

STUDY ON EUROPEAN FORESTRY INFORMATION AND COMMUNICATION SYSTEM
REPORTS ON FORESTRY INVENTORY AND SURVEY SYSTEMS

Volume 1

Austria, Belgium, Denmark, Finland, France, Germany
Greece, Iceland, Ireland, Italy, Liechtenstein

Document

CONTENTS

Foreword........ 1

I COUNTRY REPORTS 3

Country Report for Austria 5
- Austrian Forest Inventory 14
- Austrian Forest Development Plan........ 69

Country Report for Belgium 75
- Walloon Region 79
- Flemish Region........ 108
- Brussels Region 118

Country Report for Denmark 119
- National Forest Statistics 120

Country Report for Finland 145
- National Forest Inventory of Finland 156
- Permanent Field Plots in the National Forest Inventory of Finland 209

Country Report for France 227
- National Forest Survey........ 236

Country Report for Germany 277
- National Forest Inventory, NFI 288
- Forest Health Monitoring........ 336
- Survey of Forest Soil Condition........ 345
- Forest Enterprise Survey 350
- Forest Inventory of the Federal State Nordrhein-Westfalen 354

Country Report for Greece 359
- National Forest Inventory 362

Country Report for Iceland 381

Country Report for Ireland 385
- Coillte's Forest Inventory........ 392
- Inventory of Private Woodlands, 1973 412

Country Report for Italy 423
- Italian National Forest Inventory........ 451
- Lombardia Region Forest Inventory 480
- Veneto Regional Forest Inventory 494
- Friuli-Venezia Giulia Regional Forest Inventory 511
- Multipurpose Forest Inventory of Liguria Region........ 526
- Toscana Regional Forest Inventory 556
- Emilia- Romagna Regional Forest Inventory 572
- Umbria Regional Forest Inventory........ 599
- Lazio Regional Forest Inventory 613

Valle D'Aosta Regional Forest Inventory 627
Italian National Inventory of Forest Damage 638

Country Report for Liechtenstein 647
National Forest Inventory of Liechtenstein 653

Country Report for Luxembourg 675
Local Forest Inventories in Luxembourg 680
National Forest Condition Assessment 686
Community Forest Condition Assessment 702
Long Term Survey of Forest Ecosystems 705
Forest Soils Monitoring Raster 706

Country Report for the Netherlands 711
National Forest Area Survey 1980-1983 721
Other Wooded Land Survey 1983-1984 736
Forest Functions Surveys 754
Forest Condition Survey 793

Country Report for Norway 799
National Forest Inventory/Monitoring of Forest Condition 807

Country Report for Portugal 861
National Forest Inventory 871

Country Report for Spain 905
National Forest Inventory of Spain 914

Country Report for Sweden 955
The Swedish National Forest Inventory 960

Country Report for Switzerland 1019
Swiss National Forest Inventory 1029

Country Report for The United Kingdom 1095
National Inventory of Woodlands and Trees - Woodlands Survey 1116
National inventory of Woodlands and Trees Survey of Small Woodlands and Countryside Trees 1142
Forestry Commission - Crop Inventory Survey (Subcompartment Database) 1169
Forest Service (NI) Subcompartment Database 1184
Forest Service (NI) Inventory Database 1199

Country Report for the Czech Republic 1219

Country Report for Hungary 1231

Country Report for Poland 1243

II COMPARATIVE STUDY 1265

Appendices

FOREWORD OF THE COMMISSION

Council Regulation (EEC) No 1615/89, extended by Council Regulation (EEC) No 400/94, established a European Forestry Information and Communication System (EFICS). The objective of EFICS is to collect comparable and objective information on the structure and operation of the forestry sector in the Community, and thus facilitate implementation and monitoring of the Community forestry provisions in force. To that end, the system must collect, coordinate, standardize and process data concerning the forestry sector and its development. It must also make use of information available in the Member States, in particular data contained in national forestry inventories, and of any database accessible at Community and international level.

Within this context, the Commission financed in 1996 a study of which the overall objective was to analyse in detail the statistical sources on forestry resources in the Member States of the European Union and to draw up proposals for obtaining data which would be mutually compatible and comparable, so as to be able to establish a reliable and consistent forestry statistics database at European level.

The European Forest Institute (EFI) was given the task of carrying out this study. In accordance with the specifications of the study, EFI provided the Commission with a document analysing and comparing the existing forestry inventory and survey systems in the Member States.

Because of the quality of the work carried out by EFI, we consider that its results should be distributed as widely as possible. We hope this publication will contribute to this objective, and that it will benefit the whole forestry community.

C. ANZ
Head of Unit VI-F.II.2

FOREWORD OF THE EUROPEAN FOREST INSTITUTE

European Forest Institute was given a planning task which will serve as a basis for future decisions on the establishment of the system. The one-year project, funded by the Directorate General VI of the European Commission, started in January 1996. The objectives were to analyse the differences in the national systems and to study the needs and possibilities to harmonise the existing information systems regarding forest resources. The analysis covers EU and EFTA countries: Austria, Belgium, Denmark, Finland, France, Germany, Greece, Iceland, Ireland, Italy, Liechtenstein, Luxembourg, the Netherlands, Norway, Portugal, Spain, Sweden, Switzerland and the UK. In addition to these, the inventory systems in Czech Republic, Hungary and Poland are reported shortly.

The following institutions were involved in preparing the country reports:

- Federal Forest Research Institute, Austria
- University of Gembloux, Belgium
- Forest and Nature Agency, Denmark
- Finnish Forest Research Institute (regional co-ordinator)
- Agence MTDA, France
- University of Freiburg, Germany (regional co-ordinator)
- NAGREF, Forest Research Institute of Thessaloniki, Greece
- Coillte Teoranta, Ireland
- Forest and Range Management Research Institute, Italy (regional co-ordinator)
- EFOR, Luxembourg
- Institute for Forest and Forest Products, the Netherlands
- NIJOS - Norwegian Institute for Agricultural and Forest Mapping, Norway
- Instituto Superior de Agronomia, Department of Forestry, Portugal
- Escuela Tecnica Superior de Ingenieros de Montes, Spain
- Swedish University of Agricultural Sciences, Sweden
- Swiss Federal Institute of Forest, Snow and Landscape Research, Switzerland (principal researcher)
- National Remote Sensing Centre Ltd. and Forestry Commission, UK
- Czech University of Agriculture, the Czech Republic
- Forest Management Planning Service of the Ministry of Agriculture, Hungary
- Forest Research Institute, Poland

This document includes the reports on the forest inventory systems in Europe. The reports have been comprised by using common guidelines for all countries. In the second part of the document, some most relevant aspects of the inventory systems have been compared.

We believe that this document can provide valuable basic information not only for EFICS, but also for the whole professional community involved with forest inventories or their results.

Risto Päivinen
Project Leader

Michael Köhl
Principal Researcher

I COUNTRY REPORTS

COUNTRY REPORT FOR AUSTRIA

Norbert Winkler
Institute of Forest Inventory
Federal Forest Research Centre Vienna
Vienna, Austria

Summary

2.1 Nomenclature

After the nonforest/forest decision is made in accordance with the forest area definition stated in the assessment instructions of the Austrian Forest Inventory, as many as 139 attributes have to be directly assessed in the field. Grouped by a subject area, 9 attributes refer to 'Geographic Regions', 3 to 'Ownership', 24 to 'Wood Production', 14 to 'Site and Soil', 12 to 'Forest Structure', 19 to 'Regeneration', 12 to 'Forest Condition', 19 to 'Accessibility and Harvesting', 9 to 'Attributes describing Forest Ecosystems', none to 'Non-wood Goods and Services' and 18 to 'Miscellaneous'.

2.2 Data sources

The data sources utilised in the forest resource assessment are field data and maps. All kinds of maps utilised for forest inventory purposes are needed to localize the tracts. The land register maps provide information on forest ownership also.

2.3 Assessment techniques

The sampling frame of the nationwide Austrian Forest Inventory is given by the forest area definition. Sampling units are a sample plot of 300 m^2, a fixed circular plot of 21 m^2 and a Bitterlich plot which have the same centre and are located in the corners of a tract. The sampling design is a sample grid pattern where tracts are systematically distributed over Austria in a regular network. At present the 5th inventory cycle, which is the second re-assessment of the permanent plots, is being carried out. Data collection is carried out by means of handheld computers, results of the inventory period are expected at the end of the year 1997.

2.4 Data storage and analysis

The assessment data are stored in the Oracle database management system of the Federal Forest Research Centre in Vienna, where inquiries concerning data should be addressed. The databank consists of 16 tables linked by concatenated key (6 digit tract number). The data stored are not updated to a common point.

Forest area estimations are done by multiplying the total area of the reference unit by the forest area proportion of the reference unit, which is derived from the forest area proportion of the sample plots on the reference unit. Since the angle-count method is used, the aggregation of single tree data is done by weighting each single tree attribute by the represented number of trees per hectare. Total values for the plot are obtained by summing up the individual single tree attributes. The value of a concerned attribute for a tract is considered to be independent from adjacent tracts. For sampling error calculation only variance between the tracts is considered. Growth estimation is done by use of the 'gross growth' equation. Each piece of stand and area-related data found on a plot is assigned to a proportion of the plot area. Estimates are obtained independently for reference units.

The analysis software are SQL and PL/SQL procedures as well as Fortran routines. Data transfer from handheld palmtop-computers to PC via user-written interface. All raw data and derived attributes are stored in an Oracle7 database, which is installed under VMS on a DEC-VAX, and they are not available for outsiders. Forest inventory results are available for all of Austria, the federal states of Austria and the regional forest authorities, but individual results for main productive regions and growth zones can be provided if requested. Analysis can be made for any thematic subunit. There are no major links to other information sources in the whole forest resource assessment.

2.5 Reliability of data

In order to obtain information on the reliability of the data, 1-2 % of the sample plots are visited a second time by another field crew and all attributes thus assessed are analysed at the office. On the one hand, the check assessments are used for feedback to field crews to improve data quality, and on the other hand they are used to revise the assessment instructions. Immediate plausibility check in the field while data is being fed into the handheld computer and another consistency check of data is carried out when the data are loaded into the database.

2.6 Models

Cubing functions have been developed for calculation of volume as well as volume increment of the individual sample tree. DBH as one of the input variables for the model is measured for each sample tree in the field. The other variables, upper diameter and total tree height, are calculated by means of the models described in this chapter.

2.7 Inventory reports

For all finished inventory cycles printed reports are available. The inventory results of the previous cycle are also disseminated on disc. The proportion of discs sold shows that the main users of the inventory results are authorities and in most cases all standard references are requested.

2.8 Future development and improvement plans

Since the next inventory period will not be planned in detail and directions will be obtained from the BMLF until the current inventory cycle has been completed, there is no sound information available at present. For reasons of continuity, no major changes are expected.

2.9 Miscellaneous

The general guidelines for assessment show that, depending on the forest management type (FMT), different sets of attributes must be assessed according to specific information requirements. As a result of different information needs there are selected lists for tree species assessment.

Shares of tree species in age class *(Baumarten in den Altersklassen)*
b) minimum: 10%-share within age class under review, a maximum of five tree species can be specified
c) %, 10% classes, shortened selected list of 6 coniferous and 2 broadleaf tree species and additional list of 1 coniferous group and 2 broadleaves groups.
d) rounded down to closest class limit
e) ocular estimate
f) field assessment

Tree species *(Baumart)*
b) 28 tree species and 14 general groups
c) field assessment
d) shrub species not included

Woody plants *(Holzgewächse)*
b) 73 tree, 52 shrub, 17 dwarf shrub species and 4 general groups
c) field assessment
d) —-

Shares of tree species in age class *(Baumarten in den Altersklassen)*
b) minimum: 10%-share within age class under review, a maximum of five tree species can be specified
c) %, 10% classes, shortened selected list of 6 coniferous and 2 broadleaf tree species and additional list of 1 coniferous group and 2 broadleaves groups.
d) rounded down to closest class limit
e) ocular estimate
f) field assessment

Shares of tree species in protection forest *(Baumarten im Schutzwald außer Ertrag)*
b) minimum: 10%-share of area under review forest management type: (1031) protection forest without yield - forested area, a maximum of five tree species can be specified
c) %, 10% classes, shortened selected list of 7 coniferous and 3 broadleaf tree species and additional list of 1 coniferous group, 2 broadleaves groups and 1 shrub group.
d) rounded down to closest class limit
e) ocular estimate
f) field assessment

1. OVERVIEW

1.1. GENERAL FORESTRY AND FOREST INVENTORY DATA

In a mountainous and tourism-oriented country such as Austria, the forest is not only an economic factor but also renders a multitude of irreplaceable services to human beings and the environment. The following facts and figures, based on Schieler et al. (1995) and ÖSTAT (1994), provide a general view of the Austrian forest:

the total forested area	3 878 000 ha
the proportion of forested land	46.2 %
the main tree species	60.9 % Norway spruce
	4.7 % Silver fir
	6.9 % European larch
	8.5 % Scots pine
	1.4 % other conifers
	9.1 % European beech
	2.3 % Oak species
	6.2 % other broadleaves
the average volume per hectare	292 m^3/ha
the average increment per hectare	9.4 m^3/ha
the ownership proportions (public versus private forests)	85 % private 15 % state owned
the number of inhabitants per hectare forested land	2.01 inhabitants/ha
smallest management unit (for forest inventory purposes)	500 m^2

The forests are not evenly distributed over the Austrian territory. The mountainous regions of the Alps are most densely covered with forests dominated by coniferous trees where a protective function rather than a commercial function of the forest predominates on steep slopes. The areas in the East, well suited for agricultural purposes, are particulary poor in forest land. Therefore environmental and recreational function become more important.

In order to obtain information on forest resources as well as non-wood services, two main systems of nationwide forest resource assessment have been established. On the one hand there is the Austrian Forest Inventory (AFI), focusing on quantitative features such as volume and growth, which has been further developed by collecting data about forest ecosystems in the current assessment period. The AFI will be discussed in detail in chapter 2.

name of the survey	Austrian Forest Inventory - AFI *(Österreichische Waldinventur)*
time of first assessment	AFI 1961-70 (systematic grid, temporary design)
time of last assessment	AFI 1992-96 (systematic grid, permanent design)
number of inventory cycles	5 cycles (second reassessment of permanent plots)
area covered by the AFI	Austrian territory

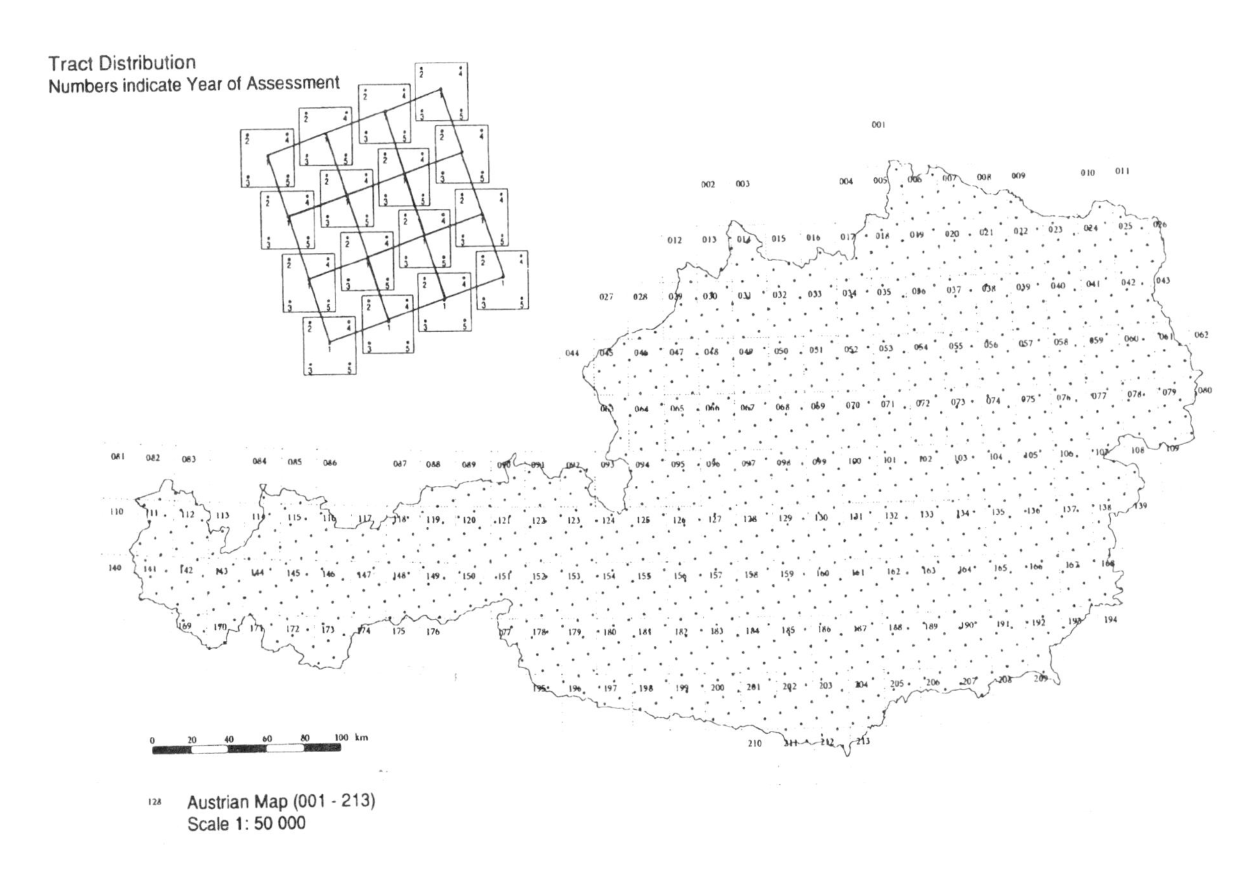

Figure 1. Outline map of the Austrian Forest Inventory - AFI

institutions and organisations involved in the assessments	Federal Forest Research Centre *(Forstliche Bundesversuchsanstalt FBVA)*, Institute of Forest Inventory
Institute of Forest Inventory tasks	nationwide Forest Resource Assessment (field assessment, data storage and analysis), provision of data for Forest Research
Institute of Forest Inventory resources	30 permanent employees
permanent staff	13 university graduates (Forestry) 12 foresters (Secondary Forestry School) 5 office personnel
temporary field crew members	7 assistants during field work period (April-October)
legal status of the assessment	Inventory is required by national forest law (adopted 1975), assessment by order of the Federal Ministry of Agriculture and Forestry *(Bundesministerium für Land- und Forstwirtschaft, BMLF)*

On the other hand is the Austrian Forest Development Plan (AFDP), designed as an instrument of forest regional planning, which is a more qualitative assessment. Basically, it is an evaluation of the importance of different forest functions, namely, 'utility' (timber production), 'protection', 'welfare' and 'recreation'. A brief outline of the AFDP will be given in chapter 3.

name of the survey	Austrian Forest Development Plan - AFDP *(Waldentwicklungsplan)*
time of first assessment	AFDP 1977-89
number of inventory cycles	2 cycles
area covered by the AFDP	Austrian territory
institutions and organisations involved in the assessments	Regional Forest Authority
Regional Forest Authority tasks and resources in connection with AFDP	field assessment for AFDP on district level done by the permanent staff
legal status of the assessment	Forest development plan is required by national forest law (adopted 1975), assessment by order of the Federal Ministry of Agriculture and Forestry *(Bundesministerium für Land- und Forstwirtschaft, BMLF)*

In addition, the Austrian Forest Damage Monotoring System (AFDMS) has been established in order to obtain information on forest condition. The surveys of forest condition (formerly called forest health assessments) are conducted according to the regulations of the International Cooperative Programme on Assessment and Monitoring of Air Pollution Effects on Forests (ICP).

1.2. OTHER IMPORTANT FOREST STATISTICS

1.2.1. Other forest data and statistics on the national level

All relevant information and data on forests and forestry on the national level which are generally available without restrictions are published in the annual Forest Report *(Österreichischer Waldbericht)* issued by the Federal Ministry of Agriculture and Forestry.

The Forest Report shows the following content:

statistics		data source
Forest Condition *(Waldzustand)*		
Forested Area Assessed by Land Register		BMLF
Land-use Class Assessed by Land Register		BEVW
Agricultural and Forest Enterprises Census		ÖSTAT-LFB
Interpretation of the Austrian Forest Inventory		AFI
Interpretation of the Austrian Forest Damage Monotoring System		AFDMS
Forest Protection and Forest Pests		BMLF
Forest Supervision		BMLF, ÖSTAT
The Economic Situation of Forestry *(Die wirtschaftliche Lage der Forstwirtschaft)*		
The Economic Situation in General		ÖSTAT, ÖIWF
Forest Returns		FBVA, HLFÖ, LBG, ÖBF, BOKU
Investments in Forests	Investments	FBVA, BMLF
	Promotion	BMLF
	Measures for the Rehabilitation of the Protection Function of Forests	BMLF
	Forest Training, Public Relations and Research	BMLF
Forest Utilization	Removals	BMLF
	Timber Prices	ÖSTAT - LFE, LHWR
	Timber Processing	ÖSTAT, FPI
	Timber Trade	ÖSTAT-AHÖ
Forest Personnel	Developments	ÖSTAT
	Accidents on the work site	SVAB, AUVA
The Torrent and Avalanches Control Service *(Wildbach- und Lawinenverbauung)*		
Investments		BMLF
Damages by Torrents and Avallanches		BMLF
The Structure of the Austrian Forestry *(Forstorganisation)*		
International Participations of the Austrian Forestry *(Internationale Agenden der österreichischen Forstwirtschaft)*		
Forest Damage Caused by Game and Domestic Animals *(Beeinträchtigung des Waldes durch Wild und Weidevieh)*		
Forest Damage Statistics on District Level		BMLF

BMLF: Bundesministerium für Land- und Forstwirtschaft, BEVW: Bundesamt für Eich- und Vermessungswesen, ÖSTAT-LFB: Österreichisches Statistisches Zentralamt, Land- und forstwirtschaftliche Betriebszählung 1980 und 1990, AFI: Austrian Forest Inventory, AFDMS: Austrian Forest Monotoring System, ÖSTAT: Österreichisches Statistisches Zentralamt, ÖIWF: Österreichisches Institut für

Wirtschaftsforschung, FBVA: Forstliche Bundesversuchsanstalt, Wien, HLFÖ: Hauptverband der Land- und Forstwirtschaftsbetriebe Österreichs, LBG - Wirtschafts- und Beratungsgesellschaft m.b.H., ÖBF: Österreichische Bundesforste, BOKU: Universität für Bodenkultur Wien, ÖSTAT-LFE: Österreichisches Statistisches Zentralamt - Land- und Forstwirtschaftliche Erzeugerpreise, LHWR: Landesholzwirtschaftsrat Steiermark, FPI: Fachverband der Papierindustrie, ÖSTAT-AHÖ: Österreichisches Statistisches Zentralamt - Der Außenhandel Österreichs, SVAB: Sozialversicherungsanstalt der Bauern, AUVA: Allgemeine Unfallversicherungsanstalt

Further information and data on national level concerning non-wood goods and services are provided by the Statistic Annual Report *(Statistisches Jahrbuch)* / Chapter Forestry issued by ÖSTAT. These are statistics on hunting areas, hunting supervisors and licenses, and game statistics.

1.2.2. Delivery of the statistics to UN and Community Institutions

1.2.2.1. Responsibilities for international assessments

The person responsible for delivery of the statistics to UN and Community Institutions is the head of the Forest Statistics Section located in the Federal Ministry of Agriculture and Forestry (BMLF), Department of Forestry (Dr. Albert Knieling, Stubenring 1, A-1010 Vienna, Tel: +43-1-21323-0). He has been responsible for providing forest resource data

a) to FAO/ECE for the 1990 Forest Resource Assessment (FRA 1990)
b) to EUROSTAT
c) to OECD (Environment Compendium) and
d) to UN-ECE for the European Timber Trends and Prospects Study V (ETTS V).

1.2.2.2. Data compilation

There is no special analysis conducted to meet the requirements of international assessments, but expert guesses are used to place a monetary value on non-wood goods and services in order to provide the requested figures.

2. AUSTRIAN FOREST INVENTORY

2.1. NOMENCLATURE

2.1.1. List of attributes directly assessed

A) Geographic regions

Attribute	Data source	Object	Measurement unit
Tract number *(Traktnummer)*	map	tract	number
Austrian map number *(Kartennummer)*	map	tract	number
Land register community *(Katastralgemeinde)*	map	tract	name
Location name *(Riedbezeichnung)*	map	tract	name
Main productive region *(Landw. Hauptproduktionsgebiet)*	map	sample plot	categorical
Growth zone *(Wuchsraum)*	map	sample plot	categorical
Height above sea level *(Meereshöhe)*	map	sample plot / intersection	m
Federal state *(Bundesland)*	map	sample plot	categorical
Regional forest authority *(Bezirksforstinspektion)*	map	sample plot	categorical

B) Ownership

Attribute	Data source	Object	Measurement unit
Ownership *(Eigentumsart)*	map	sample plot	categorical
Ownership of area being opened *(EA des Erschließungsgebietes)*	field assessment	intersection	categorical
Road ownership *(Besitzverhältnisse)*	field assessment	intersection	categorical

C) Wood production

Attribute	Data source	Object	Measurement unit
Volume estimation *(Vorratsscätzung)*	field assessment	sample plot	m^3
Diameter at breast height - DBH *(Brusthöhendurchmesser)*	field assessment	sample tree	mm
Upper diameter - D03 *(Oberer Durchmesser)*	field assessment	sample tree	mm
Forked tree *(Zwiesel)*	field assessment	sample tree	categorical
Tree height *(Baumhöhe)*	field assessment	sample tree	dm
Crown base *(Kronenansatzhöhe)*	field assessment	sample tree	dm
Tree species *(Baumart)*	field assessment	sample tree	categorical
Reserved tree / advance growth *(Überhälter / Vorwuchs)*	field assessment	sample tree	categorical
Dead standing tree *(Dürrling)*	field assessment	sample tree	categorical
Development stage *(Wuchsklasse)*	field assessment	sample tree	categorical
Age class *(Altersklasse)*	field assessment	sample tree	years
Stem damage *(Stammschädigung)*	field assessment	sample tree	categorical
Total number of peeling spots *(Anzahl der Schälwunden)*	field assessment	sample tree	number
Stem quality class *(Schaftgüteklasse)*	field assessment	sample tree	categorical
Crown position *(Baumklasse)*	field assessment	sample tree	categorical
Marking *(Auszeige)*	field assessment	sample tree	categorical
Top of crown *(Wipfelregion)*	field assessment	sample tree	categorical
Needle loss extent *(Kronenzustandsform)*	field assessment	sample tree	categorical
Type of needle loss *(Entnadelungstyp)*	field assessment	sample tree	categorical
Water sprout/mistletoe *(Mistel/Wasserreiser)*	field assessment	sample tree	categorical

Edge of stand *(Bestandesrand)*	field assessment	sample tree	categorical
Type of harvesting *(Nutzungsart)*	field assessment	sample tree	categorical
Stump rot *(Stockfäule)*	field assessment	sample tree	categorical
Underwood/overwood *(Unter-/Oberholz)*	field assessment	sample tree	categorical

D) Site and soil

Attribute	Data source	Object	Measurement unit
Slope exposure *(Neigungsrichtung)*	field assessment	sample plot	grade
Slope gradient *(Hangneigung)*	field assessment	sample plot	%
Relief *(Relief)*	field assessment	sample plot	categorical
Local climate *(Lokalklima)*	field assessment	sample plot	categorical
Water regime *(Wasserhaushalt)*	field assessment	sample plot	categorical
Soil depth *(Bodengründigkeit)*	field assessment	sample plot	cm
Fermentation layer *(Of-Horizont)*	field assessment	sample plot	cm
Humus layer *(Oh-Horizont)*	field assessment	sample plot	cm
Upper mineral soil *(Humoser Mineralboden)*	field assessment	sample plot	cm
Humustyp *(Humus type)*	field assessment	sample plot	categorical
Soil classification *(Bodengruppe)*	field assessment	sample plot	categorical
Ground vegetation type *(Vegetationstyp)*	field assessment	sample plot	categorical
Mass movements *(Bodenbewegungen)*	field assessment	sample plot	categorical
Avalanche path *(Lawinenstrich)*	field assessment	sample plot	categorical

E) Forest structure

Attribute	Data source	Object	Measurement unit
Nonforest-/forest shares *(Nicht-/Waldanteil)*	field assessment	sample plot	%
Forest management type - FMT *(Betriebsart)*	field assessment	sample plot	categorical
Forested area without timber yield *(Holzboden außer Ertrag)*	field assessment	sample plot	categorical
Development stage of stand*(Wuchklasse des Bestandes)*	field assessment	sample plot	categorical
Development stage shares *(Wuchsklasse in Zehntel)*	field assessment	sample plot	%
Stand structure *(Bestandesaufbau-vertikal)*	field assessment	sample plot	categorical
Age class of stand *(Altersklasse des Bestandes)*	field assessment	sample plot	years
Age class shares *(Altersklasse in Zehntel)*	field assessment	sample plot	%
Age sub-class *(Altersstufe)*	field assessment	sample plot	years
Tree species in age class *(Baumarten in den Altersklassen)*	field assessment	sample plot	categorical
Tree species in protection forest without yield *(Baumarten im Schutzwald außer Ertrag)*	field assessment	sample plot	categorical
Crown coverage *(Schlußgrad)*	field assessment	sample plot	categorical
Coppice forest alteration *(Ausschlagwald/Überführung)*	field assessment	sample plot	categorical

F) Regeneration

Attribute	Data source	Object	Measurement unit
Regeneration presence *(Verjüngung vorhanden)*	field assessment	sample plot	categorical
Regeneration necessity *(Verjüngungsnotwendigkeit)*	field assessment	sample plot	categorical

Reasons of absence *(Hemmfaktor für Verjüngung)*	field assessment	sample plot	categorical
Regeneration origin *(Entstehungsart der Verjüngung)*	field assessment	sample plot	categorical
Scale of regeneration area *(Flächenausmaß der Verjüngung)*	field assessment	sample plot	m^2
Regeneration distribution *(Verteilung der Verjüngung)*	field assessment	sample plot	categorical
Damage to regeneration *(Verjüngungsbeeinträchtigung)*	field assessment	sample plot	categorical
Crown coverage of regeneration *(Deckung)*	field assessment	sample plot	%
Height class *(Höhenklasse der Baumart)*	field assessment	sample plot	cm
Tree species distribution *(Mischform der Baumart)*	field assessment	sample plot	categorical
Browsing *(Flächenverbiß auf der Teilfläche)*	field assessment	sample plot	%
Dry plants *(Dürrlingsanteil)*	field assessment	sample plot	%
Number of dominant plants *(Hauptpflanzenanzahl)*	field assessment	sample plot	number
Age of dominant plant *(Pflanzenalter)*	field assessment	selected dominant tree	years
Height class of dominant plant *(Höhenklasse des Probebaumes)*	field assessment	selected dominant tree	cm
Protective measures *(Schutzmaßnahmen)*	field assessment	selected dominant tree	categorical
Browsing *(Verbiß)*	field assessment	selected dominant tree	categorical
Other damages *(Sonstige Schäden)*	field assessment	selected dominant tree	categorical
Further regeneration development *(erwartete Funktionserfüllung)*	field assessment	sample plot	categorical

G) Forest condition

Attribute	Data source	Object	Measurement unit
Shrubs inside stand *(Strauchanteil im Bestand)*	field assessment	sample plot	%
Fraying *(Fege- und Schlagschädigung)*	field assessment	sample plot	categorical
Harvesting damage *(Schäden durch Bewirtschaftung)*	field assessment	sample plot	%
Bark peeling damage *(Schälschädigung)*	field assessment	sample plot	%
Tending measures *(Pflegemaßnahmen)*	field assessment	sample plot	categorical
Disintegration phase *(Zerfallsphase)*	field assessment	sample plot	categorical
Influence on game *(Wildökologische Einflußgrößen)*	field assessment	sample plot	categorical
Forest grazing *(Aktuelle Beweidung)*	field assessment	sample plot	categorical
Tree distribution *(Bestandesform-Verteilung)*	field assessment	sample plot	categorical
Stand stability *(Bestandesstabilität)*	field assessment	sample plot	categorical
Formation phase *(Entwicklungsphase)*	field assessment	sample plot	categorical
Shares of dead standing trees *(Dürrlingsanteil der Wuchsklasse)*	field assessment	sample plot	%

H) Accessibility and harvesting

Attribute	Data source	Object	Measurement unit
Pre-logging distance *(Distanz bis Rückeweg)*	field assessment	tract main point	m
Skidding road distance *(Länge des Rückeweges)*	field assessment	tract main point	m
Forest road distance *(Länge der LKW-Straße)*	field assessment	tract main point	m
FMT of area being opened *(Betriebsart-Erschließungsgebiet)*	field assessment	intersection	categorical

Road condition *(Befahrbarkeitszustand)*	field assessment	intersection	categorical
Logging opportunities *(Wertigkeit für Holzabfuhr)*	field assessment	intersection	categorical
Road class *(LKW-Befahrbarkeit)*	field assessment	intersection	categorical
Driving surface *(Fahrbahnbreite)*	field assessment	intersection	m
Road surface *(Fahrbahndecke)*	field assessment	intersection	categorical
Road gradient *(Gefälle des Weges)*	field assessment	intersection	%
Site gradient *(Gefälle des Geländes)*	field assessment	intersection	%
Width of cut-/fillslope *(Böschungsbreite oben/unten)*	field assessment	intersection	m
Angle of cut-/fillslope *(Böschungneigung oben/unten)*	field assessment	intersection	%
Exposure of cut-/fillslope *(Böschungsexposition)*	field assessment	intersection	grade
Condition of cut-/fillslope *(Böschungszustand oben/unten)*	field assessment	intersection	categorical
Road drainage system *(Entwässerung des Weges)*	field assessment	intersection	categorical
Slides and erosion *(Rutschung und Erosion)*	field assessment	intersection	categorical
Driving surface coverage *(Bewuchs der Fahrbahn)*	field assessment	intersection	categorical
Road completion *(Baufertigstellung der Straße)*	field assessment	intersection	categorical

I) Attributes describing forest ecosystem

Attribute	Data source	Object	Measurement unit
Natural woodland community *(Natürliche Waldgesellschaft)*	map	sample plot	categorical
Feed for game *(Wildäsung)*	field assessment	sample plot	%

Mast / Litterfall *(Mast / Laubfall)*	field assessment	sample plot	%
Debris coverage *(Totholz-Flächendeckung)*	field assessment	sample plot	%
Origin of debris *(Totholzherkunft)*	field assessment	sample plot	categorical
Volume of dead lying trees *(Totholz-Volumen)*	field assessment	sample plot	m^3
Origin of dead lying trees *(Anthropogenes Totholz)*	field assessment	sample plot	%
Stage of decay *(Zersetzungsgrad)*	field assessment	sample plot	%
Woody plants *(Holzgewächse)*	field assessment	sample plot	categorical

J) Non-wood goods and services

No directly assessed attributes in this subject area.

K) Miscellaneous

Attribute	Data source	Object	Measurement unit
Year of assessment *(Jahr der Erhebung)*	field assessment	tract	year
Date of assessment *(Datum der Erhebung)*	field assessment	sample plot	date
Tract help point number *(Hilfspunkt auf Traktumfanglinie)*	field assessment	tract	number
Sample plot number *(Probeflächennummer)*	field assessment	sample plot	number
Plot centre location *(Punktort)*	field assessment	sample plot	categorical
Plot centre *(Punktart)*	field assessment	sample plot	categorical
Foreign territory *(Außer Staatsgebiet)*	field assessment	sample plot	%
Subplot number *(Teilflächennummer)*	field assessment	sample plot	number
Reason for subplots *(Teilungsgrund)*	field assessment	sample plot	categorical
Sample plot draft *(Teilungsskizze)*	field assessment	sample plot	categorical

Subplot existence *(Teilflächenexistenz)*	field assessment	sample plot	categorical
Means of age measurement *(Meßart der Alterstufe)*	field assessment	sample plot	categorical
Development stage of age sub-class *(Wuchsklasse d. Altersstufe)*	field assessment	sample plot	categorical
Actual ground vegetation *(Aktuelle Bodenvegetation)*	field assessment	sample plot	categorical
Text *(Freitext)*	field assessment	sample plot	categorical
Azimuth *(Azimut)*	field assessment	sample tree	grade
Horizontal distance *(Horizontalentfernung)*	field assessment	sample tree	m
Measurement height of DBH *(Meßhöhe des BHD)*	field assessment	sample tree	cm

2.1.2. List of derived attributes

A) Goegraphic regions

No derived attributes in this subject area.

B) Ownership

No derived attributes in this subject area.

C) Wood production

Attribute	Measurement unit	Input attributes
Tree number *(Baumnummer)*	number	in order of ascending azimuth
Ingrowth tree mark *(Einwuchskennzeichnung)*	number	comparison of current and previous azimuth and horizontal distance
Single tree volume *(Volumen)*	m^3	DBH, D03, tree height, tree species

D) Site and soil

No derived attributes in this subject area.

E) Forest structure

No derived attributes in this subject area.

F) Regeneration

No derived attributes in this subject area.

G) Forest condition

No derived attributes in this subject area.

H) Accessibility and harvesting

Attribute	Measurement unit	Input attributes
Road number *(Wegnummer)*	number	tract help points

I) Attributes describing forest ecosystems

No derived attributes in this subject area.

J) Non-wood goods and services

No derived attributes in this subject area.

K) Miscellaneous

Attribute	Measurement unit	Input attributes
Degree of difficulty *(Schwierigkeitsgrad)*	categorial	area-related data

2.1.3. Measurement rules for measurable attributes

For each of the measureable attributes the following information is given:
a) measurement rule
b) threshold values
c) measurement scale
d) rounding rules
e) instrument used for measurement
f) data source

A) Geographic Regions

Tract number (Traktnummer)
a) Co-ordinates of the tract main point on a map needed in order to find it.
b) —
c) six digit number, digits 1-3 refer to West-East location and digits 4-6 to South-North location of tract within the permanent sample grid
d) —
e) —
f) Austrian maps with grid knots drawn in

Austrian map number (Kartennummer)
a) Number of Austrian map (ÖK with scale of 1:50000) referring to a certain part of Austria where the tract under review is located.
b) —
c) three digit number (001-213)
d) —
e) —
f) outline map of Austria

Height above sea level (Meereshöhe)
a) Height above sea level of sample plot centre or of intersection of tract perimeter with road.
b) —
c) m, recorded in 100m classes above 200 m height above sea level (0-199m, 200-299m,...2400-2499m, 2500+)
d) —
e) altimeter
f) nearest triangulation point or other point where height above sea level available

B) Ownership

No measureable attributes in this subject area.

C) Wood Production

Volume estimation (Vorratsschätzung)

a) Volume of growing stock estimated by means of Denzin-formula ($V=DBH^2/1000$) in protection forest without timber yield if accessible.
b) minimum DBH of tree to be included in estimation: 20 cm
   assessed only in FMT (1031)
c) m^3
d) rounded to closest m^3
e) calliper and calculator, ocular estimation
f) field assessment

Diameter at breast height - DBH (Brusthöhendurchmesser)

a) Measured at 1.3m height above ground, on slopes measured from uphill side. One reading, calliper (right leg) pointing to plot centre, on slope calliper pointing downhill.
   Measuring point marked by drawing-pin.
b) minimum DBH: 50 mm
c) mm
d) rounded down to closest mm
e) calliper, girth measuring tape (for trees with DBH >= 600 mm)
f) field assessment

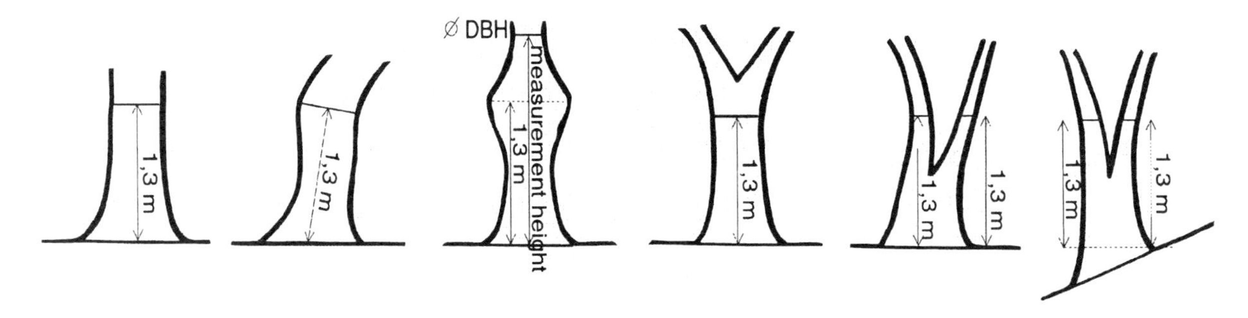

Figure 2. Measurement of DBH.

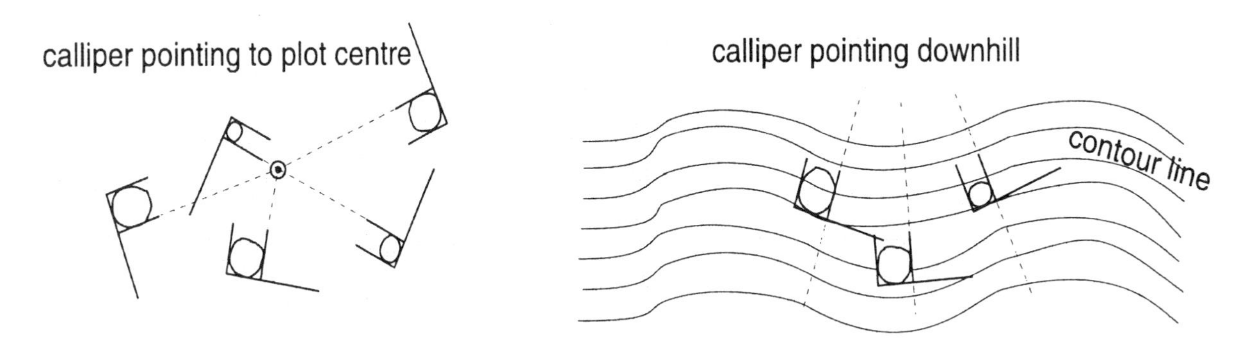

Figure 3. Measurement rule for DBH.

Upper diameter - D03 (Oberer Durchmesser)

a) Measured at 3/10 of tree height above ground, measured in same direction as DBH. One reading.
   Measured on a subset of sample trees (3 trees per plot) and each unforked ingrowth tree.
b) minimum ingrowth DBH: 105 mm
c) mm
d) rounded to closest mm
e) relascope
f) field assessment

Tree height (Baumhöhe)

a) Height of tree from ground level to top of tree. Measured on a subset of sample trees (3 trees per plot) and each ingrowth tree.
b) minimum DBH: 50 mm
c) dm
d) rounded to closest dm
e) relascope
f) field assessment

Crown base (Kronenansatzhöhe)

a) Length of stem from ground level to first living branch of crown. Only broadleaves. Measured on a subset of sample trees (3 trees per plot) and each ingrowth tree.
b) minimum DBH: 105 mm
c) dm
d) rounded to closest dm
e) relascope
f) field assessment

Figure 4. Measurement rule for crown base.

Age class (Altersklasse)

a) Age class to which tree under review belongs due to its estimated age
b) —
c) years, recorded in 20-year classes (1-20, 21-40,...,121-140,140+)
d) —
e) counting of branch whorls, annual ring counting on stumps, boring trees (but never sample trees!)
f) field assessment

Total number of peeling spots (Anzahl der Schälwunden)

a) Total number of peeling spots where not only bark is removed by bark peeling but also bast or cambium has been damaged.
b) maximum number: 12
c) number
d) —
e) counting
f) field assessment

D) Site and Soil

Slope exposure (Neigungsrichtung)

a) The exposure of a slope is the direction in which it faces.
b) —
c) grade, 50-grade classes designated in accordance with the cardinal points (north, north-east, east, ...north-west)
d) rounded to closest class middle
e) Suunto compass
f) field assessment

Slope gradient (Hangneigung)

a) Average slope gradient.
b) —
c) %, recorded in %-classes (0-5%, 6-10%, 11-20%, 21-30% ... 101-110%, > 110%)
d) —
e) Suunto clinometer
f) field assessment

Soil depth (Bodengründigkeit)

a) Soil depth assessed by digging with spade.
b) —
c) cm, three categories (0-30 cm deep, soil > 30 cm deep, others (assessment by digging impossible))
d) rounded to closest cm
e) ruler
f) field assessment

Fermentation layer (Of-Horizont)

a) Thickness of litter (F-layer), origin of most material easily identifiable.
b) —
c) cm
d) rounded to closest cm
e) ruler
f) field assessment

Humus layer (Oh-Horizont)

a) Thickness of humus (H-layer), origin of most material not easily identifiable.
b) —
c) cm
d) rounded to closest cm
e) ruler
f) field assessment

Upper mineral soil (Humoser Mineralboden)

a) Thickness of upper mineral soil (A-horizon), usually discoloured by incorporated humus.
b) —
c) cm
d) rounded to closest cm
e) ruler
f) field assessment

E) Forest structure

Nonforest / forest shares (Nicht-/ Waldanteil)

a) Total number of 10%-shares of sample plot area covered by forest.
b) Minimum: 10%-share of sample plot area
c) %, 10% classes
d) rounded down to closest class limit
e) measuring tape, Suunto compass
f) field assessment

Development stage shares (Wuchsklasse in Zehntel)

a) Total number of 10%-shares of sample plot area or subplot area with a certain development stage.
b) minimum: 10%-share of area under review
c) %, 10% classes
d) rounded down to closest class limit
e) ocular estimate, measuring tape, Suunto compass
f) field assessment

Age class of stand (Altersklasse des Bestandes)

a) Age class of stand or age class of groups formed by trees of same development stage on sample plot or subplot area.
b) maximum number of age classes on area under review: 3
c) years, recorded in 20-year classes (1-20, 21-40,...,121-140,140+)
d) —
e) counting of branch whorls, annual ring counting on stumps, boring trees (but never sample trees!)
f) field assessment

Age class shares (Zehntelanteil der angegebenen Altersklasse)

a) Total number of 10%-shares of sample plot area or subplot area covered by trees of a certain assessed age class.
b) minimum: 10%-share of area under review
c) %, 10% classes
d) rounded down to closest class limit
e) ocular estimate, measuring tape, Suunto compass
f) field assessment

Age sub-class (Altersstufe)

a) Exact age of trees in homogeneous stands.
b) —
c) years, recorded in 5years classes (1-5, 6-0,...,135-140,140+)
d) —
e) counting of branch whorls, annual ring counting on new stumps, boring trees (but never sample trees!)
f) field assessment

Tree species in age class
(Baumarten in den Altersklassen)

a) Total number of 10%-shares of a tree species within a certain assessed age class on sample plot area or subplot area.
b) minimum: 10%-share within age class under review
a maximum of five tree species can be specified
c) %, 10% classes, shortened selected list of 6 coniferous and 2 broadleaf tree species and additional list of 1 coniferous group and 2 broadleaves groups.
d) rounded down to closest class limit
e) ocular estimate
f) field assessment

Tree species in protection forest
(Baumarten im Schutzwald außer Ertrag)

a) Total number of 10%-shares of sample plot area or subplot area covered by a certain tree species.
b) minimum: 10%-share of area under review forest management type: (1031) protection forest without yield - forested area a maximum of five tree species can be specified
c) %, 10% classes, shortened selected list of 7 coniferous and 3 broadleaf tree species and additional list of 1 coniferous group, 2 broadleaves groups and 1 shrub group.
d) rounded down to closest class limit
e) ocular estimate
f) field assessment

F) Regeneration

Scale of regeneration area
(Flächenausmaß der Verjüngung)

a) For regeneration without shelter the area covered is assessed in order to get indications concerning scales of clear cuts.
b) —
c) m^2, in scale-classes (clearcut < 500 m^2, 500-1000 m^2, 1000-5000 m^2, > 5000 m^2)
d) —
e) ocular estimate, measuring tape
f) field assessment

Crown coverage of regeneration
(Deckung)

a) Ratio of area covered by crowns of young plants and total forested area.
b) —
c) %, %-categories:
1) some plants
2) numerous plants, but <= 5% coverage
3) 6 - 25%
4) 26 - 50%
5) 51 - 75%
6) 76 - 100%
d) —
e) ocular estimate
f) field assessment

Height class (Höhenklasse der Baumart)

a) Mean height of young plants seperately assessed for each tree species described on sample plot area or subplot area.
b) minimum height: 10cm, maximum height: 130cm
c) cm, cm-classes (10-30cm, 31-50cm, 51-80cm, 80-130cm)
d) —
e) ruler
f) field assessment

Browsing (Flächenverbiß auf der Teilfläche)

a) Ratio of young plants where terminal sprout is damaged by browsing to total number of young plants grouped by tree species on sample plot area or subplot area.
b) minimum of damaged trees: 1%
c) %, %-classes (1-50 %, 51-90%, > 90%)
d) —
e) counting, ocular estimate
f) field assessment

Dry plants (Dürrlingsanteil)

a) Ratio of dry plants to total number of plants grouped by tree species on sample plot area or subplot area.
b) minimum number of dry plants: 1%
c) %, in 10% classes (1-10%,11-20%,...)
d) —
e) counting, ocular estimate
f) field assessment

Number of dominant plants (Hauptpflanzenanzahl)

a) Total number of (pre-)dominant plants on sample plot area or subplot area.
b) minimum number: 1, maximum tree height: 1.3m
c) number, number-classes (1-10 plants, 11-25 plants, 26-50 plants, > 50 plants)
d) —
e) counting
f) field assessment

Age of dominant plants (Pflanzenalter)

a) Assessment of age of selected dominant plant (a maximum of 5 plants for each tree species) on sample plot area or subplot area.
b) —
c) years, in age sub-classes (1-5 years, 6-10 years, 11-20 years, 21-50 years, > 50 years)
d) —
e) counting of branch whorls
f) field assessment

Height class of dominant plant (Höhenklasse des Probebaumes)

a) Height of selected dominant plant from ground level to the top of plant (a maximum of 5 plants for each tree species).
b) minimum height: 10cm
c) cm, cm-classes (10-30cm, 31-50cm, 51-80cm, 80-130cm)
d) —
e) ruler
f) field assessment

G) Forest Condition

Shrubs inside stand (Strauchanteil im Bestand)

a) Total number of 10%-shares of sample plot area or subplot area covered by shrubs inside a stand.
b) minimum: 10%-share of area under review
c) %, 10% classes
d) rounded down to closest class limit
e) ocular estimate, (measuring tape, Suunto compass)
f) field assessment

Harvesting damage (Schädigungen durch Bewirtschaftung)

a) Damage to trees in stand under review on sample plot area or subplot area caused by felling, logging or falling rocks.
b) minimum number of damaged trees: 1
c) number, classes:
   1) one tree damaged on area under review
   2) up to 1/3 of trees damaged on area under review
   3) 1/3 to 2/3 of trees damaged on area under review
   4) more than 2/3 of trees damaged on area under review
d) —
e) counting
f) field assessment

Bark peeling damage (Schälschädigung)

a) Damage to trees in stand under review on sample plot area or subplot area caused by game as a result of bark peeling.
b) minimum number of damaged trees: 1
c) number, classes:
   1) one tree damaged on area under review
   2) up to 1/3 of trees damaged on area under review
   3) 1/3 to 2/3 of trees damaged on area under review
   4) more than 2/3 of trees damaged on area under review
d) —
e) —
f) field assessment

Shares of dead standing trees (Dürrlingsanteil der Wuchsklasse)

a) Total number of 10%-shares of sample plot area or subplot area covered by dead standing trees in protection forest without yield if accessible.
b) minimum: 10%-share of area under review
development stage: from regeneration II upwards
assessed only in FMT (1031)
c) %, 10% classes
d) rounded down to closest class limit
e) ocular estimate
f) field assessment

H) Accessibility and Harvesting

Pre-logging distance (Distanz bis Rückeweg)

a) Distance from tract main point to next skidding road, regardless of whether tract main point is forested or not.
b) minimum length: 5m
c) m, recorded in 10m classes and statement: uphill or downhill
d) rounded to closest class limit
e) measurement tape
f) field assessment

Skidding road distance (Länge des Rückeweges)

a) Distance from point where skidding road is joined to next forest road.
b) minimum length: 5m
c) m, recorded in 10m classes and statement: uphill or downhill
d) rounded to closest class limit
e) measurement tape, (car gauge)
f) field assessment

Forest road distance (Länge der LKW-Straße)

a) Distance from junction with forest road to next main road.
b) minimum length: 50m
c) m, recorded in 100m classes
d) rounded to closest class limit
e) car gauge
f) field assessment

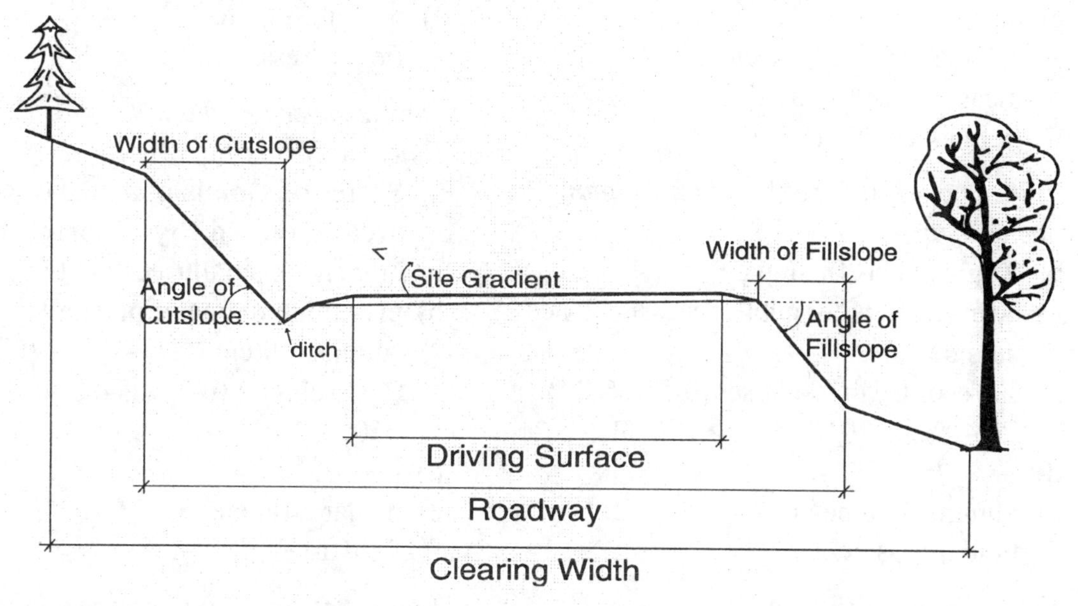

Figure 5. Road cross section

Driving surface (Fahrbahnbreite)

a) Driving surface measured at right angles to road centreline.
b) —
c) m, m-classes (width < 2m, 2-3m width, 3-5m width, > 5m width)
d) rounded down to lower category limit
e) measuring tape
f) field assessment

Road gradient (Gefälle des Weges)

a) Gradient of road measured from intersection of tract perimeter with road downhill.
b) —
c) %
d) —
e) Suunto clinometer
f) field assessment

Site gradient (Gefälle des Geländes)

a) Average gradient of site intersected by road.
b) no assessment for public roads and skidding roads
c) %
d) —
e) Suunto clinometer
f) field assessment

Width of cut-/fillslope (Böschungsbreite oben/unten)

a) Horizontal width from edge of driving surface to outside edge of cut-/fillslope.
b) no assessment for public roads and roads outside forest
c) m
d) rounded upwards to closest m
e) measurement tape
f) field assessment

Angle of cut-/fillslope (Böschungneigung oben/unten)

a) Angle of cut-/fillslope.
b) no assessment for public roads and roads outside forest
c) %, recorded in %-classes (0-5%, 6-20%, 21-40%, ... 201-220%, > 220%)
d) —
e) Suunto clinometer
f) field assessment

Exposure of cut-/fillslope (Böschungsexposition)

a) The exposure of a slope is the direction in which it faces.
b) no assessment for public roads and roads outside forest
c) grade, 50-grade classes designated in accordance with the cardinal points (north, north-east, east, ...north-west)
d) rounded to closest class midst
e) Suunto compass (400-grade scale)
f) field assessment

I) Attributes describing forest ecosystems

Feed for game (Wildäsung)

a) Total number of 10%-shares of sample plot area or subplot area covered by feed for game.
b) minimum: 10%-share of area under review
c) %, 10% classes combined stated with 12 categories of suitable feed for game
d) rounded down to closest class limit
e) ocular estimate
f) field assessment

Mast/Litterfall (Mast/Laubabfall)

a) Total number of 10%-shares of sample plot area or subplot area covered by tree species providing seasonal feed for game.
b) minimum: 10%-share of area under review
c) %, 10% classes
d) rounded down to closest class limit
e) ocular estimate
f) field assessment

Debris coverage (Totholz-Flächendeckung)

a) Shares of sample plot area or subplot area covered by debris (twigs, branches), small trees or stumps.
b) maximum diameter of small trees and stumps: 10 cm
c) %, %-classes (0-3%, 3-10%, 11-50%, >50%)
d) —
e) ocular estimate
f) field assessment

Volume of dead lying trees (Totholz-Volumen)

a) Volume of dead lying trees or parts of them, stumps, harvesting debris and/or obviously forgotten logs on sample plot area or subplot area. Mean diameter and length measured in order to calculate volume. Volume given for two classes.

b) minimum diameter: 10 cm
c) 0.01 m3, two classes (10-20 cm diameter, >= 20 cm diameter)
d) rounded to closest 0.01 m3
e) calliper, ocular estimate
f) field assessment

Origin of dead lying trees (Anthropogenes Totholz)

a) Man caused share of total volume of dead lying trees.
b) minimum diameter: 10 cm
c) %, recorded in % for two classes (10-20 cm diameter, >= 20 cm diameter)
d) —
e) calliper, ocular estimate
f) field assessment

Stage of decay (Zersetzungsgrad)

a) Stage of decay of dead lying trees or parts of them, stumps, harvesting debris and/or obviously forgotten logs on sample plot area or subplot area.
b) minimum diameter: 10 cm
c) %-shares of total dead lying tree volume within the categories:
   1) timber hard
   2) timber in peripherial parts of stem soft, stem centre hard
   3) timber in peripherial parts of stem hard,stem centre soft
   4) timber decomposed, completely soft
d) —
e) calliper, ocular estimate
f) field assessment

J) Non-wood goods and services

No measureable attributes in this subject area.

K) Miscellaneous

Year of assessment (Jahr der Erhebung)

a) Year of assessment determined by the inventory cycle.
b) —
c) actual year
d) —
e) —
f) Austrian map

NETWORK DEVELOPMENT

Numbers indicate Year of Assessment

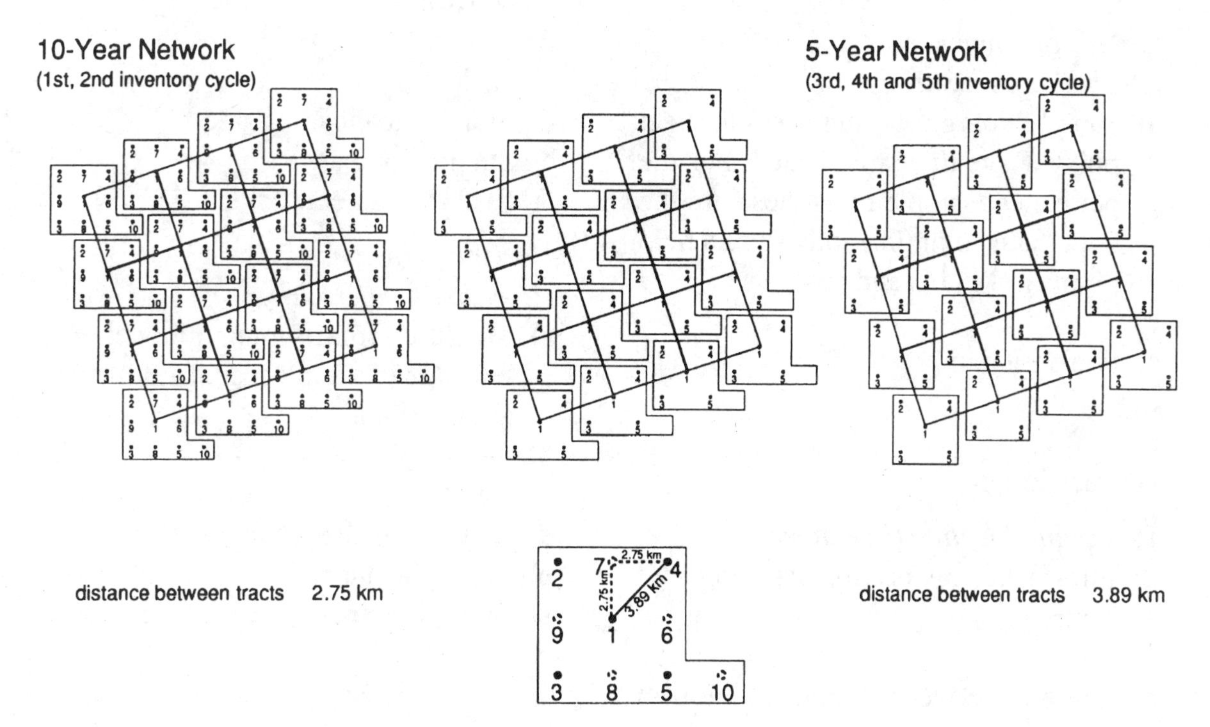

Figure 6. Year of assessment indicated by number.

Road number (Wegnummer)

a) Number of road according to its intersection with the tract perimeter.
b) —
c) number
d) —
e) —
f) road inventory data assessed

Date of assessment (Datum der Erhebung)

a) Date of assessment period predetermined by the date of former assessments.
b) —
c) actual date
d) —
e) —
f) calender

Tract help point number (Hilfspunkt auf Traktumfanglinie)

a) Number of tract help point according to point's location along tract perimeter.
b) —
c) two digit number corresponding to 25m classes (thp 01: 0.00- 24.99m,thp 02: 25-49.99m, ... thp 31: 775-799.99m)
d) rounded down to closest class limit
e) measurement tape
f) field assessment

Sample plot number (Probeflächennummer)

a) Sample plots get the number of the tract point where they are located. For the permanent sample plots these are the tract main point (00) and the tract help points (08), (16) and (24).
b) —
c) two digit number
d) —
e) —
f) tract design

Plot centre location (Punktort)

a) Information on location of sample plot centre.
b) —
c) number between 1 and 4 (subplot number)
d) —
e) —
f) field assessment

Foreign territory (Außer Staatsgebiet)

a) Total number of 10%-shares of sample plot area outside national territory.
b) minimum: 10%-share of sample plot area
c) %, 10% classes
d) rounded down to closest class limit
e) measuring tape, Suunto compass
f) field assessment

Subplot number (Teilflächennummer)

a) Subplots are subunits of a sample plot area established for different reasons.
b) maximal number of subplots on sample plot area: 4
minimum: 10%-share of sample plot area
c) number 1-4
d) —
e) ocular estimate, measuring tape, Suunto compass
f) field assessment

Azimuth (Azimut)

a) Measured from sample plot centre to left edge of sample tree at 1.3m height above ground.
b) —
c) grade
d) rounded to closest grade
e) Suunto compass
f) field assessment

Horizontal distance (Horizontalentfernung)

a) Measured from sample plot centre to left edge of sample tree at 1.3m height above ground.
b) —
c) cm
d) rounded to closest cm
e) measuring tape
f) field assessment

Measurement height of DBH (Meßhöhe des BHD)

a) Measurement height of DBH if measurement at 1.3m above ground is impossible.
b) —
c) cm
d) —
e) measuring tape
f) field assessment

Tree number (Baumnummer)

a) Number of sample tree according to its azimuth and horizontal distance.
b) —
c) number
d) —
e) —
f) sample tree data assessed

2.1.4. Definitions for attributes on nominal or ordinal scale

The following information is provided for each attribute:

a) definition
b) categories (classes)
c) data sources
d) remarks

A) Geographic Regions

Land register community (Katastralgemeinde)

a) Name of land register community/ies where tract is located.
b) list of names of land register communities
c) land register map

Location name (Riedbezeichnung)

a) Name of location which is a small area of land register community.
b) set of location names for the whole of a land register community
c) land register map

Main productive region (Landw. Hauptproduktionsgebiet)

a) List of names of 8 regions of similar characteristics within a region.
b) 8 regions
c) map
d) similar characteristics are landscape features, land-use pattern, rural settlement features etc.

Growth zone (Wuchsraum)

a) List of names of 21 zones of similar conditions for tree growth.
b) 21 growth zones
c) map
d) similar conditions are climatic conditions, geological-morphological composition and natural woodland communities within a zone

Federal state (Bundesland)

a) Name of federal states of Austria.
b) 9 federal states
c) map

Regional forest authority (Bezirksforstinspektion)

a) Name of regional forest authority
b) 86 districts
c) map
d) not always congruent with districts of administration

B) Ownership

Ownership (Eigentumsart)

a) Owner of the forest area under review.
b) 1) private ownership (< 200 ha)
   2) private ownership (200 - 1000 ha)
   3) private ownership (> 1000 ha)
   4) federal state
   5) political community (> 200 ha)
c) land register map, regional forest authority

Ownership of area being opened (EA des Erschließungsgebietes)

a) Owner of the forested area opened up by road under review.
b) categories see under 'Ownership'
c) land register map, regional forest authority

Road ownership (Besitzverhältnisse)

a) Ownership of road under review.
b) 1) public road owned by state, federal state or political community
   2) road owned by community of citizens, either with or without limited access
   3) private road, usually access controlled by the owner
c) land register map, regional forest authority

C) Wood Production

Forked tree (Zwiesel)

a) Information on whether stem of sample tree is forked or not.
b) yes/no
c) field assessment

Tree species (Baumart)

a) Selected list of scientific name of 10 coniferous and 17 broadleaf tree species and additional list of 5 coniferous groups and 9 broadleaf groups.
b) 28 tree species and 14 general groups
c) field assessment
d) shrub species not included

Reserved tree/advance growth (Überhälter/Vorwuchs)

a) Information on whether sample tree is a reserved tree or an advance growth.
b) 1) neither reserved tree nor advance growth
   2) reserved tree
   3) advance growth
c) field assessment

Dead standing tree (Dürrling)

a) Information whether sample tree is dead standing tree (almost or already dead) or not.
b) 0) none
   1) dead standing tree insignificant
   2) dead standing tree significant
c) field assessment

Development stage (Wuchsklasse)

a) Each sample tree has to be assigned to a certain development stage assessed on area under review.
b) categories see under 'Development stage of stand'
c) field assessment

Stem damage (Stammschädigung)

a) Damage to stem caused by various impacts.
b) 0) none
   1) fraying
   2) bark peeling
   3) felling/logging
   4) falling rocks
   5) either 3) or 4) if unclear
   6) top break
   7) others
   8) resin extraction
c) field assessment

Stem quality class (Schaftgüteklasse)

a) Stem quality according to given categories.
b) 1) stem upright, full-bodied, free of knots, without faults
   2) stem upright, full-bodied, with some knots or little faults
   3) stem crooked, knotty timber or stem conical, with bad faults
c) field assessment
d) assessed only for trees with DBH >= 205 mm

Crown position (Baumklasse)

a) Canopy layer to which sample tree under review belongs.
b) 5 categories (Kraft classification) for stand with one canopy layer and 9 categories for stand with more than one canopy layer.
c) field assessment

Marking (Auszeige)

a) Record of trees where removal would be recommendable.

b) 1) no marking/improvement felling
2) cleaning
3) selective thinning
4) low thinning
5) felling to initiate natural regeneration
6) removal of remaining trees
7) clearing of junk (damaged or dead standing trees)

c) field assessment

Top of crown (Wipfelregion)

a) Needle loss assessment within the top of crown of selected sample trees for tree species mentioned below.

b) 9 categories for Norway spruce, 5 categories for European silver fir and 6 categories for Scots pine and Austrian pine.

c) field assessment

d) assessed only on predominant and dominant sample trees for stands > 40 years and development stage from pole stage upwards

Needle loss extent (Kronenzustandsform)

a) Needle loss assessment for whole crown of selected sample trees for tree species mentioned below.

b) 5 categories (from no needle loss to dead crown) for Norway spruce, European silver fir and pine species

c) field assessment

d) assessed only on predominant and dominant sample trees for stands > 40 years and development stage from pole stage upwards

Type of needle loss (Entnadelungstyp)

a) Type of needle loss for whole crown of selected sample trees for tree species mentioned below.

b) 7 types of needle loss for Norway spruce and European silver fir

c) field assessment

d) assessed only on predominant and dominant sample trees for stands > 40 years and development stage from pole stage upwards

Water sprout/misteltoe (Mistel/Wasserreiser)

a) Presence and extent of water sprouts and/or misteltoe on selected sample trees for tree species mentioned below.

b) 3 categories for pine species concerning misteltoe (none, some, many), 9 categories for European silver fir (combination of categories for misteltoe and water sprouts)

c) field assessment

d) assessed only on predominant and dominant sample trees for stands > 40 years and development stage from pole stage upwards

Edge of stand (Bestandesrand)

a) Information whether selected sample tree is adjacent to open edge of stand or not. Information given for Norway spruce, European silver fir and pine species.

b) edge of stand with/without wind-firm-margin:
1)tree at the edge of stand
2) tree within 25 m of edge of stand
3) tree more than 25 m from edge of stand

c) field assessment

d) assessed only on predominant and dominant sample trees for stands > 40 years and development stage from pole stage upwards

Typ of harvesting (Nutzungsart)

a) Reason for disappearance of sample tree.

b) 1) natural disappearance
2) clear felling > 500 m^2
3) cleaning
4) selective thinning
5) low thinning
6) removal of remaining trees
7) clearing of junk (damaged or dead standing trees)
8) selection felling and clear felling < 500 m^2
9) incidental felling

c) field assessment

Stump rot (Stockfäule)

a) Supposed condition of stump when sample tree was felled.
b) 1) stump disappeared, no judgement possible
   2) no rottenness
   3) hard rottenness
   4) soft rottenness
c) field assessment

Underwood/overwood (Unter-/ Oberholz)

a) Information given in coppice forest whether sample tree belongs to underwood or overwood.
b) 1) underwood: covers the whole area under review, rotation time from 10 to 30 years
   2) overwood: some valueable trees in most cases regularly distributed over area under review, rotation time at least twice of underwood rotation time
c) field assessment
d) only assessed in coppice forest

D) Site and Soil

Relief (Relief)

a) Topography of sample plot area or its subunits.
b) 1) upper slope and ridge
   2) mid slope
   3) lower slope
   4) ephemeral areas
   5) riverside
   6) flat
   7) depression
c) field assessment

Local climate (Lokalklima)

a) Local climatic features of area under review.
b) 1) no local climatic features
   2) local climate of high humidity
   3) cold local climate
   4) duration of snow cover above average
   5) sub-average duration of snow cover
c) field assessment
d) categories 4) and 5) only assessed in subalpine regions

Water regime (Wasserhaushalt)

a) Water regime assessed by means of indicators.
b) 5 categories from dry to wet
c) field assessment
d) indicators are relief, soil features, slope exposure and indicator plants

Humus type (Humustyp)

a) Name of humus type.
b) 6 types
c) field assessment

Soil classification (Bodengruppe)

a) Name of soils derived from soil types and parent materials.
b) 26 categories
c) field assessment

Ground vegetation type (Vegetationstyp)

a) Selected list of names of 38 vegetation types and 19 additional general groups.
b) 38 specific types and 19 general groups
c) field assessment

Mass movements (Bodenbewegungen)

a) Information on directly or indirectly recognized mass movements.
b) 1) slides
   2) erosion
c) field assessment
d) assessed only in protection forest except FMT (1034)

Avalanche path (Lawinenstrich)

a) Information on occurence of avalanches by assessment of avalanche path >10 m width.
b) 1) starting zone above actual timber line
   2) starting zone underneath actual timber line
   3) avalanche path with defenses
c) field assessment
d) assessed only in protection forest except FMT (1034)

E) Forest structure

Forest management type - FMT (Betriebsart)

a) The forest management type gives information on treatment of stand considering the whole rotation time including course of regeneration.

b) key of forest management types:

1) high forest	0) ---	1) production forest	1) forested area 2) forested area without timber yield 3) shrub area
	0) ---	2) protection forest with yield	1) forested area 2) forested area without timber yield 3) shrub area
	0) ---	3) protection forest without yield	1) forested area 2) forested area without timber yield 3) shrub area 4) unsuitable for walking
2) coppice forest	1) far from stream	1) production forest	1) forested area 2) forested area without timber yield 3) shrub area
	2) adjacent to stream	1) production forest	1) forested area 2) forested area without timber yield 3) shrub area

c) field assessment

d) 4 digit FMT-number, digit 1 refers to the main type, digit 2 to specific location, digit 3 to the function and digit 4 refers to type of forested area

Forested area without timber yield (Holzboden außer Ertrag)

a) Type of forested area without timber yield.

b) Absence of timber yield due to forest management:
1) forest roads if clearing width >= 5 m
2) permanent landing
3) nursery
4) fields for game, feeding of game
5) ride
6) pre-skidding track
Absence of timber yield not found in forest management:
7) routes of mains
8) skiing grounds if width < 10 m
9) areas surrounding springs
Absence of timber yield due to natural reasons or reasons of nature conservation:
10) avalanche path if width >= 10 m
11) virgin forest, nature reserve or wildlife preserve, Heißland unter 500 m2

c) field assessment

Development stage of stand (Wuchsklasse des Bestandes)

a) Development stage of stand or group of trees on sample plot area or subplot area.

b) 1) blank (> 500 m^2)
2) blank with shrubs (50 - 500 m^2)
3) blank without shrubs (50 - 500 m^2)
4) regeneration I: tree height < 1.3 m
5) regeneration II: tree height > 1.3 m
6) pole stage: DBH 105-204 mm
7) timber I: DBH 205-354 mm
8) timber II: DBH 355-504 mm
9) big timber: DBH => 505 mm

c) field assessment

Stand structure (Bestandesaufbau-vertikal)

a) Discription of stand by total number of crown layers.
b) 1) one crown layer
   2) two crown layers
   3) no layers / more than two layers
c) field assessment

Crown coverage (Schlußgrad)

a) Ratio of area covered by tree crowns and total forested area.
b) 1) light
   2) thin
   3) close
   4) dense
c) field assessment

Coppice forest alteration (Ausschlagwald/Überführung)

a) Information on change in forest management type from coppice forest to high forest in stand under review.
b) 1) current alteration
   2) no alteration
c) field assessment
d) only assessed in coppice forest

F) Regeneration

Regeneration presence (Verjüngung vorhanden)

a) Presence of regeneration on sample plot.
b) 1) no:
   - no plants
   - only plants < 10 cm in height
   - number of plants under threshold value
   2) yes:
   number of plants exceed threshold value
c) field assessment
d) minimum number of plants varies according to plant height for area under review

Regeneration necessity (Verjüngungsnotwendigkeit)

a) Information on whether regeneration is desirable or not in current stage of development on sample plot area or subplot area.
b) yes/no decision
c) field assessment
d) natural pattern of stand development and actual stand condition have to be considered

Reasons for absence (Hemmfaktor für Verjüngung)

a) Reason for absence of regeneration.
b) 1) deficiency of light
   2) competition of grass, herbs or shrubs
   3) high litter layer, undeveloped soils (raw humus, raw soil)
   4) deficiency of mature trees producing seeds
   5) forest pasture
   6) browsing by game
   7) soil erosion
   8) unfavourable microclimate
   9) new felling area
   10) other reasons
c) field assessment
d) a maximum of 3 reasons per sample plot area or subunit (major ones)

Regeneration origin (Entstehungsart der Verjüngung)

a) Origin of regeneration.
b) 1) natural regeneration
   2) artificial regeneration
   3) natural regeneration supplemented by planted trees
   4) artificial regeneration supplemented by natural
c) field assessment

Regeneration distribution (Verteilung der Verjüngung)
a) Distribution of all plants on area under review.
b) 1) single-tree
2) groups
3) whole area
c) field assessment

Damage to regeneration (Beeinträchtigung der Verjüngung)
a) Impacts causing damage to regeneration.
b) 0) none
1) deficiency of light
2) competition of grass, herbaceous plants or shrubs
3) high litter layer, undeveloped soils (raw humus, raw soil)
4) forest grazing
5) timber harvesting
6) skiing, tourism
7) snow, frost
8) erosion (falling rocks)
9) unfavourable micro/macroclimate
10) herbicides
11) chlorotic colour changes
12) others (e.g. fungus, insects etc.)
c) field assessment
d) a maximum of 3 reasons per sample plot area or subunit (major ones)

Tree species distribution (Mischform der Baumart)
a) Distribution of certain tree species on area under review.
b) 1) single-tree
2) groups (also rows)
3) whole area
c) field assessment

Protective measures (Schutzmaßnahmen)
a) Measures in order to protect regeneration or selected tree species from damage by game or grazing animals.
b) 1) none
2) chemical (single-tree)
3) mechanical (single-tree)
4) fence (single-tree)
5) fence (area)
6) uncertain
c) field assessment

Browsing (Verbiß)
a) Damage to plants caused by browsing (terminal and axillary buds or shoots).
b) 9 categories according to severity of damage
c) field assessment

Other damages (Sonstige Schäden)
a) biotic damages other than browsing.
b) 0) none
1) fraying
2) bark peeling
3) insects, fungus
4) other
c) field assessment

Further regeneration development (Erwartete Funktionserfüllung)
a) Question whether distribution and condition of regeneration meet expectations or not.
b) 1) yes
2) uncertain
3) no (insufficient number of plants)
4) no (unsatisfactory distribution)
5) no (unsatisfactory mixture)
6) no (combination of 3-5)
c) field assessment
d) —

G) Forest Condition

Fraying (Fege- und Schlagschädigung)
a) Damage to trees by fraying on area under review.
b) 1) none
2) damage by fraying
c) field assessment

Tending measures (Pflegemaßnahmen)
a) Proposal of tending measures in order to reach silvicultural targets.
b) 1) weeding
2) cleaning, tending
3) selective thinning
4) low thinning
5) felling to initiate natural regeneration
6) removal of remaining trees

7) clearing of junk (damaged or dead standing trees)
c) field assessment

Disintegration phase (Zerfallsphase)
a) Information on stage of disintegration assessed by means of features indicative of starting or progressive disintegration.
b) 1) no disintegration
2) beginning disintegration
3) advanced disintegration
c) field assessment
d) assessed only in high forest - production forest (1011)

Influence on game (Wildökologische Einflußgrößen)
a) Measurements on forest area under review suitable to affect the behaviour of game.
b) 0) none
1) forest grazing
2) area under review inside fenced wintering area
3) area under review fenced
c) field assessment

Forest grazing (Aktuelle Beweidung)
a) Actual influence on stand and soil by forest grazing.
b) 1) none
2) forest grazing
c) field assessment
d) assessed only in protection forest except FMT (1034)

Tree distribution (Bestandesform-Verteilung)
a) Spatial distribution of trees on the area under review
b) 1) uniform distribution
2) trees in small clusters or groups
3) individual trees
c) field assessment
d) assessed only in protection forest except FMT (1034)

Stand stability (Bestandesstabilität)
a) Stand stability concerning the next twenty years must be assessed.
b) 1) stable
2) stable-labile
3) labile-critical
4) critical-unstable
5) areas covered by Dwarf pine or Green alder
c) field assessment
d) assessed only in protection forest except FMT (1034)

Formation phase (Entwicklungsphase)
a) Information on formation phase in protection forest.
b) 1) juvenile phase
2) initial phase
3) optimum phase
4) terminal phase
5) disintegration phase
6) overmature phase
7) regeneration phase
8) areas covered by dwarf pine or green alder
c) field assessment
d) assessed only in protection forest except FMT (1034)

H) Accessibility and harvesting

Forest management type of area being opened (Betriebsart des Erschließungsgebietes)
a) Forest management type of the forested area opened up by the road
b) categories see under 'FMT'
c) field assessment

Road condition (Befahrbarkeitszustand)
a) Information on road condition to show changes.
b) 0) similar to last review
1) road abondoned
2) old road replaced by new road
3) road unpassable due to erosion, slides etc.
c) field assessment

Logging opportunities (Wertigkeit für Holzabfuhr)

a) Information on road location and resulting logging opportunities.
b) 1) road inside forest, logging to road possible from both sides along road
2) road at the edge of forest, logging to road possible only from one side
3) road outside forest, but logging to road possible
c) field assessment
d) maximal distance from edge of forest to outside road: 75 m

Road class (LKW-Befahrbarkeit)

a) Information whether road is passable for lorries or not.
b) 1) skidding road (not passable for lorries)
2) road passable for lorries
c) field assessment

Road surface (Fahrbahndecke)

a) Kind of road surface and required base.
b) 1) subgrade soil
2) gravel base and gravel surface
3) gravel base and bituminous surface / concrete pavement
c) field assessment
d) —

Condition of cut-/fillslope (Böschungszustand oben/unten)

a) Condition of cut-/fillslope of road under review.
b) 1) without vegetation
- controlled (bioengineering)
- rock
- bare soil
2) with vegetation
- grass
- grass and herbaceous plants
- only young trees
- grass, herbaceous plants and young trees
c) field assessment
d) no assessment for public roads and roads outside forest

Road drainage system (Entwässerung des Weges)

a) Drainage facilities in order to avoid erosion and slides.
b) 0) none
1) open drain
2) ditch
3) combination of 1 + 2
4) road on bank or road outsloped
c) field assessment
d) no assessment for public roads and roads outside forest

Slides and erosion (Rutschung und Erosion)

a) Information on road related slides or erosion.
b) 0) none
1) slides or erosion
c) field assessment
d) no assessment for public roads and roads outside forest

Driving surface coverage (Bewuchs der Fahrbahn)

a) Information whether driving surface is covered by plants or not.
b) 1) no plants on driving surface
2) driving surface covered by grass-herb vegetation, but tracks free of vegetation
3) driving surface entirely covered by vegetation
c) field assessment
d) no assessment for public roads and roads outside forest

Road completion (Baufertigstellung der Straße)

a) Information on time of road completion.
b) 1) road new (up to two years)
2) road older than two years
c) field assessment
d) no assessment for public roads and roads outside forest

I) Attributes describing forest ecosystems

Natural woodland community (Natürliche Waldgesellschaft)
a) List of natural woodland communities found in Austrian forest.
b) 26 woodland communities
c) map

Origin of debris (Totholzherkunft)
a) Origin of debris, small trees and stumps found on area under review.
b) 1) predominantly harvesting slash
   2) predominantly natural debris
c) field assessment
d) maximum diameter of small trees and stumps: 10 cm

Woody plants (Holzgewächse)
a) Scientific name of selected list of 14 coniferous and 59 broadleaf trees, 49 shrubs, 17 dwarf shrubs species, 1 perennial herbaceous plant and 4 additional classes (other coniferous, other broadleaf-hard wood ,other brosdleaf-soft wood and other shrub species)
b) 73 tree, 49 shrub, 17 dwarf shrub species and 4 general groups (see chapter 2.9.2).
c) field assessment

J) Non-wood goods and services

No attributes in this subject area, neither on nominal nor on ordinal scale.

K) Miscellaneous

Degree of difficulty (Schwierigkeitsgrad)
a) Information on degree of tract difficulty.
b) 15 categories
c) data assessed, map
d) information for field work planning

Plot centre (Punktart)
a) Information on category of sample plot centre.
b) 1) forest
   2) nonforest
   3) foreign territory
c) field assessment

Reason for subplots (Teilungsgrund)
a) Reasons for division of sample plot area into subunits called subplots.
b) 0) none
   1) age class
   2) stand structure
   3) tree species
   4) tree species in age class
   5) forest management type
   6) regional forest authority
   7) soil classification
   8) ownership
   9) slope gradient
   10) slope exposure
   11) null-division
   12) relief
   13) crown coverage
   14) shrub area
   15) vegetation type
   16) feed for game
   17) development stage
c) field assessment

Sample plot draft (Teilungsskizze)
a) Information on whether old sample plot drawing has to be replaced or not
b) 1) unchanged
   2) new
c) field assessment
d) sample plot drawing shows bounderies and location of subplots

Subplot existence (Teilflächenexistenz)
a) Information on whether subplot should be maintained or not.
b) 1) subplot maintained
   2) subplot abandoned
   3) new subplot added
c) field assessment

Means of age measurement (Meßart der Altersstufe)

a) Information on how age of tree was assessed.
b) 1) counting of annual rings on stump
   2) growth drill
   3) counting of branch whorls
   4) obtained information (from owner or his staff)
   5) project up to a subsequent date
c) field assessment

Development stage of age sub-class (Wuchsklasse der Altersstufe)

a) Development stage of age sub-class stated for sample plot area or subplot area.
b) categories see under 'Stage of development of stand'
c) field assessment

Actual ground vegetation (Aktuelle Bodenveg.)

a) Information on how ground vegetation type has been assessed on area under review.
b) 1) indirect (from beside area under review due to absence of ground vegetation on it)
   2) direct (presence of ground vegetation on area under review)
c) field assessment

Text (Freitext)

a) Relevant information concerning forest resources assessment.
b) 5 groups (correction, data, data register/record, information, observation)
c) field assessment

2.1.5. Forest area definition and definition of 'other wooded land'

Forest area definition according to assessment instructions of the Austrian Forest Inventory (AFI) which is slightly different to definition given by a national forest law. Independent of land register, all areas of the country covered by woody plants specified in the appendix of forest law 1975 are forest area if some criteria are met. These criteria have to be proved for the area entirely or partly surrounding the particular sample plot under review:

a) minimum area of 500 m^2 covered by woody plants and
b) width >= 10 m of area covered by woody plants and
c) minimum crown coverage of 30 % (ocular estimation).

Note:

Width >= 10 m is required for both stream side areas covered by woody plants if stream width > 3 m.

No crown coverage criterion for temporary blanks and forested areas without timber yield treated under 2.1.4.

There is no explicit definition of other wooded land in the AFI instructions since they are considered as nonforest areas (e.g. parks, wind protection belts).

There is no explicit definition of forest boundaries in the AFI instructions, but the procedure used to fix the forest edge is the following: the forest boundary is determined by the vertical crown projection of the trees at the edge of the forest.

2.2. DATA SOURCES

A) field data: sample plots, treated under 2.3.

B) questionnaire: no questionaire utilised

C) aerial photography: no aerial photography used in the current forest resource assessment. (However, during the first permanent designed forest inventory (1981-85) aerial photography was used for nonforest/forest decisions as well as on an estimation of a total number of forested 10%-shares of sample plots where plots are unsuitable for walking)

D) spaceborne or airborne digital remote sensing: no spaceborne or airborne digital remote sensing used

E) maps: Austrian map *(Österreichische Karte)* issued by the Federal Land Surveyor's Office *(Bundesamt für Eich- und Vermessungswesen (Landesaufnahme))*. Maps continuously revised by a surveyor's office, but some maps have not been updated for up to twenty years. Austrian maps used by the forest inventory are printed topographic maps with a scale of 1:50 000, where boundaries of growth zones are drawn in by hand.

Land register map *(Mappenblatt)* issued by the Federal Land Surveyor's Office *(Bundesamt für Eich- und Vermessungswesen (Landesaufnahme))*. Land register maps used during the first inventory period (1960-70) are still in use and they are replaced if needed. Land register maps used by the forest inventory are printed maps of registered lots with a scale of 1:10 000.

A district map *(Revierkarte)* is supplied by forest enterprises on a voluntary basis if available. District maps are printed topographic maps with boundaries of property and management units, normally with a scale of 1:10 000.

F) other geo-referenced data: no other geo-referenced data used

2.3. ASSESSMENT TECHNIQUES

2.3.1. Sampling frame

The sampling frame of the Austrian Forest Inventory is given by the forest area definition. The entire Federal territory is covered by the assessment. However, roughly 9 percent of the tracts are not visited, as they are considered constantly unproductive (surfaces of water, wasteland, built-up areas). Some sample plots (roughly 4 percent of forested ones) are not accessible or unsuitable for walking due to terrain conditions in the Alps (FMT 1034).

2.3.2. Sampling units

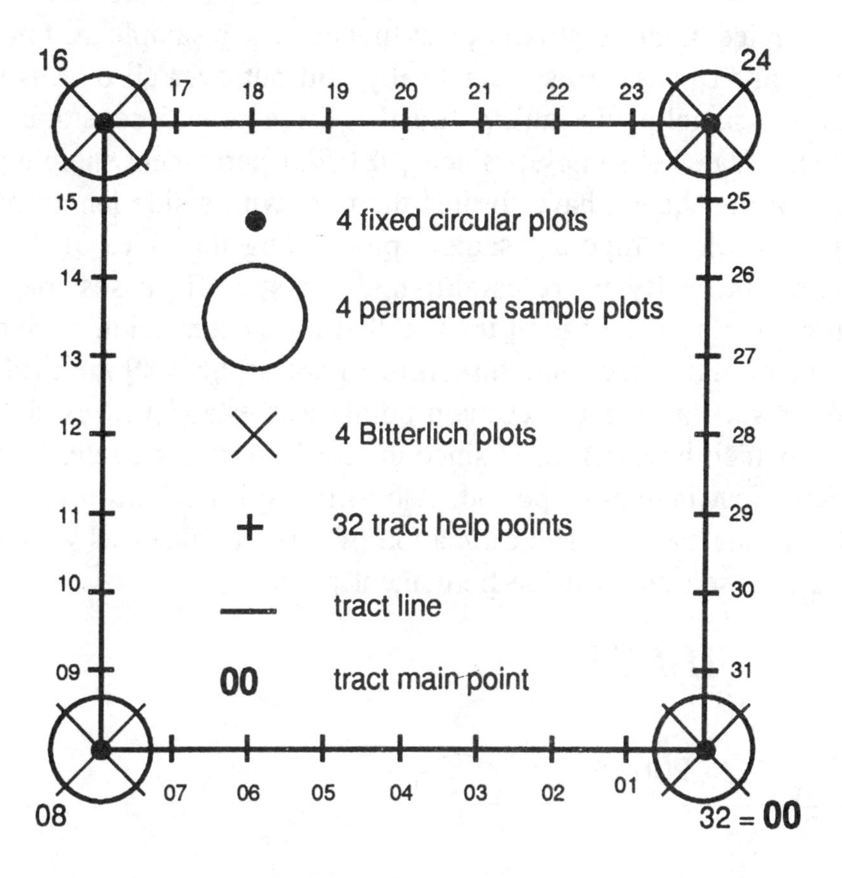

Figure 7. Sampling units of a tract.

Sample plot: Circular plot of 300 m^2 (radius 9.77 m) for assessments of area and stand related data. Slope correction is done by increasing plot radii according to the slope. All sample plots are permanently marked, but these markings are hidden from the landowners and other people in order to keep forest management on permanent plots unbiased. Four permanent sample plots per tract are located in the corners of the square.

Bitterlich plot: horizontal point sampling for trees with DBH >= 105 mm using a basal area factor four. Slope correction is automatically done by the relascope accordding to slope gradient. Observation points for the angle-count sampling are the permanently marked centres of the sample plots. Azimuth and distance of the sample trees from observation point are recorded. The radius (2.60 m) of the fixed circular plot corresponds to the distance of a tree with DBH = 104 mm from the observation point using the angle-count method to consider this tree a sample tree. Sample tree counting by accurate measurements in doubtful cases.

Fixed circular plot: Smaller circle of 21 m^2 (radius 2.60 m) for trees with DBH >= 50 mm and DBH < 105 mm. Fixed circular plots established in order to enhance counting probability of smaller trees. Same centre as sample plot and Bitterlich plot. Azimuth and distance of tallied trees from observation point recorded.

Line survey: a road inventory is carried out along the sides of the tract.

2.3.3. Sampling designs

The sampling design applied in the Austrian Forest Inventory is a sample grid method where the assessment units (tracts) are systematically laid out over all of Austria in a regular network (cluster sampling design). A 'tract' is a square with several circular plots along the perimeter (satellite sample). Since 1981 four permanent sample plots, located in the corners of the square, have formed the tract with a side-length of 200 m. In former inventory cycles temporary sample plots along the sides of the tract were established and can easily be re-established for specific assessments or purposes. The south-eastern corner of each tract, called tract main point, is fixed by Gauß-Krüger co-ordinates and at the same time it is a knot of the 3.89 km grid. The distance of 3.89 km between adjacent tract main points was established by deriving the five-year network, which has been used since the 3 rd inventory cycle, from the 10-year network of the last inventory period. Along the sides of the tract a road inventory is carried out (line survey). The estimation of current values, as well as the estimation of change is based only on the permanent plots.

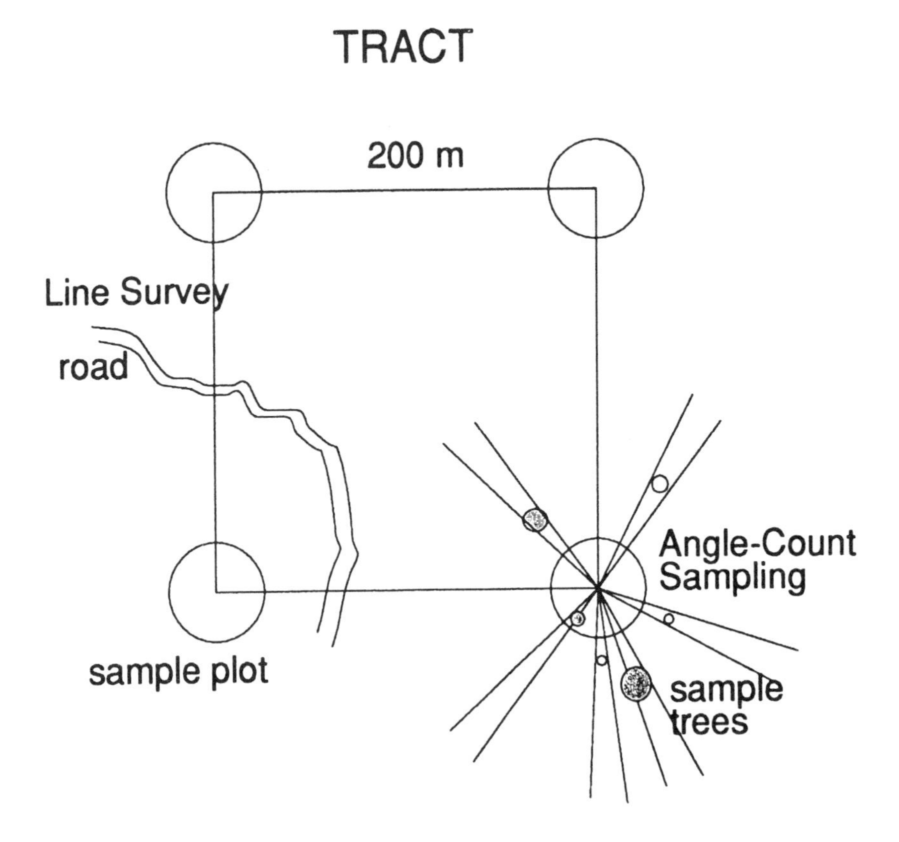

Figure 8. Tract design of current inventory.

2.3.4. Techniques and methods for the combination of data sources

Techniques and methods for the combination of data sources are not applied in the survey since there is, in a sense, only one data source.

2.3.5. Sampling fraction

Data source and sampling unit	Proportion of forested area covered by sample	Represented mean area per sampling unit
field assessment / sample plots	0.008	377.14 ha

2.3.6. Temporal aspects

Inventory cycle	time period of data assessment	publication of results	time period between reference dates	reference date
1st AFI	1961-1970	1973/74	-	1. July 1965
2nd AFI	1971-1980	1985	10 years	1. July 1975
3rd AFI	1981-1985	1986	8 years	1. July 1983
4th AFI	1986-1990	1993	5 years	1. July 1988
5th AFI	1992-1996	1997 (exp.)	6 years	1. July 1994

2.3.7. Data collection techniques in the field

The data is recorded in the field by handheld computer with an immediate plausibility check, the data are then transferred to a computer via an interlink cable to a LAN (local area network) -integrated PC.

2.4. DATA STORAGE AND ANALYSIS

2.4.1. Data storage and ownership

The assessment data are stored in a database management system at the Federal Forest Research Centre in Vienna, Austria. The person responsible for the permission to submit data to outside users is the director of the Federal Forest Research Centre. Inquiries concerning data should first be addressed to the head of the Institute of Forest Inventory (DI. K. Schieler, Seckendorff-Gudentweg 8, A-1131 Vienna, Tel: +43-1-87838-0, Fax: +43-1-8775907, email: fbva@forvie.ac.at).

2.4.2. Database system used

The data are stored in an ORACLE database.

2.4.3. Databank design

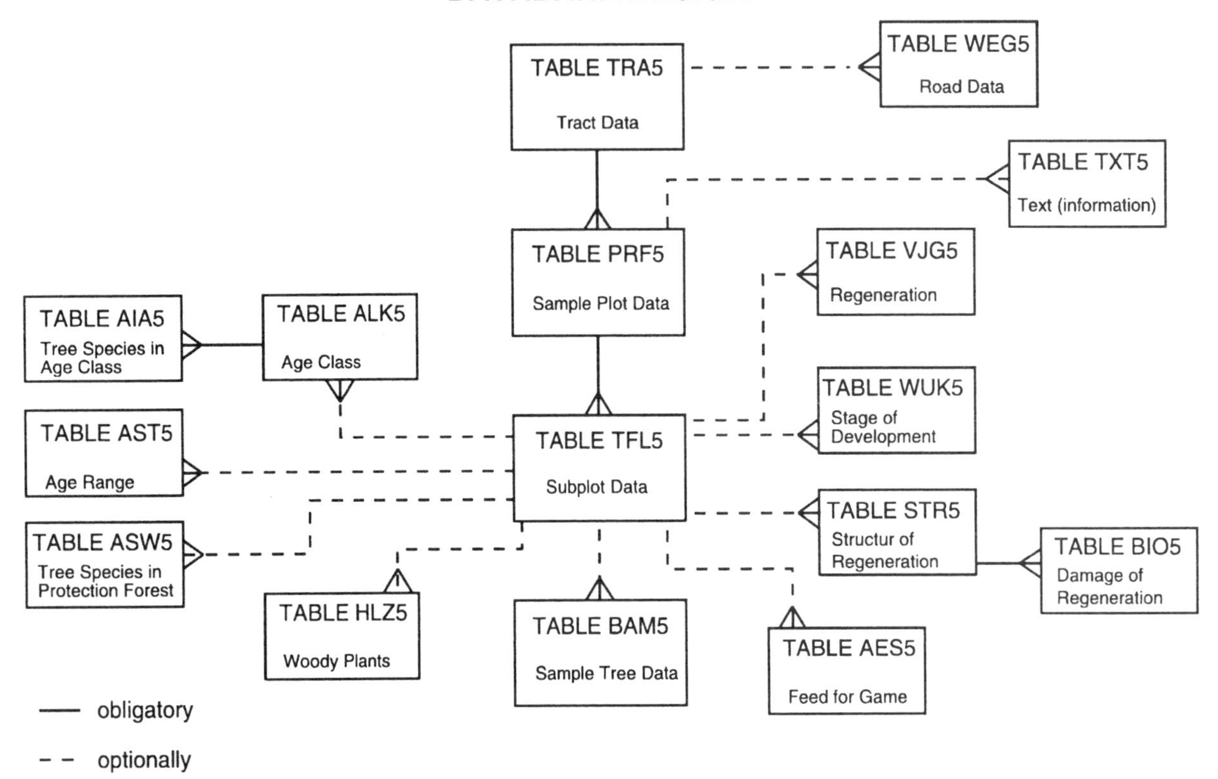

Figure 9. Databank design.

2.4.4. Update level

For each record the date when the data were assessed is available. The data are not updated to a common point in time and thus reflect the situation at the time of assessment.

2.4.5. Description of statistical procedures used to analyse data including procedures for sampling error estimation

The following in chapter 2.4.5. is based on SCHIELER (1988).

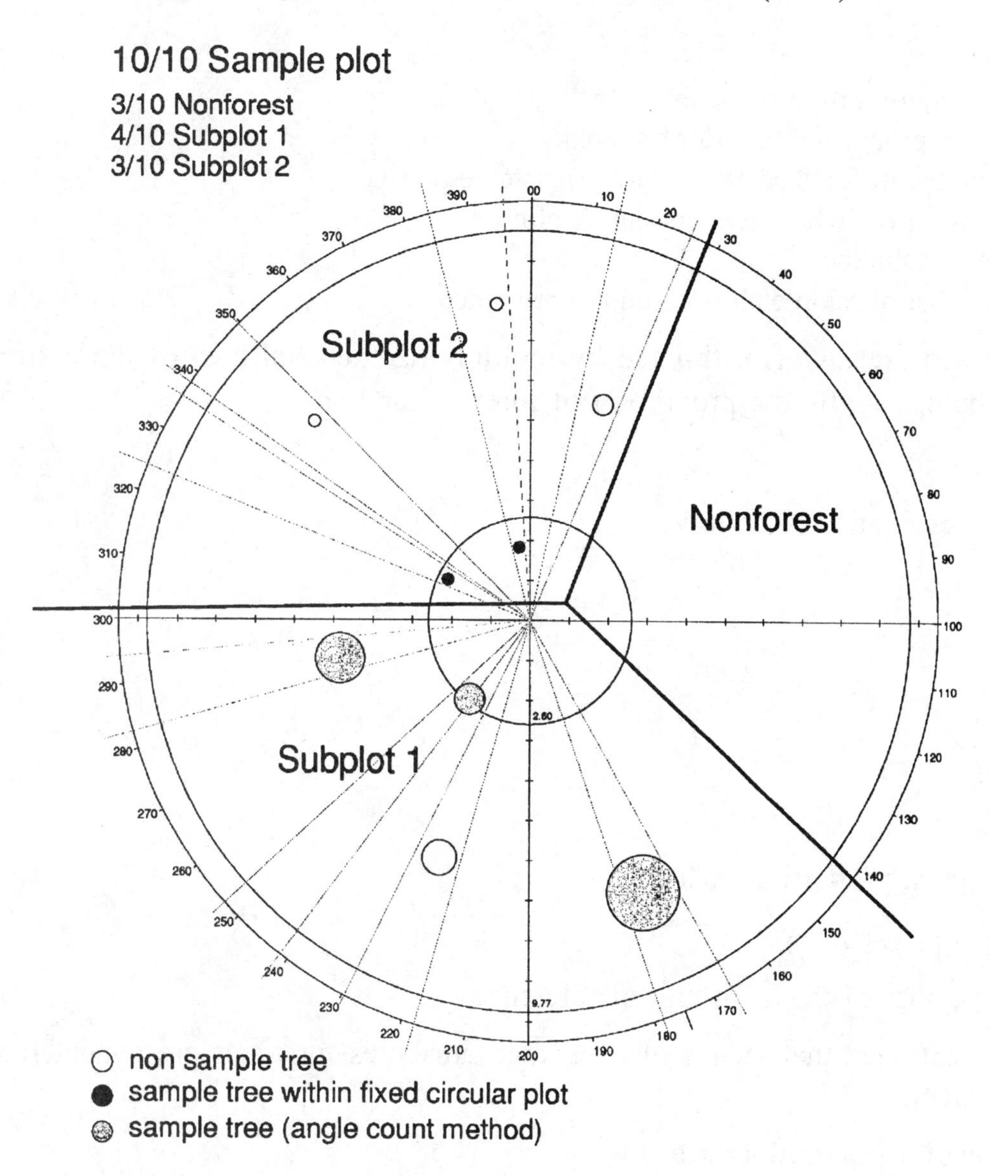

Figure 10. Sample plot draft

a) area estimation

A systematic sample grid is laid over the entire country and for each sample plot of 300 m^2 the 10%-shares of sample plot area covered by forest are assessed in the field. The estimation of the proportion of forested area is done as follows:

$$p_i = \frac{f_i}{w_i}$$

$$\overline{p} = \frac{\sum_{i=1}^{w} p_i w_i}{\sum w_i}$$

where

p_i = forest area proportion of sample plot
p = forest area proportion of unit of reference
f_i = total number of forested 10%-shares on sample plot
w_i = total number of 10%-shares on sample plot
i = sample plot number
w = total number of sample plots on unit of reference

The total forested area A_w is estimated by multiplying the total area of the entire country or subunits, A, by the proportion of forested land, $\overline{p}$.

$$A_w = A\overline{p}$$

Sampling error estimation as follows:

$$p_{t_i} = \frac{\sum f_i w_i}{\sum w_i}$$

$$s_{\overline{p}} = \sqrt{\frac{(p_{t_i} - \overline{p})^2}{n_t - 1}}$$

where

p_{t_i} = forest area proportion of tract

$s_{\overline{p}}$ = standard error of $\overline{p}$

n_t = total number of tracts on unit of reference

It should be mentioned that values of concerned attributes for tracts are considered to be independent.

b) aggregation of tree and plot data

The aggregation of single tree data is done by weighting each single tree attribute Yij by

$$w_{ij} = \frac{k}{DBH_{ij}^2} \qquad \text{(represented number of trees per hectare)}$$

where ij stands for tree i on plot j and

$$k = \frac{baf}{\pi} 4$$

where

baf = basal area factor applied for angle-count method.

Procedure for plots not entirely in the forested area is treated below under e) 'Sampling at the forest edge'.

Once the single tree attributes have been related to unit area by weighting with the represented number of trees per hectare, the total values for plot i, Y_i, can be calculated by summing up the individual single tree attributes.

$$y_i = \sum y_{ij} w_{ij}$$

c) estimation of total values

A = the area of the unit of reference is assumed to be known without error, n = total number of sample plots on the unit of reference (forested and nonforest plots). On the nonforest area the attributes under review are assigned the value zero. For estimation of area related attributes and error estimation the variance of A is considered to be zero ($v_A = 0$).

Sampling error estimation according to procedure treated under a) 'area estimation' and where f_i = total number of forested 10%-shares with considered attribute on sample plot.

d) estimation of ratios

Each angle-count observation provides area related values, e.g. basal area per hectare, volume per hectare or number of trees per hectare. Ratios are estimated as means of ratios as shown in the following example:

$$V_i = \sum_{j=1}^{n_i} \frac{v_{ij} nrep_{ij}}{f_i}$$

$$V_{/ha} = \frac{\sum_{i=1}^{n} V_i f_i}{\sum_{i=1}^{n} f_i}$$

where

V_i = volume per hectare of sample plot
$V_{/ha}$ = volume per hectare of unit of reference
v_{ij} = volume of tree j on plot i
nrep = number of trees represented per hectare
i = sample plot number
j = sample tree number
n_i = total number of sample trees on concerned sample plot
n = total number of sample plots

Sampling error estimation according to LOETSCH-HALLER (1964) as well as HENGST (1966) as follows:

$$s_{(V/ha)} = \sqrt{\frac{n}{(n-1)(\sum f_{t_i})^2}\sum(V_{t_i} - \frac{\sum V_{t_i}}{\sum f_{t_i}} f_{t_i})^2}$$

where

$s_{(V/ha)}$ = standard error of $V_{/ha}$
V_{t_i} = volume per hectare of tract
f_{t_i} = total number of forested 10%-shares of tract
i = tract number
n = total number of tracts

e) sampling at the forest edge

The angle-count sampling at the forest edge is projected to full sample plot area by using the number of forested 10%-shares on the sample plot area as one can see from the following formula already stated under d) 'estimation of ratios':

$$V_i = \sum_{j=1}^{n_i} \frac{v_{ij} nrep_{ij}}{f_i}$$

where

V_i = volume per hectare of sample plot
v_{ij} = volume of tree j on plot i
nrep = number of trees represented per hectare
f_i = total number of forested 10%-shares on sample plot
i = sample plot number
j = number of sample tree
n_i = total number of sample trees on concerned sample plot

f) estimation of growth and growth components (mortality, cut, ingrowth)

The growth of a forest stand is made up of the components survivor growth, ingrowth, mortality and cut. Ingrowth is the number or volume of trees periodically growing into a measurable size, mortality is the number or volume of trees periodically dying from natural causes, cut is the growth related to trees observed on the first and ongoing occasions.

Successive tree volumes are paired to determine the growth contribution of each tree.

Type of growth	individual tree growth figures
Gross growth	$= V_{s2}+I+M+C-V_{s1}-M-C = G_s+I$

where

V_{s1} = the initial volume of survivor trees
V_{s2} = the final volume of survivor trees
G_s = survivor growth
M = the initial volume of trees dying during the period between inventories
C = the initial volume of trees cut during the period between inventories
I = the volume of trees at the second inventory that were below the measurable size on the first occasion

Equations for survivor growth and ingrowth on sample plot:

$$G_{si} = \sum_{i=1}^{n_1} (v_{2_i} - v_{1_i})k + \sum_{i=1}^{n_2} (v_{2_i} - v_{1_i})nrep_1$$

$$I_i = \sum_{i=1}^{n_3} v_{2_i} k$$

where

v_1 = the volume of trees measured on the first occasion
v_2 = the volume of trees measured on the second occasion
n_1 = total number of trees with DBH < 105 mm (fixed circular plot)
k = plot expansion factor of fixed circular plot
n_2 = total number of trees with DBH >= 105 mm (angle count method)
$nrep_1$ = initial number of trees represented per hectare
n_3 = total number of trees on fixed circular plot with DBH < 50 mm on the first occasion and with DBH > 50 mm on the second occasion
i = sample plot number

and gross increment on sample plot:

$$Incr_i = G_{s_i} + I_i$$

g) allocation of stand and area related data to single sample plots/ sample points

Each class of stand and area related data which can be found on the plot is recorded and assigned to a proportion of the plot area.

h) Hierarchy of analysis (treatment of subunits)

Estimates are obtained independently for the entire area and regional subunits. Sampling error estimation must be carried out for each unit of reference as well.

2.4.6. Software applied

The analysis software are SQL and PL/SQL procedures as well as FORTRAN routines which are fed by Oracle data selected using the Oracle tool SQL*Plus. Data transferred via an interlink cable from a handheld computer to a LAN-integrated PC are sent to VAX via user-written interface based on DCL-scripts (disc command language is a part of VMS environment).

2.4.7. Hardware applied

Handheld palmtop-computers (MS-DOS) are used for data capture in the field. The Oracle7 data-base is installed under VMS on a DEC-VAX 4000-600. The analysis is done on above-mentioned DEC-VAX via terminals and LAN-integrated PC's.

2.4.8. Availability of data (raw and aggregated data)

All raw data and derived attributes stored in the data-base are not available for outside persons. All costs for extraction data from the database and preliminary analysis must be reimbursed. Exceptions are possible. Analysis according to special requests is possible, but the Institute reserves the right to reject any request for analysis. Data usually submitted on disc. Inquiries concerning data should be adressed to the head of the Institute of Forest Inventory, Federal Forest Research Centre (DI. K. Schieler, Seckendorff-Gudentweg 8, A-1131 Vienna, Tel: +43-1-87838-0, Fax: +43-1-8775907, email: fbva@forvie.ac.at).

2.4.9. Subunits (strata) available

In general, the forest inventory results can be obtained since they are published for all of Austria, the nine federal states of Austria and for regional forest authorities. In addition, results for main productive regions and growth zones can be provided without major expenditure. Analysis can be made for any thematic subunit if requested.

2.4.10. Links to other information sources

Topic and spatial structure	Age/period availability	Responsible agency	Kind of data-set
outline map of land register communities *(Übersichtskarte der Katastralgemeinden mit Mappenblatteinteilung)*	since 1956 available	BEV	map
geological outline map of Austria *(Geologische Übersichtskarte der Republik Österreich mit tektonischer Gliederung)*	since 1961 available (1:1000000)	BEV	map
growth zones map of Austria *(Wuchsraumkarte)*	publ. 1994 available	FBVA	map

BEV: Bundesamt für Eich und Vermessungswesen (Landesaufnahme), Krotenthallergasse 3, 1080 Vienna (Tel: +43-1-438 935-464), FBVA: Federal Forest Research Centre, Seckendorff-Gudentweg 8, A-1131 Vienna (Tel: +43-1-87838-0)

2.5. RELIABILITY OF DATA

2.5.1. Check assessments

a) Sample plots: 1-2 % of the sample plots are visited a second time by field crews. The composition of the field crews varies, but they are all well-trained and experienced members of the permanent staff of the Institute of Forest Inventory, usually dealing with regular field work. The sample plots to be checked are randomly selected from the list of already assessed plots. All attributes are assessed a second time independently from the assessment of the first field crews, which means that data recorded by the field crews are not availible for the checking crews. The field crews do not know in advance which plots will be checked. The checking crews visit the plots after the field crews have assessed them. The organisation and the analysis of the check assessments is done at the office.

Analysis of the check assesssments:
All data that are assessed by both regular field crews and checking crews are analysed. For measured variables the mean difference deviation between regular and checking crews and its standard deviation is calculated and visually represented in graphs. Attributes on nominal or ordinal scale are presented in contingency charts with the classes for regular and checking crews. Graphs and contingency tables are for official use only.

b) The check assessments are used for the following purposes:
- feedback to field crews to improve the data quality
- results are used to revise the assessments instructions
- results will not be used to set up error budgets and to quantify the total error
- results will not be used to correct any raw data or results.

2.5.2. Error budgets

For the publication of inventory results only sampling errors, already treated under 2.4.5., are concerned. No procedures other than those described for the calculation of sampling errors which might enable one to quantify error sources such as measurement errors, prediction errors, grouping errors or other non-statistical errors are applied.

2.5.3. Procedures for consistency checks of data

Data assessed in the field:
The data are put directly into a handheld computer on the field plot. While the data is being fed into the computer a first consistency check is carried out for selected attributes, using data from the last assessment and cross-checks of different attributes for plausibility tests. When the data are loaded on the database, a second, more formal check is done by the Oracle7 data base kernel. Of special interest is the assignment of tree numbers, which allow the pairing of information from different assessments for individual trees as well as the assignment of road numbers.

2.6. MODELS

2.6.1. Volume (functions, tariffs)

The following in chapter 2.6.1. is based on BRAUN (1969) and SCHIELER (1988).

a) Outline of the model

Cubing functions *(Kubierungsfunktionen)* have been developed for the calculation of volume as well as the volume increment of the individual sample tree.

Four basic types of cubing functions for specific species groups were found to be adequate for a nationwide forest resource assessment in Austria.

Within the basic types for trees, with DBH >= 105 mm, constant and regression coefficients vary for specific tree species.

For trees with DBH < 105 mm one basic type with different parameters for 28 tree species and specific species groups is applied.

Input variables for the model are DBH, upper diameter (D03) and total tree height, in case of broadleaves the height of crown base in addition.

Upper diameter (D03) and total tree height are calculated by means of models whereas DBH and crown base are measured for each selected sample tree.

b) Overview of prediction errors

The relative standard error of the cubing functions is between 4 and 6 % for different tree species (after Braun 1969).

c) Data material for derivation of the model

Input data were 17000 single tree measurements from the most important tree species cubed by sections from plots distributed over the entire country.

d) Methods applied to validate the model

The cubing funtions have been developed by a set of single tree data. The functions obtained have been verified using an independent set of data. There is no current verification.

2.6.2. Assortments: no model for assortments applied.

2.6.3. Growth components: no model for growth components applied.

2.6.4. Potential yield: no model for potential yield applied.

2.6.5. Forest functions: no model for forest functions applied.

2.6.6. Other models applied

Upper diameter model

a) outline of the model

Upper diameter (D03) model *(D03 - Modell)* has been derived using initial DBH and final DBH as well as initial D03 of individual surviver trees as input variables. One basic type with different parameters for 28 tree species and specific species groups is applied.

b) overview of prediction errors

Growth estimation unbiased.

c) data material for derivation of the model

Input data were a large number of single tree measurements from the most important tree species from temporary plots which were located along the sides of a tract assessed by the forest inventory between 1981-1990.

d) methods applied to validate the model

The upper diameter (D03) model has been developed by a set of single tree data. The model obtained has been verified using a subset of 6650 single trees. A subset of sample trees is used to readjust the parameters of the regression model by current forest inventory measurements (1992-1996).

Height growth model

a) outline of the model

Height growth model *(Höhenzuwachsmodell)* has been derived using initial DBH and final DBH as well as initial tree height of individual surviver trees as input variables. One basic type with different parameters for 28 tree species and specific species groups is applied.

b) overview of prediction errors

Growth estimation unbiased.

c) data material for derivation of model

Input data were 28035 single tree measurements from the most important tree species taken from temporary plots which were located along the sides of a tract assessed by the forest inventory between 1981-1990.

d) methods applied to validate the model

The height growth model has been developed by a set of single tree data. The model obtained has been verified using an independent set of data. A subset of sample trees is used to readjust the parameters of the regression model by current forest inventory measurements (1992-1996).

2.7. INVENTORY REPORTS

2.7.1. List of published reports and media for dissemination of inventory results

Inventory	Year of publication	Citation	Language	Dissemination
1st AFI	1973	BRAUN R., 1973: Österreichische Forstinventur 1961/1970. Zehnjahres-Ergebnisse für das Bundesgebiet. Mitteilungen der Forstlichen Bundesversuchsanstalt, Wien, Nr. 1o3/I, 117 p.	German	printed
1st AFI	1974	BRAUN R., 1974: Österreichische Forstinventur 1961/1970. Zehnjahres-Ergebnisse für das Bundesgebiet. Mitteilungen der Forstlichen Bundesversuchsanstalt, Wien, Nr. 1o3/II, 217 p.	German	printed
2nd AFI	1985	FBVA, 1985: Österreichische Forstinventur 1971-1980. Zehnjahresergebnis. Mitteilungen der Forstlichen Bundesversuchsanstalt, Wien, Nr. 154/I, 216 p.	German	printed
2nd AFI	1985	FBVA, 1985: Inventurgespräch. Mitteilungen der Forstlichen Bundesversuchsanstalt, Wien, Nr. 154/II, 91 p.	German	printed
3rd AFI	1986	FBVA, 1986: Österreichische Forstinventur 1981-90. Auswertung 1981-85. Forstliche Bundesversuchsanstalt, Wien,	German	microfiche

3rd AFI	1989	HASZPRUNAR J., NIEBAUER O., REITTER A., 1989: Auszüge aus den Bundes- und Landesergebnissen der Auswertung 1981-85 (Österreichische Forstinventur). Forstliche Bundesversuchsanstalt, Wien, 206 p.	German	printed
4th AFI	1993	FBVA, 1993: Forstinventur 1986-90. Forstliche Bundesversuchsanstalt, Wien	German	disc
		FBVA, 1993: Österreichische Forstinventur 1986/90. Erläuterung zur Auswertung und Handhabung der Ergebnisdiskette. Österreichische Forstinventur,. Wien, 24 p.	German	printed
4th AFI	1996	FBVA, 1995: Österreichische Forstinventur. Ergebnisse 1986/90. Forstliche Bundesversuchsanstalt - Waldforschungszentrum, Wien, Nr. 92, 262 p.	German	printed
1st AFI	1985	BRAUN R., 1985: Über die Bringungslage und den Werbungsaufwand im österreichischen Wald. Mitteilungen der Forstlichen Bundesversuchsanstalt, Wien, Nr. 155, 243 p.	German	printed
1st/2nd AFI	1982	MILDNER H., HASZPRUNAR J., SCHULTZE U., 1982: Weginventur im Rahmen der Österreichischen Forstinventur. Mitteilungen der Forstlichen Bundesversuchsanstalt, Wien, Nr. 143, 114 p.	German	printed

2.7.2. List of contents of the latest report and update level

Reference of latest report:

FBVA, 1996: Österreichische Forstinventur. Ergebnisse 1986/90. Forstliche Bundesversuchsanstalt - Waldforschungszentrum, Wien, Nr. 92, 262 p.
Update level of inventory data used in this report:
Field data: 1986-1990
Content:

Chapter	Title	Number of pages
1.	Introduction	1
2.	Methods	1
2.1	Methods of data assessment	
2.2	Methods of data analysis	
3.	Results	
3.1	Forest area	7
3.2	Growing stock	3
3.3	Increment and Harvesting	
3.3.1	Increment	2
3.3.2	Harvesting	1
3.4	Damages	
3.4.1	Bark peeling	3
3.4.2	Browsing	1
3.4.3	Harvesting	2
3.4.4	Falling rocks	1
3.5	Forest opening	
3.5.1	Forest roads	2
3.5.2	Logging conditions	3
3.6	Tending measures	1
4	Glossary	2
5	Literature	1
6	Tables	
	Austria	44
	Federal states	162
7	Appendix	3

2.7.3. Users of the results

A user profile of the forest inventory results can be obtained by the proportion of discs sold to different user groups, the users' interests by the proportion of reference units requested.

Grouped by users		Grouped by reference units requested	
Research institutions	8 %	Austria, all federal states	8 %
Schools	3 %	Austria, all federal states, specific regional forest authority/ies	23 %
Special interest groups	8 %		
Authorities	77 %	Austria, all federal states, all regional forest authorities	69 %
Private use	4 %		

2.8. FUTURE DEVELOPMENT AND IMPROVEMENT PLANS

2.8.1. Next inventory period

The next inventory period will be planned in detail after the current inventory cycle is completed and directions are obtained from the BMFL. Hence there is no sound information concerning changes availabe at present.

2.8.2. Expected or planned changes

For reasons of continuity, a permanently designed forest inventory development is evolutional and changes occur in details rather than in the whole system. Hence no major changes are expected.

2.8.2.1. Nomenclature

There are no changes planned up to now.

2.8.2.2. Data sources

Re-establishment of tempory sample plots for specific assessments in order to keep permanent plots unbiased if required.

2.8.2.3. Assessment techniques

There are no changes planned up to now.

2.8.2.4. Data storage and analysis

There are no changes planned up to now.

2.8.2.5. Reliability of data

There are no changes planned up to now.

2.8.2.6. Models

There are no changes planned up to now.

2.8.2.7. Inventory reports

There are no changes planned up to now.

2.8.2.8. Other forestry data

There are no changes planned up to now.

2.9. MISCELLANEOUS

2.9.1 General guidelines for assessment of attributes

High forest - production forest:
all attributes listed in chapter 2.1.1 where the object assigned is not 'sample tree' must be directly assessed in the field if no restriction can be found under b) 'threshold value' for measureable attributes treated under chapter 2.1.3 or under d) 'remarks' for attributes on nominal or ordinal scale treated under chapter 2.1.4.

High forest - protection forest with yield:
the same attributes as in high forest - production forest must be assessed, except the attribute 'disintegration phase'. In addition, the following attributes must be directly assessed in the field:

- 'forest grazing',
- 'tree distribution'
- 'stand stability'
- 'mass movements'
- 'formation phase'
- 'avalanche path'

High forest - protection forest without yield if accessable:
the same attributes as in protection forest with yield must be assessed, except 'tree species in age class'. In addition, the following attributes must be directly assessed in the field:

- 'shares of dead standing trees'
- 'tree species in protection forest'
- 'volume estimation'

High forest - protection forest without yield unsuitable for walking:
only FMT must be stated, no further assessments necessary.

Coppice forest:

the same attributes as in high forest - production forest must be assessed, except the attributes: 'fraying', 'harvesting damage', 'tending measures', 'disintegration phase'. No 'regeneration assessment' will be assessed, since the sucker formation normally appears after coppicing.

In addition the following attribute must be directly assessed in the field:

- 'coppice forest alteration'

Sample tree assessment:
Sample trees must be assessed on every sample plot of a tract if the sample plot area or parts of it (subunits) have been designated one of the following forest management types (FMT):

- high forest / production forest / forested area (FMT 1011)
- high forest / protection forest with yield / forested area (FMT 1021)
- coppice forest / far from stream / production forest / forested area (FMT 2111)
- coppice forest / adjacent stream / production forest / forested area (FMT 2211)

The following can never become sample trees:

- shrubs, even if they show tree-like growth which meets the DBH - threshold value
- trees on shrub areas (FMT 1013, 1023, 2113, 2213)
- trees on forested areas without timber yield
- trees on forested areas designated protection forest without yield
- trees under environmental protection or connected with an historical event.

2.9.2. List of woody plants

The list below has been compiled from different lists of the 'Assessment Instructions of the Austrian Forest Inventory' (FBVA 1994); for translation of tree species VAUCHER (1986) has been used.

List of woody plants		*Liste der Holzgewächse*				
Tree species - Conifers		***Nadelbaumarten***				
Spruce	*Picea*	*Fichte*	w	s		
Norway Spruce	*Picea abies*	*Fichte*	w	s	a	p
Fir	*Abies*	*Tanne*	w	s		
Silver Fir	*Abies alba*	*Weißtanne*	w	s	a	p
Larch	*Larix*	*Lärche*	w	s		
European Larch	*Larix decidua*	*Europäische Lärche*	w	s	a	p
Pine	*Pinus*	*Kiefer*	w	s		
Scots Pine	*Pinus silvestris*	*Weißkiefer*	w	s	a	p
Corsican Pine	*Pinus nigra*	*Schwarzkiefer*	w	s	a	p
Swiss Stone Pine	*Pinus cembra*	*Zirbe*	w	s	a	p
Eastern White Pine	*Pinus strobus*	*Strobe*	w	s		
Mountain Pine	*Pinus mugo*	*Berg- und Moorkiefer*	w	s		
Douglas Fir	*Pseudotsuga*	*Douglasienarten*	w	s		
Other conifer		*Sonstiger Nadelbaum*	w	s	a	p
English Yew	*Taxus baccata*	*Eibe*	w	s		
Tree species - Broadleaves		***Laubbaumarten***				
European Beech	*Fagus silvatica*	*Rotbuche*	w	s	a	p
Oak	*Quercus*	*Eiche*	w	s	a	p
European White Oak	*Quercus robur*	*Stieleiche*	w			
Sessile Oak	*Quercus petraea*	*Traubeneiche*	w			
Pubescent Oak	*Quercus pubescens*	*Flaumeiche*	w			
Turkey Oak	*Quercus cerris*	*Zerreiche*	w			
Northern Red Oak	*Quercus rubra*	*Roteiche*	w			
Sweet Chestnut	*Castanea sativa*	*Edelkastanie*	w	s		
European Hornbeam	*Carpinus betulus*	*Hainbuche*	w	s		
Elm	*Ulmus*	*Ulme*	w	s		
Mountain Elm	*Ulmus glabra*	*Bergulme*	w			
European White Elm	*Ulmus laevis*	*Flatterulme*	w			
Field Elm	*Ulmus carpinifolia*	*Feldulme*	w			
Plum	*Prunus*	*Prunus*	w	s		
Wild Cherry	*Prunus avium*	*Vogelkirsche*	w	s		

Bird Cherry	*Prunus padus*	*Traubenkirsche*	w			
Mountain Ash	*Sorbus*	*Sorbus*	w	s		
European Mountain Ash	*Sorbus aucuparia*	*Eberesche*	w	s		
Wild Service Tree	*Sorbus torminalis*	*Elsbeere*	w	s		
Service Tree	*Sorbus domestica*	*Speierling*	w			
Whitebeam	*Sorbus aria*	*Mehlbeere*	w			
Locust species	*Robinia species*	*Robinienarten*	w	s		
Maple	*Acer*	*Ahorn*	w	s		
Sycamore	*Acer pseudoplatanus*	*Bergahorn*	w			
European Maple	*Acer platanoides*	*Spitzahorn*	w			
Field Maple	*Acer campestre*	*Feldahorn*	w			
Box Elder	*Acer negundo*	*Eschenblättriger Ahorn*	w			
Ash	*Fraxinus*	*Esche*	w	s		
European Ash	*Fraxinus excelsior*	*Waldesche*	w			
Flowering Ash	*Fraxinus ornus*	*Blumenesche*	w			
Narrow-leaved Ash	*Fraxinus angustifolia*	*Quirlesche*	w			
Other broadleaf (hard wood)		*Sonstiger Laubbaum (Hartholz)*	w	s	a	p
Hop Hornbeam	*Ostrya carpinifolia*	*Hopfenbuche*	w	s		
European Nettle Tree	*Celtis australis*	*Zürgelbaum*	w			
Mulbeery-tree species	*Morus species*	*Maulbeerbaumarten*	w			
European Walnut	*Juglans regia*	*Walnuß*	w	s		
Black Walnut	*Juglans nigra*	*Schwarznuß*	w	s		
Plane species	*Platanus species*	*Platanenarten*	w			
Apple-tree species	*Malus species*	*Apfelarten*	w	s		
Pear-tree species	*Pyrus species*	*Birnenarten*	w	s		
Tree of Heaven	*Ailanthus peregrina*	*Götterbaum*	w			
Birch	*Betula*	*Birke*	w	s		
European White Birch	*Betula pendula*	*Sandbirke*	w			
Pubescent Birch	*Betula pubescens*	*Moorbirke*	w			
Alder	*Alnus*	*Erle*	w			
Black Alder	*Alnus glutinosa*	*Schwarzerle*	w	s		
Speckled Alder	*Alnus incana*	*Grauerle*	w	s		
Poplar	*Populus*	*Pappel*	w			
European Aspen	*Populus tremula*	*Aspe*	w	s		
Silver Poplar	*Populus alba*	*Silberpappel*	w	s		
Black Poplar	*Populus nigra*	*Schwarzpappel*	w	s		
Poplar species (X)	*Populus species (X)*	*Hybridpappeln (can)*	w	s		

Willow	*Salix*	*Weiden*	w	s		
White Willow	*Salix alba*	*Silberweide*	w			
Crack Willow	*Salix fragilis*	*Bruchweide*	w			
Common Sallow	*Salix caprea*	*Salweide*	w			
Lime tree	*Tilia*	*Linde*	w	s		
Large-leaved Lime	*Tilia platyphyllos*	*Sommerlinde*	w			
Small-leaved Lime	*Tilia cordata*	*Winterlinde*	w			
Other broadleaf (soft wood)		*Sonstiger Laubbaum (Weichholz)*	w	s	a	p
Horse Chestnut species	*Aesculus species*	*Roßkastanienarten*	w			
Shrubs		*Sträucher*				
Shrubs		*unbestimmte Sträucher*	w			
Shrubs		*strauchflächenfähige Sträucher*			a	p
Green Alder	*Alnus viridis*	*Grünerle*	w			p
Dwarf Mountain Pine	*Pinus mugo ssp.*	*Latsche*	w			p
Snowy Mespilus	*Amelanchier ovalis*	*Felsenbirne*	w			
Barberry species	*Berberis species*	*Berberitzenarten*	w			
Common Box	*Buxus sempervirens*	*Buchsbaum*	w			
Common Bladder Senna	*Colutea arborescens*	*Blasenstrauch*	w			
Cornelian Cherry	*Cornus mas*	*Gelber Hartriegel*	w			
Blood-twig Dogwood	*Cornus sanguinea*	*Roter Hartriegel*	w			
Hazel	*Corylus avellana*	*Hasel*	w			
Wig Tree	*Cotinus coggygria*	*Perückenstrauch*	w			
Cotoneaster species	*Cotoneaster species*	*Steinmispelarten*	w			
Hawthorn species	*Crataegus species*	*Weißdornarten*	w			
Oleaster	*Elaeagnus angustifolia*	*Ölweide*	w			
Spindle Tree	*Euonymus europaeus*	*Gemeiner Spindelstrauch*	w			
Broad-leaved Spindle Tree	*Euonymus latifolius*	*Breitblätteriger Spindelstrauch*	w			
Warted Spindle Tree	*Euonymus verrucosus*	*Warziger Spindelstrauch*	w			
Alder Buckthorn	*Frangula alnus*	*Faulbaum*	w			
Sea Buckthorn species	*Hippophae species*	*Sanddornarten*	w			
Holly Tree	*Ilex aquifolium*	*Stechpalme*	w			
Juniper species	*Juniperus species*	*Wacholderarten*	w			

Laburnum species	*Laburnum species*	*Goldregenarten*	w
Privet species	*Ligustrum species*	*Ligusterarten*	w
Honeysuckle species	*Lonicera species*	*Heckenkirschenarten*	w
Medlar	*Mespilus germanica*	*Echte Mispel*	w
German Tamarisk	*Myricaria germanica*	*Deutsche Tamariske*	w
Rockcherry	*Prunus mahaleb*	*Steinweichsel*	w
Sloe	*Prunus spinosa*	*Schlehe*	w
Buckthorn species	*Rhamnus species*	*Kreuzdornarten*	w
Rose species	*Rosa species*	*Rosenarten*	w
Willow species	*Salix species*	*Strauchweiden*	w
Elder	*Sambucus nigra*	*Schwarzer Holunder*	w
European Red Elder	*Sambucus racemosa*	*Roter Holunder*	w
Bladdernut	*Staphylea pinnata*	*Pimpernuß*	w
Common Lilac	*Syringa vulgaris*	*Flieder*	w
Wayfaring Tree	*Viburnum lantana*	*Wolliger Schneeball*	w
Water Elder	*Viburnum opulus*	*Gemeiner Schneeball*	w
Shrubs		*Nicht strauchflächengähige Sträucher*	
Dwarf Birch	*Betula nana*	*Zwergbirke*	w
Clematis species	*Clematis species*	*Waldrebenarten*	w
Spurge Laurel	*Daphne laureola*	*Lorbeer Seidelbast*	w
February Daphne	*Daphne mezereum*	*Gewöhnlicher Seidelbast*	w
Broom species	*Genista species*	*Ginster*	w
Common Ivy	*Hedera helix*	*Efeu*	w
Savin Juniper	*Juniperus sabina*	*Sadebaum*	w
Box Thornspecies	*Lycium species*	*Bocksdornarten*	w
Currant species	*Ribes species et ssp.*	*Johannisbeerarten*	w
Wild Gooseberry	*Ribes uva-crispa*	*Stachelbeere*	w
Blackberry	*Rubus fructicosus agg.*	*Brombeerarten*	w
Wild Raspberry, Dewberry, ---	*Rubus idaeus, caesius, saxatilis*	*Him-, Kratz-, Steinbeere*	w
Furze species	*Ulex species*	*Stechginsterarten*	w
Dwarf-shrub		*Zwerg- und Kleinsträucher*	
Common Heather	*Calluna vulgaris*	*Besenheide*	w

Striated Mezereon	*Daphne striata*	*Steinröschen*	w
Black Crow Berry	*Empetrum nigrum*	*Krähenbeere*	w
Snow Heath	*Erica carnea*	*Schneeheide*	w
Dwarf Juniper	*Juniperus communis ssp. nana*	*Zwergwacholder*	w
Marsh Ledum	*Ledum palustre*	*Sumpfdorst*	w
Alpine Azalea	*Loiseleuria procumbens*	*Gemsheide*	w
European Loranth	*Loranthus europaeus*	*Riemenblume*	w
Rusty-leaved Alpenrose	*Rhododendron ferrugineum*	*Rostrote Alpenrose*	w
Hairy Alpenrose	*Rhododendron hirsutum*	*Behaarte Alpenrose*	w
Dwarf Alpenrose	*Rodothamnus chameacistus*	*Zwergalpenrose*	w
Bitter Nightshade	*Solanum dulcamara*	*Bittersüßer Nachtschatten*	w
Whortleberry	*Vaccinium myrtillus*	*Heidelbeere*	w
Moosberry	*Vaccinium oxycoccus*	*Moosbeere*	w
Bog Bilberry	*Vaccinium uliginosum*	*Rauschbeere*	w
Red Bilberry	*Vaccinium vitis idaea*	*Preiselbeere*	w
Mistletoe species	*Viscum species*	*Mistelarten*	w
Perennial herbaceous plants		*Stauden*	
Hop	*Humulus lupulus*	*Wilder Hopfen*	

w -indicates tree species and groups for assessment of attribute 'woody plants'
s - indicates selected tree species and groups for 'sample tree' assessment
a - indicates selected tree species and groups for assessment of attribute 'tree species in age class'
p - indicates selected tree species and groups for assessment of attribute 'tree species in protection forest without yield'

3. AUSTRIAN FOREST DEVELOPMENT PLAN

The following in chapter 3 is based on OTTITSCH (1994).

3.1. NOMENCLATURE

Criteria for the Evaluation of Protection

Criteria Present Within Function Area	Value
wind protection belts	3
protection against natural force on sites of special public interest	3
sand-drift	3
erosion	3
rocky, precipitous or shallow soiled sites, where reforestation is difficult	3
dangerous land slides	3
area covered with woody plants between timber line and tree line	3
areas close below the timber line	3
protection of infrastructure against natural forces - area under forest ban	3
protection of traffic and electricity infrastructure - area under forest ban	3
protection of military infrastructure - area under forest ban	3
protection of forests against forest damage - area under forest ban	3
protection against natural forces on sites of increased public interest	2
potential erosion	2
rocky, precipitous or shallow soiled sites	2
land slides	2
protection of infrastructure against natural forces	2
protection of traffic and electricity infrastructure	2
protection of military infrastructure	2
protection of forests against forest damage	2

Criteria for the Evaluation of Welfare

Criteria Present Within Function Area	Value
importance for balance of climate and water on sites of special public interest	3
filtering and cleaning of air and water on sites of special public interest	3
protection against noise on sites of special public interest	3
protection against emissions of air pollution - area under forest ban	3
protection of mineral springs and spas - area under forest ban	3
protection of water resources- area under forest ban	3
importance for balance of climate and water	2
filtering and cleaning of air and water	2
protection against noise	2
protection against emissions of air pollution	2
protection of mineral springs and spas	2
protection of water resources	2

Criteria for the Evaluation of Recreation

Criteria Present Within Function Area	Value
importance of the forest as recreation area - special public interest	3
demand for recreation in the vicinity of urban areas - need for regulation of demand	3
demand for recreation in tourism regions - need for regulation of demand	3
importance of the forest as recreation area - increased public interest	2
demand for recreation in the vicinity of urban areas - no regulation needed	2
demand for recreation in tourism regions - no regulation needed	2
area closed to the public	0

Additional Function Area Information

groups	classes
Terrain factors	slope
	morphology
	exposition
Geology and site conditions	geology
	soil depth
	water supply
	nutrition
Forest stand characteristics	stand types (deciduous/conifers/mixed,...)
	structure
	crown coverage
	development phase (initial, optimal, terminal)
	deciduous tree species
	conifer species
	site appropriateness of stand
Forest type and special sites	forest type
	special sites
Area characteristics	hydrology
	presence of function area < 10 ha within area
	protection forest characteristics
Additional information	water reserves
	nature reserves
	wild water and avalanche control danger zone plans
	closed areas (e.g. military zones)

Groups and Classes of Impairments

group	classes
Abiotic impairments	immissions
	site specific climate
	use of non-renewable resources (mining, quarries,...)
	infrastructure (electric or traffic)
	uprooting pressure (for development areas)
Damage by deer	peeling (of bark)
	browsing
	rubbing
Forest pasture	general
	stomping (of plants)
	browsing
	soil compaction
Recreation, Tourism	general
	winter recreation
	summer recreation
	extreme tourism
	recreation around urban areas
Improper forest management	general
	silvicultural failures
	unstable stands
	insufficient regeneration
	infrastructure (e.g. too few or too many forest roads)
Legal factors	ownership structure
	legal rights of non-owners (e.g. forest pasture)
	nature reserves
	water reserves
Hydrologic factors	ground water
	wetlands
	floods and wildwater
Forest protection (biotic factors)	insects
	fungi
	parasitic plants
	others (viruses, bacteria, salt damages, drought)
	loss of leaves or needles
Other impairments	forest fires
	agriculture

Groups and Classes of Necessary Actions

group	classes
Abiotic impairments	immissions control
Damage by deer	reduction
	total elimination
	prohibition of feedings
	fencing of regeneration sites
	protection of single plants
Forest pasture	division of forest and pasture
Tourism and recreation	regulation for wintersport infrastructure
	regulation for other types of recreation
Forest management	silvicultural actions
	protection forest improvement
	forest infrastructure planning and repair of damages
	special management techniques
Legal impairments	regulation of ownership structure
	regulation of non-owners' rights
	compensation for impairments from nature reserves
	compensation for impairments from water reserves
	subsidising
	other legal regulations
	(environmental sensitivity assessments)
Hydrologic impairments	groundwater
	wetlands
	flood- and wild water control
Forest protection actions against:	general
	insects
	fungi
	parasitic plants
	others (according to responsible factor)
Other actions	preservation of forest lands (against uprooting)

Note: The information on 'additional function area information', 'impairments' and 'necessary actions' is divided into groups, classes (both stated above) and types. The urgency of 'necessary actions' is also stated for the function area as follows:

- very urgent (now or within the next ten years)
- urgent (within the next 10 to 30 years)
- desireable (no time frame)

3.2. DATA SOURCES

The only data source used for the assessment are field data.The field work and inventory for the AFDP are done by staff members of the Regional Forest Authorities.

3.3. ASSESSMENT TECHNIQUES

The Austrian Forest Development Plan (AFDP), designed as an instrument of regional forest planning, is the only nationwide mapping of the conditions in Austrian forests. To describe the different goods and services provided by forests, the AFDP uses the terms 'forest functions' and 'function values'. The term 'impairments' is used to describe possible difficulties in these functions due to natural conditions, as well as various forms of human impact. The spatial entities for which function assessments are stated are called function areas and they are assigned functions values ranging from 1 to 3 for 'protection', 'welfare' and 'recreation'. The function 'utility' (timber production) is not evaluated, since it is considered a prerequisite for the provision of the other functions. The AFDP consists of partial plans for each of the nine federal states; the partial plans on federal state level consists of forest development plans at the district level of the Regional Forest Authority. The FDP for each district consists of a set of Austrian maps (scale 1:50000), in which boundaries of function areas and their assigned function values are drawn. The function that has been given the value 3, which is the highest level of importance, becomes the 'leading function'. If all functions are assigned the same value in the function area, then 'protection' comes before 'welfare' and 'welfare' comes before 'recreation'. Each function area is coloured in accordance with its leading function (red stands for protection, blue for welfare, yellow for recreation). The green colour of the forest layer on the Austrian map indicates timber production, which is generally the leading function if none of the other functions is assigned the value 3. The revision for the first district FDPs, finished in 1981, are already being carried out, as each district has to be revisioned after a period of ten years.

3.4. DATA STORAGE AND ANALYSIS

The Ministry of Agriculture and Forestry (BMLF) performed a country wide digitization of the AFDP, using AUTOCAD (Autosketch) as a digitization tool and GKS-based customary software to plot the results. The thematic data are stored in an ORACLE database. In addition to this, the Federal Forest Authoritiy of several federal states have started individual activities using a GIS-equipment. In order to coordinate the activities at federal sate level, a general guideline for data structure of a digitized AFDP has been developed and published by the Ministry of Agriculture and Forestry.

3.5. RELIABILITY OF DATA: —

3.6. MODELS: —

3.7. INVENTORY REPORTS

The AFDP-maps are the Austrian maps at a scale of 1:50000 with coloured function areas drawn on them. Function areas smaller than 10 hectares are not mapped individually, but occasionally they are mapped as symbols if necessary. These areas smaller than 10 hectares are then included in the function area description of the surrounding larger function area.

3.8. FUTURE DEVELOPMENT AND IMPROVEMENT PLANS: —

3.9. MISCELLANEOUS

The suggested 'necessary actions' are of no legal relevance. They are a guideline for evaluating whether any private or public activity is in accordance with the statements and prescriptions made in the Forest Development Plan.

REFERENCES

Ahlsved, K-J. et al. 1979. Lexicon forestale. Werner Söderström, Osakeyhtiö

BMLF,1995. Österreichischer Waldbericht 1994. Bundesministerium für Land- und Forstwirtschaft, Wien, 121 p.

Braun, R., 1969. Österreichische Forstinventur - Methodik der Auswertung und Standardfehler-Berechnung. Forstliche Bundesveruchsanstalt, Wien, Nr. 84, 60 p.

FBVA, 1994. Instruktion für die Feldarbeit der Österreichischen Waldinventur 1992-19996. Forstliche Bundesveruchsanstalt, Wien, überarbeitete Fassung 1994, 193 p.

Hengst, M. 1966. Einführung in die Mathematische Statistik und ihre Anwendung. B.I. Hochschultaschenbücher, Bd. 42, Bibliographisches Institut Mannheim

Loetsch, F. & Haller, K. 1964. Forest Inventory; Vol. 1. München Basel Wien; BLV Verlagsges. München

ÖSTAT, 1994. Österreichischer Amtskalender 1994/95. Das Lexikon der Behörden und Institutionen. Österreichische Staatsdruckerei, Wien, Jg. 62

Ottisch, A. 1994. Geo-referenced Forest Information in Austria. In „Designing a system of nomenclature for European forest mapping", International Workshop Proceedings, European Forest Institute, Joensuu, Finland 1994, p. 159-173

Schieler, K. 1988. Methodische Fragen in Zusammenhang mit der österreichischen Forstinventur. Diplomarbeit, Univ. für Bodenkultur, Wien

Schieler, K., Buchsenmeister, R. & Schadauer, C. 1995. Österreichische Forstinventur. Ergebnisse 1986/90. Forstliche Bundesversuchsanstalt - Waldforschungszentrum, Wien, Bericht Nr. 92, 262 p.

Vaucher, H. 1986. Elsevier's dictionary of trees and shrubs. Elsevier, Amsterdam, 413 p.

COUNTRY REPORT FOR BELGIUM

Hugues Lecomte, Jacques Hébert and Jacques Rondeux,
Faculté Universitaire des Sciences Agronomiques
Gembloux, Belgium

Summary

1. Overview

Belgium is divided into 3 regions: Walloon, Flemish and Brussels regions. Each one has its own political autonomy in various fields, in particular the natural resources management.

Wooded area in the Walloon region covers 530,600 ha. A first temporary inventory has been realised in the beginning of the 1980s. A new permanent forest inventory is carried out since the beginning of the 1994.

The Flemish region will start the first permanent inventory of its 128,600 ha of wooded area in 1997.

With only 2,000 ha the Brussels region does not organise a forest inventory up to now.

The method used in the Walloon region is summarised below as the Flemish inventory method is quite similar.

2.1 Nomenclature

Forest inventory of the Walloon Region covers more than 500,000 ha of wooded area. These includes productive area and non-productive surfaces (firebreaks, roads, meadows, waste lands, ...). A lot of information is collected, namely, based upon a systematic rectangular grid (distances between the points are 1,000 m and 500 m):

- general and administrative information
- soil and vegetation observations
- dendrometric data (stands, trees)
- description of the stands (quality, health, ...).

The main general and administrative information is collected from maps; this concerns geographic region, ecological territory, ownership, ...

All the observations on soil and vegetation are field observations: topography (slope, altitude, relief, exposure, ...), pedology (humus, soil texture, drainage, ...), vegetation (at four different levels).

Information concerning the stands is: structure, type, age, silviculture (clear cutting, complete soil depth and profile, gap, stand regularity, regeneration, ...), forest conditions (health, quality, game damage), accessibility and harvesting.

Following measurements and observations are realised on each sample tree: species, girth at breast height (gbh), total height, dominant height, height at different levels (only for hardwood with gbh 120 cm and softwood with gbh 90 cm), health state and quality of wood.

The main derived values are density, basal area, volumes (for trees and stands, extrapolated to ha), mean girth, dominants girth and height, site index.

Stand type is defined according to the percentage of each species. Regeneration state is estimated and described in terms of percentage or stems numbers.

2.2 Data sources

The main data sources are sample plot on the field and maps. Aerial photographs are used only when available. No airborne digital remote sensing is used.

2.3 Assessment techniques

Walloon forest inventory is based upon a systematic grid of points where sample plots (single plots, no clusters) are located. The sample unit is composed of four circular plots which radii are related to the size of the sample trees.

The distances between points are 0.5 km (North-South) and 1 km (West-East).

No a priori stratification is defined.

2.4 Data storage and analysis

The data is stored in an ACCESS database, installed on a PC Computer.

Area estimation is made by points counting (1 point per 50 ha).

Area estimation error is determined by the BOUCHON's method (1975) derived from the MATTERON's theory (regionalised variables).

Error estimation for the other parameters and variables (basal area, volume, ...) is calculated by the random sampling error estimation method. Total error is a combination of both errors (area and variables).

The software is developed from ACCESS software and the hardware is a PC Pentium 100 Mhz with 16 Mega RAM and 1 Giga HD.

2.5 Reliability of data

Several procedures exist in the data treatments so that mistakes are as reduced as possible.

2.6 Models

13 different volume tables are used in order to determine the volumes of all the threes (solid wood of the stem).

Secondary species volumes are calculated as similar main species.

2.7 Inventory reports

Many results and papers have been published about the first forest inventory of the Walloon forest. They describe inventory methodology and present many results.

The latest results concerning the whole Wallonia were published in 1984. The new forest inventory has started two years ago and covers up to now only 20 % of the sample plots. Only theoretical and methodological papers are available for this new IFW.

Users of the results are mainly industry, forestry administration and scientific institutes.

2.8 Future development and improvement plans

The new forest inventory of Wallonia has started in 1994 and must be finished in 2003. No important changes will be made in a next future.

INTRODUCTION

This country report for Belgium on analyses of the existing forest inventory and survey systems is prepared as a contribution for the study of the European Forestry Information and Communication System (EFICS). The objective of this report is to provide a standardised description of the Belgian forest inventory systems according to the guidelines prepared by the European Forest Institute.

Belgium is divided into three parts (regions): Walloon region, Flemish region and Brussels region.

	Total area (km^2)	Population (1994) (1000)
Walloon region	16 845	3 313
Flemish region	13 521	5 866
Brussels region	162	952
Belgium	30 528	10 131

Forest policy is an attribute of each region so that Belgian forests are also divided into three parts: Flemish forests, Brussels forests and Walloon forests. The following table gives the distribution of forest area.

	Area (ha)	% of Belgian forest
Walloon region	530 600	80.4
Flemish region	128 600	19.3
Brussels region	2 000	0.3
Belgium	661 200	100 %

Reference year : 1st April 1996 (Poplar excluded).

Abbreviations :

IGN : "Institut Géographique National"
National Geographic Institute (maps author)

DNF : "Division de la Nature et des Forêts"
Forestry administration of the Walloon Region

IFW : "Inventaire Forestier Wallon"
Forest inventory of the Walloon Region

Definition of typical forest expressions:

"Semis": height < 150 cm

"Fourrés": height ≥ 150 cm with girth at 1.5 m < 10 cm

THE WALLOON REGION

1. OVERVIEW

1.1. GENERAL FORESTRY AND FOREST INVENTORY DATA

Walloon Region

Total forest area : 530,600 ha
Proportion of forested land : 31.5 %
Main tree species : spruce, oak, beech
Average volume per hectare : 258 m^3
Average increment per hectare : N.A.([1])
Ownership proportions : public : 48.5 %
private : 51.5 %
Number of inhabitants per hectare forested land : 6
Average size of a stand : N.A.

Name of the survey : **Forest inventory of the Walloon Region**
"Inventaire des Ressources ligneuses de Wallonie"
First assessment : 1979 - 1983 (temporary inventory)
Second assessment : 1994 - 2003 (first cycle of the permanent inventory)
Area covered : Walloon Region (South Part of Belgium).

Institution and organisation :
First assessment : Faculté universitaire des Sciences agronomiques de Gembloux
Second assessment :
1994 - 1996 Faculté universitaire des Sciences agronomiques de Gembloux
1996 - 2003 Division de la Nature et des Forêts - Ministère de la Région Wallonne (collect of data) and Faculté universitaire des Sciences agronomiques de Gembloux (scientific follow up).
Forest act : Décret du Gouvernement Régional Wallon du 16-2-1995 (Moniteur belge du 7/4/96).

([1]) N.A. = Not Available

1.2. OTHER IMPORTANT FOREST STATISTICS

1.2.1. Other forest data and statistics on the national level

1.2.1.1. Non-woods goods and services

-

1.2.1.2. Removals

-

1.2.1.3. Other important forest statistics

"Recensement décennal de l'Agriculture et des Forêts".
Source : Questionnaire completed by private and public owners.
Realised by the "INS" (Institut National des Statistiques) every 10 years.

Tables are published in a vulgarisation book. Last complete publication (public and private owners: 1976).

For the statistics specific to public forests "Ministère de la Région Wallonne, Division de la Nature et des Forêts".

1.2.2. Delivery of the statistics to UN and Community institutions

1.2.2.1. Responsibilities for international assessments

Mr Parmentier
Inspecteur en Chef-Directeur
Administration des Relations économiques
Ministère des Affaires économiques
Rue Général Leman, 60
B-1040 BRUXELLES
Phone : + 32 2 230 90 43
Fax : + 32 2 230 95 65

1.2.2.2. Data compilation

Joined evaluation with the help of

Regional forest administrations
Office Central du Commerce Extérieur (OCCE)
Institut Natonal des Statistiques (INS)
Fédération Nationale des Scieries et des Industries Connexes (FNS)
Fédération Belge des Exploitants Forestiers (FEDEMAR)
Fédération Belge des Entreprises de la Transformation du Bois (FEBELBOIS)
Fédération Nationale des Négociants en Bois (FNN)
Fédération Belge du Commerce d'Importation de Bois
Fédération Belge des Producteurs de Pâte, Papier et Cartons (COBELPA)

2. FOREST INVENTORY OF THE WALLOON REGION

The whole information (descriptions and presentations) concerning the forest inventory of the Walloon Region is included in the "Guide méthodologique de l'Inventaire forestier wallon" (Fac. Univ. Sc. Agron. de Gembloux, Février 1996, 208 p.).

2.1. NOMENCLATURE

2.1.1. List of attributes directly assessed

A) Geographic regions

Attribute	Data Source	Object	Measurement unit
Administrative forest unit (*Cantonnement*)	Map	Sample plot	categorical
Administrative forest sub-unit (*Triage*)	Map	Sample plot	categorical
Province (*Province*)	Map	Sample plot	categorical
Vegetal association (*Association végétale*)	Field observation	Stand	categorical
Ecological zoning (*Territoire écologique*)	Map	Sample unit	categorical
Forest Region (*Région forestière*)	Map	Sample unit	categorical
Locality (*Localité*)	Map	Sample unit	categorical

B) Ownership

Attribute	Data source	Object	Measurement unit
Kind of owner (*Type de propriétaire*)	Map	Sample unit	categorical

C) Wood production

Attribute	Data source	Object	Measurement unit
Species (*Espèce forestière*)	field observation	tree	categorical
Girth at breast height (*Circonférence à 1.5 m*)	field assessment	tree	cm
Total height (hardwood) (*Hauteur totale*)	field assessment	tree	m

Dominant height (softwood) *(Hauteur dominante)*	field assessment	tree	m
Height at different level (hardwood gbh > 120 cm) *(Hauteur à différents niveaux)*	field assessment	tree	m
Quality of wood (hardwood gbh > 120 cm) (softwood gbh > 90 cm) *(Qualité du bois)*	field assessment	tree	categorical
Health state *(Etat sanitaire)*	field assessment	tree	categorical

D) Site and soil

Attribute	Data source	Object	Measurement unit
Topography			
- relief *(Relief)*	field observation	stand	categorical
- exposure *(Exposition)*	field assessment	stand	categorical
- slope *(Pente)*	field measurement	sample unit	degree
- altitude *(Altitude)*	map	sample unit	m
Pedology			
- humus *(Humus)*	field observation	sample unit	categorical
- texture *(Texture)*	field observation	sample unit	categorical
- stone type *(Charge caillouteuse)*	field observation	sample unit	categorical
- stone quantity *(Abondance de la charge caillouteuse)*	field observation	sample unit	categorical
- drainage (*Drainage*)	field observation	sample unit	categorical
- soil profile (*Type de sol*)	field observation	sample unit	categorical
- soil depth (*Profondeur du sol*)	field observation	sample unit	categorical

E) Forest structure

Attribute	Data source	Object	Measurement unit
Age (only for softwood) (*Age*)	field measurement	stand	year

Attribute	Data source	Object	Measurement unit
Species (*Espèce*)	field observation	tree	categorical
Structure (*Structure*)	field observation	stand sample unit	categorical
Type (*Type*)	field observation	stand sample unit	categorical
Clear cutting (*Mise à blanc*)	field observation	stand	categorical
Complete gap (*Vide total*)	field observation	stand	categorical
Stand regularity (*Régularité*)	field observation	stand	categorical

F) Regeneration

Attribute	Data source	Object	Measurement unit
Species (*Espèce*)	field observation	stand	categorical
Realisation (*Réalisation*)	field observation	stand	categorical
Species (*Espèce*)	field observation	sample unit	categorical
Number (*Abondance*)	field measurement	sample unit	number per m^2

G) Forest condition

Attribute	Data source	Object	Measurement unit
Health state (*Etat sanitaire*)	field observation	stand	categorical
Stand quality (*Qualité du peuplement*)	field observation	stand	categorical
Game damage (*Dégâts de gibier*)	field observation	stand	categorical

H) Accessibility and harvesting

Attribute	Data source	Object	Measurement unit
Harvesting condition (*Conditions d'exploitation*)	field observation	stand	categorical

I) Attributes describing forest ecosystems

Attribute	Data source	Object	Measurement unit
Humus (*Humus*)	field observation	sample unit	categorical
Soil study (*Pedologie*)	field observation	sample unit	categorical

Vegetation observation (*Phytosociologie*)	field observation	sample unit	categorical
- trees : high level			
median level			
low level			
- plants			

J) Non-wood goods and services

-

K) Miscellaneous

Attribute	Data source	Object	Measurement unit
Map number (*Carte IGN*)	map	sample unit	-
Sample plot number (*Numéro de placette*)	map	sample unit	-
Date (*Date*)	-	sample unit	ddmmyy
Beginning hour (*Heure de début*)		sample unit	hhmm
End hour (*Heure de fin*)		sample unit	hhmm
X and Y co-ordinates (*Coordonnées X et Y*)	map	sample unit	hm (= 0.1 km)

2.1.2. List of derived attributes

A)

-

B)

-

C) Wood production

Attribute	Measurement unit	Input attributes
Single tree basal area (*Surface terrière individuelle*)	m^2	girth at 1.5 m
Single tree volume (*Volume individuel*)	m^3	girth at 1.5 m, total tree height or dominant height by species
Volume at different levels (*Volumes à différents niveaux*)	m^3	height level individual tree volume
Density/ha (*Densité*)	-	number of stems (number extended/ha)

Basal area/ha (*Surface terrière/ha*)	m^2	Sum of individual tree basal area extended to ha
Volume/ha([1]) (*Volume/ha*)	m^3	Sum of individual tree volume extended to ha
Basal area and volume by categories (trade, miscellaneous, ...) (*surface terrière et volume par catégorie*)	m^2 or m^3	tree basal area, tree volume
Mean girth (*Circonférence moyenne*)	cm	individual girths of trees
Dominant girth (*Circonférence dominante*)	cm	individual girths of dominant trees (mean)
Dominant height (*Hauteur dominante*)	m	individual height of dominant trees (mean)
Site index (*Indice de fertilité*)	categorical	age - dominant height (spruce, Douglas fir)
Basal area percentage by species (*Proportion de chaque essence en surface terrière*)	%	basal area by species total basal area

D)

-

E) Forest structure

Attribute	Measurement unit	Input attributes
Stand type (*Type de peuplement*)	categorical	basal area by species (proportion)

F) Regeneration

Attribute	Measurement unit	Input attributes
seedling (*semis*([2]))	%	percentage of area covered by regeneration
thicket (*fourrés*([2]))	%	percentage of area covered by regeneration
	-	number of stems
sapling (*gaulis*([2]))	%	percentage of area covered by regeneration
	-	number of stems

G)

-

([1]) Volumes are always solid wood volume over bark (volume of the stem to 7 cm diameter)

([2]) The definitions are given in introduction

H)

-

I)

-

J)

-

K)

-

2.1.3. Measurements rules for measurable attributes

For each of the measurable attributes the following information is given:

a) measurement rule
b) threshold values
c) measurement scale
d) rounding rules
e) instrument used for measurement
f) data source

C) Wood production

Girth at breast height (1.5 m) (Circonférence à 1,5 m)

a) measured at 1.5 m above ground, on slopes measured from uphill side. One reading with tape
b) minimum girth : 20 cm
c) cm, no classes
d) rounded to lower cm (0.0 to 0.4), to upper cm (0.5 to 0.9)
e) tape
f) field assessment

Total tree height (Hauteur totale individuelle)

a) length of the tree from ground level to the top of the tree, on slopes measured from equal side or uphill side
b) minimum tree height : 3 m
c) m
d) rounded to closest level : 0.00, 0.25, 0.50, 0.75
e) Blume-Leiss dendrometer
f) field assessment

Tree height at different levels (Hauteur de la grume à différents niveaux)

a) length of the tree from ground level to these levels (first default, trade level), on slopes measured from equal site or uphill side
b) minimum tree height: 2 m
c) m
d) rounded to closest level: 0.00, 0.25, 0.50, 0.75
e) Blume-Leiss dendrometer
f) field assessment

Age (Age)

a) measured at 0.30 m above ground level by radial boring
b) minimum age: 1 year
c) year
d) not rounded
e) Pressler borer
f) field assessment

D) Site and soil

Slope (*Pente*)
a) mean of downward and upward measurements
b) minimum: 0°
c) degree
d) rounded to nearest degree
e) Blume-Leiss dendrometer
f) field assessment

Stone quantity (*Abondance de la charge caillouteuse*)
a) estimate of the percentage of stones with a soil sampler
b) no threshold
c) in four classes: < 5 %, 5-14 %, 15-49 %, ≥50 %
d) rounded to closest class limit
e) ocular estimate
f) field assessment

Soil depth (*Profondeur du sol*)
a) measured by soil penetration with a soil sampler
b) no threshold
c) in five classes: 0-19 cm, 20-39 cm, 40-59 cm, 60-79 cm and ≥80 cm
d) rounded to closest class limit
e) soil sampler
f) field assessment

Altitude (Altitude)
a) altitude of the unit sample above the sea level
b) no threshold
c) m
d) not rounded
e) ocular observation
f) map

E) Forest structure

Stand type (*Type de peuplement*)
a) ratio of basal area covered by each tree species and total tree species
b) no threshold
c) limit between pure and mixed stand :
66.7 % (hardwoods)
80.0 % (softwoods)
d) not rounded
e) ocular estimate (based on the number of trees by species)
f) field assessment

F) Regeneration

Area covered by regeneration (Surface occupée par la régénération)
a) ratio of the area covered by regeneration (stems with girth at 1.5 m < 20 cm) and the area of the small sampling plot (radius = 2.25 m)
b) maximum girth: 19 cm at 1.5 m
c) % for the different regeneration levels seedling, thicket, sapling, number of stems (per ha) for thicket and sapling
d) not rounded
e) ocular estimate (area) and counting for thicket and sapling
f) field assessment

G) Forest condition

Health state (Etat sanitaire)
a) number of trees (diseased or dead)
b) no threshold
c) %, in four classes :
< 25 %,
25-49 %,
50-74 %,
≥75 %
d) not rounded
e) ocular estimate
f) field assessment

Stand quality (Qualité du peuplement)
a) number of trees with defaults
b) only for trees with girth above 120 cm (hardwood) or 90 cm (softwood)
c) %, in four classes :
< 25 %,
25-49 %,
50-74 %,
≥75 %
d) not rounded
e) ocular estimate and trees counting
f) field assessment

Game damage (Dégâts de gibier)

a) number of trees with defaults caused by game
b) for all trees with girth above 20 cm
c) %, in five classes:
 0 %,
 1-24 %,
 25-49 %,
 50-74 %,
 ≥75 %
d) not rounded
e) ocular estimate and trees counting
f) field assessment

H) Accessibility and harvesting

a) distance between the point sampling and the nearest way for truck
b) no threshold
c) m
d) rounded to the nearest 10 m
e) graduate ruler
f) map

2.1.4. Definitions for all attributes or nominal or ordinal scale

The following information is provided for each attribute:

a) the definition
b) the categories (classes) and
c) the data sources
d) remarks

A) Geographic regions

Forest unit (Cantonnement)

a) territory subdivision in administrative units, specific to the "Division Nature et Forêts", managed by a forest engineer who controls the respect of the forest laws, hunting, fishing and environment rules and manages the public forests
b) 36 different units grouped in seven local centres
c) specific maps of the DNF

Forest region (Région naturelle)

a) subdivision of the Walloon Region according to the soil types and geology substratum
b) seven regions; one of them (Ardenne) is subdivided into three sub-regions according to the altitude
c) maps obtained by the classification of each village or city in one of the regions based upon the "field knowledge" of the forest engineer

Ecological zoning (Territoire écologique)

a) hierarchical classification of the forest station based upon climate, geology stone type and soil
b) 27 ecological territories
c) ecological territories map (DELVAUX and GALOUX (1962) Les territoires écologiques du Sud-Est de la Belgique, Travaux hors série, Groenendael, Belgium, 2 vol. 315 pp.)

Vegetal association (Association végétale)

a) ecosystem characterised by the common presence of a community of herbaceous and woody species which makes it possible the classification of the forest station in terms of aptitude and productivity
b) 27 vegetal associations and 43 subdivisions
c) definition from the vegetation included in the sample unit
d) missing data for some sample units (according to the inventory season)

Province (Province)
a) administrative division of the Walloon Region
b) five provinces
c) administrative maps of Belgium

Forest sub-unit (Triage)
a) subdivision of the "*cantonnement*" including 300 to 400 ha of public forests managed by a forest ranger (technical level)
b) 500 *triages* in Wallonie
c) "*triages*" map (management map of the DNF)

B) Ownership

Kind of owner (Nature du propriétaire)
a) owner of the forest area
b) 1) region
2) province
3) commune
4) social assistance centre (CPAS - "*Centres Publics d'Aide Sociale*")
5) church patrimony and other public owners
6) natural reserves
7) national defence
8) private owners
c) specific map of public owners
d) confirmation of the owner by the forest ranger

C) Wood production

Stand quality (Qualité du peuplement)
a) stand classification according to the presence and frequency of the common defects
b) tree categories: good, normal or poor according to the damaged trees number:
< 25 %,
25-49 %,
50-74 %,
≥75 %
main defects:twisted fibre, bad conformation suckers presence, crutches, frost split dead branches, harvesting damages, rotten other defects
c) field observations
d) only for the high forests

Game damages (Dégâts de gibier)
a) stand classification among the frequency of game defects: (*abroutissements*), (*frottures*), peelings,
b) five categories:
stands with no damages,
low (< 25 %, damaged trees),
median (25 to 49 %),
high (50 to 74 %),
generalised (≥ 75 %)
c) field observations
d) the observations take into account the damage depth

Cuttings (Coupes)
a) estimation of the frequency and the intensity of the clearing or thinning
b) six classes: no cutting, slight cuttings, normal cuttings, hard cuttings, premature cuttings or late cuttings
c) field observations
d) concerns mainly softwoods stands in private ownership

Individual quality of trees (Qualité individuelle des grumes)
a) quality trees estimation based upon several criteria (form conformation, defects, damages, ...)
b) four classes for hardwoods :
1 (cabinet making),
2 (joinery),
3 (industry),
4 (rejects)
three classes for softwoods :
1 (framework),
2 (industry),
3 (rejects)
c) field observations
d) only for high forests

Health state of the trees (individual observations) (Etat sanitaire des arbres)
a) observation of the health state of each tree according to the vegetation aspect (defoliation, yellowing, sap flow, ...)
b) five classes :
1 (healthy tree),
2 (affection beginning),

3 (clearly declining),
4 (nearly dead),
5 (dead)

c) field observation

d) difficult observations for the hardwoods during the autumn, winter and the beginning of the spring

D) Site and soil

Relief (Relief)

a) survey of topography type at the site of the sampling unit

b) eight classes :
plateau = ground without slope,
dome = flat ground with slopes on each side,
side = ground with a regular slope,
depression = small valley between two sides,
terrace = not extended flat area in a slope,
complex relief = many types of relief are present,
revised relief = place where the ground has been modified,
abrupt = ground with a very hard slope (rock)

c) field observation

d) observation possible on map

Exposure (Exposition)

a) side orientation from the North

b) nine classes :
no orientation,
North,
N-E,
E,
S-E,
S,
S-W,
W,
N-W

c) field observation (compass)

d) if the slope is complex, the axe of the more important slope is only considered

Humus (Humus)

a) description and classification of the humus considered as station factor

b) seven categories :
(*mull calcique*),
mull,
(*moder mulleux*),
moder,
dysmoder,
mor,
turf

c) field observation

d) voluntarily simplified analysis

Soil texture (Texture du sol)

a) soil description and classification based upon its granulometric composition

b) nine classes :
limon (A)
sanded limon (L)
light sanded limon (P)
sand with limon (S), sand (Z)
light clay (E), heavy clay (U)
stay limon (G)
turf (V)

c) field observation (based upon a core)

d) classes similar to those used in soils cartography

Stone type (Nature de la charge caillouteuse)

a) description of the type of stones found in the soil

b) nine types :
limestone (k),
schist (f)
(*phyllade*) (fi),
(*grès schisteux*) (r),
sandstone (g)
(*psammite*) (Lp),
chalk (n),
(*dragées de quartz*) (o),
(*graviers de terrasses fluviales*) (t)

c) field observations (based upon a core)

d) classes similar to those used in soils cartography

Drainage (Classe de drainage)

a) soil classification according to water economy; observation based upon the depth where the first signs of gley and pseudogley are visible

b) nine classes of drainage :
excessive (a)
normal (b)
moderated (c): pseudogley deeper than 80 cm
imperfect (d): pseudogley from 50 cm
quite poor (h): pseudogley from 30 cm
poor (i): pseudogley before 30 cm
quite poor to poor (e): gley deeper than 80 cm
quite poor to very poor (f): gley from 30 cm
very poor (g): gley before 30 cm

c) field observations (based upon a core)

d) classes similar the those used in soils cartography

Soil profil (Type de sol)

a) classification of the soil profil in the localisation of the sample unit

b) nine types :
degraded brown soil = *sol brun lessivé* (a)
brown soil = *sol brun* (b)
degraded soil = *sol lessivé, dégradé* (c)
degraded calcareous soil = *sol d'altération de calcaire* (d)
little developed podzol = *sol podzolique peu développé* (f)
podzol = *podzol* (g)
undefined profile soil = *sol à profil non défini* (x)
undeveloped profile soil = *sol sans développement de profil* (p)

c) field observations at the sample unit

d) classes similar to those used in the soils cartography

Vegetation observations (Observations phytosociologiques)

a) observation and identification of the woody, herbaceous and moss species in the sample unit

b) three lagers: wood subdivided in
high level (> 10 m)
median level (3 - 10 m)
low level (< 3 m)
herbaceous
moss

c) field observation

d) observations in the 18 m radius plot for woody species and in the 9 m radius plot for herbaceous and moss species

E) Forest structure

Stand structure (Structure du peuplement)

a) classification of the structure of the sampled stand

b) eight types :
high forest with 1 level (layer)
high forest with 2 levels (layers)
uneven forest
plantation
young forest
coppices with standard
coppices
undefined structure

c) field observations

d) observations realised in the neighbourhood of the sample unit (in the stand which includes the samples plot), to be related to the observations on the aerial plots.

Stand type (Type de peuplement)

a) classification based upon species in the stand

b) ten types: stand with
beech
oak
precious hardwoods
mixed or others hardwoods
spruce
Douglas fir
larch
pine
mixed softwoods or other softwoods
poplar

c) field observations

d) classification based upon an estimation of the proportion of different species which are growing in the stand.

Clear cutting (Mise à blanc)

a) specific observations for the clear cuttings
b) collected data:
   time since cutting
   stand age at the cutting
   cutting causes:
      normal age (exploitability age)
      premature cutting
      cutting after storm damages
      other causes (expropriation, ...)
c) field observations

Complete gap (Vide total)

a) observation realised when the sample unit is completely located in a gap (= clearing in a stand)
b) collected data:
   species of the surrounding stand
   stand age
   probably causes of the gate:
   silviculture (bad cuttings)
   storm damages
   diseases
   other causes
c) field observations

Stand regularity (Régularité du peuplement)

a) uniformity and homogeneity of the stand from the point of view of the stems distribution on the field
b) four classes:
   regular distribution of the stand (good distribution)
   irregular density (bad distribution of the stems in the stand = gaps, high densities, ...)
   gaps in the stand
   high density in the stand
c) field observations
d) data collected in the stand where the sample unit is located

Sample unit structure (Structure de l'unité d'échantillonnage)

a) classification of the sample unit structure
b) eight classes :
   high forest with one level (layer)
   high forest with two levels (layers)
   uneven forest
   plantation
   young forest
   coppice with standards
   coppice
   undefined structure
c) field observations (on the sample unit)
d) structure independent of the surrounding stand structure

Sample unit type (Type de l'unité d'échantillonnage)

a) classification of the sample unit based upon the forest species which are identified on the sample unit (measured trees)
b) ten types: sample unit with
   beech,
   oak,
   precious hardwoods,
   mixed or other hardwoods,
   spruce,
   Douglas fir,
   larch,
   pine,
   mixed or other softwood,
   poplar
c) classification based upon computer definitions (basal area species)
d) classification independent of the type of the surrounding stand

Forest species (Espèces forestières)

a) forest species identification of each measured tree
b) identification of 33 hardwood species and 15 softwood species
c) field observations
d) rare or non-economic species classified as "other softwoods" or "other hardwoods"

F) Regeneration

Stand regeneration (Régénération dans le peuplement)

a) description of the natural regeneration in the stand and estimation of the possible necessity to regenerate the stand

b) four categories:
regeneration in progress (good for the future)
realised (existing regeneration)
none (no regeneration in the stand)
should be present (old stand without regeneration)

c) field observation

d) observation for the three main regeneration species

Regeneration on the sample unit (Régénération sur l'unité d'échantillonnage)

a) study of the natural regeneration on the area of the sample unit

b) three steps of development for the natural regeneration
seedling *(semis)* (height less than 1.5 m)
thicket *(fourrés)* (height more than 1.5 m, girth at 1.5 m < 10 cm)
sapling *(gaulis)* (height more than 1.5 m, girth between 10 and 19 cm)

c) field observation

d) observations and measurements for the three main regeneration species

G) Forest condition

Stand health (Etat sanitaire du peuplement)

a) observation of the global health state of the whole stand, according to the presence, the frequency and the decline degree of trees

b) six stand categories based upon the diseased trees number :
- no diseased tree
- diseased trees: rare and dispersed
- diseased trees: rare and grouped
- diseased trees:
  25 to 50 % of the stand
  50 to 75 % of the stand
  75 % and more of the stand

c) field observations

d) observations to correlate with aerial photos which were realised for forest decline study; identification of the decline causes (insects, fungi, undefined) and of the decline development state (initial, progressive, final, all states), ...

H) Accessibility and harvesting

Exploitation conditions (Conditions d'exploitation)

a) estimation of the stand harvesting conditions based upon the mechanisation possibilities and the distance to a road which can be used by a truck

b) three types of conditions:
easy (flat field, hard soil)
difficult because of the slope or the soil
impossible
estimation of the mean distance to access to a road for truck

c) field observation, map measurement (m)

d) the distance is measured from the plot centre

I) Attributes describing forest ecosystems

-

J) Non-wood foods and services

-

K) Miscellaneous

-

2.1.5. Forest area definition and definition of "Other wooded land"

Wooded area are considered and inventoried :

- forests (more than 100 ha)
- woods between 10 and 100 ha
- small woods with area between 10 ares and 10 ha.

Here is no crown cover specification: as the inventory is based upon a systematic grid which is applied on coloured topographic IGN map, all the sampling points included in the green coloured area are considered as wooded points.

Forest edge is defined as the right line between edge trees and does not follow the crowns covers.

The general aspect of the point and his localisation (town, village, near a big house or a castle) tend to consider this point as "park".

A sampling point located in forest area is considered as productive or not productive point if there is growing trees or not on the sampling unit.

The non-productive forest points are :

- firebreaks
- forestry ways and roads
- unforested area under electric lines
- natural or artificial meadows
- waste area
- heaths, peat bags, miry and marshy area
- slopes and banks of ways, roads, train lines located in forest area
- quarries in forest area
- nurseries
- arboreta
- ponds and rivers.

The stands with a width lower than 9 m are considered as lines and inventoried with a specific methodology.

Poplar stands (clumps or lines) are inventoried separately. Forest species lines are also considered in the inventory.

Are not sampled the following wooded area, which have no productive function: grounds, ornamental alignments, hedges, lonely trees, Christmas trees plantations.

2.2. DATA SOURCES

A) Field data

sample plot: general and administrative data
soil and vegetation data
dendrometric data
area estimation

B) Questionnaire
not used.

C) Aerial photography
(only used if available)
- company Walphot S.A.
- scale 1:25000 or orthophotoplan 1:10000
- infrared colour
- last survey: 1989
- instrument: common stereoscope
- incomplete air cover
- cost per photo: ± 1200 Belgian francs

D) Spaceborne or airborne digital remote sensing
not used

E) Map
- "Institut géographique national" (IGN),
Abbaye de la Cambre, 13, B-1050 Bruxelles
. date : from 1985 to 1996
. scale : 1:25000
. type : topographic map printed (multicolour)
- "Division Nature et Forêts" (Ministère de la Région Wallonne),
Avenue Prince de Liège, 7, B-5100 Jambes
. date: from 1980 to 1995
. scale: 1:25000
. type: forest map for ownership

F) Other georeferenced data
not used

2.3. ASSESSMENT TECHNIQUES

2.3.1. Sampling frame

The inventory covers the whole forest of Wallonia, independently of ownership and forest kind. As there is one point per 50 ha wooded land, nearly 10,600 sample units are located in Walloon forest.

Some of these points (about 1 percent) are not accessible for different reasons : relief (high slopes), or impossible approach (point located on an island).

2.3.2. Sampling units

Field sample plots

- Sample unit composed of four fixed area circular concentric plots :
radius of 2.25 m : regeneration (girth < 20 cm)

radius of 4.5 m :	high forest with girth from 20 to 69 cm (at the level 1.5 m) coppice (all girths)
radius of 9 m :	high forest (girth between 70 to 119 cm)
radius of 18 m :	high forest with girth of ≥120 cm

- Each plot area is calculated from the relation $S = \pi r^2 \cos \alpha$
  S = area (m^2)
  r = plot radius (m)
  α = slope (degree)

These sample plots are circular on the field: no slope correction is realised on the field but is calculated by computer procedures.

- Determination of an "plot expansion factor" to ha for each plot
  Fext = 10 000/S
  S = plot area (m^2)
- Permanent sample unit materialised by a metallic stake in the soil located at the plot centre; 4 trees colour marked ("witness trees"); numbered from 1 to 4 from North to West. All the measured trees are identified by their distance and azimuth from the centre.
- Systematic distribution of the sample units based upon a rectangular gird 1 km*0.5 km (1 km W-E and 0.5 km N-E axes).
- Sample unit centre located by a compass approach from a permanent field point; the azimuth is measured by compass and the distance by a "lost cotton yarn measurer". The covered distances are corrected according to the slope if necessary.

2.3.3. Sampling designs

The sampling used in the frame of the Forestry inventory of Wallonia is a systematic sampling. No stratification has been defined.

It is important to note that no clusters are used but single plots localised at the knots of the grid.

Aerial photos study is independent of the field work and will only be realised if the necessary documents are available.

The main objectives of the photos study are: stands cartography, area estimation.

2.3.4. Techniques and methods for combination of data sources

At this time, there is no combination between data obtained by aerial photos and by maps and field measurements.

GIS is nevertheless used for combination data found on maps with these collected in forest:

- province
- forest region

- ecological zoning
- coordinates X and Y
- city or village
- ownership (for public forests).

2.3.5. Sampling fraction

Data source and sampling unit	proportion of forested area covered by sample	represented mean area per sampling unit
field assessment area sample plots	0.1 %	50 ha

2.3.6. Temporal aspects

Inventory cycle	Time period of data assessment	Publication of results	Time period between assessments	Reference data
1st IFW	1979 - 1983	1984	-	1980
2nd IFW (permanent)	1994 - 2003	2004 (expected)	10 years	1994

2.3.7. Data capturing techniques in the field

Data recorded on tally sheets in the field and edited by hand into the computer at the office.

2.4. DATA STORAGE AND ANALYSIS

2.4.1. Data storage and ownership

The data are stored in a data-base system at the "Ministère de la Région Wallonne" "Direction Générale des Ressources Naturelles et de l'Environnement", "Division de la Nature et des Forêts", Namur (Jambes) who is the owner of the data.
(Phone: +32 81 32 12 11, Fax: +32 81 32 12 63).

It is possible to ask for the results of the first inventory at the FUSAGx, Gembloux, Prof. J. Rondeux, Phone/Fax: + 32 81 62 23 01; Email: "rondeux@fsagx.ac.be".

2.4.2. Database system used

The data are stored in an ACCESS database (installed on a PC computer).

2.4.3. Data bank design

The Data Bank Design is presented here.

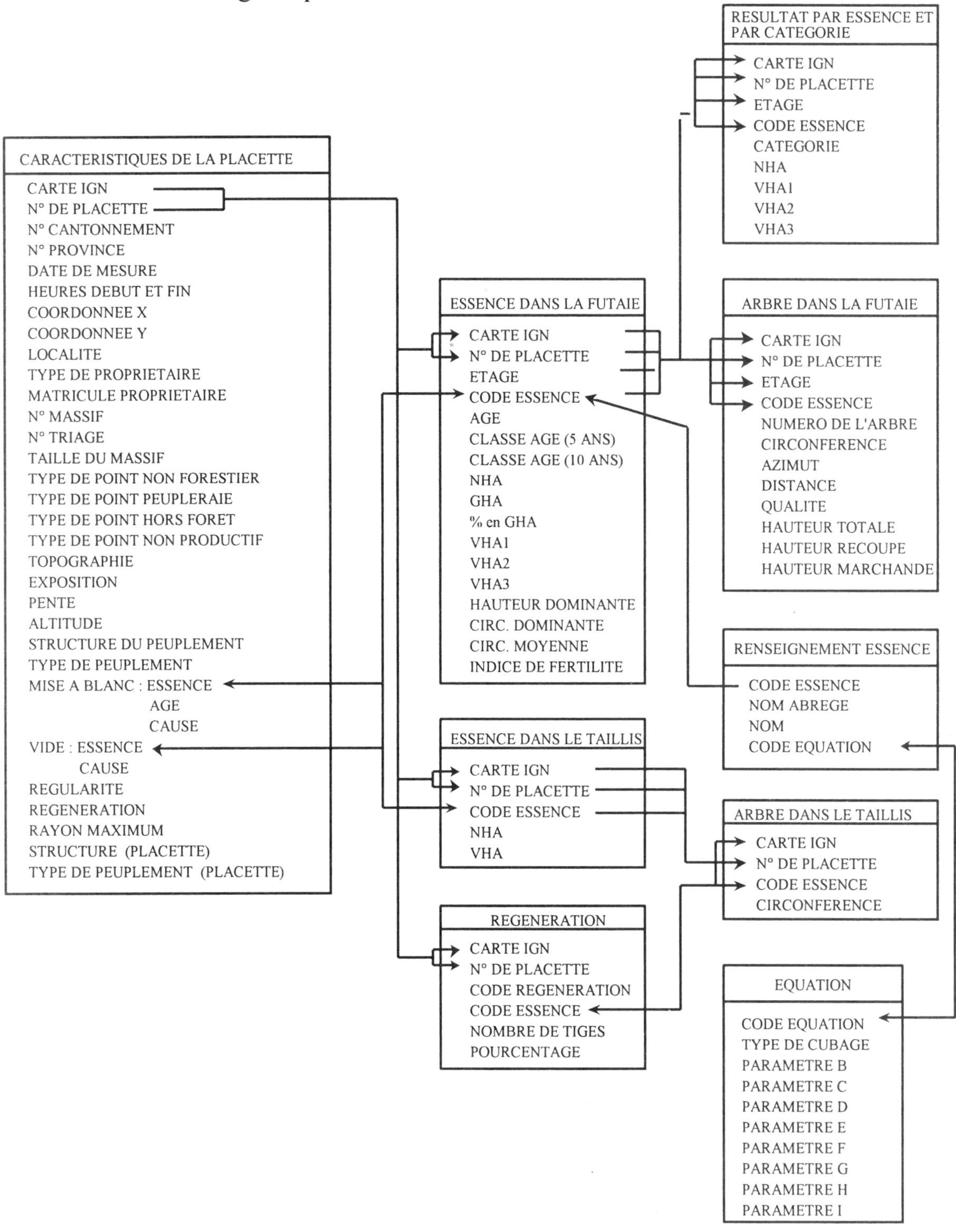

2.4.4. Update level

For each sample unit, the date of data collection is available. Until now there is no reference date and the data are not updated to a common point in time. They reflect the situation at the time of assessment (the new inventory has started only two years ago). Error estimate is not published.

2.4.5. Description of statistical procedures used to analyse data including procedures for sampling error estimation

a) Area estimation

Area estimations are dealing with forest units identified on a map or a photo but also with some particular results like the area distribution according to the age. For this reason, the adopted area estimation method is dot-counting or point-counting.

$$S_n = n\,a\,b$$

where S_n = area estimation (ha)
n = number of dots within the area
a and b = lengths of the grid unit (hm = 0.1 km)

As we use a systematic dot-grid, the standard error cannot be estimated according to the general rules of the random and simple sampling. We have to take into account the form of the area but often we have only the coordinates of the point located inside the area to be estimated. We use the BOUCHON's method (1975) derived from the MATTERON's theory (regionalised variables) and confirmed by PERROTTE in 1981. (See especially BOUCHON, J. and TOMIMURA, S. (1979) Comparaison des mesures de surfaces par comptage de points. Annales des Sciences Forestières 36(4): pp.321-330)

$$\hat{\sigma} = \frac{100}{n}\sqrt{\frac{N_2}{12} + 0.305\frac{N_1^2}{N_2}}$$

where= $\hat{\sigma}$ =standard error (%)
n = number of dots within the selected area
N_i = values obtained by separate addition of the number of horizontal and vertical grid unit sides located along the perimeter of the area formed by all the grid units build around the selected dots. N_1 is the larger of both values.

The best way is to give an example.

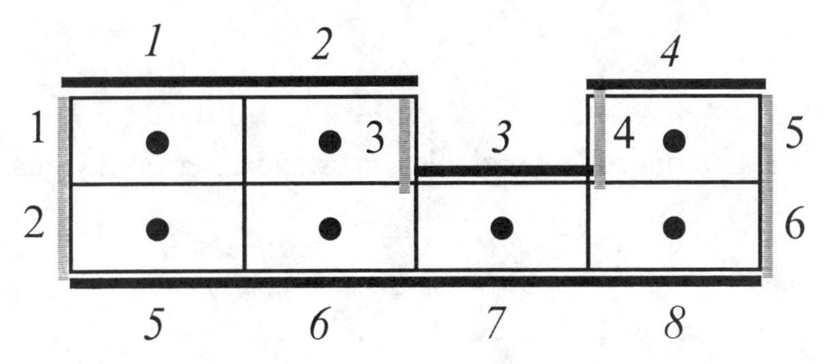

In this case, we have

n = 7
S_n = 7 a b
N_1 = 8
N_2 = 6

$$\hat{\sigma} = \frac{100}{7}\sqrt{\frac{6}{12} + 0.305\frac{8^2}{6}} = 13\%$$

b) Aggregation of tree and plot data

The aggregation of single tree data is done by weighting each single tree attribute by its plot expansion factor. As we use concentric sample plots, the plot expansion factors are different for each sub-plot and take into account the slope a.

Girth at 1.5 m (c)	Sample plot radius	Plot expansion factor (w)
< 20 cm	2.25 m	628.8 / cos(α)
20 - 69 cm	4.5 m	157.2 / cos(α)
70 - 119 cm	9 m	39.30 / cos(α)
> 120 cm	18 m	9.824 / cos(α)

The total values for one plot can be calculated by summing the single tree attributes multiplied by their plot expansion factor. As an example, the estimation of the volume per ha VHA for the sample plot i is computed as follows

$$VHA_i = \sum_j v_{ij} \, w_{ij}$$

where v_{ij} is the volume estimation for the tree j of the sample plot i and w_{ij} is the plot expansion factor for the tree j of the sample plot i.

In the same way, we can compute the arithmetic mean girth for the sample plot i :

$$\bar{c}_i = \sum_j c_{ij} \, w_{ij} \Big/ \sum_j w_{ij}$$

c) Estimation of total values

From n points selected from our database according to different criteria, we can compute the various mean values (VHA,GHA, NHA) with their standard errors

$$\bar{y} = \frac{1}{n}\sum_{i=1}^{n} y_i \quad \text{and} \quad \hat{\sigma}_{\bar{y}} = \sqrt{\frac{\sum_{i=1}^{n}(y_i - \bar{y})^2}{n\,(n-1)}}$$

as each estimation gained from the different n points has the same weight. The total value for the area S_n is estimated as follows

$$y_{tot} = \bar{y} S_n$$

As far as we can consider $\bar{y}$ and S_n as independent variables, the standard error of y_{tot} can be computed as

$$\hat{\sigma}_{y_{tot}} = \sqrt{\hat{\sigma}^2_{\bar{y}} \, \hat{\sigma}^2_{S_n} + \hat{\sigma}^2_{\bar{y}} \, S_n^2 + \hat{\sigma}^2_{S_n} \, \bar{y}^2}$$

and generally simplified as

$$\hat{\sigma}_{y_{tot}} = \sqrt{\hat{\sigma}^2_{\bar{y}} \, S_n^2 + \hat{\sigma}^2_{S_n} \, \bar{y}^2}$$

d) Estimation of ratios

As all our estimations (volume per ha, number of stems per ha, ...) are linked to the centre of the plot (point sampling), we never have an estimation of ratio. As an example, volume per ha is a random variable which takes a different value for each plot depending on the coordinates of the centre of the plot. It is not the estimation of a ratio. For the other kinds of ratio like proportion of two areas, the information on standard value is given only for the area, not for the ratio.

e) Sampling at the forest edge

If necessary, the centre of the sample plot is shifted in such a way that the whole plot is just included in the stand where the centre of the plot was initially located. When it is not possible (stand too narrow), the plot area is adapted and the plot expansion factors are corrected (very rare situation).

f) Estimation of growth and growth components (mortality, cut, in growth)

No definitive decision has been taken on this subject. In fact, our permanent forest inventory has started in 1994 with a 10 years periodicity. The next occasion at an old place is planned in 2003.

g) Allocation of stand and area related data to single sample plots/sample points

We have chosen the "point decision": as we can move the centre of the sample plot in order to have the whole plot in the same stand, it is not necessary to assign information to a proportion of the plot area. In case of several species (mixed stand), information is distributed in proportion of the basal area of each species.

h) Hierarchy of analysis: how are sub-units treated?

Many selections of points can be produced according to the species, the kind of owner, the administrative division, ... but also the age, the altitude, the slope, the mean distance to the nearest road or the importance of damage. All these kinds of selections are treated as a whole. The only problem is to aggregate enough points to get relevant estimations. Generally, 30 sample points are considered as a minimum.

2.4.6. Software applied

Home made software developed with MS-Access

2.4.7. Hardware applied

Data capture: paper sheet
Data storage: PC pentium 100 Mhz, 16 Mega RAM, 1 Giga HD
Data analysis: idem
Operating system: Window

2.4.8. Availability of data (raw and aggregated data)

All data can be available for anyone with the authorisation of the Walloon Region. The database cannot be consulted directly but results tables can be obtained from the IFW service.

For special requests with specific analysis and/or program, costs will be added to the user.

2.4.9. Subunits (strata) available

Results are available for the Walloon Region, the five provinces and for the "cantonnements" (units of the forest administration). It is possible to obtain results for forest regions, cities or villages (if these contains enough sample units).

Results are also available for other criteria, by example: ownership, structure and type stand, forest species, age, slope classes, soil types, wood quality, girth and height or productivity classes, ...

All the information, observations and variables collected may be used as selection criteria to present results.

2.4.10. Links to other information sources

Topic and spatial structure	Age/period availability	Responsible agency	Kind of data-set
maps for public ownership and cantonnements	permanent	DNF	map
maps for provinces, cities, villages and forest regions	permanent	IGN	map

2.5. RELIABILITY OF DATA

2.5.1. Check assessments

No field check.

2.5.2. Error budgets

Only statistic errors are considered and estimated. Classification and measurement errors are not included in the work.

2.5.3. Procedures for consistency checks of data

When the data are collected on the field plot, a first check is realised by the field team, in order to verify if all the information, observations and measures are collected and are correct.

A new control is done at the office when the tally sheets arrive from the forest; data is recorded in the database and just after this operation, a new complete checking is realised by another operator so that all the data are verified one by one to eliminate recording errors.

Some procedures have been introduced in the software in order to detect possible errors (especially for measurement attributes) by searching limit values and by plotting correlated variables.

2.6. MODELS

2.6.1. Volumes

a) Outline of the model

Volumes are calculated from volume tables created for 13 main species. For the other species, the most suitable is chosen. The volume is the solid wood volume of the stem over bark (until 7 cm diameter of the stem). Branch volumes are not included.

Three volume tables are used :

volume table based upon the girth at 1.5 m level
volume table based upon the girth and the dominant height
volume table based upon the girth and the total height for each tree

The single tree volume equations have been published in 1985 by "*Les presses agronomiques de Gembloux*" under the title "*Tarifs de cubage des arbres et des peuplements forestiers*" and signed by DAGNELIE P., PALM R., RONDEUX J. and THILL A.

N°	Main species	Species treated as main species
1	Beech	Hornbeam, others hardwoods, hazel, elders
2	Birch	Aspen, Willows
3	Scots pine	All pines
4	Elm	
5	Wild cherry	Walnut, crab tree, pear tree
6	Larch	
7	Ash	Sorb
8	Sycamore maple	All maples, lime
9	Spruce	Stick spruce, Tsuga, Thuya, Cypress, other coniferous
10	Douglas fir	All firs
11	Red oak	
12	Oak	Alders
13	Poplar	All poplars

b) Overview of prediction errors

No model.

c) Data material for deviation of model

Volumes tables constructed from measurements of nearly 7,000 (exactly 6,800) trees in the Walloon forests (DAGNELIE *et al.*, 1985: p 15).

d) Method applied to validate the model

Due to the large quantity of measurements, no validation has been conducted with external data.

2.6.2. Assortments

At stand level, assortment volume table is available for spruce (according to the stand mean girth) (DAGNELIE, PALM, RONDEUX and THILL., 1988. Tables de production relatives à l'épicéa commun, Les presses agronomiques de Gembloux. 123 p.). At tree level, various assortment volume tables are available for all the main species (12) (DAGNELIE et al., 1985)

2.6.3. Growth components

Site index for two species is identified, based upon age and dominant height of the stand (reference age : 50 years).

Yield tables are used for spruce and Douglas fir, as "stands models".

References:

for spruce: Dagnelie, P., Palm, R., Rondeux, J. & Thill, A., 1988. Tables de production relatives à l'épicéa commun, Les presses agronomiques de Gembloux. 123 p.

for Douglas fir: Rondeux, J., Laurent, C. and Thibaut, A., 1991. Construction d'une table de production pour le douglas (Pseudotsuga menziesii (Mirb.) Franco) en Belgique, Les Cahiers forestiers de Gembloux 3. 23 p.

2.6.4. Potential yield

The method is defined for species treated in even-aged stands. The needed information are area and volume per ha for all the age classes, growth model like yield table and a silviculture model giving the proportion of clean cutting for each age class.

Our method is described in HEBERT, J. and LAURENT, C., 1995. Estimation de la disponibilité de la ressource forestière. Revue Forestière Française 47 (5): pp.572-580.

2.6.5. Forest functions

-

2.6.6. Other models applied

Single tree model is available for beech.

2.7. INVENTORY REPORTS

2.7.1. List of published reports and media for dissemination of inventory results

Inventory	Year of publication	Citation	Language	Dissemination
First IFW	1981	Principaux résultats relatifs à l'inventaire des massifs forestiers de la province de Liège. FSAGx, Gembloux. 47p.	French	printed
	1982	Principaux résultats relatifs à l'inventaire des massifs forestiers de la province de Luxembourg. FSAGx, Gembloux. 64 p.	French	printed
	1983	Principaux résultats relatifs à l'inventaire des massifs forestiers de la province de Namur. FSAGx, Gembloux. 65 p.	French	printed
	1983	Principaux résultats relatifs à l'inventaire des massifs forestiers de la province du Hainaut et du Brabant Wallon. FSAGx, Gembloux. 47 p.	French	printed
	1984	La pessière wallonne en chiffres. Bull. Soc. R. For. Belg. 91 (3): pp. 89-98	French	printed
	1984	Principaux résultats relatifs à l'inventaire des massifs forestiers de la Région Wallonne. FSAGx, Gembloux. 66 p.	French	printed
	1984	Guide méthodologique de l'Inventaire Forestier Wallon, FSAGx. 170 p.	French	printed
	1985	Quelques considérations chiffrées sur la forêt feuillue wallonne. Annales de Gembloux 92: pp. 111-125	French	printed

	1986	Quelques statistiques récentes sur la forêt wallonne. Bull. Soc. R. For. Belg. 93 (1)1-22	French	printed
	1988	Considération sur la structure actuelle des hêtraies en Ardenne et Région jurassique (Implications en matière de traitement sylvicole). Bull. Soc. Roy. For. Belg. 95 (6):279-293	French	printed
	1991	La pessière wallonne : son évolution entre 1980 et 1990. Sylva Belgica 1992, 99 (4):7-14	French	printed
Second IFW (new since 1994)	1994	Comparaison de plusieurs types d'unités d'échantillonnage dans la perspective d'un inventaire forestier régional. Forestry Chronicle 70 (3):304-310	French	printed
	1994	L'inventaire forestier régional wallon: brève présentation méthodologique. Sylva Belgica 101 (6):9-16	French	printed
	1995	L'inventaire forestier régional: un outil de développement régional. Wallonie, 34:3-8	French	printed
	1996	Inventaire des Ressources Ligneuses de Wallonie. Guide méthodologique. Faculté Universitaire des Sciences Agronomiques de Gembloux, Gembloux. 208 p.	French	printed

2.7.2. List of contents of latest report and update level

Reference of the latest complete report (which contains results about the whole Wallonia): Principaux résultats relatifs à l'inventaire des massifs forestiers de la Région Wallonne. Faculté des Sciences agronomiques de Gembloux. 66 pp. (1984).

Chapitre	Titre	Nombre de pages
1	Présentation générale	2
2	Caractéristiques générale de l'inventaire	1
3	Nature des données récoltées	1
4	Traitement et exploitations des données	2
5	Détermination et contrôle des surfaces boisées	13
6	Quelques résultats types relatifs à l'inventaire	41
7	Conclusions	2
8	Bibliographie	2

2.7.3. Users of the results

Main users of the results of the inventory:

- "Division de la Nature et des Forêts" (forestry administration)
- industry:
  paper mills,
  panel factories,
  sawmills
- members of the forest sector
- private experts
- scientific members (research centres and teaching faculties).

Main questions are put by the industry and the forestry administration.

2.8. FUTURE DEVELOPMENT AND IMPROVEMENT PLANS

The new Walloon forest inventory is planned to be realised between 1994 and 2003. The data collected and analyses procedures will be slightly modified but no important change will be made in a next future. The checking of the field assessment by an independent team will be organised.

The realisation of the inventory is guaranteed by a regional act.

2.9. MISCELLANEOUS

-

THE FLEMISH REGION

1. OVERVIEW

1.1. GENERAL FORESTRY AND FOREST INVENTORY DATA

Total Flemish area: 152,730 ha with poplar
128,577 ha without poplar
Proportion of forested land: 11.3 % (9.5 without poplar)
Main tree species: hardwoods 49,562 ha (32.4 %)
poplar 25,153 ha (16.5 %)
softwoods 53,608 ha (35.1 %)
mixed 15,339 ha (10.0 %)
bare terrain 9,069 ha (6.0 %)
Average volume per hectare: N.A.
Average increment per hectare: N.A.
Ownership proportion: public forests: 35,856 ha (23.5 %)
private forests: 116,874 ha (76.5 %)
Number of inhabitants per hectare forested land: 38.4 (45.6 without poplar)
Average size of a stand: N.A.

Name of the survey: **Forest inventory of the Flemish Region (*Bosinventarisatie Vlaams Gewest*)**
First assessment: First field data collected 1 January 1997
Area covered: Flemish Region
Institutes and organisations involved in the assessments:
Faculty agricultural and applied biological Sciences, University of Gent (Belgium)
(*Faculteit Landbouwkundige en Toegepaste biologische wetenschappen Rijk Universiteit te Gent*)

1.2. OTHER IMPORTANT FOREST STATISTICS

See item 1.2 in the report for the Walloon Region.

2. FOREST INVENTORY OF THE FLEMISH REGION

(BOSINVENTARISATIE VLAAMS GEWEST)

The following information comes out the publication "Concept bosinventarisatie Vlaams Gewest" (Fac. Landbouwkundige en Toegepaste biologische wetenschappen, Gent, Juli 1995. 80 p.)

2.1.1. List of attributes directly assessed

Attribute	Data source
Date of the assessment (*opnametijdstip*)	(field)
year (*jaar*)	
month (*maand*)	
day (*dag*)	
Administrative data (*administratieve gegevens*)	(map)
province and locality (*provincie en gemeente*)	
administrative district (*bestemming*)	
function (*bestemming*)	
Ownership (*eigenaarscategorie*)	
state (*staat*)	
region (*gewest*)	
province (*provincie*)	
locality (*gemeente*)	
other public owners (*overige publiekrechtelijke eigenaars*)	
private (*privé*)	
Stand type (*opstand type*)	(field)
hardwoods (*loofhout*)	
softwoods (*naaldhout*)	
mixed hardwoods (*gemmengd loofhout*)	
bare terrain (covers recently harvested areas, as well as fire-devastated blocks or simple bare soils) (*te bebossen oppervlakten*)	
open area (includes areas of a forest complex having no timber production function) (*overige oppervlakten behorend tot het bosdomein*)	
Age and development state (*leeftijd and ontwinkkelingsfase*)	(field)
age classes:	
0	
1-10	
11-20	
21-40	

41-60
61-80
81-100
101-120
121-140
141-160
> 160
uneven aged

development steps:

null step or deforested area (*onbebost*)
first step (*jongwas*) tot complete cover (height≥ 2m)
second step (*dichtwas*) from complete cover to 7 cm dbh
third step (*stookhout*) 7 cm < dbh≤14 cm
forward step (*boomhout*) dbh > 14 cm

Structure *(structuur)*

high forest (*hooghout*) vertical: 1 or more levels
horizontal: individual, in groups, homogeneous
coppice with standards (*middelhout*)
coppice (*hakhout*)
to be determined (= bare terrain) (*te bepalen*)
not applicable (= open areas) (*niet van toepassing*)

Crown cover (*sluitingsgraad*) (field)

<1/3
1/3-2/3
>2/3
not applicable (*niet van toepassing*)

Dendrometric measures and data (field)
(*dendrometrische gegevens*)

tree coordinates (*coordinaten van de boom*)
tree species (*boomsoort*)
girth at 1.5 m (*omtrek op 1,5 m*)
level (*stade*)
height (*totale hoogte*)
bark thickness (*schorsdikte*)
branch free height (*takvrije stamlengte*)

Coppice (*hakhout*)

tree species (*boomsoort*)
girth at 1.5 m (*omtrek op 1,5* m)

Coppice with standards (*middelhout*)

same data as high forest and coppice

Regeneration (*verjonging*)

tree species (*boomsoort*)
regeneration type (*verjongingswijze*):
natural
artificial
mixed

- number of stems (*aantal bomen*):
  - < 500
  - 500 - 1499
  - 1500 - 2499
  - 2500 - 4999
  - 5000 - 9999
  - > 10000
- mean height (*gemiddelde hoogte*):
  - 0 - 50
  - 50 - 100
  - 100 - 150
  - 150 - 200
  - > 200
- distribution: individual, in groups, homogeneous

Quality (*kwaliteit*) (field)

- four classes:
  - A cabinet making (*schil- en fineerhout, meubelhout*)
  - B joinery (*constructiehout*)
  - C industry (*industriehout*)
  - D rejects (*pulphout, spaanders, brandhout*)

Vegetation (*vegetatie type*) (map)

- data collected on vegetation map

Site conditions (*standplaats*)

- soil type
  - texture (*textuurklasse*)
  - drainage (*drainageklasse*)
  - profil (*profielontwikkeling*)
  - data collected from soil map (*gegevens van bodemkaart*)
- topography (*topografie*)
  - altitude (*hoogteligging*) — on map (topographic map)
  - slope (*helling*) — on field
  - exposure (*expositie*) — on field

Health state (*gezondheidstoestand*)

- data collected by "*Instituut voor Bosbouw en Wildbeheer*" (grid of 4*4 km)

Social function of the forest

- accessibility (*toegankelijkheid*)
- opening (*opening*)
- relaxation infrastructure (*recreatieve infrastructuur*)
- waste (*afval*)
- penetration of the forest out of the paths (*betreding buiten de paden*)
- enclosures (*omheiningen*)
- proximity of coffee shops, riding schools, sport clubs, ... (*nabijheid van horeca, maneges, sportclubs*)

2.1.2. List of derived attributes

Tree parameters (*boomparameters*)
mean girth (*gemiddelde omtrek*)
mean height (*gemiddelde hoogte*)
mean bark thickness (*gemiddelde schorsdikte*)
mean branch free height (*gemiddelde takvrije stamlengte*)

Stand parameters (*bestandsparameters*)
number of stems (N) (*stamtal*)
girth distribution (*stamtalverdeling*)
< 19 cm
20-29 cm
30-39 cm
...
140 - 149 cm
150 - 159 cm
> 159 cm

Data collected from
basal area (G) (*bestandsgrondvlak*)
dominant height (*dominante hoogte*) (Hdom)
timber volume (*werkhout volume*)
calculated from volume tables

Dead trees (*dood hout*)
number (*aantal*)
mean girth (*gemiddelde omtrek*)
mean height (*gemiddelde hooghte*)

2.1.3. Measurement rules for measurable attributes

See above or not available.

2.1.4. Definitions for attributes on nominal or ordinal scale

See above or not available.

2.1.5. Forest area definition and definition of "Other Wooded Land"

Forest area :

area ≥ 50 ares
width ≥ 25 m
cover ≥ 20 % (excepted for clear cutting or forest damages).

2.2. DATA SOURCES

A) field: see above

B) Questionnaire: not used

C) Photos: orthophotoplans scale 1:5000 infra-red colour

D) Spaceborne or airborne digital remote sensing: not used

E) Maps : maps realised by the Flemish Region mapping ("*Boskartering van het Vlaamse Gewest*") :

1. map (base map) with administrative boundaries, roads, ways, ...
2. stand type map
3. stand development map with development stage, crown cover, structure
4. ownership map
5. topographic map

Map available in digital or analogue format (scale for 1:5000 to 1:25000).

F) Other georeferenced data :

soil maps (1:20000),
topographical maps (1:50000),
land use planning map (1:25000) (*gewestplannen*) (maps with indications of the function/destination (*bestemming*) of the land, for example agriculture, forestry, residential area, nature reserve),
vegetation map (1:25000) (*biologische waarderingskaart*)

2.3. ASSESSMENT TECHNIQUES

2.3.1. Sampling frame

The inventory covers the whole forest of Flemish Region, independently of ownership.

2.3.2. Sampling units

4 circular plots :

1) r=2.25 m
   seedlings with total height < 2 m
2) r = 4.5 m
   coppiceand trees with girth at 1.5 m < 22 cm and total height ≥ 2 m
3) r = 9 m
   trees with girth at 1.5 m ≥ 22 cm and < 122 cm
4) r = 18 m
   trees with girth at 1.5 m ≥ 122 m

2.3.3. Sampling designs

Two phases are proposed :

phase 1: wooded/non wooded area classification from "Vlaams boskartering" maps
phase 2: field work on the points localised in forested area. Grid of 1 km*0.5 km.

2.3.4. Techniques and methods for combination of data sources

Use of a GIS ARCINFO and a database management system.

2.3.5. Sampling fraction

Grid of 1 km*0.5 km
Proportion of forested area covered by sample: 0.1 %
Represented mean area per sampling unit: 50 ha

2.3.6. Temporal aspects

Three field teams together for a three-years data collection.

2.3.7. Data capturing techniques in the field

A Microflex 9500 field computer will be used.

2.4. DATA STORAGE AND ANALYSIS

2.4.1. Data storage and ownership

Data are stored in a data base system.

Ownership: "Afdeling Bos en Groen" (Flemish Forestry Commission).

2.4.2. Database system used

Relational database.

2.4.3. Data bank design

See table next page.

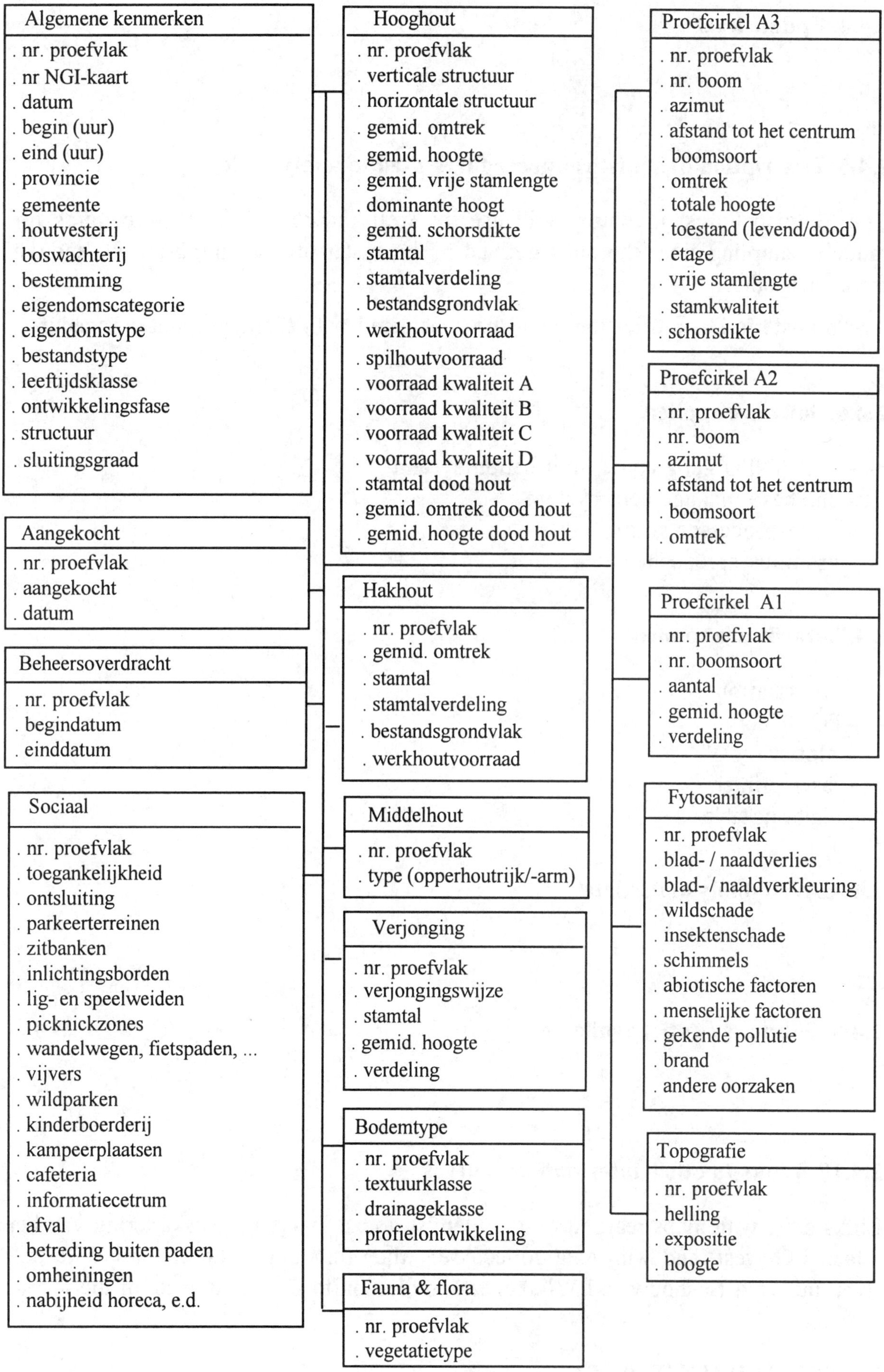

Algemene kenmerken
. nr. proefvlak
. nr NGI-kaart
. datum
. begin (uur)
. eind (uur)
. provincie
. gemeente
. houtvesterij
. boswachterij
. bestemming
. eigendomscategorie
. eigendomstype
. bestandstype
. leeftijdsklasse
. ontwikkelingsfase
. structuur
. sluitingsgraad

Aangekocht
. nr. proefvlak
. aangekocht
. datum

Beheersoverdracht
. nr. proefvlak
. begindatum
. einddatum

Sociaal
. nr. proefvlak
. toegankelijkheid
. ontsluiting
. parkeerterreinen
. zitbanken
. inlichtingsborden
. lig- en speelweiden
. picknickzones
. wandelwegen, fietspaden, ...
. vijvers
. wildparken
. kinderboerderij
. kampeerplaatsen
. cafeteria
. informatiecetrum
. afval
. betreding buiten paden
. omheiningen
. nabijheid horeca, e.d.

Hooghout
. nr. proefvlak
. verticale structuur
. horizontale structuur
. gemid. omtrek
. gemid. hoogte
. gemid. vrije stamlengte
. dominante hoogt
. gemid. schorsdikte
. stamtal
. stamtalverdeling
. bestandsgrondvlak
. werkhoutvoorraad
. spilhoutvoorraad
. voorraad kwaliteit A
. voorraad kwaliteit B
. voorraad kwaliteit C
. voorraad kwaliteit D
. stamtal dood hout
. gemid. omtrek dood hout
. gemid. hoogte dood hout

Hakhout
. nr. proefvlak
. gemid. omtrek
. stamtal
. stamtalverdeling
. bestandsgrondvlak
. werkhoutvoorraad

Middelhout
. nr. proefvlak
. type (opperhoutrijk/-arm)

Verjonging
. nr. proefvlak
. verjongingswijze
. stamtal
. gemid. hoogte
. verdeling

Bodemtype
. nr. proefvlak
. textuurklasse
. drainageklasse
. profielontwikkeling

Fauna & flora
. nr. proefvlak
. vegetatietype

Proefcirkel A3
. nr. proefvlak
. nr. boom
. azimut
. afstand tot het centrum
. boomsoort
. omtrek
. totale hoogte
. toestand (levend/dood)
. etage
. vrije stamlengte
. stamkwaliteit
. schorsdikte

Proefcirkel A2
. nr. proefvlak
. nr. boom
. azimut
. afstand tot het centrum
. boomsoort
. omtrek

Proefcirkel A1
. nr. proefvlak
. nr. boomsoort
. aantal
. gemid. hoogte
. verdeling

Fytosanitair
. nr. proefvlak
. blad- / naaldverlies
. blad- / naaldverkleuring
. wildschade
. insektenschade
. schimmels
. abiotische factoren
. menselijke factoren
. gekende pollutie
. brand
. andere oorzaken

Topografie
. nr. proefvlak
. helling
. expositie
. hoogte

Structure database for the forest inventory in the Flemish region.

2.4.4. Update level

-

2.4.5. Description of statistical procedures used to analyse ate

The Flemish Forest Inventory will use the well-known statistical procedures for random sampling since the plots defined in the systematic sampling are far enough from each other.

An "a posteriori" stratification could be considered with the appropriate procedures.

2.4.6. Software applied

- ARC/INFO geographic information system
- database management system
- word processor software
- graphics spreadsheet

2.4.7. Hardware applied

- workstation
- PC 80586
- plotter
- laser printer
- digitising table.

2.4.8. Availability of the data

-

2.4.9. Subunits (static) available

-

2.4.10. Links to other information sources

Links exist with maps realised by the Flemish forest mapping (Boskartering van het Vlaams Gewest) and with data collected by the "Instituut of Forest and Wildlife" (Instituut voor Bosbouw and Wildbeheer) in relation to the health state of the forest.

2.5. RELIABILITY OF DATA

Information not available.

2.6. MODELS

Information is only available for volume estimation.

Volume tables are used for determination of timber volume.

DAGNELIE, PALM, RONDEUX and THILL., 1985. Tables de cubage des arbres et des peuplements forestiers. Les Presses Agronomiques de Gembloux, Gembloux. 148 p. For birch, oak, red oak, maple, ash, beech, wild cherry, elm, spruce, Douglas fir, larch, scots pine.

FABER and DIK., 1968. De samenstelling van inhouds- en opbrengsttabellen voor Pinus nigra Arm. in Nederland. Stichting Bosbouwproefstation "De Dorschkamps", Wageningen. 78 p. For Corsica pine.

FABER and TIEMENS., 1975. "De opbrengstniveaus van populier". Rijksinstituut voor onderzoek in de bos- en landschapsbouw "De Dorschkamp", Wageningen. 117 p. Birch volume tables are used also for willow and alder.

Ash volumes tables are used for other hardwoods and spruce volumes tables are used for other softwoods.

2.7. INVENTORY REPORTS

Vandenbil, V. & Janssens, F., 1989. Een inventarisatianet voor de bossen in Vlaanderen, Een eerst survey. Ministerie van de Vlaamse Gemeenschap. Dienst Waters en Bossen. 120 p.

Waterinckx, M., 1990. Simulatie van een systematische bemonstering tbv de bosinventarisatie in het Vlaamse Gewest. Afstudeenwerk. Faculteit van de Landbouwwetenschappen. R.U.G. 84 p.

Waterinckx, M. & Goossens, R., 1995. Concept bosinventarisatie Vlaams Gewest. Faculteit van Landbouwwetenschappen, R.U.G. 80 p.

2.8. FUTURE DEVELOPMENT AND IMPROVEMENT PLANS

A forest inventory in the Flemish Region is planned for the next months. The report which has been used to write this paper contains a lot of propositions in order to carry out this inventory.

No more information is available at this time.

2.9. MISCELLANEOUS

-

THE BRUSSELS REGION

The Brussels region has a total forest autonomy but because of its size, there is no statistical inventory.

1. OVERVIEW

1.1. GENERAL FORESTRY DATA

Total forest area: 2,008 ha
Proportion of forest land: 12.4 %
Main tree species: beech

Ownership proportions:	public forest	1,940 ha (97 %)
	private forest	68 ha (3 %)

No other data available.
Information source:

Institut Bruxellois pour la Gestion de l'Environnement, Service Espaces Verts, Eaux et Forêts. X. LEJEUNE
Gulledelle 100, 1200 Bruxelles
Phone : 32 2 775 77 37 ; Fax : 32 2 775 76 11

1.2. OTHER IMPORTANT FOREST STATISTICS

See report for the Walloon Region.

2. NO FOREST INVENTORY

COUNTRY REPORT FOR DENMARK

Peter Munk Plum
National Forest and Nature Agency
Division of Forest Management Planning
Copenhagen, Denmark

1. OVERVIEW

Forest inventories have not been conducted in Denmark, as e.g. in Finland, Sweden, Norway and many other countries. The reason is that nearly all Danish forests are cultivated forests on former agricultural land. They predominantly consist of tree-species and provenances from the Northern hemisphere. Therefore, almost every Danish forest stand is small - less than 1 ha in average - with a great diversity between stands as regards to tree-species, provenances, age, stand-structure, etc. This situation makes the forest inventories difficult to conduct.

The reason for the current situation is entirely historical. After a number of lost wars between about 1600 until 1815 it was necessary to have material for warships, houses, tools, fuel, and enough agricultural land in a diminishing country with an increasing population. Around the year 1800 the country, which was fully covered by natural (broad-leaved) forest during the bronze-age about 1500 years BC, it only had about 2 % of forest which was in a very poor condition.

A that time a huge afforestation scheme was put into force and - after some changes - this policy is still essential to Danish forest policy.

At the same time it was decided to establish forest statistics based on questionnaires to different forest owners. During the period 1810 to 1990 thirteen such forest statistics have been made.

According to the Danish forest law of 1989 the Government, National Forest and Nature Agency's duty is to provide new national forest statistics at least every ten years. Therefore, a new national forest statistic shall be finished before the year of 2000.

The national forest statistics are naturally part of the total Danish statistical system. Besides these statistics there are statistics on agriculture, wood-production and consumption, working conditions, number of forest owners, studies on forest health, biodiversity, use of forests for recreational purposes, etc. (see below).

2. NAME OF THE SURVEY

The main source for this presentation is:

Skove og plantager 1990 / Forests 1990
Miljøministeriet, Skov- og Naturstyrelsen / Ministry of the Environment, National Forest and Nature Agency, and
Danmarks Statistik / The National Statistical Office of Denmark.

2.1. NOMENCLATURE

2.1.1. List of attributes directly assessed

The following attributes are addressed by the survey:

- the name of the forest
- the cadastral conditions for the forest
- the nearest road
- the commune or communes the forest are placed in
- the type of ownership
- the number of employees
- area of the different tree-species according to the list shown below
- supplementary area to the forestry ("other wooded areas")
- areas divided in ten-years-classes splitted out to the tree-species according to the list shown below
- average production-class of each of these ten-years-classes

2.1.2. List of derived attributes

There are no actual derived attributes to the forest statistic apart from (besides in many cases) supplementary information on Christmas-tree- and greenery-area on noble fir and Caucasian fir split up to five-year-classes for the 20 youngest years. Also a specific survey has been conducted on the annual cut based on the information from a small number of forests.

2.1.3. Measurement rules for measurable attributes

Its is important to underline that the national forest statistical bodies do not measure private land. Consequently the information concerning these areas is therefore entirely based on questionnaires.

The information on different types of ownership is therefore based on the following sources:

Type of ownership	Area
State forests	184 000 ha
Danish Heath Society (Danish Land Develop Company) *(Hedeselskapet)*	100 000 ha
Other private plans with forest management plans	100 000 ha
Private forest areas without forest management plans	76 000 ha

In other words about 84 % of the forest statistic should be based on forest management plans with stand-level information. The forest owner is not obliged to follow specific rules for the measurable attributes. Therefore, the forest statistic relies very much on the reliability of each forest owner.

However, specific principles in forest management plans are generally used:

- each stand over 0.1 hectare is normally measured individually
- the age is normally registered yearly in the forest management plan or elsewhere and this information is taken from earlier management plans
- the density of the stand is normally measured sample areas
- the diameter of trees is normally measured in no more than 10 % of the stand when the trees are thicker than about 25 cm. If a stand is very valuable more trees are measured.
- the average height together with the diameter is measured of 10 - 20 trees if the height is more than about 10-15 meters. On the basis of the information on age and height the production class is calculated.

2.1.4. Definitions for attributes on nominal or ordinal scale

Information on forest structure or regeneration methods is not included in the forest statistics. It is possible that some information in this field will be included in the forest statistics around year 2000. This could be a consequence of the efforts that have been made on the conservation of “natural forests”, i.e. mainly old broad-leaved stands.

2.1.5. Forest area definition and definition of “other wooded land”

In the forest statistics “other wooded land” are areas which, in a cadastrial sense are connected to forest land. But these areas are be a minor part of the forest ownership, up till 20 - 30 percent.

2.2. DATA SOURCES

Normally the cadastral system. The following definitions for forest areas are used:

- the forest is to be or should be planted with tree-species which can develop on the location into high trees with stems, that would say at least, to the height of 6 meters
- the area should be more than 0.5 hectares
- the area should be wider than 20 meters (at least on average)

2.3. ASSESSMENT TECHNIQUES

2.3.1. Sampling frame

Every ownership bigger than 0.5 hectares was asked to inform the Statistical Bureau about the mentioned figures. The forest statistic is in any way not followed by public samplings. The private owners can independently make samples and many field sampling systems is probably used (n.b.: if samples at all are used).

Normally, sample plots are not put into a specific sampling frame, but measurements are mostly made in the most valuable stands.

The forests included in the statistic have differed a little bit during the time. A rough overview is as follows:

Forest Statistics, year of:	Number of forests, sizes of forests, main or all tree species, type of ownership	Age class and site index distributions, and for the last statistics, standing volume, annual increment and annual cuttings
1880-1896	Only entailed estates, bigger land-development plantations and community-owned estates	Made by the national forest research station
1907	Forest area and tree species: all forests over 0.5 ha, in principle. Ownership groups: all ownerships over 50 ha	Age-class distributions: all ownerships over 50 ha
1923	As in 1907	As in 1907, plus removal in the period of 1917-1923
1931	As in 1923	As in 1923
1951	All forests: number of forests, forest area, distribution between classes of tree species and ownership structure	For forests over 50 ha: as in 1931
1965	As in 1951	As in 1951, plus questionnaires from about 20% of forest ownerships under 50 ha and over 0.5 ha
1976	As in 1965	As in 1965
1990	As in 1976	As in 1976

2.3.7. Data capturing techniques in the field

As mentioned, data capturing in the field is only made in the management planning, not by national forest inventories. Each owner can decide which system he wishes to use. Normally the data capturing is more intense in more valuable stands than in less valuable. In general, one to ten measurements are made in the stands measured, according to the general measurement-capacity in the planning process and the expected value of the specific stand. To design with an overall view of a specific stand will be made including a judgement of the density of the stand.

This is followed by tree-measurements. Many techniques are used here, for instance:

- measurement of heights and diameter of a number of trees on a typical part of the stand, or
- the same in "Bielefelt-circles", or
- measurement of each tree or a part of the trees in a part of the rows, or
- measurement of all trees in sample plots in triangles (for instance with 30 or 40 meter of
  sides according to the size of the trees) or in circles, or
- measurement of all trees in the stand (extremely valuable stands)

2.4. RELIABILITY OF DATA

The reliability is expected to be very good. Formally, no check is made. There are no specific check-assessments or error budgets.

2.6. MODELS

2.6.1. Volume (functions, tariffs)

Volume and increment functions exists for all major tree-species in Denmark as follows:

- Norway spruce, most of the country
- Norway spruce, mid and west Jutland
- Norway spruce, south eastern part of the country
- Sitka spruce
- Douglas fir
- silver fir
- mountain pine
- Scots pine, Western part of Jutland
- Scots pine, most of the country
- contorta pine
- Japanese larch
- beech
- oak
- ash

- hybrid aspen
- birch
- red alder

Up till now most of the table are based on the measurements done in the forests.

Computerised models have been made from these tables to be used in computer systems for forest management planning.

A project will begin in 1996 concerning computerised, dynamic yield tables.

2.6.2. Assortments

The yield tables are divided in a different group of yield-classes.

2.6.3. Growth components

Normally the following growth components are incorporated in the yield tables:

- age
- number of trees per hectare
- height of an average tree with a medium ground surface of one tree at the height of 1.3 meter
- diameter of the same tree
- total ground surface of one hectare
- proportion between the real tree and a cylinder with the same height and diameter as the
  medium tree in the stand
- saleable wood over 5 centimetres

2.6.4. Potential yield

The potential yield is simply calculated from the figures.

2.6.5. Forest functions

Other forest functions such as wood production is not included in the national forest statistics.

Anyhow, a number of analyses describe other forest functions of the wood. But these analyses are not made on a regular basis but either as investigations or as a part of the physical planning for open land which regularly are carried out by the counties.

The main "other functions" which, in this way, are described are the following:

- nature conservation interests such as flora, fauna, nature types (heather, lakes, strands, meadows, strand-meadows, prehistoric remains, etc.)
- the use of forests for leisure-purposes
- groundwater quality under forests

2.7. INVENTORY REPORTS

2.7.1. List of published reports and media for dissemination of inventory reports

The following forest statistics can be compared (please see the literature list):

Author or institution	Year of the statistic
August Niemann, 1809	1809
Christian Olufsen	1811
Adolph Frederik Bergsøe	1847
Peter Erasmus Müller	1876
Danmarks Statistik (the National Statistical Office of Denmark)	1907
Danmarks Statistik	1923
Danmarks Statistik	1931
Danmarks Statistik	1951
Danmarks Statistik	1965
Danmarks Statistik	1976
Danmarks Statistik and Skov- og Naturstyrelsen (National Forest and Nature Agency)	1990

A summary of these statistics is shown in the figure below. The figure is central in the understanding of the Danish forest history, problems in the past and big efforts to build up new forest land.

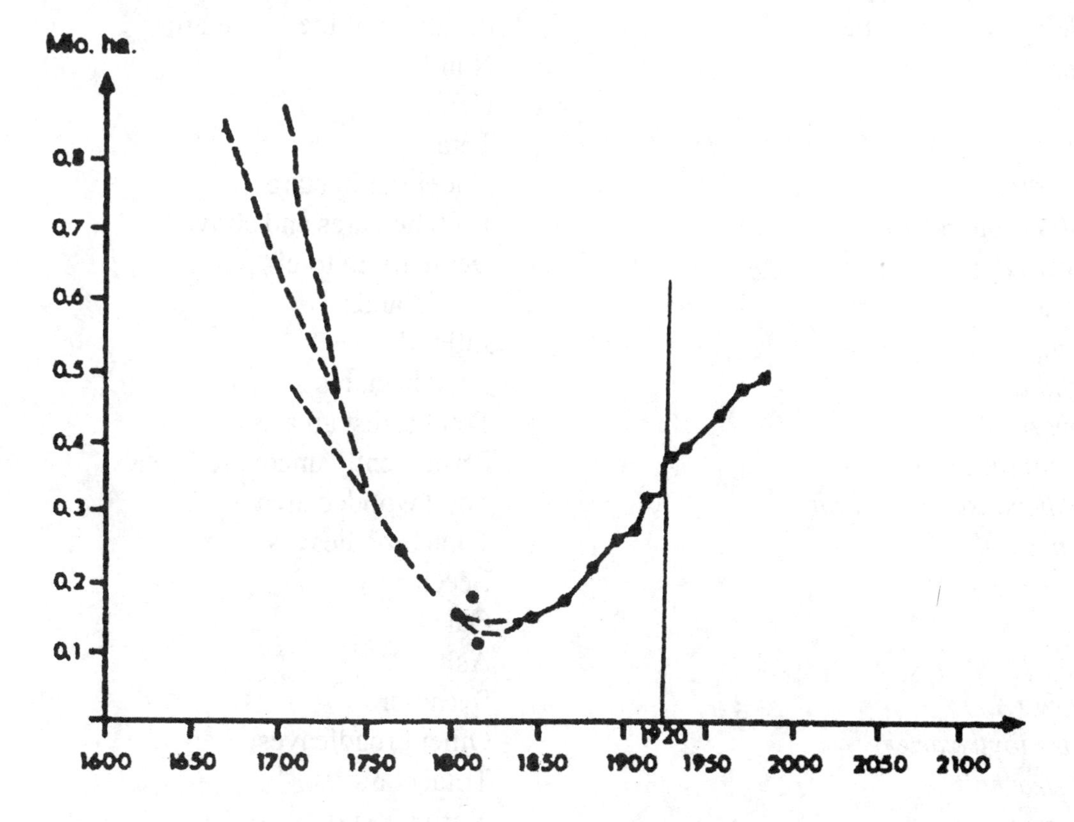

Figure 1. The development of the forest area in Denmark. (Plum 1988). The line at 1920 represent year when The Kingdom of Denmark took over The Northern part of Schleswig.

2.7.2. Users of the results

The forest statistics are used by a very broad group of individuals, amongst them school and university students, etc., organisations, institutions, politicians, etc. It is difficult to estimate where the greatest use of the statistics is (and there have not been made any analysis in this field).

2.8. THE CONTENT OF THE LATEST FOREST STATISTIC

The contents of the latest forest statistics is listed below. The forest statistics are enclosed.

PLEASE REMARK:

Tables on age-class distributions are always calculated with the intervals: year 1980 - 1989, 1970 - 1979, 1960 - 1969 and so on. How many years the ageclasses go back in time is mentioned in the list below.

Tables on site-class distributions are always calculated as the medium productivity class - defined by increment over bark at average felling age of the group.

Structure in the forestry sector. 1990

Andel af driftsenheder pct.	Share of forestry units pct.
Andel af skovareal pct.	Share of forest area pct.
Andel af driftsenheder i forhold til andelen af skovarealet	Share of forestry units compared to the share of the forest area
Antal	Number
Ha	Hectares
I alt	Total
Under 0,5 ha	Under 0.5 hectare
1000 ha og derover	1000 hectares and above
Hele landet	Denmark in total
Øerne	The Islands
Jylland	Jutland
1000 ha	1000 hectares
Skovareal i alt	Total forest area
Ubevokset areal	Permanently uncovered area
Skovbevokset areal i alt	Total wooded area
Løvtræ i alt	Total broadleaves
Bøg	Beech
Eg	Oak
Ask	Ash
Ahorn (Ær)	Sycamore
Andre løvtræarter	Other broadleaves
Nåletræ i alt	Total conifers
Rødgran	Norway spruce
Sitkagran mv.	Sitka spruce and other spruce
Ædelgran mv.	Silver fir and other fir
Nobilis	Noble fir

Nordmannsgran	Caucasian fir
Bjergfyr mv.	Mountain pine and lodgepole pine
Andre nåletræarter	Other conifers
Midlertidigt ubevokset	Temporarily uncovered area

Wooded area by species. 1990

Bøg	Beech
Eg	Oak
Ask	Ash
Ahorn (Ær)	Sycamore
Andre løvtræarter	Other broadleaves
Nåletræ i alt	Total conifers
Rødgran	Norway spruce
Sitkagran mv.	Sitka spruce and other spruce
Ædelgran mv.	Silver fir and other fir
Nobilis	Noble fir
Nordmannsgran	Caucasian fir
Bjergfyr mv.	Mountain pine and lodgepole pine
Andre nåletræarter	Other conifers
Midlertidigt ubevokset	Temporarily uncovered area

Area of broadleaves. 1965, 1976 and 1990

Bøg	Beech
Eg	Oak
Ask	Ash
Andre løvtræarter	Other broadleaves
1000 ha	1000 hectares

Area of conifers. 1965, 1976 and 1990

Gran og ædelgran	Spruce and fir
Bjergfyr	Mountain pine
Andre nåletræarter	Other conifers
1000 ha	1000 hectares

Forest area by species and county. 1990

Skovpct.	Percentage of land area
Skovareal i alt	Total forest areal
Hjælpearealer	Auxiliary areas
Skovbevokset areal i alt	Total wooded area
Løvtræ i alt	Total broadleaves
Nåletræ i alt	Total conifers
Midlertidigt ubevokset	Temporarily uncovered area
Pct.	Percentage
Ha	Hectares
Hele landet	Denmark in total
Amt	County

Age-class distribution of beech. 1990

Ha	Hectares
Før 1840	Pre- 1840

Age-class distribution of oak. 1990

Ha	Hectares
Før 1840	Pre- 1840

Age-class distribution of Norway spruce. 1990

Ha	Hectares
Før 1870	Pre- 1870

Age-class distribution of Sitka spruce and other spruce. 1990

Ha	Hectares
Før 1870	Pre- 1870

Age-class distribution of all fir (including Noble fir and Caucasian fir). 1990

Ha	Hectares
Før 1870	Pre- 1870

Average yield class by planting year classes: Beech, oak, ash and sycamore. 1990

Løvtræ	Total
Bøg	Beech
Eg	Oak
Ask	Ash
Ahorn (Ær)	Sycamore
Ha	Hectares
PK	Yield class
I alt	Total
Før 1840	Pre- 1840

Average yield class by planting year classes: Norway spruce, Sitka spruce and other spruce, silver fir and other fir. 1990

I alt	Total
Rødgran	Norway spruce
Sitkagran mv.	Sitka spruce and other spruce
Ædelgran mv.	Silver fir and other fir
Ha	Hectares
PK	Yield class
I alt	Total
Før 1870	Pre- 1870

Average yield class by size classes. Beech, oak, Norway spruce and Sitka spruce and other spruce. 1990

Bøg	Beech
Eg	Oak
Rødgran	Norway spruce
Sitkagran mv.	Sitka spruce and other spruce
PK	Yield class
Ha	Hectares

Estimated, standing volume overbark 1990 and expected mean annual volume increment 1990-2000

Mio. m^3 stående vedmasse	Standing volume in millions of cubic metres overbark
Gnst. årlig tilvækst mio. m^3	Mean annual volume increment in millions of cubic metres overbark
Bøg	Beech
Eg	Oak
Ask	Ash
Ahorn (Ær)	Sycamore
Andre løvtræarter	Other broadleaves
Rødgran	Norway spruce
Sitkagran mv.	Sitka spruce and other spruce
Ædelgran mv.	Silver fir and other fir
Nobilis, nordmannsgran	Noble fir, Caucasian fir
Bjergfyr mv.	Mountain pine and lodgepole pine
Andre nåletræarter	Other conifers

Wooded area by ownership classes. 1965, 1976 and 1990

Offentlig sektor	The National Forest and Nature Agency, other stateowned forests, benefices, counties and municipalities
Hedeselskabet	Danish Land Development Service
Fonde mv.	Foundations, organisations and independent institutions
Selskaber mv.	Joint-stockcompanies, partnerships, cooperative societies and other societies
Privatejede	Privately owned forests

Forest area by species and ownership classes. 1990

I alt	Total
Privatejede	Privately owned forests
Fonde mv.	Foundations, organisations and independent institutions
Hedeselskabet	Danish Land Development Service
Selskaber og foreninger	Joint-stock companies, partnerships, cooperative societies and other societies

Skov- og Naturstyrelsen	The National Forest and Nature Agency
Staten iøvrigt	Other stateowned forests
Præsteembeder	Benefices
Amter og kommuner	Counties and municipalities
Ha	Hectares
Skovareal i alt	Total forest area
Hjælpearealer	Auxiliary areas
Skovbevokset areal	Wooded area
Midlertidigt ubevokset areal	Temporarily uncovered area
Løvtræ i alt	Total broadleaves
Bøg	Beech
Eg	Oak
Ask	Ash
Ahorn (Ær)	Sycamore
Andre løvtræarter	Other broadleaves
Nåletræ i alt	Total conifers
Rødgran	Norway spruce
Sitkagran mv.	Sitka spruce and other spruce
Ædelgran mv.	Silver fir and other fir
Nobilis	Noble fir
Nordmannsgran	Caucasian fir
Bjergfyr mv.	Mountain pine and lodgepole pine
Andre nåletræarter	Other conifers

Forests and forest area by ownership classes and forest administration classes. 1990

I alt	Total
Privatejede	Privately owned forests
Fonde mv.	Foundations, organisations and independent institutions
Hedeselskabet	Danish Land Development Service
Selskaber og foreninger	Joint-stock companies, partnerships, cooperative societies and other societies
Skov- og Naturstyrelsen	The National Forest and Nature Agency
Staten iøvrigt	Other stateowned forests
Præsteembeder	Benefices
Amter og kommuner	Counties and municipalities
Antal	Number
Ha	Hectares
I alt	Total
Eget personale	Own forestry staff
Hedeselskabet	Danish Land Development Service
De Danske Skovdyrkerforeninger	The Association of Danish Forest Owner Co-operations
Anden bistand	Other forestry administration
Ingen bistand	No forestry administration

Forests and forest area by forestry administration classes. 1990

Eget personale	Own forestry staff
Hedeselskabet	Danish Land Development Service
De Danske Skovdyrkerforeninger	The Association of Danish Forest Owner Co-operations
Anden bistand	Other forestry administration
Ingen bistand	No forestry administration
Antal	Number
Ha	Hectares
Skovareal (ha)	Forest area (hectares)
Antal skove	Number of forests

Annual cut by species. 1978 - 1991

Bøg	Beech
Eg	Oak
Andre løvtræarter	Other broadleaves
Nåletræ	Conifers
1000 m^3	Thousands of cubic metres

Annual cut by assortments. 1978 - 1991

1000 m^3	Thousands of cubic metres
I alt	Total
Løvtræ i alt	Total broadleaves
Gavntræ	Sawnwood
Brænde	Fuelwood
Nåletræ i alt	Total conifers

Age-class distribution of Noble fir and Caucasian fir. 1990

Hele landet	Denmark in total
Øerne	The Islands
Jylland	Jutland
Nobilis	Noble fir
Nordmannsgran	Caucasian fir
Ha	Hectares
I alt	Total
Før 1870	Pre- 1870

Age-class distribution of Noble fir and Caucasian fir. 1990

Nobilis	Noble fir
Nordmannsgran	Caucasian fir
Ha	Hectares

Forests and forest area by region and size classes. 1965, 1976 og 1990

Hele landet	Denmark in total
Øerne	The Islands

Jylland	Jutland
Antal	Number of forests
Ha	Hectares
I alt	Total
Under 0,5 ha	Under 0.5 hectare
1000 ha og derover	1000 hectares and more

Forests and forest area by size classes and county. 1990

Skovens størrelse i ha	Size of forest (hectares)
I alt	Total
1000 og derover	1000 hectares and more
Antal	Number of forests
Ha	Hectares
Hele landet	Denmark in total
Amt	County
Øerne	The Islands
Jylland	Jutland

Forests 25 ha and above with area in 2 or more municipalities. By size classes and county. 1990

Skovens størrelse i ha	Size of forest (hectares)
I alt	Total
1000 og derover	1000 hectares and more
Kommuner	Municipalities
Antal	Number of forests
Hele landet	Denmark in total
Amt	County

Forest area by species and region. 1881-1990

1000 ha	1000 hectares
Hele landet	Denmark in total
Skovareal i alt	Total forest areal
Ubevokset areal	Permanently uncovered area
Skovbevokset areal i alt	Total wooded area
Løvtræ i alt	Total broadleaves
Bøg	Beech
Eg	Oak
Ask	Ash
Ahorn (Ær)	Sycamore
Andre løvtræarter	Other broadleaves
Nåletræ i alt	Total conifers
Rødgran	Norway spruce
Sitkagran mv.	Sitka spruce and other spruce
Ædelgran mv.	Silver fir and other fir
Nobilis	Noble fir
Nordmannsgran	Caucasian fir

Bjergfyr mv.	Mountain pine and lodgepole pine
Andre nåletræarter	Other conifers
Blandet bevoksning	Mixed forest
Midlertidigt ubevokset	Temporarily uncovered area
Øerne	The Islands
Jylland	Jutland

Forest area by species and municipality. 1990

Skovareal i alt	Total forest area
Hjælpearealer	Auxiliary areas
Skovbevokset areal i alt	Total wooded area
Midlertidigt ubevokset areal	Temporarily uncovered area
Løvtræ	Broadleaves
Løvtræ i alt	Total broadleaves
Bøg	Beech
Eg	Oak
Ask	Ash
Ahorn (Ær)	Sycamore
Andre løvtræarter	Other broadleaves
Nåletræ	Conifers
Nåletræ i alt	Total conifers
Rødgran	Norway spruce
Sitkagran mv.	Sitka spruce and other spruce
Ædelgran mv.	Silver fir and other fir
Nobilis	Noble fir
Nordmannsgran	Caucasian fir
Bjergfyr mv.	Mountain pine and lodgepole pine
Andre nåletræarter	Other conifers
Ha	Hectares
Hele landet	Denmark in total
Amt	County
I alt	Total

Forest area by size classes, species and region. 1990

I alt	Total
Skovens størrelse i ha	Size of forest (hectares)
1000,0 og derover	1000 hectares and more
Ha	Hectares
Hele landet	Denmark in total
Skovareal i alt	Total forest area
Hjælpearealer	Auxiliary areas
Skovbevokset areal i alt	Total wooded area
Løvtræ i alt	Total broadleaves
Bøg	Beech
Eg	Oak
Ask	Ash

Ahorn (Ær)	Sycamore
Andre løvtræarter	Other broadleaves
Nåletræ i alt	Total conifers
Rødgran	Norway spruce
Sitkagran mv.	Sitka spruce and other spruce
Ædelgran mv.	Silver fir and other fir
Nobilis	Noble fir
Nordmannsgran	Caucasian fir
Bjergfyr mv.	Mountain pine and lodgepole pine
Andre nåletræarter	Other conifers
Midlertidigt ubevokset areal	Temporarily uncovered area
Øerne	The Islands
Jylland	Jutland

Forest area by size classes, species and county. 1990

I alt	Total
Skovens størrelse i ha	Size of forest (hectares)
1000,0 og derover	1000 hectares and more
Ha	Hectares
Hele landet	Denmark in total
Amt	County
Skovareal i alt	Total forest area
Hjælpearealer	Auxiliary areas
Skovbevokset areal i alt	Total wooded area
Løvtræ i alt	Total broadleaves
Bøg	Beech
Eg	Oak
Ask	Ash
Ahorn (Ær)	Sycamore
Andre løvtræarter	Other broadleaves
Nåletræ i alt	Total conifers
Rødgran	Norway spruce
Sitkagran mv.	Sitka spruce and other spruce
Ædelgran mv.	Silver fir and other fir
Nobilis	Noble fir
Nordmannsgran	Caucasian fir
Bjergfyr mv.	Mountain pine and lodgepole pine
Andre nåletræarter	Other conifers
Midlertidigt ubevokset areal	Temporarily uncovered area

Forests and forest area by size classes, ownership classes and region. 1990

I alt	Total
Skovens størrelse i ha	Size of forest (hectares)
1000,0 og derover	1000 hectares and more
Antal	Number
Ha	Hectares
Hele landet	Denmark in total

Alle ejerforhold	All types of ownership
Privatejede	Privately owned forests
Fonde mv.	Foundations, organisations and independent institutions
Hedeselskabet	Danish Land Development Service
Selskaber og foreninger	Joint-stock companies, partnerships, cooperative societies and other societies
Skov- og Naturstyrelsen	The National Forest and Nature Agency
Staten i øvrigt	Other state-owned forests
Præsteembeder	Benefices
Amter og kommuner	Counties and municipalities
Øerne	The Islands
Jylland	Jutland

Forests and forest area by size classes, ownership classes and county. 1990

I alt	Total
Skovens størrelse i ha	Size of forest (hectares)
1000,0 og derover	1000 hectares and more
Antal	Number
Ha	Hectares
Amt	County
Alle ejerforhold	All types of ownership
Privatejede	Privately owned forests
Fonde mv.	Foundations, organisations and independent institutions
Hedeselskabet	Danish Land Development Service
Selskaber og foreninger	Joint-stock companies, partnerships, cooperative societies and other societies
Skov- og Naturstyrelsen	The National Forest and Nature Agency
Staten i øvrigt	Other state-owned forests
Præsteembeder	Benefices
Amter og kommuner	Counties and municipalities

Forest area by species, ownership classes and region. 1990

Skovareal i alt	Total forest area
Hjælpearealer	Auxiliary areas
Skovbevokset areal i alt	Total wooded area
Midlertidigt ubevokset areal	Temporarily uncovered area
Løvtræ	Broadleaves
Løvtræ ialt	Total broadleaves
Bøg	Beech
Eg	Oak
Ask	Ash
Ahorn (Ær)	Sycamore
Andre løvtræarter	Other broadleaves
Nåletræ	Conifers

Nåletræ i alt	Total conifers
Rødgran	Norway spruce
Sitkagran mv.	Sitka spruce and other spruce
Ædelgran mv.	Silver fir and other fir
Nobilis	Noble fir
Nordmannsgran	Caucasian fir
Bjergfyr mv.	Mountain pine and lodgepole pine
Andre nåletræarter	Other conifers
Ha	Hectares
Hele landet	Denmark in total
Alle ejerforhold	All types of ownership
Privatejede	Privately owned forests
Fonde mv.	Foundations, organisations and independent institutions
Hedeselskabet	Danish Land Development Service
Selskaber og foreninger	Joint-stock companies, partnerships, cooperative societies and other societies
Skov- og Naturstyrelsen	The National Forest and Nature Agency
Staten i øvrigt	Other state-owned forests
Præsteembeder	Benefices
Amter og kommuner	Counties and municipalities
Øerne	The Islands
Jylland	Jutland

Forest area by species, ownership classes and county. 1990

Skovareal i alt	Total forest area
Hjælpearealer	Auxiliary areas
Skovbevokset areal i alt	Total wooded area
Midlertidigt ubevokset areal	Temporarily uncovered area
Løvtræ	Broadleaves
Løvtræ i alt	Total broadleaves
Bøg	Beech
Eg	Oak
Ask	Ash
Ahorn (Ær)	Sycamore
Andre løvtræarter	Other broadleaves
Nåletræ	Conifers
Nåletræ i alt	Total conifers
Rødgran	Norway spruce
Sitkagran mv.	Sitka spruce and other spruce
Ædelgran mv.	Silver fir and other fir
Nobilis	Noble fir
Nordmannsgran	Caucasian fir
Bjergfyr mv.	Mountain pine and lodgepole pine
Andre nåletræarter	Other conifers
Ha	Hectares
Amt	County

Alle ejerforhold	All types of ownership
Privatejede	Privately owned forests
Fonde mv.	Foundations, organisations and independent institutions
Hedeselskabet	Danish Land Development Service
Selskaber og foreninger	Joint-stock companies, partnerships, cooperative societies and other societies
Skov- og Naturstyrelsen	The National Forest and Nature Agency
Staten i øvrigt	Other state-owned forests
Præsteembeder	Benefices
Amter og kommuner	Counties and municipalities

Forests and forest area by size classes, forestry administration classes and region. 1990

I alt	Total
Skovens størrelse i ha	Size of forest (hectares)
1000,0 og derover	1000 hectares and more
Antal	Number
Ha	Hectares
Hele landet	Denmark in total
I alt	Total
Eget forstligt personale	Own forestry staff
Hedeselskabet	Danish Land Development Service
De Danske Skovdyrkerforeninger	The Association of Danish Forest Owner Co-operations
Anden forstlig bistand	Other forestry administration
Ingen forstlig bistand	No forestry administration
Øerne	The Islands
Jylland	Jutland

Forests and forest area by ownership classes, forest administration classes and region. 1990

Alle ejerforhold	All types of ownership
Privatejede	Privately owned forests
Fonde mv.	Foundations, organisations and independent institutions
Hedeselskabet	Danish Land Development Service
Selskaber og foreninger	Joint-stock companies, partnerships, cooperative societies and other societies
Skov- og Naturstyrelsen	The National Forest and Nature Agency
Staten i øvrigt	Other state-owned forests
Præsteembeder	Benefices
Amter og kommuner	Counties and municipalities
Antal	Number
Ha	Hectares
Hele landet	Denmark in total
I alt	Total

Eget forstligt personale	Own forestry staff
Hedeselskabet	Danish Land Development Service
De Danske Skovdyrkerforeninger	The Association of Danish Forest Owner Co-operations
Anden forstlig bistand	Other forestry administration
Ingen forstlig bistand	No forestry administration
Øerne	The Islands
Jylland	Jutland

Average yield class by species, planting year classes and region. 1990

Alle træarter	All species
Løvtræ i alt	Total broadleaves
Bøg	Beech
Eg	Oak
Ask	Ash
Ahorn (Ær)	Sycamore
Andre løvtræarter	Other broadleaves
Nåletræ i alt	Total conifers
Rødgran	Norway spruce
Sitkagran mv.	Sitka spruce and other spruce
Ædelgran mv.	Silver fir and other fir
Nobilis	Noble fir
Nordmannsgran	Caucasian fir
Bjergfyr mv.	Mountain pine and lodgepole pine
Andre nåletræarter	Other conifers
Areal	Area
PK	Yield class
Ha	Hectares
Hele landet	Denmark in total
I alt	Total
Før 1840	Pre- 1840
Før 1870	Pre- 1870
Øerne	The Islands
Jylland	Jutland

Average yield class by species, planting year classes and county. 1990

Alle træarter	All species
Løvtræ i alt	Total broadleaves
Bøg	Beech
Eg	Oak
Ask	Ash
Ahorn (Ær)	Sycamore
Andre løvtræarter	Other broadleaves
Nåletræ i alt	Total conifers
Rødgran	Norway spruce
Sitkagran mv.	Sitka spruce and other spruce
Ædelgran mv.	Silver fir and other fir

Nobilis	Noble fir
Nordmannsgran	Caucasian fir
Bjergfyr mv.	Mountain pine and lodgepole pine
Andre nåletræarter	Other conifers
Areal	Area
PK	Yield class
Ha	Hectares
Amt	County
I alt	Total
Før 1840	Pre- 1840
Før 1870	Pre- 1870

Average yield class by species, size classes and region. 1990

Alle træarter	All species
Løvtræ i alt	Total broadleaves
Bøg	Beech
Eg	Oak
Ask	Ash
Ahorn (Ær)	Sycamore
Andre løvtræarter	Other broadleaves
Nåletræ i alt	Total conifers
Rødgran	Norway spruce
Sitkagran mv.	Sitka spruce and other spruce
Ædelgran mv.	Silver fir and other fir
Nobilis	Noble fir
Nordmannsgran	Caucasian fir
Bjergfyr mv.	Mountain pine and lodgepole pine
Andre nåletræarter	Other conifers
Areal	Area
PK	Yield class
Ha	Hectares
Hele landet	Denmark in total
I alt	Total
1000 ha og derover	1000 hectares and more
Øerne	The Islands
Jylland	Jutland

Average site index for beech and oak by planting year classes and region. Forests above 50 ha. 1976 and 1990.

Bøg	Beech
Eg	Oak
Ha	Hectares
Bonitet	Site index
Hele landet	Denmark in total
0-10 år	0-10 years
121 år og derover	121 years and above
Øerne	The Islands
Jylland	Jutland

Average site index for ash/sycamore by planting year classes and region. Forests above 50 ha. 1976 and 1990

Ask	Ash
Ahorn	Sycamore
Ha	Hectares
Bonitet	Site index
Hele landet	Denmark in total
0-10 år	0-10 years
101 år og derover	101 years and above
Øerne	The Islands
Jylland	Jutland

Average site index for Norway spruce/sitka spruce and silver fir by planting year classes and region. Forests above 50 ha. 1976 and 1990.

Rødgran og sitkagran mv. spruce	Norway spruce, sitka spruce and other
Ædelgran mv.	Silver fir and other fir
Ha	Hectares
Bonitet	Site index
Hele landet	Denmark in total
0-10 år	0-10 years
71 år og derover	71 years and above
Øerne	The Islands
Jylland	Jutland

Estimated standing volume overbark 1990 and expected mean annual volume increment 1990-2000. By region and species

Stående vedmasse	Standing volume overbark
Gnst. årlig tilvækst 1990-2000	Mean annual volume increment 1990-2000
Hele landet	Denmark in total
Øerne	The Islands
Jylland	Jutland
Mio. m^3	Millions of cubic metres overbark
Mio. m^3/år	Millions of cubic metres overbark per year
I alt	Total
Løvtræ i alt	Total broadleaves
Bøg	Beech
Eg	Oak
Ask	Ash
Ahorn (Ær)	Sycamore
Andre løvtræarter	Other broadleaves
Nåletræ i alt	Total conifers
Rødgran	Norway spruce
Sitkagran mv.	Sitka spruce and other spruce
Ædelgran mv.	Silver fir and other fir
Nobilis, nordmannsgran	Noble fir, Caucasian fir
Bjergfyr mv.	Mountain pine and lodgepole pine
Andre nåletræarter	Other conifers

Annual cut by species and assortments. 1978-1991

Hugst i alt	Total cut
Bøg	Beech
Eg	Oak
Andre løvtræarter	Other broadleaves
Nåletræ	Conifers
Gavntræ	Sawnwood
Brænde	Fuelwood

Annual cut by species, assortments and county. 1978-1991

I alt	Total
Bøg	Beech
Eg	Oak
Andre løvtræarter	Other broadleaves
Nåletræ	Conifers
Hugst i alt	Total cut
Hele landet	Denmark in total
Øerne	The Islands
Jylland	Jutland
Amt	County
Gavntræ	Sawnwood
Brænde	Fuelwood

Annual cut per ha wooded area by species and county. 1978 and 1990

I alt	Total
Bøg	Beech
Eg	Oak
Andre løvtræarter	Other broadleaves
Nåletræ	Conifers
Hugst i alt	Total cut
Hele landet	Denmark in total
Øerne	The Islands
Jylland	Jutland
Amt	County

Annual cut per ha wooded area by size classes and species. 1978 and 1990

I alt	Total
Skovens størrelse i ha	Size of forest (hectares)
Under 50 ha	Less than 50 hectares
250 ha og derover	250 hectares and above
Hugst i alt	Total cut
Bøg	Beech
Eg	Oak
Andre løvtræarter	Other broadleaves
Nåletræ	Conifers

2.9.2. Delivery of the statistics to UN and Community institutions

Delivery of data to these institutions is made on the basis of the forest statistics by filling in the different questionnaires which are sent to the national correspondent.

2.9.2.1. Responsibilities for international assessments

Forestry affairs the national Forest and Nature Agency is the only responsible correspondent in Denmark.

2.9.2.2. Data provided for FRA 1990

As mentioned in chapters 2.9.2. to 2.9.2.1.

2.10 FUTURE DEVELOPMENT AND IMPROVEMENT PLANS

2.10.1. Next inventory period

The National Forest and Nature Agency is legally obligated to initiate new forest statistics before year 2000 (that would say 10 years after the last forest statistics of 1990).

A board which will follow this work is being established at the moment.

2.10.2. Expected or planned changes

As a consequence of the agreements of the conferences in Strassbourg, 1989, Rio de Janeiro, 1992, and Helsinki, 1993, as well as a consequence of a number of EEC-regulations it is expected that new forest statistics will be much broader than the previous ones.

It would be natural to include a number of social, environmental, economical, biological, cultural and other topics in a new forest statistic.

On the other hand, the line from the last seven forest statistics must not be broken. Therefore, it is believed that a new forest statistic will be carried through more or less in the same way as the former seven statistics supplemented by a number of analysis etc. concerning different nonponderable factors. The board mentioned in chapter 2.10.1 shall discuss this matter.

2.10.2.1. Nomenclature

See chapter 2.9.1.

2.10.2.2. Data sources

See chapter 2.9.1.

2.10.2.8. Other forest data

A very large inventory of leisure activities in forests was carried out in 1976 - 77. This analysis was repeated in 1995 but was not been published yet.

These two similar inventories have at least tree main topics:

- How much is these leisure activities large Danish forests?
- How expensive are the leisure activities and what is a typical Danish leisure pattern ?
- What would the preferences of different groups of Danes be as regards to the structure, development, forest types, stand types, etc. ?

These analyses make part of the basis for forest management planning and administration in private as well as in public forest enterprises.

REFERENCES

Andersen, K.F., 1976. Efterladte papirer fra skovpolitikken. Den Kgl. Veterinær- og Landbohøjskole, Skovbrugsinstituttet. 167 p.

Bergsøe, Adolph Frederik, 1847. Om Skovdyrkningen. In.: Den danske Stats Statistik. Pp. 197 - 204

Danmarks Statistik, 1909. Skovbruget i Danmark 1907. Statistiske meddelelser.

Danmarks Statistik, 1925. Skovbruget i Danmark 1923. Statistiske meddelelser.

Danmarks Statistik, 1931. Landbrugstællingen 1931. Statistiske Meddelelser.

Danmarks Statistik, 1954. Skove og Plantager 1951. Statistiske Meddelelser.

Danmarks Statistik, 1964. Skove og Plantager 1965. Statistiske Meddelelser.

Danmarks Statistik, 1967. Skove og Plantager 1965. Statistiske Meddelelser.

Danmarks Statistik, 1979. Skovtællingen 1976. Statistiske Meddelelser.

Koch, Niels Elers, 1978. Skovenes friluftsfunktion i Danmark, I del. Befolkningens anvendelse af de danske skove. Forest recreation in Denmark. Part I: The udse of the country's forest by the population. Det forstlige forsøgsvæsen i Danmark. The Danish Forest Experiment Station. Vol. 35, pp. 285 - 451.

Koch, Niels Elers, 1980. Skovenes friluftsfunktion i Danmark, II del: Anvendelsen af skovene, regionalt betragtet. Forest recreation in Denmark, part II: The use of the forests considered regionally. Ib., vol 37, pp. 73 - 383.

Koch, Niels Elers, 1984. Skovenes friluftsfunktion i Danmark, III del: Anvendelsen af skovene, regionalt betragtet. Forest recreation in Denmark, part III: The use of the forests considered locally. Ib. Vol. 39, pp. 121 - 362.

Lütken, Chr., 1870. Statistisk Beskrivelse af de danske Statsskove, Kjøbenhavn.

Miljøministeriet, Skov- og Naturstyrelsen, og Danmarks Statistik, 1994. Skove og Plantager 1990, 131 p.

Müller, P. E., 1876. Optegnelser om vort skovbrug i 1876. Tidsskrift for Skovbrug, vol. 1, pp 151 - 166.

Niemann, A., 1809. Forststatistik der dänischen Staten, Altona, 88 p.

Olufsen, Chr., 1811. Danmarks Brændselsvæsen, physikalsk, cameralistisk og oeconomisk betragtet. Kjøbenhavn. 352 p.

Plum, Peter Munk, 1988. Denmark: Forest Policy and Legislation. Environmental Policy and Law, vol. 18/4, pp. 111-115

Statens Forstlige Forsøgsvæsen, 1979. Skovbrugstabeller, 170 p.

COUNTRY REPORT FOR FINLAND

Erkki Tomppo, Jari Varjo, Kari Korhonen, Arto Ahola, Antti Ihalainen, Juha Heikkinen, Hannu Hirvelä, Heli Mikkelä, Eero Mikkola, Sakari Salminen and Tarja Tuomainen

Finnish Forest Research Institute

National Forest Inventory of Finland

Executive Summary

Chapter 2 in this report describes two branches in the National Forest Inventory. They are separated whenever notable differences exist. The *Field inventory* includes a traditional field sample based approach (85672 field plots) for inventory and an estimation of results. *Multi-source inventory* is a remote sensing and digital map information aided extension of the field inventory, allowing an accurate estimation of the results for small areas. In addition to these two inventory systems, the National Forest Inventory of Finland includes a separate permanent field plot inventory (3009 field plots), mainly for the purposes of the forest health assessment. This part of the inventory is described in Chapter 3.

2.1 Nomenclature

Altogether 123 different variables are directly assessed in the field inventory. Most of the variables are measured in the field. Two thirds of the variables are categorical ones, and one third are continuous ones.

2.2 Data sources

In the traditional field inventory, the field assessment, maps and area statistics are the only data sources. In the multi-source inventory both satellite images and various digital map information are applied, in addition to the information collected in field inventories.

2.3 *Assessment*

The National Forest Inventory is mainly based on temporary angle gauge sample plots (Bitterlich plots). During the 8th inventory, permanent field plots have been established, starting in Northern Finland. Fixed radius field plots are applied for some special purposes, such as surveying the seedling storey or reindeer foddering conditions in Northernmost Finland.

2.4 *Data storage and analysis*

The calculation of the field inventory results is based on the application of the Bitterlich field plots. The satellite image analysis, the multi-source inventory, is applied by generalizing the field information for smaller areas than it is possible when using field measurements only. The estimation of the sampling error of the field inventory estimates is based on the results of Matérn (1960) and Ranneby (1981). The estimation of the errors of multi-source inventory is, so far, based on empirical tests.

2.5 *Reliability of data*

About 5% of the field tracts are remeasured during each field season, and immediate feedback is given about possible measurement errors. Consistency checks are applied after the field season and the field group leaders correct the detected errors.

2.6 *Models*

The sample tree specieswise polynomial taper curves are applied to derive the total volumes and the timber assortment volumes, which are then generalized to tally trees. The timber assortment volumes are transformed to describe the contribution of the trees measured from the Bitterlich plot, and these transformed assortment volumes are generalized to tally trees. The increment is estimated by computing the current volume and volume 5 years ago. An automated stand simulator based on development of individual trees, combined with linear optimization package for selecting optimal future treatments, is applied for the estimation of the cutting possibilities and future yield.

2.7 *Inventory reports*

The inventory results have been published in scientific forest series. The National Forest Inventories have been published in Folia Forestalia, Communicationes Instituti Forestalis Fenniae and Acta Forestalia Fennica. The results of the NFI form the basis for the forest policy and for the planning of long term forest management in large areas. After the multi-source inventory was made operational, the results have been available for the purposes of operative forest management planning, as well.

2.8 Future development and improvement plans

Research is going on continuously for improving the National Forest Inventory. Applications of the improving spaceborne remote sensing material and the new airborne remote sensing material will have great potential in future inventories. Field sampling is being developed continuously; it will contribute to new information needs, such as assessment of biodiversity and the multiple use of forests.

1. OVERVIEW

1.1 GENERAL FORESTRY AND FOREST INVENTORY INFORMATION

The National Forest Inventories of Finland (Valtakunnan metsien inventointi) cover the whole Finland. The first assessment was carried out in 1921–24, and the latest, the 8th national forest inventory, was completed in 1994. The responsible organization of the National Forest Inventories is the Finnish Forest Research Institution (Metla) (Statistical yearbook... 1995).

Finland is located between the latitudes of 59.5 and 70.2 degrees. The total area, including inland watercourses, is 33.8 mill hectares and the land area is 30.5 mill. hectares. The forestry land comprises 26.3 mill. ha, forest and other wooded land (FOWL) 23.0 mill ha and forest land (FL) 20.0 mill. ha. The other categories of principal land use are: agricultural land 3.0, built-up land 0.8 and transport routes 0.4 mill ha.

The total growing stock is 1937 mill. m^3 (1989–94) of which Scots pine (*Pinus silvestris* L.), comprises 890 mill. m^3 , Norway spruce (*Picea abies* Karst) 701 mill m^3, two birch species (*Betula pendula* Roth., *Betula pubescens* Ehrh.) and other deciduous trees (aspen, grey alder etc.) 346 mill m^3. The mean annual increment of the 5 years preceding the measurement year of the 8th national inventory was 75.4 mill. m^3 while the mean annual removals of the period 1986–1994 were 53 mill. m^3. The mean growing stock on forest and other wooded land is 82.0 m^3/ha, on forest land 94 m^3/ha (see Table 1) (Tomppo and Henttonen 1995).

Table 1. Some forest characteristics in South Finland, North Finland and in the whole country.

Characteristic	South	North	Total
Land area	16074	14386	30459
Forestry land (1000 ha)	12544	13732	26276
FOWL (1000 ha)	12081	10923	23004
FL (1000 ha)	11499	8534	20032
FOWL/inhabitant (ha)	2.8	16.7	5.3
Vol pine (mill. m^3)	552	338	890
Vol spruce (mill. m^3)	582	118	701
Vol birch (mill. m^3)	196	90	286
Vol others (mill. m^3)	51	9	60
Vol total (mill. m^3)	1381	555	1937
Mean vol FOWL (m^3/ha)	114.3	50.9	84.2
Mean vol FL (m^3/ha)	119	60.70	94.00
Increment (mill. m^3/a)	57.9	17.5	75.4
Privately owned (%)	75	40	58
State owned (%)	8	51	29
Other owners (%)	17	9	13

FOWL = forest and other wooded land, FL = forest land

The average size of a forest stand is about 1 ha in Southern Finland but it increases towards Northern Finland.

Figure 1 shows the distribution of the growing stock across the country. Figure 2 shows the development of the land use classes, Figure 3 the development of the growing stock by main tree species groups and Figure 4 the development of the increment and removals in the main tree species groups since the beginning of 1920's.

Note that the land area of Finland in the first inventory (1921–24) was 34.36 mill. hectares, of which 30.3 mill ha was forestry land. Of this, 25.3 mill. ha belonged to forest and other wooded land. After the second world war, Finland lost 3.9 million hectares of its land area and 3.4 million hectares of FOWL.

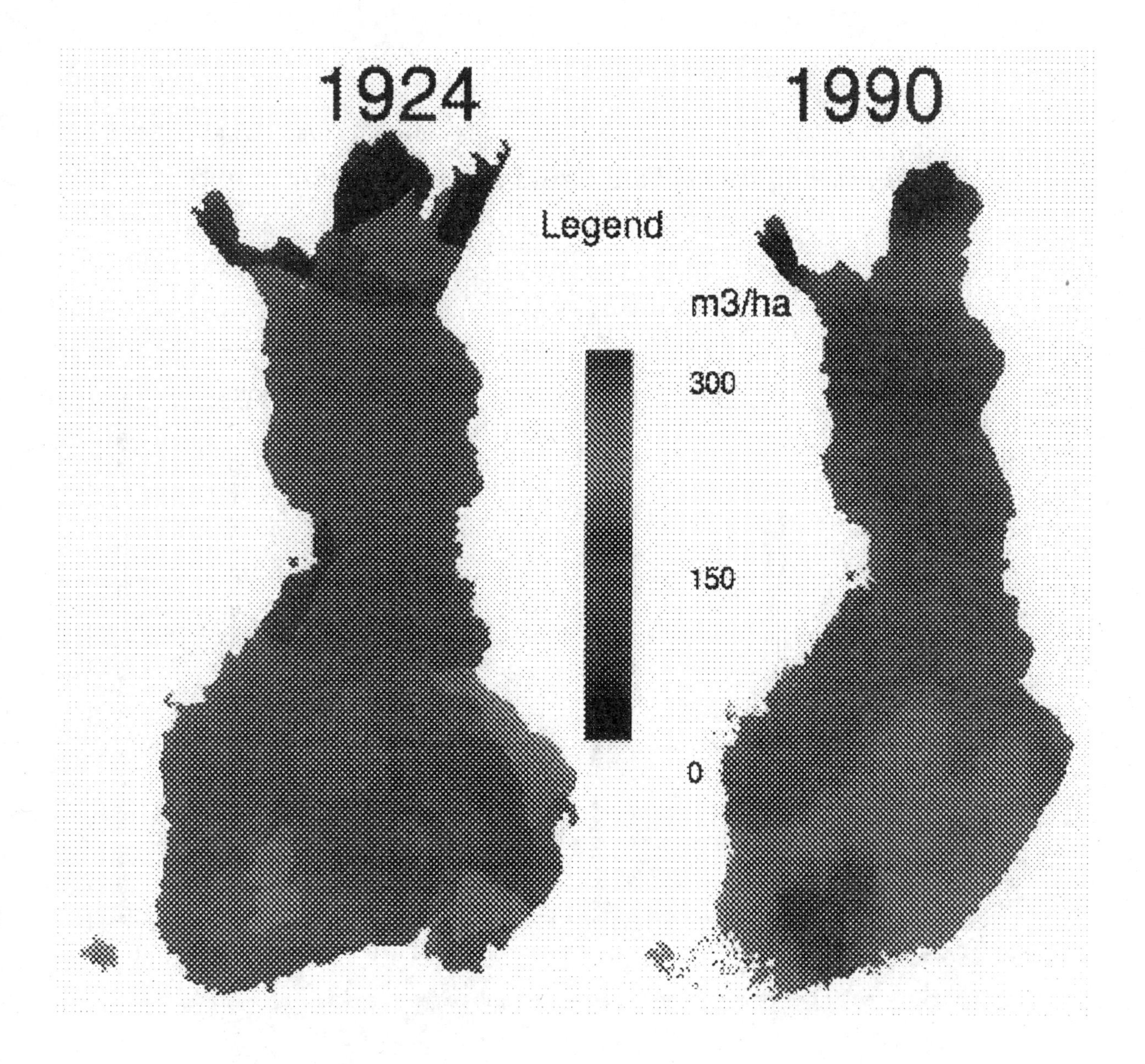

Figure 1. Mean volume of the growing stock in 1990 and in 1924 (first NFI) on forest and scrub land.

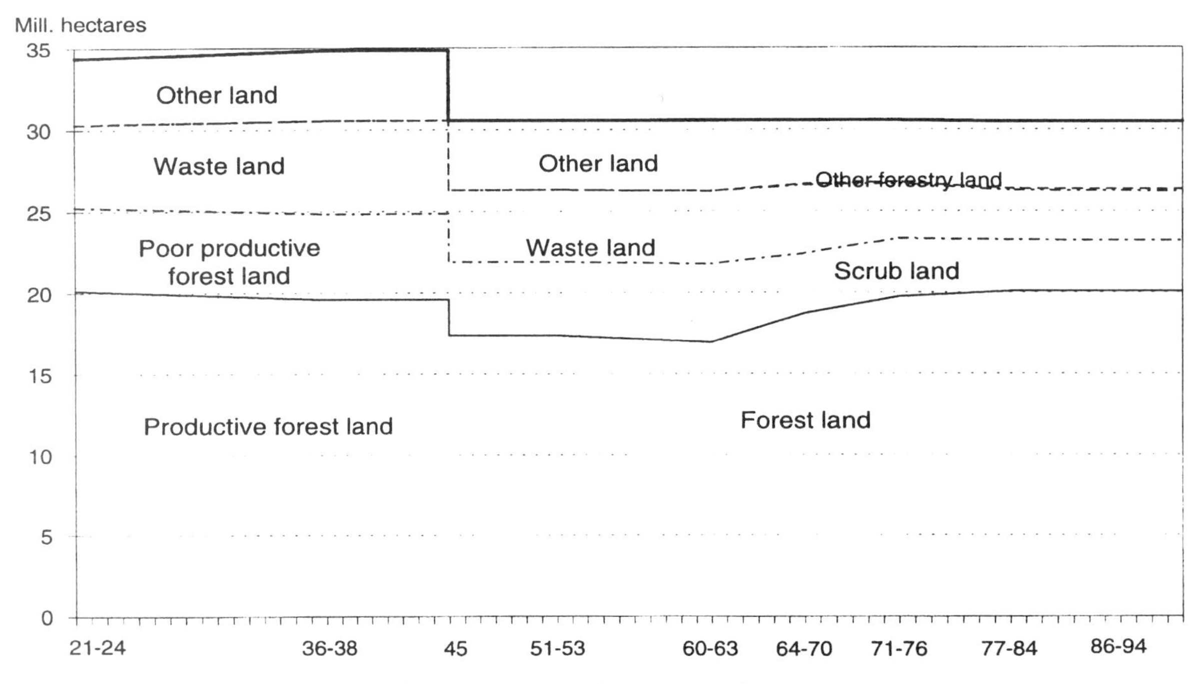

Figure 2. Development of the land use classes in the inventories.

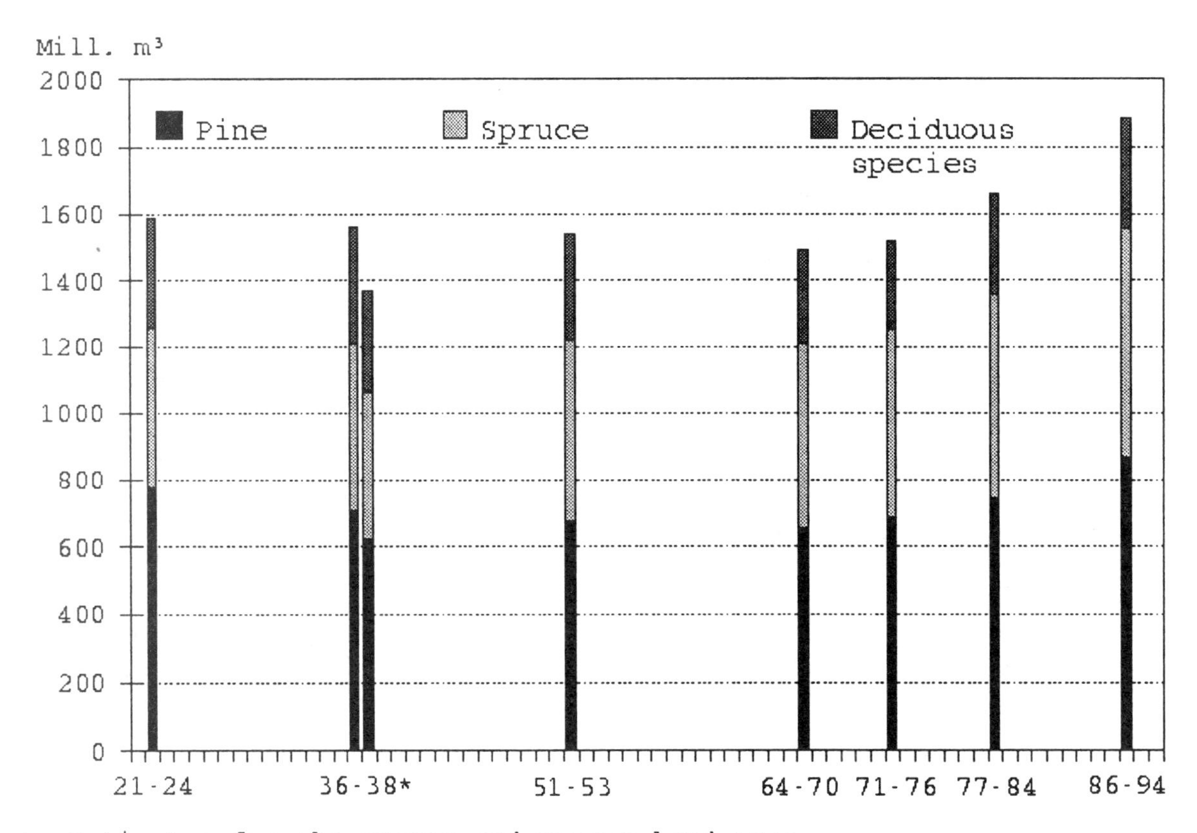

Figure 3. Trends in the growing stock volumes by the main tree species groups in the inventories in 1921–94. The volume in the second inventory is given for both the old and the current land areas.

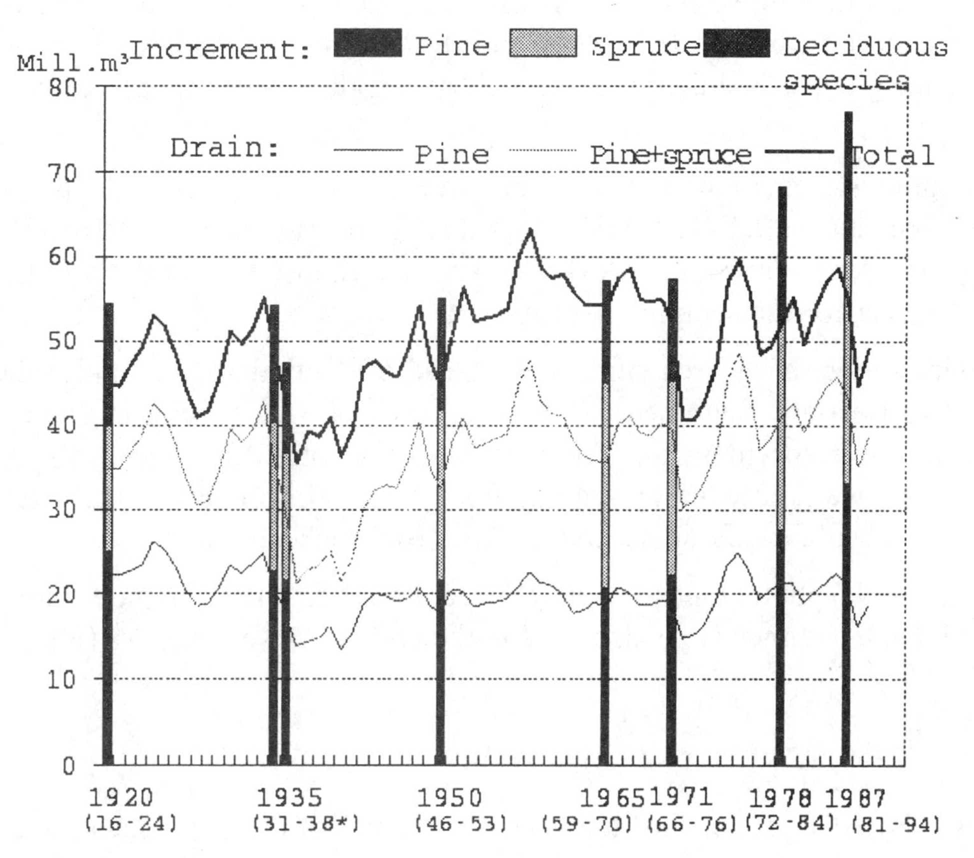

Figure 4. Trends in the growing stock increment and total drain by tree species (1916–1994). The increment in the second inventory is given for both the old and the current land areas.

Background

Slash-and-burn agriculture, the burning of wood for tar-making, the household use of timber, ship building and, from the second half of the nineteenth century onwards, the expanding forest industry exhausted the forest resources of Finland. The further growth of forest industries into a large-scale industry in 1870's and the fact that the cuttings proceeded deeper into the wilderness stimulated the interest in forest resources. According to the pre-inventory documents, the forest resources of Finland were at their lowest when the increase in crops on burnt-over clearings halted around the turn of the century.

Old Inventories

The National Forest Inventory of Finland (NFI) has produced large-area forest resource information for 75 years. The traditional role has been to produce objective and up-to-date information about the forest resources, the forest health conditions and their development for national and regional decision making. The information has been used for the planning of forest management in large areas, for making decisions concerning forest industry investments, and as a basis for forest income taxation. The NFI covers all the forests, and the information has also been used by the ownership groups for justifying and calibrating their own standlevel inventory results.

The sampling system of the First National Inventory (1921–24) was the line survey sampling (Ilvessalo 1927). The methods in the following second (1936–38), third (1951–53) and fourth inventories (1960–63) were almost the same as in the first inventory, except for the sampling density.

Detached clusters have been employed instead of continuous lines since the fifth inventory (Kuusela and Salminen 1969). This design is statistically more effective. From the fifth inventory onwards, the work has been carried out each year, and the measurements have proceeded by regions from the south to north. Fixed sized field plots were also replaced by angle gauge plots (Bitterlich plots).

The use of aerial photographs was introduced, for the first time, in the 5th inventory in North Finland, and they were also used in the 6th and 7th inventories (Poso 1972).

Current National Forest Inventory in Finland

The Finnish Forest Research Institute started to develop a new inventory system in 1989, during the 8th inventory, in order to receive geographically localized, up-to-date information, as well as information for smaller areas than previously. Sampling design and field plot measurements have been changed, to some extent. One fifth of the field plots have been made permanent in Northern Finland. In addition to the ground measurements, the method exploits satellite image data, digital map data (arable land, roads, built up areas, digital terrain model) and, in the future, geographical data, e.g. soil and meteorological data. Image analysis methods have been chosen in such a way that estimates of all variables of the inventory can be computed for each pixel.

The system is now operative. The inventory results for the whole country involving thematic maps and statistics for large and small areas will be ready during 1996. Statistics for a forest holding level are available by order.

A lot of information other than on the growing stock and increment is collected in the National Forest Inventories. Examples include information on forest health and ground vegetation and, in some inventories, observation of animals for estimating the distribution of species, and bryophyte samples for analysing the distribution and concentrations of sulfur and heavy metals, etc.

In 1995, the grid of 3000 permanent field plots, established for forest health monitoring purposes 1985–86, was measured for the third time.

The ninth national forest inventory of Finland started in 1996. New features include for example the measurement of additional characteristics describing biological diversity of the forests.

Ongoing research topics in the inventory are the testing of radar data (ERS-1 SAR, RADARSAT) and the introduction of airborne imaging spectrometer (AISA) into the system. Methods for monitoring the biodiversity of forests are being developed.

The financing of the first National Forest Inventories was granted by the Finnish Government in a separate law given by the Parliament. Today, the NFI is one of the public tasks imposed by the Finnish Government on the Finnish Forest Institute in the statute (law) no: 374/91 of the Institute,

1.2. OTHER IMPORTANT FORESTRY STATISTICS

1.2.1 Other forest data and statistics on the national level

Forest Statistics Information Service is other main research project involved with forest statistical work in the Finnish Forest Research Institute (Metla). The major part of the statistics on forestry and the forest industries in Finland have been compiled by the Forest Statistics Information Service.

The forest statistical data originate from the Forest Statistics Information Service's own statistical enquiries, as well as from outside statistical authorities and organisations in the forest sector (i.e. Statistics Finland, the Board of Customs and the Finnish Forest Industries Federation). On the next list is represented main statistics and sources compiled by the Forest Statistics Information Service:

Statistics	Data Source
Silvicultural and forest improvement work	Forestry Development Centre Tapio, The Finnish Forest Industries Federation & the Finnish Forest and Park Service
Forest fires	Ministry of the Interior
Logging and transportation volumes and costs of roundwood	The Finnish Forest Industries Federation
Roundwood volumes marked for felling in private forests	Forestry Development Centre Tapio
Roundwood purchases and prices in private forests	The Finnish Forest Industries Federation
Roundwood stocks of the forest industries	The Finnish Forest Industries Federation
Roundwood fellings	Forest Industries & The Finnish Forest and Park Service
Removals	The Finnish Forest Research Institute
Growing stock drain	The Finnish Forest Research Institute
Wood consumption	The Finnish Forest Industries Federation & Forest Industries
Production of the forest industries	The Finnish Forest Industries Federation
The labour force in the forestry sector	The Finnish Forest Research Institute & Statistics Finland

Forest taxation	The Finnish Forest Research Institute
Forest sector in national economy	Statistics Finland
Foreign trade of roundwood and forest industry products	The Board of Customs
International forest statistics	FAO

Statistics on non-wood goods and services:

Statistics	**Data Source**
Private-owned nature protection areas, threatened species	The Ministry of Environment
State-owned nature protection areas	The Finnish Forest and Park Service
Market information of wild berries and edible mushrooms	Food and Farm Facts
Hunting	The Finnish Game and Fisheries Research Institute
Reindeers	The Association of Reindeer Herding Cooperatives
Nature-based recreation sites	Nation-wide sports site system

The Forest Statistics Information Service utilizes three communication channels:

- *The yearbook of forest statistics*, consisting of thirteen main chapters, covers the whole Finnish forest sector, ranging from forest resources to exports of forest industry products. Detailed information is presented for the last two years, together with extended time series, many of which date back to the 1950's.
- *Forest statistical bulletins* covering the following topics on a monthly basis: roundwood sales, prices, removals, foreign trade in roundwood, and exports of forest industry products. Approximately 70 bulletins are published yearly.
- *METINFO* is computer-based forest statistical information system. This system makes it possible for users to have an on-line access to actual forest statistics. The first phase of the information system comprising monthly statistics on roundwood markets.

1.2.2. Delivery of statistics to the UN and institutions of the European Community

1.2.2.1. Responsibilities for international assessments

a) FAO/ECE Forest Resource Assessment information:

Prof. Erkki Tomppo
The Finnish Forest Research Institute
Unioninkatu 40 A
00170 HELSINKI
FINLAND

b) OECD (Environment Compendium):

Statistics Finland
Työpajakatu 13
00022 TILASTOKESKUS
FINLAND

c) Information for Helsinki Ministerial Conference

The Finnish Forest Research Institute
Unioninkatu 40 A
00170 HELSINKI
FINLAND

d) CORINE Land Cover

Tapani Säynätkari
Finnish Environment Institute
P.O. Box 140
00251 HELSINKI
FINLAND

e) Statistics for EUROSTAT

Statistics Finland
Työpajakatu 13
00022 TILASTOKESKUS
FINLAND

F) Forest health monitoring EEC Regulation 3528/86

Prof Eino Mälkönen
The Finnish Forest Research Institute
PB 18
01301 VANTAA
FINLAND

1.2.2.2. Data compilation

The required data is based on the existing statistics.

2. THE NATIONAL FOREST INVENTORY OF FINLAND

2.1. NOMENCLATURE

2.1.1. List of the directly assessed attributes

Note: In order to keep this presentation short, all the identification numbers of the measurement objects such as tract, plot, tally tree etc. are excluded from the following lists of variables. In the description of the measurement rules and categories, only one description is given if the same rules and categories are used for several different objects such as stand, tally trees, sample tree etc.

A) Geographic regions

Attribute	Data source	Object	Measurement unit
The Finnish Uniform Coordinates of the tract (*Rypään koordinaatit*)	1:20000 map	tract	m
Deviation of the tract line *(linjan siirtymä*)	field assessment /geographical map	field plot	categorical
Distance and bearing of the stand boundary (*kuvion raja*)	field assessment	stand boundary, measured if existing closer than 40 m from the centre of the field plot	categorical
Percentage of the field plot covered by the stand *(kuvion koko koealalla)*	field assessment	field plot	categorical
Field plot size (*kuvion koko*)	field assessment	field plot	categorical
Completed measurements (*mittaustapa*)	field assessment	field plot	categorical
Land use class *(maaluokka)*	field assessment	stand	categorical
Land use class specification *(maaluokan tarkennus*)	field assessment	stand	categorical
Change in the land use class (*maaluokan muutos*)	field assessment	stand	categorical
Time of the change in the land use class (*maaluokan muutoksen aika*)	field assessment	stand	categorical

B) Ownership

Attribute	Data source	Object	Measurement unit
Owner (*omistaja)*	owner register	field plot	categorical

C) Wood production

Attribute	Data source	Object	Measurement unit
Saw-timber stage (*tukkipuukokoisuus)*	field assessment	stand	categorical
Pole size of the growing stock *(järeys*)	field assessment	stem of mean basal area or dominant trees	cm
Proportion of the saw-timber *(tukkikokoisten puiden osuus päätehakkuussa)*	field assessment	stand	categorical
Quality distribution of the saw-timber *(tukkikokoisten puiden laatujakauma päätehakkuussa*)	field assessment	stand	categorical
Stand basal area *(kuvion pohjapinta-ala)*	field assessment	stand	m^2/hectare
Stand basal area of the 2nd tree storey *(2. jakson pohjapinta-ala)*	field assessment	stand	m^2/hectare
Breast height diameter *(läpimitta)*	field assessment	tally tree	mm
Timber assortment class (*puuluokka)*	field assessment	tally tree	categorical
Timber assortment specification *(puuluokan tarkennus)*	field assessment	tally tree	categorical
Crown storey *(latvuskerros)*	field assessment	tally tree	categorical
Height to the lowest dead branch *(kuivaoksaisuusraja*)	field assessment	sample tree	dm
Height of the base of living crown *(elävän latvuksen raja)*	field assessment	sample tree	dm
Height *(pituus)*	field assessment	sample tree	dm
Height increment of the past five years *(viiden vuoden pituuskasvu)*	field assessment	sample tree	dm
Height increment of the current growing season *(inventointikesän pituuskasvu)*	field assessment	sample tree	dm

Diameter increment *(läpimitan kasvu)*	field/laboratory assessment	sample tree	mm
Revision of the timber assortment *(puuluokan muutos)*	field assessment	sample tree	categorical
Revision of the timber assortment specification *(puuluokan tarkennuksen muutos)*	field assessment	sample tree	categorical
Length of the broken part *(katkenneen osan pituus)*	field assessment	sample tree	dm
Condition of the growing stock *(laatu)*	field assessment	sample tree	categorical
Lenght of saw-timber or cull *(tukki tai raakkiosan pituus)*	field assessment	sample tree	dm
Cause for revision of the condition of the growing stock *(laadunalenemisen tai pakkokatkaisun syy)*	field assessment	sample tree	categorical

D) Site and soil

Attribute	Data source	Object	Measurement unit
Main site class *(kasvupaikan päätyyppi)*	field assessment	stand	categorical
Main site class specification for peat land *(suon sekatyyppi)*	field assessment	stand	categorical
Site fertility class *(kasvupaikkatyyppi)*	field assessment	stand	categorical
Site fertility class specification for peat land *(suotyypin lisämääre)*	field assessment	stand	categorical
Drainage stage *(ojitustilanne)*	field assessment	stand	categorical
Completed drainage *(tehty ojitus)*	field assessment	stand	categorical
Time of completed drainage *(ojituksen ajankohta)*	field assessment	stand	categorical
Need for drainage *(ojitustarve)*	field assessment	stand	categorical

Urgency of the required drainage *(ojituksen kiireellisyys)*	field assessment	stand	categorical
Soil type *(maalaji)*	field assessment	stand	categorical
Type of the organic layer (*orgaanisen kerroksen laatu)*	field assessment	stand	categorical
Taxation class *(veroluokka)*	field assessment	stand	categorical
Taxation class specification (*veroluokan tarkennus)*	field assessment	stand	categorical

E) Forest structure

Attribute	Data source	Object	Measurement unit
Number of the tree storeys (*jaksojen määrä)*	field assessment	stand	categorical
Tree storey position *(jakson asema)*	field assessment	stand	categorical
Development class *(kehitysluokka)*	field assessment	stand	categorical
Origin of the stand *(perustamistapa)*	field assessment	stand	categorical
Dominant tree species *(pääpuulaji)*	field assessment	stand	categorical
2nd tree species *(1. sivupuulaji)*	field assessment	stand	categorical
3rd tree species *(2. sivupuulaji)*	field assessment	stand	categorical
Proportion of the coniferous species of the total volume *(havupuuosuus 1)*	field assessment	stand	categorical
Proportion of the coniferous species of the total number of stems *(havupuuosuus 2)*	field assessment	stand	categorical
Number of stems *(runkoluku)*	field assessment	stand	number/hectare
Total number of seedlings *(taimien kokonaismäärä)*	field assessment	stand	1000/hectare
Number of seedlings capable for further development *(kehitys-kelpoisten taimien määrä)*	field assessment	stand	100/hectare

Mean age of growing stock at breast height *(rinnankorkeusikä)*	field assessment	stand/tree storey	a
Birth type *(syntytapa)*	field assessment	sample tree	categorical
Upper diameter *(yläläpimitta)*	field assessment	sample tree	cm
Thickness of bark *(kuoren paksuus)*	field assessment	sample tree	mm
Tree height *(pituus)*	field assessment	sample tree	dm
Age at breast height *(rinnankorkeusikä)*	field assessment	sample tree	a
Proportional size of the regeneration plot *(pienpuukoealan koko)*	field assessment	regeneration plot	1/10
Bearing of the tally tree from the field plot centre *(suunta)*	field assessment	tally tree	angle
Distance of the tally tree from field plot centre *(etäisyys)*	field assessments	tally tree	cm
Tally tree species *(puulaji)*	field assessment	tally tree	categorical

F) Regeneration

Attribute	Data source	Object	Measurement unit
Tree species *(puulaji)*	field assessment	seedlings	categorical
Capability for development *(kehityskelpoisuus)*	field assessment	seedlings	categorical
Maximum height of seedlings *(kuvausositteen suurimman puun pituus)*	field assessment	seedlings by strata	dm
Maximum d.b.h *(kuvaus-ositteen suurimman puun läpimitta)*	field assessment	seedlings by strata	cm
Minimum height *(kuvaus-ositteen pienimmän puun pituus)*	field assessment	seedlings by strata	dm
Minimum d.b.h. *(kuvaus-ositteen pienimmän puun läpimitta)*	field assessment	seedlings by strata	cm

G) Forest condition

Attribute	Data source	Object	Measurement unit
Description of the damage *(tuhon ilmiasu)*	field assessment	stand/growing stock	categorical
Age of the damage *(tuhon syntyajankohta)*	field assessment	stand	categorical
Causing agent of the damage *(tuhon aiheuttaja)*	field assessment	stand	categorical
Damage degree *(tuhon aste)*	field assessment	stand	categorical
Defoliation *(harsuuntuminen)*	field assessment	stand	categorical
Abundance of beard lichens *(naavamaiset jäkälät)*	field assessment	tally trees on permanent field plot	categorical
Abundance of leaf lichens *(lehtimäiset jäkälät)*	field assessment	tally trees on permanent field plot	categorical
Abundance of *Scoliciosporum clorococum* and *Desmococcus olivaeus* *(vihersukkulajäkälät)*	field assessment	permanent field plot stand	categorical
Stand quality *(metsikön laatu)*	field assessment	stand/growing stock	categorical
Reason for the decreased quality *(laadun alenemisen syy)*	field assessment	stand/growing stock	categorical
Description of the damage *(tuhon ilmiasu)*	field assessment	sample tree	categorical
Age of the damage *(tuhon syntyajankohta)*	field assessment	sample tree	categorical
Causing agent of the damage *(tuhon aiheuttaja)*	field assessment	sample tree	categorical
Damage degree *(tuhon aste)*	field assessment	sample tree	categorical
Defoliation *(harsuuntuminen)*	field assessment	sample tree	categorical
Diameter of the rotten part of the stem *(lahon pituus)*	field assessment	sample tree	cm
Stage of decay *(lahon laatu)*	field assessment	sample tree	categorical
Location of the decay *(lahon sijainti)*	field assessment	sample tree	categorical

H) Accessibility and harvesting

Attribute	Data source	Object	Measurement unit
Applied cutting methods *(tehdyt hakkuut)*	field assessment	stand	categorical
Time of applied cuttings *(tehtyjen hakkuden ajankohta)*	field assessment	stand	categorical
Applied soil preparations methods *(maapinnan käsittely)*	field assessment	stand	categorical
Time of applied soil preparation *(maanpinnan käsittelyn ajankohta)*	field assessment	stand	categorical
Applied silvicultural treatment *(tehdyt metsänhoitotyöt)*	field assessment	stand	categorical
Time of the applied silvicultural treatments (*tehtyjen metsänhoitotöiden ajankohta)*	field assessment	stand	categorical
Proposed cuttings *(hakkuuehdotus)*	field assessment	stand	categorical
Urgency of proposed cuttings *(ehdotetun hakkuun ajankohta)*	field assessment	stand	categorical
Proposed soil preparation *(maapinnan käsittelyehdotus)*	field assessment	stand	categorical
Proposed silvicultural treatments *(ehdotetut metsänhoitotyöt)*	field assessment	stand	categorical
Tree species *(puulaji)*	field assessment	tallied stump	categorical
Stump age *(kannon ikä)*	field assessment	tallied stump	categorical
Type of felling *(kaatotapa)*	field assessment	tallied stump	categorical
Stump diameter *(kannon läpimitta)*	field assessment	tallied stump	cm
Stump height *(kannon korkeus)*	field assessment	tallied stump	cm

J) Non-wood goods and services

Attribute	Data source	Object	Measurement unit
Location of the reindeer fodder field plot *(porokoealan sijainti)*	field assessment	reindeer fodder field plot	categorical
Coverage of *Deschampsia flexuosa (metsälauhan peittävyys)*	field assessment	reindeer fodder field plot	%
Coverage of grasses and herbs *(heinämäisten ja ruohomaisten kasvien peittävyys)*	field assessment	reindeer fodder field plot	%
Coverage of *Stereocaulon sp. (tinajäkälien peittävyys)*	field assessment	reindeer fodder field plot	%
Coverage of reindeer lichen *(poronjäkälien peittävyys)*	field assessment	reindeer fodder field plot	%
Height of reindeer lichen *(poronjäkälien korkeus)*	field assessment	reindeer fodder field plot	mm
Location of the square plot *(näyteruudun sijainti)*	field assessment	square fodder plot	categorical
Coverage of *Deschampsia flexuosa (metsälauhan peittävyys)*	field assessment	square fodder plot	%
Coverage of grasses and herbs *(heinämäisten ja ruohomaisten kasvien peittävyys)*	field assessment	square fodder plot	%
Coverage of *Stereocaulon sp. (tinajäkälien peittävyys)*	field assessment	square fodder plot	%
Coverage of *Cladonia alpestris (palleroporon-jäkälän peittävyys)*	field assessment	square fodder plot	%
Coverage of reindeer lichens except *Cladonia alpestris (muiden poronjäkälien peittävyys)*	field assessment	square fodder plot	%
Height of reindeer lichen *(poronjäkälien korkeus)*	field assessment	square fodder plot	mm
Soil condition *(kasvualustan tila)*	field assessment	square fodder plot	categorical

Condition of the vegetation *(kasvillisuuden tila)*	field assessment	square fodder plot	categorical
Multiple use *(monikäyttö)*	cadaster map	stand	categorical
Multiple use specification *(monikäytön tarkennus)*	cadaster map/field assessment	stand	categorical

2.1.2. List of derived attributes

Note: Please note that the following derived variables include the variables presented in the general results, including the characteristics of the forest area. Thus, in addition to the tree, plot and stand characteristics, examples of the characteristics of the forest area (computation unit) are given. In addition to the physical measurement units, the proportion of many of these variables is presented in the results. Many other measured variables are applied to deliver special results based on orders.

A) Geographic regions

Attribute	Measurement unit	Input attributes
Land use class *(maaluokat)*	km^2	Land use class, number of field plots
Areal gain of the land use classes of the present forestry land during the previous 10-year-period *(10-vuotiskauden pinta-alasiirtymät nykyisen matsätalousmaan maaluokkiin)*	km^2	Change in land use class, number of field plots, area of the computation unit
Land use classes on non-forestry land and changes to and from forestry land during the previous 10-year-period *(muun maan maaluokat, ja niiden 10-vuotiskauden pinta-ala-siirtymät kohti metsätalousmaata)*	km^2	Change in land use class, number of field plots, area of the computation unit
Area by dominant tree species and age class *(puulajin vallitsevuus)*	km^2	number of field plots, tree species, age
Forest land area by development class *(metsämaan ala kehitysluokittain)*	km^2	number of field plots, development class

B) Ownership

Attribute	Measurement unit	Input attributes
Forest and forestry land ownership *(omistajaerittely metsä- ja metsätalousmaalla)*	% of the class area	Ownership, number of field plots

C) Wood production

Attribute	Measurement unit	Input attributes
Single tree volume including bark *(tilavuus)*	m^3	d.b.h., diameter at 6m height, length, tree species
Single tree volume including bark by timber assortment *(puutavaralajeittainen tilavuus)*	m^3	d.b.h., diameter at 6m height, length, tree species, length of saw timber or cull
Single tree volume increment including bark *(yksittäisen puun kasvu)*	m^3/a	d.b.h., thickness of the bark, height, diameter increment, height increment, diameter at 6 m height, tree species, taxation class
Volume by dominant tree species and age class *(puuston tilavuus puulajivaltaisuuksittain ja ikäluokittain)*	m^3/ha	volume, tree species, age, number of field plots
Volume by tree species *(tilavuus puulajivaltaisuuksittain)*	m^3/ha	Volume, tree species, number of field plots, area of the computation unit
Proportional volume by diameter class *(suhteellinen tilavuus läpimittaluokittain)*	%	Volume, tree species, d.b.h, number of field plots, area of the computation unit
Timber assortment volume *(puuston puutavaralajirakenne)*	m^3	Timber assortment volume, number of field plots on, area of the computation unit
Diameter distribution of the saw-timber stock *(tukkipuuston järeysrakenne)*	stems/ha	d.b.h, number of field plots, area of the computation unit
Increment *(kasvu)*	m^3	diameter increment, height increment
Increment by tree species *(kasvu puulajeittain)*	m^3/ha	Increment, tree species, number of field plots, area of the computation unit
Basal area by dominant tree species and age class *(ikäluokittaiset pinja-alat puulajivaltaisuuksittain)*	m^3	Basal area, tree species, age, number of field plots, area of the computation unit

D) Site and soil

Attribute	Measurement unit	Input attributes
Forest land by subsidiary land class and site class *(metsämaan jakaantuminen alaryhmiin ja kasvupaikkatyyppeihin)*	%	Main site class, site fertility class, number of field plots, area of the computation unit
Scrub land by subsidiary land class and site class *(kitumaan jakaantuminen alaryhmiin ja kasvupaikkatyyppeihin)*	%	Main site class, site fertility class, number of sample, area of the computation unit
Waste land by subsidiary land class and site class *(joutomaan jakaantuminen alaryhmiin ja kasvupaikkatyyppeihin)*	km^2	Main site class, site fertility class, number of field plots, area of the computation unit
Forest land by taxation class *(veroluokkien osuudet metsämaan kankailla ja soilla)*	%	Taxation class, number of field plots
Division of peatlands according to peat depth *(suoalan jakaantuminen turvekerroksen paksuuden perusteella)*	km^2 and %	Soil type, land use class, number of field plots, area of the computation unit
Completed drainage on mineral soils and peatlands *(ojitustilanne maaluokittain kankailla ja soilla)*	km^2 and %	Completed drainage, land use class, number of sample, area of the computation unit
Drainage stage of forest land peatlands *(ojitustilanne ja ojistusten jakaantuminen kuivumisasteen mukaan metsämaasoilla)*	km^2 and %	Drainage stage, land use class, number of field plots, area of the computation unit
Area drained during the previous 10-year-period *(kymmenvuotiskauden metsäojitukset)*	km^2	Completed drainage, number of field plots, area of the computation unit
Area suitable for forest drainage *(metsäojitukseen soveltuvat pinta-alat)*	km^2	Main site class, need for drainage, number of field plots, area of the computation unit

E) Forest structure

Attribute	Measurement unit	Input attributes
Dominance of tree species on forest and scrub land *(puulajien vallitsevuus metsä- ja kitumaalla)*	%	Dominant tree species, number of field plots
Dominance of deciduous species on forest and scrub land *(lehtipuulajien vallitsevuus metsä- ja kitumaalla)*	%	Dominant tree species, number of field plots

Tree species composition on forest land, areas based on the volume percentage of the dominant tree species in the main storey *(puulajikoostumus metsämaalla, pinta-alajakauma vallitsevan puulajin tilavuusosuudesta vallitsevassa puujaksossa)*	km^2 and %	Volume percentage of dominant tree species , land use class, number of field plots, area of the computation unit
Tree species composition on forest land, areas based on the volume percentage of coniferous/deciduous tree species in the main storey *(puulajikoostumus metsämaalla pinta-alajakauma havu-/lehtipuuston tilavuusosuudesta vallitsevassa puujaksossa)*	km^2 and %	Volume percentage of coniferous species, number of field plots, area of the computation unit
Tree storey distribution *(puujaksot)*	km^2	Tree storey position, number of field plots, area of the computation unit
Proportions of tree species *(puulajien osuudet)*	number/ha	Stem number, tree species, basal area, number of field plots, area of the computation unit
Proportional stem number distribution *(metsämaan puuston suhteellinen runkolukusarja)*	%	Tree species, stem number, d.b.h, number of sample, area of the computation unit
Origin of the stand methods *(perustamistapa)*	km^2 and %	Birth type, capability for development, number of field plots, area of the computation unit
Seedling stand areas and seedling densities *(taimikoiden pinta-alaosuudet ja taimimäärät)*	%	Tree species, Capability for development, number of field plots, area of the computation unit

G) Forest condition

Attribute	Measurement unit	Input attributes
Stand quality with the reason for decreased quality *(metsikön laatu alennussyineen)*	km^2	Stand quality, reason for the decreased quality, number of field plots, area of the computation unit

H) Accessibility and harvesting

Attribute	Measurement unit	Input attributes
Cuttings during the previous 10-year-period *(kymmenvuotiskauden hakkuut)*	km^2	Applied cutting methods, Time of completed cuttings, number of field plots, area of the computation unit
Area recommended for cutting during the next 10-year-period, based on the silvicultural stage of the stand *(kuvioittainen hakkuuehdotus)*	km^2	Proposed cuttings, number of field plots, area of the computation unit
Time of the latest cutting *(viimeisen hakkuun ajankohta)*	%	Time of completed cuttings, number of field plots, area of the computation unit
Silvicultural treatments during the previous 10-year-period *(kymmenvuotiskauden metsänhoitotoimenpiteet)*	km^2 and %	Applied silvicultural treatment, number of field plots, area of the computation unit
Area recommended for silvicultural treatments during the next 10-year-period *(metsänhoitotoimenpide-ehdotukset)*	km^2 and %	Proposed silvicultural treatments, number of field plots, area of the computation unit
Soil preparation during the previous 10-year-period *(kymmenvuotiskauden maanmuokkaukset)*	km^2 and %	Applied soil preparation methods, number of field plots, area of the computation unit
Area recommended for soil preparation during the next 10-year-period *(maanmuokkausehdotukset)*	km^2 and %	Proposed soil preparation, number of field plots, area of the computation unit

J) Non-wood goods and services

Attribute	Measurement unit	Input attributes
Forest land areas under obligatory restrictions *(velvottavien käyttörajoitusten alaiset pinta-alat metsätalousmaalla)*	km^2	Multiple use, multiple use specification, number of field plots

2.1.3. Measurement rules for measurable attributes

For each of the measureable attributes the following information is given:

a) measurement rule
b) threshold value
c) measurement scale
d) rounding rules
e) instrument used for measurement
f) data source

A) Geographic regions

Coordinates of the field plot

a) Location of the field plot expressed with the Finnish Uniform Coordinates
b) –
c) 10 m classes
d) rounded to the closest 10 m class
e) -
f) map assessment

C) Wood production

Pole size of the growing stock

a) Mean diameter of the trees in the main storey on the centre point stand, except on the development classes 2 and 3. It is measured as a mean breast height diameter of tally trees of the main storey. On development classes 2 and 3 the pole size is defined as height of the dominant/codominant trees in the main storey.
b) Defined as the mean height for the development classes 2 and 3, and as a mean diameter for other development classes.
c) 1 cm classes for diameter and 1 dm classes for height
d) rounded off to the nearest class centre.
e) calliper
f) field assessment

Stand basal area and basal area of the second tree storey (non dominant)

a) Measured from three points 20 m from the centre point to the cardinal points
b) –
c) 1 m^2/ha
d) rounded to the nearest 1 m^2
e) relascope, relascope factor 1.5
f) field assessment

Bearing of the tally tree

a) The bearing is measured from the plot centre to the pith of the tally tree at breast height
b) Tally trees selected by relascope using relacope factor 1.5. Trees further than 12.45 m from the plot centre are excluded
c) compass scale from 0 to 400
d) –
e) compass
f) field assessment

Distance to the tally tree

a) Horizontal distance from the plot centre to the pith of the tally tree at breast height
b) Tally trees selected by relascope using relascope factor 1.5. Trees further than 12.45 m from the plot centre are excluded
c) 1 cm classes
d) rounded to the closest class centre
e) metallic tape measure
f) field assessment

Breast height diameter (d.b.h.)

a) Measured at 1.3 m height from the ground level . One measure across the radius of the plot
b) Tally trees selected by relascope using relascope factor 1.5. Trees further than 12.45 m from the plot centre are excluded
c) 1 mm classes
d) rounded to the closest class centre
e) caliper
f) field assessment

Height to the lowest dead branch

a) Distance from the ground level to the lowest dry branch which affects the quality the of saw-timber.
b) For sample trees with a diameter less than 16.5 cm, branches of 10 mm and thicker are taken into account. For larger sample trees, branches of 15 mm and thicker are taken into account. Measured from the sample trees belonging to timber assortments 1–8.
c) 1 dm classes
d) rounded to the closest class centre
e) suunto altimeter
f) field assessment

Height of the base of living crown

a) Distance from ground level to the beginning of living crown
b) Measured for all living trees. Single living branches separated from the living crown by two or more dead branches are not counted as the beginning of the living crown
c) 1 dm classes
d) rounded to the closest class centre
e) suunto altimeter
f) field assessment

Height

a) Measured from ground level to the top of the crown
b) Measured for all living trees and for standing dead trees
c) 1 dm classes
d) rounded to the closest class centre
e) suunto altimeter
f) field assessment

Length of saw-timber or cull

a) length of different quality parts of the stem beginning from the stump
b) trees belonging to timber assortment classes 5–9 and pulpwood from which a part does not fulfill pulpwood qualifications
c) 1 dm classes
d) rounded to the closest class centre
e) suunto altimeter
f) field assessment

Height increments of the past five years

a) Before 31st July, the growth of the five previous years, and form 1st August on the growth of the present season and four previous years.
b) Measured from coniferous trees and counted with models for deciduous trees
c) 1 dm classes
d) rounded to the closest class centre
e) measurement bar or binoculars with measure scale and tripod
f) field assessment

Height increments of the present growing season

a) Growth of the present growing season
b) measured for coniferous trees
c) 1 dm classes
d) rounded to the closest class centre
e) measurement bar or binoculars with measure scale and tripod
f) field assessment

Diameter increment

a) Increment core borred at the breast height 90 degrees towards the plot radius
b) measurement of the past five years from living trees
c) one year increment applying 0.01 mm accuracy and five year total in 1 mm classes
d) rounded to the closest 1 mm class centre
e) borer and growth ring microscope
f) laboratory measurement

Length of the broken part

a) Length of the broken part
b) For all the living and standing death trees except for the death trees of which the broken part can be found and it is still usable
c) 1 dm classes
d) rounded to the closest class centre
e) ocular estimation
f) field assessment

Length of saw-timber or cull

a) Length of the saw timber or cull made because of insufficient quality. Not measured if the saw-timber extends up to the minimum diameter
b) For all saw timber logs where there is need for bucking because of varying quality
c) 1 dm classes
d) rounded to the closest class centre
e) suunto altimeter
f) field assessment

E) Forest structure

Number of stems

a) Measured in three fixed radius circles (young tree plots), r=2.3m, located in the plot centre and 20 m before and after the centre on the tract line
b) measured only for the development classes 2 and 3 when the tree storey position is dominant or usable under storey.
c) number/ha
d) –
e) visual observation
f) field assessment

Distance of the tally tree from the field plot centre

a) horizontal distance from the plot centre to the side of the tree
b) only on permanent plots
c) 1 cm classes
d) rounded to the closest class centre
e) measuring tape
f) field assessment

Mean age at the breast height and the breast height age

a) Measured from the mean basal area tree at 1.3 m height applying increment borer, or counting the whorls in the case of a young tree
b) Trees higher than 1.3 m
c) 1 year scale
d) –
e) increment borer
f) field assessment

Upper diameter

a) Measured at the height of 6m from the ground level. One measurement across the radius of the plot.
b) Trees over the height of 8m
c) 1 cm classes
d) rounded to the closest class centre
e) caliper
f) field assessment, measured only at every 9th cluster

Thickness of the bark

a) Measured at the height of 1.3m.The sum of two measurements taken from the opposite sides of a tree across the radius of the plot
b) trees over the height of 1.3m
c) 1 mm scale
d) rounded to the closest class centre
e) bark meter
f) field assessment, measured only at every 9th cluster

F) Regeneration

Total number of seedlings

a) The number of seedlings is counted from a fixed radius field plot r = 2.3 m
b) diameter < 5.3 cm
c) 1000/hectare classes
d) rounded to the closest class centre
e) measuring tape
f) field assessment

Number of seedlings capable for further development

a) The number of seedlings is counted from a fixed radius field plot r = 2.3 m
b) diameter < 5.3 cm, trees capable of reaching final cutting
c) 1000/hectare classes
d) rounded to the closest class centre
e) ocular estimation
f) field assessment

Maximum height of seedlings

a) Measured from ground level to the top of the crown
b) Measured on the small tree field plots
c) 1 dm classes
d) rounded to the closest class centre
e) suunto altimeter or height bar
f) field assessment

Maximum d.b.h.

a) Measured at 1.3 m height from the ground level. One measurement across the radius of the plot
b) Measured on the small tree field plots
c) 1 cm classes
d) rounded to the closest class centre
e) caliper
f) field assessment

Minimum height

a) Measured from the ground level to the top of the crown
b) Measured on the small tree field plot
c) 1 dm classes
d) rounded to the closest class centre
e) suunto altimeter or height bar
f) field assessment

Minimum d.b.h.

a) Measured at the height of 1.3 m from the ground level . One measurement across the radius of the plot
b) Measured on the small tree field plot
c) 1 cm classes
d) rounded to the closest class centre
e) caliper
f) field assessment

G) Forest condition

Diameter of the rotten part of the stem

a) Length of the decay in the stem
b) Determined by additional borings when decay is detected from the increment core
c) 1 dm classes
d) rounded to the closest class centre
e) increment borer
f) field assessment

H) Accessibility and harvesting

Stump diameter

a) Measured at the height of the highest root collar, or if this does not exist, at the ground level. One measurement across the plot radius
b) Stumps of the previous growing season are measured on all the field plots, older stumps only on permanent plots. The stumps are selected by relascope sampling, based on the stump diameter, with the relascope factor of 1.5. The stumps further than 12.45 m from the plot centre are excluded
c) 1 cm classes
d) rounded to the closest class centre
e) caliper
f) field assessment

Stump height

a) Height from the ground level to the pith at the cutting level
b) Stumps under 1 m, when the height of over 99 cm is recorded
c) 1 cm classes
d) rounded to the closest class centre
e) metallic tape measure
f) field assessment

J) Non-wood goods and services

Height of reindeer lichen

a) Mean height from the ground level to the top of the thallus
b) measured when the coverage is over 0.5%
c) 1 mm classes
d) rounded to the closest class centre
e) metallic tape measure
f) field assessment

2.1.4. Definitions for attributes on nominal or ordinal scale

The following information is provided for each attribute:

a) the definition
b) the categories (classess)
c) the data sources
d) remarks

A) Geographic regions

Deviation of the tract line

a) Plot wise observation which reveals the difference between the plot centre spotted on the map and the one defined through the terrain measurement. Defined by objects which can be unambiguously determined from both the map and the terrain
b) 10 m classes towards the four main cardinal points
c) field assessment

Distance and bearing of the stand boundary

a) The distance between the field plot and the nearest landscape boundary is defined as: the border between the land use classes, the border between the mineral soil and the peat land, the border between two stand development classes, or the border between the composition of tree species
b) 5m classes beginning form 0 to 40 meters form the plot centre towards 8 cardinal points
c) field assessment

Field plot size

a) Area of the 10 % sectors of the field plot, which intersect forest land
b) 10 % classes
c) field assessment

Percentage of the field plot covered by the stand

a) Proportion of the stand from the area of the field plot when the relascope plot, according to the largest tree (rmax=12.45m), extends to several stands
b) 10 % classes up to the full size of the field plot
c) field assessment

Measurement category

a) Expresses the category of objects which have been measured from the plot
b) 0) no trees, young trees or stumps measured
1,A) trees measured on the plot
2,B) trees and stumps measured
3,C) stumps measured
4) young trees measured
5) plot not visited, obviously there are no tally trees on the plot
6) plot not visited, obviously there are tally trees on the plot
c) field assessment

Classes of land use

a) Expresses the land use form from the point of view of forestry
b) 1) potential forest land used for timber production; mean annual increment at least 1m3/ha
2) scrub land where the potential mean annual increment is from 0.1 to 1m3/ha
3) waste land is a domain of forestry where the potential mean annual increment is less than 0.1m3/ha
4) other forestry land i.e. forestry roads, forest depots and camp lots, small grav

5) agricultural land
6) build up land
7) land claimed by traffic lines
8) land claimed by power lines
A) fresh water
B) salt water

c) field assessment

Specification of the land use classes

a) Defines special characteristics of the land use classes
b) 0) no specification
1) small stand (0.25–1 ha) belonging to the land use classes 1–4 in the middle of other land use classes
2) small stand (under 0.25 ha) belonging to land use classes 1–4 but recorded as part of another class because of its small size or unextrable shape
3) stand located in a small island smaller than 1 ha
4) agricultural land but not field
5) farm centre
6) forested area which does not belong to land usage classes 1–4 such as a forested building plot
c) field assessment

Change in land use class

a) Changes within land use classes 1–4 and from classes 1–4 to the other classes are recorded from the past 10-year-period. Changes from the other land use classes to classes 1–4 are recorded from the past 30-year-period
b) 0) no change or change between classes 5–8
1) change from the land use classes 1 to present class during the past 10-year-period
2) change from the land use classes 2 to present class during the past 10-year-period
3) change from the land use classes 3 to present class during the past 10-year-period
4) change from the land use classes 4 to present class during the past 10-year-period
5) change from the land use class 5 to classes 1–4 during the past 30-year-period
6) change from the land use class 6 to classes 1–4 during the past 30-year-period
7) change from the land use class 7 to classes 1–4 during the past 30-year-period
8) change from the land use class 8 to classes 1–4 during the past 30-year-period
9) stand is changing to land use classes 1–4 for example agricultural land which has not been
used for agriculture during the past 10 years
A) change from the land use class A to classes 1–4
B) change form the land use class B to classes 1–4
c) field assessment

Time of the change in land use class

a) The time when the change in land use class has taken place
b) 0) inventory year
1) previous year
2) 2–5 years ago
3) 6–10 years ago
4) 11–30 years ago
c) field assessment

B) Ownership

Owner

a) The ownership recorded from cadaster maps
b) 0) farm forest owners
1) other private forest owners
2) forest company
3) other company
4) Finnish Forest and Park Service (state owned)
5) other areas than 4 owned by the state
6) jointly owned forest
7) municipalities, churches, and other communities
8) ownership not shared, such as heirs
c) map assessment

C) Wood production

Saw-timber stage

a) A tree storey is considered as saw-timber if 25% of the dominant and codominant trees have reached the dimensions of saw-timber. The saw timber stage is estimated for the tree storeys of the development classes 5–9.

b) 1) non saw-timber stage
2) saw timber stage

c) field assessment

The proportion of saw-timber

a) An estimate of the proportion of trees which will grow to the saw timber stage within rotation time

b) 1) under 25 % of the dominant trees which are capable for further development in the storey will reach the saw-timber stage. The technical quality of the storey will not fulfill the saw-timber qualifications
2) under 25 % of the dominant trees which are capable for further development in the storey will reach the saw-timber stage. The technical quality of the storey will fulfill the saw-timber qualifications
3) 25–35% of the dominant trees which are capable for further development in the storey will reach the saw-timber stage
..
+) 95–100% of the dominant trees which are capable for further development in the storey will reach the saw-timber stage

c) field assessment

The quality distribution of saw-timber

a) An estimate of the quality distribution of saw-timber logs at the final cutting

b) 1) at least 1/3 belongs to the 1st quality class and at least 2/3 fulfill the saw-timber
qualifications
2) at least 1/3 belongs to the 1st quality class and under 2/3 fulfill the saw-timber qualifications
3) at least 1/10 belongs to the 1st quality class and at least 2/3 fulfill the saw-timber
qualifications
4) at least 1/3 belongs to the 1st quality class and at least 1/3 fulfills the saw-timber qualifications
5) at least 1/3 belongs to the 1st quality class and under 1/3 fulfills the saw-timber qualifications
6) at lest 2/3 fulfill the saw-timber qualifications
7) at least 1/3 fulfills the saw-timber qualifications
8) under 1/3 fulfill the saw-timber qualifications

c) field assessment

Tree species

a) Scientific names of 2 coniferous and 7 deciduous tree species and two additional classes: other coniferous and other deciduous species

b) 1) *Pinus sylvestris* Scots pine
2) *Picea abies* Norway spruce
3) *Betula pendula* silver birch
4) *Betula pubescens* downy birch
5) *Populus tremula* European aspen
6) *Alnus incana* grey alder
7) *Alnus glutinosa* black alder
8) *Sorbus aucuparia* rowan
9) *Salix caprea* goat willow
A) other coniferous
B) other deciduous

c) field assessment

Timber assortment

a) The timber assortment class defines saw-timber and pulpwood trees and their quality based on the observed assortment class for saw-timber which has already reached the final cutting stage, and based on the estimated stage at the future final cutting for other trees.

b) 1) wastewood, a tree which will never reach even pulpwood qualifications
2) a good quality young tree which is expected to reach class 5 saw-timber qualifications

3)a good quality young tree with some quality defect which is expected to reach class 6 saw-timber qualifications
4) a normal quality young tree which is expected to reach class 7 saw-timber qualifications
5) good quality saw-timber, at least 31 dm belongs to the 1st quality class and at least 80% of
the part of stem with saw-timber dimensions fulfills the saw-timber qualifications
6) good quality saw-timber with some quality defects, at least 31 dm belongs to the 1st quality class but under 80% of the part of stem with saw-timber dimensions fulfills the saw-timber qualifications
7) normal saw-timber, at least 40 dm fulfills the saw-timber qualifications and at least 80% of the part of stem with saw-timber dimensions fulfills the saw-timber qualifications
8) saw-timber with quality defects, at least 40 dm fulfills the saw-timber qualifications but under 80% of the part of stem with saw-timber dimensions fulfills the saw-timber qualifications
9) current pulpwood, the quality of tree allows development to saw-timber but it will
never reach the saw-timber stage because of site and/or tree species or health status
A) poor quality pulpwood, the poor quality prevents development to the saw-timber stage
B) Poor quality constraint pulpwood which has reached the saw-timber dimensions, the poor
quality prevents development to the saw-timber stage
C) usable standing death tree
D) usable fallen death tree
E) stub, at least half of the tree is usable for construction purposes, diameter at least 20cm

c) field assessment
d) the death trees are considered usable if they can be used as fuelwood

Revision of timber assortment

a) defines the cause for the quality defect in case the timber assortment class is 2, 4, 6, 8 or B
b) 1) green branch or general branchiness
2) death, rotten or vertical branch, branch bump or branch hole
3) crook
4) curve
5) fork
6) decayed
7) stemwood damage
8) crown damage
9) other defect or damage
c) field assessment

Revision of the timber assortment specification

a) If the possible increment boring or other following measurement gives reason to revise timber assortment class, revised class is recorded. Otherwise previously recorded class is accepted
b) as in height increment above
c) field assessment

Condition of the growing stock

a) Logs belonging to the timber assortment classes 5–9 are divided according to the saw-timber quality classes announcing the class length beginning from stump. Pulpwood is divided if a part of the stem does not fulfill pulpwood qualifications.
b) 1) High quality butt log
2) green branch log
3) dead branch log
4) cull suitable for pulpwood
5) jump cut in deciduous tree
6) wastewood
7) saw-timber part of fork stem
8) obligatory cut in the middle of saw-timber part
c) field assessment

Cause for revision of the condition of the growing stock

a) defines the reason for lowering quality or obligatory cut in the middle of saw-timber part
b) 1) green branch or general branchiness
   2) death, rotten or vertical branch, branch bump or branch hole
   3) crook
   4) curve
   5) fork
   6) decayed
   7) stemwood damage
   8) crown damage
   9) other defect or damage
c) field assessment

Crown storey

a) Defines the crown storey of trees capable of further development in development classes 2–6 and 9.
b) 2) dominant and codominant trees of the dominant tree storey
   3) dominated tree of the dominant tree storey
   4) suppressed tree of the dominant tree storey
   5) dominant and codominant trees of the over storey
   6) dominated and suppressed trees of the over storey
c) field assessment

D) Site and soil

Main site class

a) divides forest, scrub and waste land to mineral soil and main peatland classes
b) 1) mineral soil
   2) spruce/hardwoods peatland
   3) pine peatland
   4) open bog
c) field assessment

Main site class specification for peatland

a) Divides main site classes 2–4 into pure peatland classes and combinations of peatland classes
b) 0) pure peatland class
   1) includes features of mineral soil forest
   2) includes features of spruce/hardwoods peatland forest
   3) includes features of pine peatland forest
   4) includes features of open bog
   5) includes fen-like features
c) field assessment

Site fertility class

a) Defines site fertility class
b) 1) pure rich grass-herb mineral soil and corresponding rich fens, mires
   2) lower nutrition level grass-herb forest, grass mires
   3) mineral soils and corresponding mires with Myrtillus vegetation
   4) mineral soils and mires with Vaccinium vegetation and sedgerich mires
   5) dry upland forest sites and swarf shrub and cottong-grass-sedge mires
   6) poor mineral soil and sphagnaceous mires
   7) rocky or sandy soil or alluvial land
   8) summit and field forest
c) field assessment

Site fertility class specification for peatland

a) Used for specifying the site fertility class on peatlands and for defining the factors affecting the potential yield
b) 0) no specification
   1) flarky, flarks or corresponding hollows cover over 30 % of the stand
   2) purple moor-grassy, purple moor-grass belong to dominant under vegetation species
   3) brown bog moss dominated, brown bog moss covers over 30% of the stands and the site
   quality class is not 6
   4) alluvial soil, stand under water part of the year
   5) affected by surface water
c) field assessment

Drainage stage

a) Defines possible drainage and the effect of draining on forest yield
b) 0) undrained
   1) drained mineral site
   2) drained peatland, the drainage effect cannot yet be discovered in ground vegetation or in trees
   3) drained peatland where the drainage effect is clearly visible although the ground vegetation is still characterized by the original peatland type
   4) drained peatland where the drainage effect has resulted in appearance of some mineral soil under vegetation. Water relations does not prevent the canopy closure
c) field assessment

Completed drainage

a) Defines the type of possible drainage
b) 0) undrained
   1) initial drainage
   2) reparation of old ditches
   3) complementary drainage
   4) other than forestry drainage, for example field ditches on afforested old agricultural land
c) field assessment

Time of completed drainage

a) Defines the time when the detected drainage was accomplished
b) 0) inventory year
   1) previous year
   2) 2–5 years ago
   3) 6–10 years ago
   4) 11–30 years ago
   5) more than 30 years ago
c) field assessment

The need for drainage

a) Defines the possibility of increasing the yield by drainage
b) 0) no drainage need
   1) initial drainage on peatland suitable for draining or mineral soil under peat formation
   2) reparation of old ditches
   3) complementary drainage
   4) initial drainage of reparation of ditches implemented on peatland not suitable for timber production, further draining activities not recommended
c) field assessment
d) based on the soil type and day degrees used to define productive peatland where water relations prevent timber production

Urgency of the required drainage

a) Defines urgency of the proposed drainage
b) 0) not urgent, can be combined for example with the next silvicultural or cutting activity
   1) urgent, water relations have affected the yield negatively, proposed to be drained immediately
c) field assessment

Soil type

a) defined at the depth of 30 cm, or on top of the mineral soil layer if the depth of the soil layer on top of the mineral soil is under 30 cm
b) 1) organic, mean depth of the organic layer over 30 cm, normally peat
   2) solid rock, free soil layer (organic + mineral) is under 30 cm
   3) rock material, depth of the layer of the rock blocks (diameter > 20 cm) and stones(diameter 2–20 cm) over 30 cm
   4) rough gravel mineral soil
   5) medium rough mineral soil, fine sand or rough silt and corresponding moraine
   6) fine mineral soil, medium and fine silt, clay and corresponding moraine
c) field assessment

Type of the organic layer

a) Type of the organic layer
b) 0) peat
   1) raw humus
   2) other, such as mull
   3) organic layer does not exist

Thickness of the soil mantle

a) Thickness of the organic layer on top of the mineral soil layer
b) 0–10 cm in one cm classes
10-100 cm in ten centimeter classes
over 100 cm
c) field assessment

Taxation class

a) Assessed for the forest land and for agricultural land which has not been used for agriculture for ten years
b) 0) very rich and rich sites
1) damp sites on mineral soil
2) sub-dry sites on mineral soil
3) dry upland and poor mineral soil and spruce/hardwood peatland
4) pine peatland
c) field assessment

Taxation class change

a) If the taxation class does not follow the site fertility class on mineral soils, or the main site class on peatlands, the reason for change is defined by this specification
b) 0) no change
1) closeness of solid rock or rocky soil
2) alluvial land
3) raw humus
4) location, such as low day degree area, windy area, repetitive snow damages along the sea or lake side
5) other decreasing site characteristics or surrounding phenomena
6) increase of the taxation class on drained peatland above main site class, site fertility class or draining class
7) drained peatland where the drainage effect is clearly visible although the ground vegetation is still characterized by the original peatland type but the productivity is comparable with some site quality class mineral soil
c) field assessment

E) Forest structure

Number of tree storeys

a) Tree storeys are separated if the age difference between the storeys is 40 years, or the basal area estimation requires assessment in different storeys
b) 0) uneven-aged
1) one storey
2) two storey
c) field assessment

Tree storey position

a) Defines the position of the tree storey in stand
b) 1) dominant, no nurse crop
2) standard
3) nurse crop
4) understorey, capable for further development during the next 10-year-period, the final cutting is estimated to take place within this period
5) understorey, incapable for further development because there is no final cutting estimated to take place within the next 10-year-period
6) understorey, incapable for further development due to unsuitable tree species or due to damages
7) young plants showing possibility to natural regeneration which is, at the moment, prevented by dense dominant storey
c) field assessment

Development class

a) Defines the development class, the definition mature stands are based on optimal rotation defined by the site fertility class and geographic region
b) 1) treeless regeneration area
2) young seedling stand
3) advanced seedling stand
4) young thinning stand
5) advanced thinning stand
6) mature stand
7) shelter tree stand
8) seed tree stand
c) field assessment

The origin of the stand

a) Definition is based on the amount and location of the seedling material
b) 1) natural seedling regeneration or open area without manmade regeneration
   2) sprout forest
   3) planted
   4) sown
   5) unsuccessful planted
   6) unsuccessful sown
c) field assessment

Dominant tree species

a) Dominant tree species are defined on the basis of the dominance of coniferous or deciduous species. If the proportions of coniferous and deciduous species are equal the dominant tree species is the one for which the silvicultural treatments are meant to favour
b) 0) treeless
   1) Scots pine
   2) Norway spruce
   3) silver birch
   4) downy birch
   5) European aspen
   6) grey alder
   7) black alder
   8) rowan
   9) willow
   A) other coniferous
   B) other deciduous
c) field assessment
d) The proportion of the dominant tree species is recorded in 10% classes

2nd tree specie

a) the tree species which is the most important after the dominant tree species
b) as for dominant tree species
c) field assessment
d) proportion of the 2nd tree species is recorded in 10% classes

3rd tree species

a) the tree species which is the most important after the dominant and 2nd tree species, recorded in the development classes 2-3 and 4-9.
b) as for dominant tree species
c) field assessment
d) the proportion is not recorded

Proportion of the coniferous species from the total volume

a) defines the proportion of coniferous species from the total volume of the storey
b) 0) <5 %
   1) 5-15%
   2) 15-25%
   .
   9) 85-95%
   +) >95%
c) field assessment

Proportion of the coniferous species from the total number of stems

a) defines the proportion of coniferous species in development classes 2-3 from the total number of tree of the storey
b) 0) <5 %
   1) 5-15%
   2) 15-25%
   .
   9) 85-95%
   +) >95%
c) field assessment

Birth type

a) Defines the birth type of a sample tree
b) 0) not possible to detect
   1) naturally sown
   2) natural sprout
   3) planted
   4) sown
c) field assessment

Proportion of the regeneration plot

a) Defines the proportion of the centre stand of the plot
b) 10 % classes
c) field assessment

Capability of development

a) Defines the expected timber assortment class for young tally trees
b) 1) capable of development
   2) suitable for pulp tree
c) field assessment

G) Forest condition

Description of the damage

a) defines the visually observed description of the damage
b) 0) no damage
   1) dead standing tree
   2) fallen tree or standing stem broken below the crown
   3) rotten scarf in the stem
   4) stem damage or root damage in 1 m radius from the stem
   5) broken or dry top on the upper part (50 % of crown) of crown
   6) top with multiple leaders
   7) defoliation
   8) coloration of the needles of the leaves
c) field assessment

Age of the damage

a) Defines when the damage has originated
b) 0) less than 5 years ago
   1) more than 5 years ago
c) field assessment

The damaging agent

a) Defines the cause of damage
b) 0) not known
   1) wind
   2) snow
   3) other meteorological cause such as frost
   4) soil factors such as water and nutrition availability
   5) competition between plants
   6) harvesting
   7) other human activity
   8) mole
   9) elk or other vertebrate
   A) insect A0) not recognized
   A1) Tomicus species
   A2) pine weevil
   A3) sawfly, wespen
   A4) spruce bark beetle
   b) fungus B0) not recognized
   B1) pine canker
   B2) decay fungus
   B3) pine twisting rust
   B4) Scots pine blister rust
   B5) needle cast
   B6) other rust fungi
   B7) other fungi disease
c) field assessment

Degree of the damage

a) defines the significance of the damage
b) 0) light damage which does not affect the development of a tree
   1) damage which decreases growth but does not damage the produced timber
   2) damage decreasing the amount of timber produced
   3) lethal
c) field assessment

Defoliation (tree)

a) Assessed in the development classes 4–9 from the dominant trees and in the development classes 2–3 from standards. Assessment is based on the estimated normal trees in the corresponding condition
b) 0) 0–10% of the needles lost
   1) 11–20 % of the needles lost
   .
   9) 91–100% of the needles lost
   E) upper part of the crown not visible
c) field assessment

Defoliation (stand)

a) Defoliation observation considering whole stand
b) E) not observed
   0) no defoliation
   1) 6-20% of the trees have at least 40% defoliation
   2) 21-50% –"–
   3) 51-100% –"–
   4) 6-20% of the trees have 20-40 % defoliation

5) 21-50% –"–
6) 51-100% –"–
7) defoliation observed on the sample plot but does not affect the whole stand.
c) field assessment

Abundance of beard lichen

a) In the permanent field plots the existence of beard lichens (*Alectoria, Bryoria and Usnea*) is assessed.
b) e) not assessed
0) does not exist
1) small on single trees, most of the trees do not have
2) moderate, almost every tree has lichens but only some in each
3) rich in existence, almost every tree has several types, often long growth
-
5) exceptional code, lichen exists only above the 2m level. Used in reindeer management areas where grazing affects the amount of lichens
c) field assessment

Abundance of leaf lichens

a) In the permanent field plots the proportion of the stem and branches at the height of 0.5–2.0 m covered with leaf lichens (*Hypygymnia, Parmelia and Pseudevernia*) is estimated
b) 0) no leaf lichen coverage under 1%
1) small, coverage 1–10%
2) moderate, coverage 10–40%
3) rich existence, coverage over 40% however not continuous (see 4)
4) *Hypogymnia physodes* covers the trees almost completely
5) exceptional code for rich existence where little existence under 2m level but rich existence above 2m level.
c) field assessment

Abundance of Scoliciosporum clorococum and Desmococcus olivaeus

a) In the permanent field plots the proportion of the stem and branches at the height of 0.5–2.0 m covered with *Scoliciosporum clorococum* and *Desmococcus olivaeus* is estimated
b) 0) no existence, coverage under 1%
1) 2) moderate, coverage 10–40%
3) rich existence, coverage over 40%
4) trees covered almost completely
c) field assessment
d) not estimated if beard lichen is not estimated

Stand quality

a) Stands are divided to ones capable of further development and to under-productive ones. A stand is considered under-productive if the yield without special treatment is low enough to give reason for regeneration before maturity
b) 1) good
2) satisfactory
3) adequate
4) under-productive
c) field assessment

Reason for the decreased quality

a) defines the cause of decreased stand quality
b) 1) age when the age has exceeded the normal rotation time
2) unfavorable tree species compared with the site fertility class
3) over density
4) delayed silvicultural treatment such as clearing or standard removal
5) natural low density
6) unsuccessful treatment
7) unfavorable spatial distribution
8) technical quality
9) damages
c) field assessment

Stage of decay

a) defines the stage of decay from the increment core
b) 0) soft decay
1) hard decay
c) field assessment

Location of the decay

a) defines location of the decay from the increment core
b) 1) heart decay
2) other than heart decay
c) field assessment

H) Accessibility and harvesting

Applied cutting methods

a) The cuttings during the past ten-year-period are recorded
b) 0) no cuttings
1) tending of seedling stand
2) standard removal
3) first commercial thinning
4) other commercial thinning or preparatory cut
5) selection cutting
6) special cutting such as cuttings for opening drainage or road construction areas or cuttings
for repairing detected forest damages
7) artificial regeneration
8) natural regeneration
c) field assessment

Time of completed cuttings

a) Time of the completed cuttings including cuttings older than ten years, estimated according the cutting seasons, beginning the first of June each year
b) 0) inventory summer
1) previous cutting season
2) previous cutting seasons 2–5
3) " 6–10
4) " 11–30
5) no cuttingss or cuttingss older than 30 cutting seasons
c) field assessment

Applied soil preparation methods

a) the last completed soil preparation activity during the past 30 years
b) 0) no soil preparation
1) harrowing
2) ploughing
3) ploughing with draining effect
4) hummocking
5) controlled burning
6) controlled burning + harrowing
7) controlled burning + ploughing
8) controlled burning + hummocking
c) field assessment

Time of completed soil preparation

a) Year of the completed soil preparation
b) 1) inventory year
2) previous year
3) 2–5 years ago
4) 6–10 years ago
5) 11–30 years ago
c) field assessment

Type of completed silvicultural treatment

a) the last completed silvicultural treatment during the past 30 years
b) 0) no treatment
1) regeneration
2) repair regeneration
3) pruning
c) field assessment

Time of completed silvicultural treatment

a) Year of the completed soil preparation
b) 1) inventory year
2) previous year
3) 2–5 years ago
4) 6–10 years ago
5) 11–30 years ago
c) field assessment

Proposed cuttings

a) Generally cutting is proposed if the basal area exceeds the proposition at least by 6 m3/ha
b) 0) no cuts
1) noncommercial thinning
2) standard removal
3) first commercial thinning
4) other commercial thinning or preparatory cutting
5) selection cutting
6) special cutting such as cuttings for opening drainage or road construction areas, or cuttings for repairing detected forest damages
7) artificial regeneration
8) natural regeneration
A) no proposal, uncut strip in strip cutting area
c) field assessment

The urgency of proposed cuttings
a) Made for a 10-year-period
b) 1) treatment already delayed
   2) first 5-year-period
   3) last 5-year-period
c) field assessment

The proposed soil preparation
a) Proposed for the regenerated areas. Soil preparation can be proposed immediately or after the cutting proposed for the ten-year-period
b) 0) no soil preparation
   1) harrowing
   2) ploughing
   3) ploughing with a draining effect
   4) hummocking
c) field assessment

Proposed silvicultural treatments
a) Proposal for silvicultural treatment
b) 0) no proposal
   1) regeneration
   2) repair regeneration
   3) preventing overgrowth of grass and herb
   4) clearing
c) field assessment

Stump age
a) on normal field plots always code 1, on permanent field plots may also be 1
b) 0) inventory summer after first of June
   1) previous year
c) field assessment

Type of felling
a) Estimated from stumps originating from the year preceding the inventory
b) 1) merchantable trees felled using a harvester
   2) merchantable trees felled using a chain saw
   3) non merchantable trees felled using a chain saw or a clearing saw
   4) other type of felling
c) field assessment

J) Non-wood goods and services

Location of the reindeer fodder plot
a) distance and bearing of the reindeer fodder plot
b) 2) 20m east of the field plot centre
   4) " south "
   6) " west "
   8) " north "
   9) other location
c) field assessment
d) assessed only in the reindeer management area

Coverage of Deschampsia flexuosa, coverage of grasses and herb, coverage of Stereocaulon spp., coverage of reindeer lichens
a) Describes the proportion of the plot area covered with the living parts of the vegetation on the fixed radius field plot
b) one % classes
c) field assessment
d) can be over 100% if the species grow on top of each other

Location of the square plot
a) all the reindeer fodder square plots are preferably located towards the same bearing from the plot centre, in the bearing with the smallest bearing code, preferably according to the cardinal points
b) 1) north-east
   2) east
   3) south-east
   4) south
   5) south-west
   6) west
   7) north-west
   8) north
c) field assessment
d) assessed only in the reindeer management area

Coverage of Deschampsia flexuosa, coverage of grasses and herb, coverage of Stereocaulon spp., coverage of different reindeer lichen species

a) Describes the proportion of the plot area covered with the living parts of the vegetation on the square field plot
b) one % classes
c) field assessment using an estimation scale

The soil condition

a) To be recorded, the condition has to dominate at least one half of the square plot
b) 0) narrow water 0-5m
 1) normal
 2) rock or rocky
 3) raw humus
 4) swampy
 5) soil surface broken
 6) old controlled burning area
 7) underscrub spot
 8) trampled area
 9) tree or stump
c) field assessment
d) assessed only in the reindeer management area

Condition of the vegetation

a) includes all species of scrub, grass, herb, moss and lichen
b) 1) normal
 2) mechanical stress
 3) physiologic stress
 4) darkening of *Cladonia rangiferina*
c) field assessment
d) assessed only in the reindeer management area

Multiple use

a) Defines potentials for multiple use, areas under restrictions based on land used planning (codes 1–7) are detected from the database in advance
b) 0) no restrictions
 1) areas where all cuttings are forbidden by law
 2) areas where cuttings are forbidden by authorities responsible for management
 3) areas where cuttings are restricted by law because of nature protection or amenity values
 4) as 3) but restriction based on the decision of authority responsible for management
 5) areas where special permission is required for felling trees
 6) areas with temporal cutting restrictions
 7) areas which have been decided to be protected but the protection has not yet been put into effect
 8) do not belong to classes 1–7 but the special location or amenity values should be taken into account in the forest management planning
c) field assessment

Multiple use specification

a) Multiple use specification
b) 1) strict nature reserve
 2) national park
 3) wilderness forest
 4) peatland reserve
 5) limit forest
 6) reserved stand
 7) privately reserved stand
c) field assessment
d) if multiple use registered is 8, specification is 6 or 7

2.1.5. Forest area definition and definition of "Other wooded land"

Forest land has the potential capacity to produce a mean annual increment of at lest 1 m^3/ha stemwood, over bark, given an optimum tree species mixture, growing stock volume and prescribed rotations. *Scrub land* has the potential capacity to produce a mean annual increment of at lest 0.1 m^3/ha but less than 1.0 m^3/ha given an optimum tree species mix. Forest land and scrub land combined are called *forested land. Waste land*, if not naturally treeless, is not given an optimum tree species mix,and it is not able to produce annually more than 0.1 m^3/ha. (Salminen 1993).

2.2 DATA SOURCES

A) Field data

Field plots defined in section 2.3.2.

B) Questionnaire

Not applicable

C) Aerial photography

Not applicable

D) Space borne digital remote sensing

Monotemporal Landsat TM images (a-products), preferably from the field inventory period, are rectified to the Finnish Uniform Coordinate System and rectified to 25 m pixel size. The digital terrain models are used for correcting original spectral values in order to avoid confusion in the image analysis caused by land morphology (Tomppo 1996). All the Landsat TM channels except the thermal channel are applied. The classification results are verified by comparing them with field inventory results from large areas and by means of jackknifing (Tomppo 1992). In the 8th inventory, 50 Landsat images were applied.

E) Maps

Printed topographic base maps at the scale 1:20 000.

F) Other georeferenced data

Commercial digital maps produced by the National Land Survey of Finland for digital terrain model, arable land, urban areas, single buildings, roads and peatlands are applied. Ownership and multiple use information are partly in map form.

2.3. ASSESSMENT TECHNIQUES

2.3.1 Sampling frame

The National Forest Inventory assessment covers the entire Finland including land area and inland watercourses up to the inner boundary of territorial waters. The observations which would be located on lakes are registered and 85672 field plots are observed on land.

2.3.2. Sampling units

Field inventory

The sampling design has been changed during the 8th inventory, at the administrative border of North Finland, see Figure 6.

In South Finland, all the field plots were temporary angle gauge plots, Bitterlich plots. The relascope factor was 2. If a field plot intersects on several stands (on a stand border), stand characteristics are given for each stand and stand codes are given to each tree in order to attach trees to the relevant stands. The sizes of the intersections are estimated but not used in the computations, see Chapter 2.4.5. Similarily, if a plot is located on a border of land use classes, the subplot located on forestry land is measured.

If the tract line is 'parallel' to the land use class border, the parallelism means that the tract line does not intersect the land use class border within the field plot. If the center point of the plot is located on forestry land, the size of field plot is measured and the measured size is applied in the computations, instead of the whole size, see Chapter 2.4.5. Correspondingly, if the center point in this case is located on non-forestry land, the size of the plot is zero and the possible trees on the plot are not measured. The whole idea in this system is to avoid measuring small plot parts e.g. on lake sides or sides of arable land if the plot center is on non-forestry land. Correspondingly, the measured parts will get higher weight in the computations, in order to avoid bias. This system will not be applied any more in the 9th inventory.

In North Finland, in the areas of Forestry Board District 16–19, two different types of field plots were applied. In the Districts 16–18 and in the southern part of the District 19 (19 A), the field plot was a restricted angle gauge plot with a relascope factor 1.5. A maximum radius of 12.46 m was set to the field plot in order to avoid too many subplots and substand descriptions. One fifth of the plots were established as permanent for future measurements. Other plots are temporary. Subfield plots were treated as in South Finland.

In the northernmost part of Finland, in the areas of the municipalities of Enontekiö, Inari and Utsjoki, a fixed size field plot was utilized. The plot consists of three subplots, each with a radius of 5.15 m. The three plots are located in the corners of a triangle with equal sides of 11.3 m, see Figures 5 and 6. One fourth of the plots were permanent.

Stand: The stand on which the centre point of the field plot is located is referred to as the centre point stand.

Young tree field plot: Young tree information is recorded from all the normal field plots in which the centre point stand is on forest land. A young tree field plot is a fixed radius field plot, the radius is 2.3m and the size is 16.6 m^2.

Stump field plot: All the stumps from the inventory summer are recorded from the forested normal field plots. In addition, all the stumps from the previous felling season are measured from the permanent field plots.

Fixed radius reindeer fodder field plot: In the reindeer management area a fixed radius reindeer fodder field plot is located 20 m from the centre of the permanent field plot. The fodder plot is located towards the centre point stand and the preference of the bearing is firstly along the tract line and secondly directly towards east, south, west, north and other bearings in descending order. The applied radius is 10 m and the size 314.2 m^2.

Reindeer fodder sample square: In the reindeer management area, five sample squares are measured around the permanent field plots. The squares are located in the centre point stand. The majority of the tree squares are located in one direction. The preference of locations are three squares along the tract line to the east and the remaining two west of the field plot. If this is not possible, the preference of bearings is east-west, north-south, north-east, south-east, south-west and north-west.

Multi-source inventory

The applied pixel size after image rectification and resampling is 25 m × 25 m, that is 625 m^2. Each pixel is analysed in image processing.

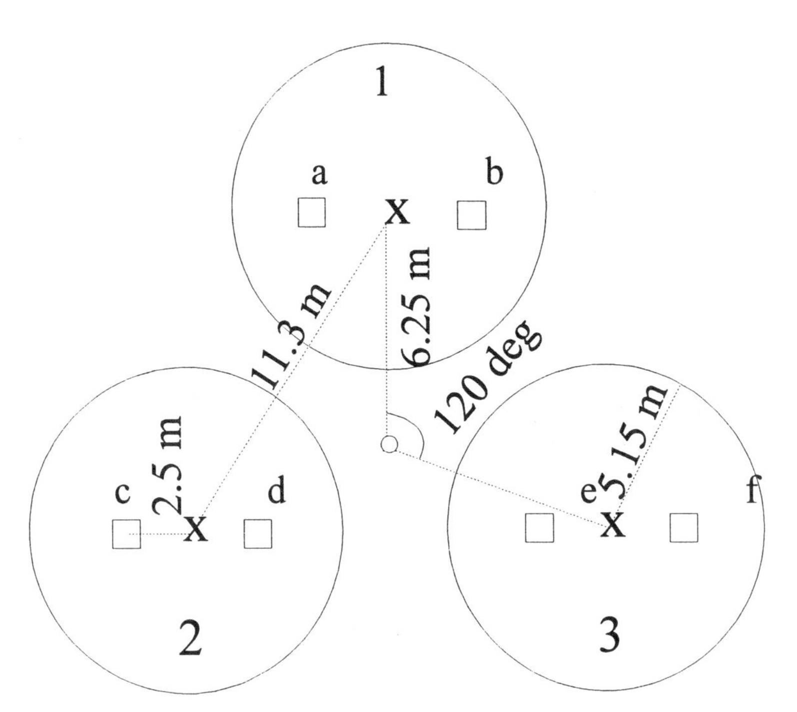

Figure 5. Field plots in northernmost Finland. 1, 2 and 3 are the tally tree plots and a-f are the vegetation sample squares.

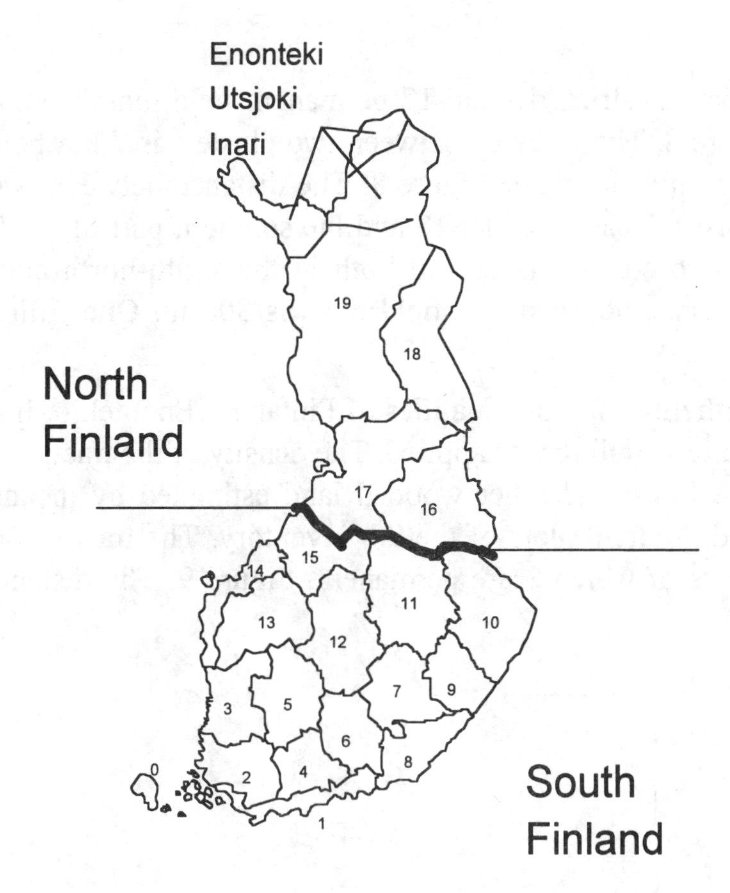

Figure 6. The division of the country used in the National Forest Inventory, 1–19 are the Forestry Board Districts.

2.3.3 Sampling designs

The field sampling design has been changed a few times between inventories. The method of the first four inventories, linewise survey sampling, was changed into a systematic cluster sampling in the fifth inventory (1964–1970). This method has been modified a few times afterwards. A bigger change was carried out during the eighth rotation, from the administrative border of North Finland northwards. One fifth of the field plots was established permanent. Until that, all the plots had been temporary. Totally, four different sampling designs have been used in the 8th inventory, 1) South Finland, in the province of Ahvenamaa (Åland) and in the areas of the15 southernmost Forestry Board Districts (FBD), 2) the areas of FBDs 16–17, 3) the area of FBDs 18 and 19, except for the three northernmost municipalities Enontekiö, Inari and Utsjoki and 4) the areas of those three municipalities, see Figure 6.

South Finland (FBD 1–15 and Ahvenanmaa):

The sampling unit is a cluster that is called a tract. There has been some variation in the cluster shape during the 8th inventory described here. Direction of the tract sides are south - north, and from the north end of the line towards east, see Figure 7. One tract has 21 angle gauge field plots. The distance of two tracts is 7 km in the west - east direction and 8 km in the south - north direction; see Figure 7. The distance between two field plots is 200 m.

North Finland:

In the Forestry Board Districts 16 and 17 permanent field plots were applied for the first time; see Figure 8. The distance between two clusters is 7 km both in the south-north and west-east directions; see Figure 8. The distance between two field plots is 300 m. In the Forestry Board District 18 and the southern part of the District 19, the distance between two clusters is 10 km both in the south-north and the west-east directions. The distance between two field plots is 300 m. One fifth of the plots is permanent.

In the three northernmost municipalities of Finland – Enontekiö, Inari and Utsjoki –stratified systematic sampling was applied. The density of the tracts was proportional to the share of the forest and other wooded land estimated by means of a satellite image analysis and the field plots of the 7th inventory. The tract has a square shape and has 8 field plots of which 2 are permanent, Figure 9. The distance between the plots is 750 m.

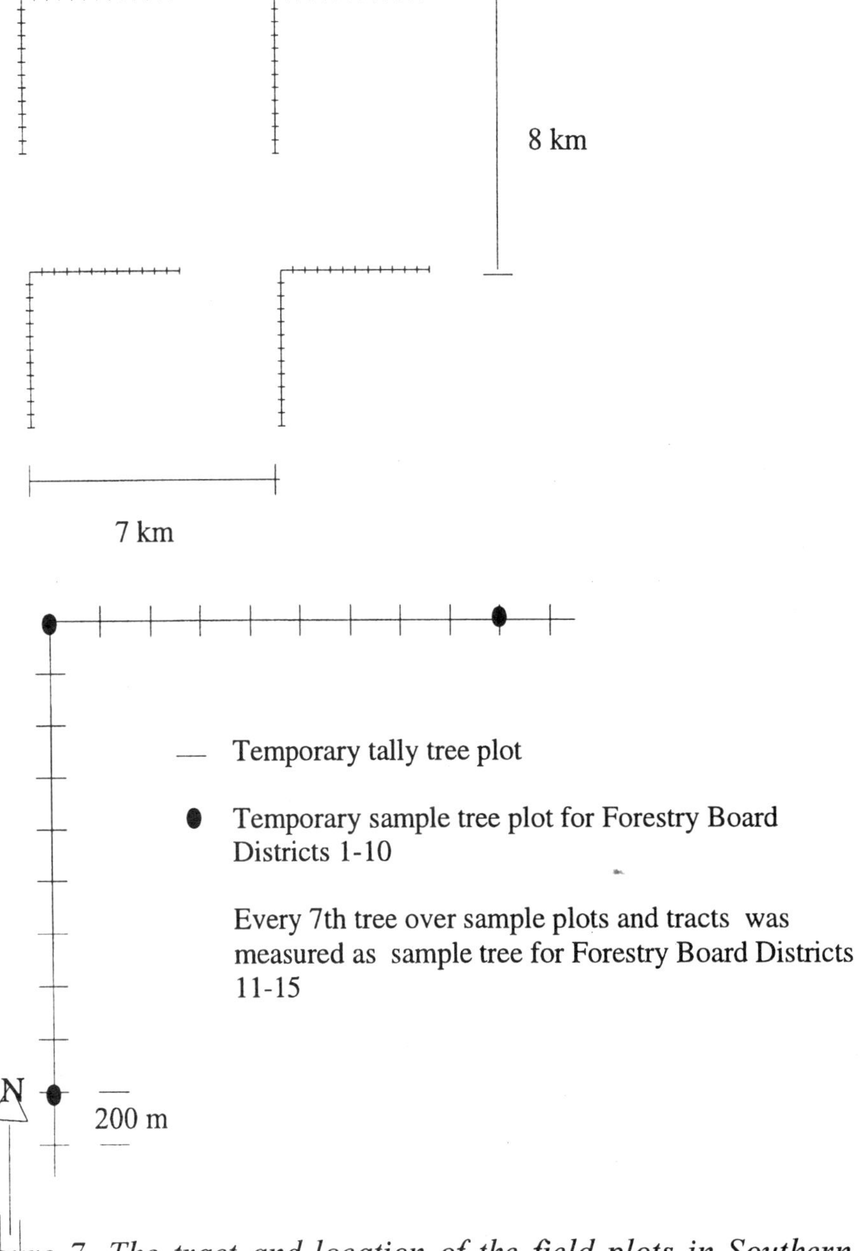

Figure 7. The tract and location of the field plots in Southern Finland.

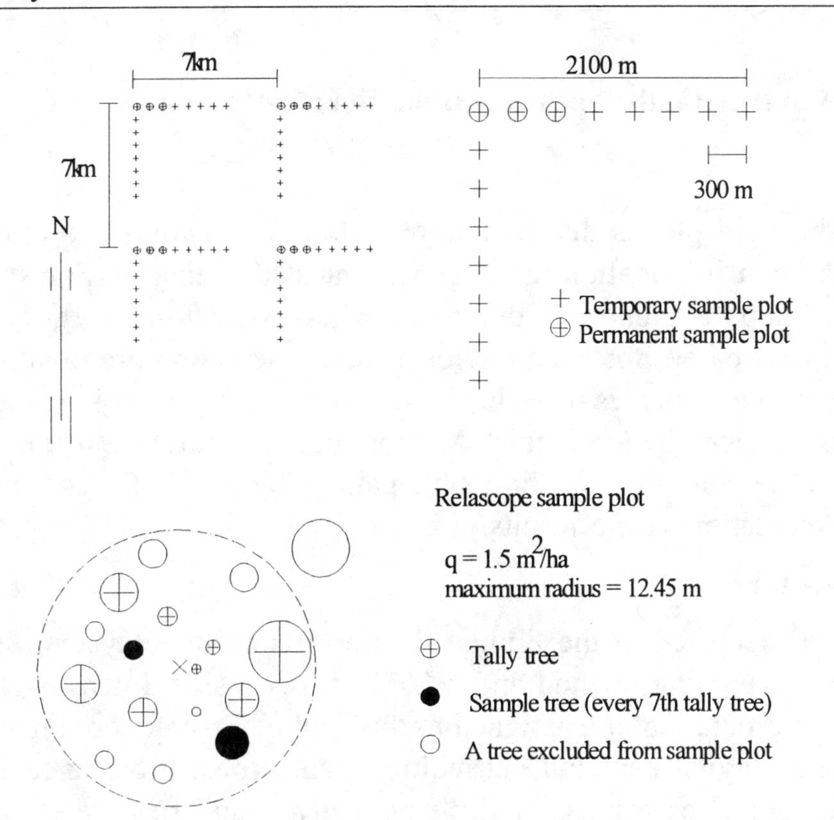

Figure 8. The tract and location of the field plots in Kainuu and Pohjois-Pohjanmaa.

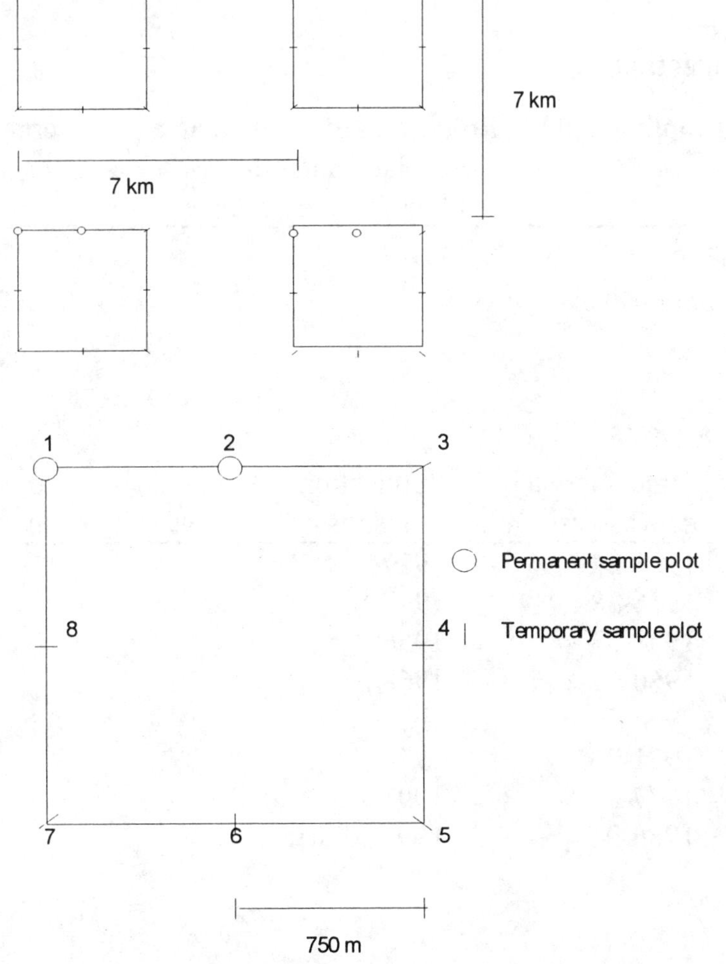

Figure 9. The tract and location of the field plots in the Northernmost Finland. Number of measured tracts in a group of four tracts varied depending on the share of forestry and other wooded land.

2.3.4. Techniques and methods for the combination of data sources

Field inventory

The altitude of the field plot is detected from a digital elevation model and it is combined with the plot information using geographical coordinates. Forest holding boundaries from maps are used to determine the ownership category. Wood production constraints based, for example, on nature reservation are obtained from maps or other information sources as well. Cumulative sum of the day degrees of the growing season is provided by the Finnish Meteorological Institute and it is attached to the field plots. The land areas of the municipalities by the Statistics Finland are applied in the computation of the results.

Multi-source inventory

Satellite images are rectified to the Finnish Uniform coordinate system using the nearest neighbour resampling method and a 25*25m pixel size. Intensity values of the pixels closest to a field plot are assigned to the field plot using coordinates of the plot and the pixels. Digital peatland, agricultural land, road, urban area and peat digging areas are applied as well. Digital municipality boundaries are employed to separate the computation units.

2.3.5. Sampling fraction

Data source and sampling unit	Proportion of forested area (forest land and other wooded land) covered by sample	represented mean area per sampling unit
Field plots (r=12.46m)	0.00014	352 ha
Landsat TM 25*25m pixels	100	0.0625 ha

2.3.6. Temporal aspects

Inventory cycle	Time interval of data assessment	Publication of results	Time interval (a) between assessments	Reference date
1st	1921-24	1927	-	1921
2nd	1936-38	1942	varying	1936
3rd	1951-53	1956	"	1951
4th	1960-63	1968	"	1960
5th	1964-70	1972	"	1964
6th	1971-76	1978	"	1971
7th	1977-84	1991	"	1977
8th	1986-94	1993 (first)	"	1986

2.3.7. Data capturing techniques in the field

Data is mainly recorded with field computers which perform some plausibility checks during the recording. The information is transferred to the main frame to the Inventory Department of The Finnish Forest Research Institute Helsinki via cellular phone every second or third working day.

If the hand computers are malfunctioning or not available, manual entry to the inventory forms is used. The forms are mailed to the Inventory Department of The Finnish Forest Research Institute Helsinki for recording.

2.4. DATA STORAGE AND ANALYSIS

The inventory data (inventories from 5th to 8th) are stored in the Finnish Forest Research Institute. The data from the four first inventories are stored in the State Archieves. The person responsible for the information is the Professor of forest inventory (Dr. Erkki Tomppo, Helsinki Research Centre, The Finnish Forest Research Institute, Tel: +350 9 857051, Fax: +358 9 85705717, email: erkki.tomppo@metla.fi)

2.4.2. Data base system used

The data is stored in self-created ASCII file system organized by inventory tracts.

2.4.3. Data bank design

Not used.

2.4.4. Update level

The inventory date of a plot is stored in the file system. The results are computed for the computation units measured during the same field season.

2.4.5. Description of statistical procedures used to analyze data including procedures for sampling error estimation

Statistics are produced within computation units. In the traditional field sample based inventory the computation unit normally consists of one Forestry Board District or, for example, of the forest land of one Forestry Board District. In the multi-source inventory usable accuracy has been obtained for smaller computation units; typical units are municipalities or large forest holdings.

a) Area estimation

Proportional and absolute areas of strata are estimated within 2 computation unit. Typical strata are land use classes, tree species classes, forest treatment classes etc.

The field inventory:

The area estimation is based on the proportions of the field plots belonging to different strata within the calculation area. The municipality area information from the National Land Survey of Finland is used to derive the areas of computation units and thus to estimate the absolute areas of strata.

Let y_i be the number of field plot centres within tract i belonging to a computation unit, and x_i be the number of the field plots within the tract i belonging to the computation unit and to a stratum. Then the proportional area estimate of the stratum within the computation unit is

$$m = \frac{\sum x_i}{\sum y_i}$$

where the summation is over the tracts intersecting the computation unit. The absolute area is estimated by multiplying the proportional estimate by the land area of the computation unit (Tomppo and Korhonen 1996).

The error estimation is based on local quadratic forms. For above area estimators the tract level residuals

$$z_i = x_i - my_i$$

are calculated and groups g of four tracts i1, i2, i3, i4 are formed. The tract groups are located in the corners of square:

i1 upper left corner
i2 upper right corner
i3 lower left corner
i4 lower right corner

The estimate of the standard error of m is then

$$e_m = \frac{\sqrt{b\sum_g T_g}}{\sum_i y_i}$$

where $T_g = \frac{1}{4}(z_{i1} - z_{i2} - z_{i3} + z_{i4})^2$,

and b is the number of tracts represented by one tract group. Details are given by Matérn (1960), Ranneby (1981) and Salminen (1985).

The multi-source inventory:

The division between forestry land use classes (classes 1–4) and other land use classes is based on the digital base map produced by the National Land Survey of Finland and on the housing register provided by the Civil Register of Finland. The discrimination of water bodies is based on satellite image analysis. In addition, the division between mineral soil forestry land and peatland forestry land is based on the stratification of field plots and the satellite image by means of a digital peatland mask. Note that the mask is used only for stratification of the field plots applied in the image process, not categorically. This approach avoids the bias caused by the peatland mask errors. The area of water and non-forestry land use classes are estimated by computation units from produced digital maps by multiplying the number of pixels classified in the land use class by the size of the pixel.

The area estimation within the forestry land is based on the combination of satellite data and the field plot information. The ground truth information is generalized to all pixels so that the ground truth of K spectrally closest (i.e. having most similar image data) field plot pixels is used. Normally K varies from 5 to 10.

More formally, the Euclidean distance $d_{p(i),p}$ is computed in the spectral space of the satellite image channels from the pixel p to be classified to each field plot pixel p(i) within 50 to 100 km radius. Also the difference in elevation is restricted to less than 50– 100 m. Then K nearest field plot pixels denoted by p(1), ... , p(k) are selected and weights

$$w_{i,p} = \frac{d_{p(i),p}^{-2}}{\sum_{j=1}^{k} d_{p(j),p}^{-2}}$$

are calculated. The area estimation of a stratum is based on applying weight

$$c_i = \sum_p w_{i,p}$$

for ground truth of the field plot i, where the summation is taken over the pixels within the computation unit.

The area estimates for computation unit are calculated by multiplying the pixel size by the sum of the weights c_i of the field plots i, that belong to the stratum. This can be interpreted in such a way that each pixel in the computation unit is divided to K parts each having the proportion $w_{i,p}$ of the area and each part is classified according to ground truth data of the corresponding field plot i. The area estimate of the stratum is calculated by summing the areas of the pixel parts labeled to the class.

The accuracy of the multi-source inventory method has been estimated by empirical experiments on test areas (Päivinen et al. 1993). Analytical method for error estimation is under development.

b) Aggregation of the tree and plot data

Sample tree information is transferred to tally trees by Vu values described in section d. The sample tree Vu-timber assortment values are used to calibrate the Vu functions which are applied to estimate the Vu values of the tally trees in the 1 cm diameter classes by the Forestry Board District, the timber assortment class and the taxation class and possibly the dd-levels.

c) Estimation of the total values

The field inventory:

The total values are obtained by multiplying the areas (see section a) of strata by estimates of the mean value (see following section d). The standard error of the total values, e_{total}, is obtained as

$$e_{total} = \sqrt{e_{area}^2 + e_{meanvalue}^2}$$

where e_{area} is the standard error of the area estimator and $e_{mean\ value}$ is the standard error of the mean value estimator.

The multi-source inventory:

The total values are obtained multiplying the continuous variable of interest on the field plot i by the weight c_i, and summing the resulting values over the computation unit.

d) Estimation of the ratios

The field inventory:

The mean value estimates are calculated for the diameter, basal area and volume, and the stem number per 1 ha unit area. For one field plot the mean basal area estimate is calculated by counting the trees in the Bitterlich plot and multiplying the result by the relascope factor q. For the computation unit, the mean basal area estimate is the mean of the field plot estimates.

One tree in the Bitterlich plot represents n trees per unit area where

$$n = \frac{1}{\pi\left(\frac{d}{2}\right)^2} q = \frac{4q}{\pi d^2}$$

where: d = diameter (m) of the tree.

The estimate for the mean stem number per area unit for a Bitterlich plot is obtained by summing the n values and the final estimate is the mean of the plot estimates.

The mean volume is estimated by summing up the Vu values describing the contribution of the trees measured in the Bitterlich plot. The Vu value of a field plot is

$$Vu = nv$$

where v is the volume of the tree.

The mean diameter D (weighed by the basal area) for the computation unit, when N is the number of stems measured in the Bitterlich plots of the computation unit, is

$$D = \frac{\sum_{i=1}^{N} d_i}{N}$$

The mean basal area (BA) for the computation unit is derived by

$$BA = 10q \frac{\sum_{j=1}^{K} N_j}{\sum_{j=1}^{K} A_j}$$

where N_j is the number of stems in the field plot j, K is the number of the field plots in the computation unit, A_j is the proportional area of the field plot j (10=full field plot) and q is the relascope factor.

The multi-source inventory:

The ratio estimated are formed by dividing the total estimate from section c by the sum of the weights c over the computation unit.

For ratio estimators the standard errors are estimated using the local quadratic forms explained in section a.

e) Sampling at the forest edge

The partial field plots are measured at the forest edge. The size of the section measured is taken in the consideration as presented in previous section.

f) Estimation of growth and the growth components

Growth is estimated as the annual mean of the past five-year-period. It is divided into two categories, the growth of the stock which has remained in the inventory, and the growth of the removed stock . The growth of the remaining stock is estimated by extracting the volume in the beginning of the five year interval from the present volume using the Vu values for each sample tree (Kujala 1980), and extended to tally trees by 1 cm diameter classes. The ingrowth is not estimated as a separate component but it is included in the growth of the remaining stock by calculating the growth from all the trees which exceed the length of 1.35 m. The growth of the removed trees includes the growth of the harvested stock and the growth of the dead stock during the five-year-interval. The increment percentage applied for the increment estimation of the removed stock is supposed to be 70 % of the increment percentage of the remaining stock (Salminen 1993), and its estimate is based on the annual drain tables and the time of removal.

g) Allocation of stand and area related data to single field plots/sample points

Each class of stand and area data, which can be found on the plot, is recorded and assigned to a proportion of the plot area.

h) Hierarchy of the analysis: how are the subunits treated

The field inventory:

Results are calculated normally for the Forestry Board Districts, South Finland, North Finland and for the entire country. When producing estimates for large areas such as South Finland, North Finland or for the whole country, the results are calculated directly for these blocks if possible. When inventory methods have varied within these large units the results are first calculated in as large entities as possible (inventored in the same way) and finally, for example, results for the entire country are produced as a combination of these results. However, the computation units can be formed freely for other purposes provided that they are large enough to allow acceptable estimation errors, which are determined by the sampling design.

The multi-source inventory:

Results can be estimated in any computation unit.

2.4.6. Applied software

The software used for calculating the results for large areas is developed in the Finnish Forest Research Institute applying VAX and Dec FORTRAN programming language. SAS statistical package is applied especially for modelling and compiling tables. Satellite image processing procedures are programmed in the Finnish Forest Research Institute, partly applying the modules provided by the DISIMP image processing software.

2.4.7. Applied hardware

Rufco 900 hand computers are used for logging the field information. The data is entered into a VAX/VMS system and consistency checks are applied. The analysis is done in VAX/VMS and DEC/OSF-1 systems. Multi-source data is mainly processed by means of DEC/OSF-1. Hard copies are produced by Versatec electrostatic plotter.

2.4.8. Availability of data (raw and aggregated data)

The raw data is owned by the Finnish Forest Research Institute. A specific agreement is needed to make the data available for other organisations. Usually reimbursement of all the costs is demanded. Special analyses based on the National Forest Inventory data are executed on commercial basis but the Finnish Forest Research Institute reserves all rights to reject any request for analysis.

2.4.9. Subunits (available strata)

The results based on the field sample are available for the forestry board districts, Southern Finland, Northern Finland and for the whole country. Results are usually computed by ownership groups, e.g. for private forests or state owned forests.

The multi-source inventory results are available for any area in Finland.

2.4.10. Links to other information sources

Topic and spatial structure	Age/period availability	Responsible agency	Kind of data set	Availability
Base map	since 1921	Land Survey of Finland	map	available
Borders of forest holdings	since 1921	Land Survey of Finland	map	available
Ownership category information	since 1921	Civil Register of Finland	data file	available
Cumulative day degree information	since 1921	Finnish Meteorological Institute	data file	available
Land area	since 1921	Statistics Finland	data file	available
Digital base map	beginning from 1990	Land Survey of Finland	raster map	available
Digital elevation model	since 1993	Land Survey of Finland	raster map	available
Profession of forest owner	1986	National Board of Income Taxation	data file	available

2.5. RELIABILITY OF DATA

2.5.1 Check assessments

Each inventory field crew controls a part of the tracts measured by other crews during the field work season. The control tracts are selected randomly from the 100 km circle around the working area of the crew. The control tracts correspond to about 5% percent of the tracts measured during the field work season. After the control measurement, the control crew and the original measurement crew compare measurements and have a meeting to find reasons for discrepancies so that they can be corrected in the remaining work. Main purpose of these checks is also the training of the field crews. The field crews have 1–3 general meetings per field work season to solve problems raised by the control measurements. It should be noted that most of the inventory field crew leaders work on permanent basis and therefore these controls take place repeatedly.

2.5.2 Error budgets

The sampling error estimation is given in section 2.4.5. Other possible error sources than those described in 2.4.5. are not taken into consideration.

2.5.3. Procedures for consistency checks of data

When the data is received, the possible transformation errors are controlled and corrected. Complete plausibility checks are carried out on the basis of the possible relationships between the recorded variables after the field work season. The confusing combinations are detected, controlled and finally corrected by the field group leaders.

2.6. MODELS

2.6.1. Volume (functions, tarifs)

a) outline of the models

The volume of the sample trees is calculated applying the volume functions by the tree species (Laasasenaho 1982) based on breast height diameter, length and the diameter at the height of 6 m. When the diameter at the height of 6 m is not measured, it is estimated applying the measured data and appriori information from Vähäsaari (1989) models. For small trees (height under 8.1 m) separate volume functions based on height and breast height diameter are used. The volumes of the sample trees are transformed to the Vu values describing the volume which the sample tree represents per basal area and per area unit.

b) overview of prediction errors

The taper curve models produce unbiased volume estimates. The standard error of total volume estimates applying d.b.h., diameter at 6 m height and length varies from 3.5 % to 5.0 % depending on tree species (*Pinus sylvestris, Picea abies* and *Betula* spp) (Laasasenaho 1982).

c) data material for derivation of the models

The study material was a resample from the 5th National Forest Inventory single tree data and it covered the whole country.

d) methods applied to validate the models

The reliability estimates of the models were based on theoretical studies and empirical tests. Empirically the models were tested by applying separately measured control material (Laasasenaho 1982).

2.6.2. Assortments

a) outline of the models

Saw-timber sized sample tree stems were divided into three saw-timber quality classes in the field. The final bucking of stems was based on value optimization. The timber assortment volumes were calculated by polynomial taper curve models (Laasasenaho 1982) and transformed to timber assortment Vu values. The classification to pulpwood and saw timber is based on scaling during the field measurements. There are separate classes for timber which does not meet even the pulpwood requirements. The division between pulpwood and unmerchantable top is based on the taper curve models using 6 cm minimum diameter. The assortment volume estimates are scaled to the level of the total volume estimated by volume functions.

b) overview of prediction errors

Comparable with section 2.6.1. b)

c) data material for derivation of the model

The study material was a resample from the 5th National Forest Inventory single tree data and it covered the whole country.

d) methods applied to validate the model

As in section 2.6.1. d)

2.6.3. Growth components

a) outline of the models

The growth of the remaining growing stock is estimated by applying increment functions based on the diameter and height increment assessments (Kujala 1980). The stem form at the beginning of the increment calculation period (n=5 years) is estimated and the corresponding volume is derived. The mean annual increment is estimated applying the volume difference in the beginning and at the end of the period.

More formally this can be expressed as follows. The development of the stem form and bark density is based on the ratio between the stem volume v and the barkless basal area of stem g. This ratio r for each tree species is

$$r = \frac{v}{g}$$

The r values are estimated for the height classes and thus r can be presented as continuous function of the tree height h. For growth estimation the estimate of r n years ago is formed

$$r_{-n} = r - (\hat{r} - \hat{r}_{-n})$$

where $\hat{r}$ is r as a function of h and $\hat{r}_{-n}$ is r_{-n} as a function of h.

The volume n years ago is estimated by

$$V_{-n} = r_{-n} g_{-n}$$

where g_{-n} is the basal area n years ago.

Finally the annual growth i_a is estimated by

$$i_a = \frac{v - v_{-n}}{n}$$

The growth of the removed stock i_r is estimated by applying the annual drain tables. The increment percentage applied for increment estimation of the removed stock is supposed to be 70 % of the increment percentage of the remaining stock. The drain is estimated as the mean of drain tables of the ongoing year and the previous year.

b) overview of prediction errors

The form increment is comparable within height classes because it is mainly affected by the thickness of bark and by the stem form. The bark is normally thicker than on rich sites but the stem form is worse on poor sites. These two factors offset each other when calculating the form increment

c) data material for the derivation of the model

The sample tree data from the 6 th National Forest Inventory.

d) methods applied to validate the model

The form growth in height classes is compared between different site classes and geographic location beset on the 6th NFI data.

2.6.4. Potential yield

a) Outline of the potential yield estimation

Potential yield estimates are based on the MELA forest management planning tool (Siitonen 1983). Mela consists of two parts: an automated stand simulator based on the development of individual trees in Finnish conditions, and a linear optimization package for selecting optimal future treatments. Tree level growth models by tree species are used to simulate alternative future development scenarios. The simulation includes growth estimation which is based on linear models presented by Ojansuu et al. (1991) and production of the alternative future treatments. For each Forestry Board District, linear optimization is applied to select from the produced alternatives simultaneously both a production program for the whole forestry unit and the management for the stands based on the actual goals of the decision maker (Lappi 1992). The goals and constraints are based on commonly accepted principles such as sustainability. The estimates for the whole country are produced as aggregates of the Forestry Board Districts.

b) prediction errors

The prediction errors can be divided into two parts: possible sampling errors due to the NFI data and the individual models errors (Ojansuu et al. 1991).

c) Data and material used for deriving the models

The growth models were estimated on the 7th NFI sample trees and the field plot information (Ojansuu 1991) and on permanent growth field plot (INKA) information (Vuokila 1986).

d) Methods used to validate the models

The models were validated by comparing the Mela growth estimates with changes in repeated field measurements (Ojansuu et al. 1991).

2.6.5. Forest functions

Not applied

2.6.6 Other applied models

The 6 m diameter models

a) When a diameter at the height of 6 m is not measured it is estimated by using the measured data and appriori information from Vähäsaari (1989) models, which are calibrated by means of measured trees.
b) The root mean square errors of the models varies from 7.79% to 10.04%.
c) The sample trees over the height of 7.5 m from the 7th and 8th National Forest Inventories were used in modeling.
d) Models were tested using subsamples from the estimation data by the Forestry Board Districts.

Thickness of the bark

a) Regression models were estimated using length of the tree and the d.b.h. Models were estimated separately for each Forestry Board District.
b) not published
c) Sample trees from the 7th and 8th National Forest Inventory
d) not published

2.7 INVENTORY REPORTS

2.7.1. List of published reports and media for dissemination of inventory results

Inventory	Year of publica tion	Citation	Language	Dissemi nation
8th NFI	1996	Suomen metsävarat 1989-1994 ja niiden muutokset vuodesta 1951 lähtien. Metsätilastotiedote 354. 18 p.	Finnish	Printed
8th NFI	1993	Eteläisimmän Suomen metsävarat 1986-1988, Forest resources of Sourthernmost Finland 1986-1988, Folia Forestalia 825. 111 p.	Finnish/ English	Printed
7th NFI	1991	Suomen metsävarat 1977-1984 ja niiden kehittyminen 1952-1980, Forest Resources of Finland in 1977-1984 and their development in 1952-1980. Acta Forestalia Fennica 220. 84 p.	Finnish/ English	Printed
7th NFI	1986	Metsävarat piirimetsälautakunnittain Pohjois-Suomessa 1982-1984, Folia Forestalia 655	Finnish/ English	Printed
7th NFI	1980	Ahvenanmaan maakunnan ja maan yhdeksän eteläisimmän piirimetsälautakunnan alueen metsävarat 1977-1979. Folia Forestalia 446. 90 p.	Finnish/ English	Printed
7h NFI	1983	Metsävarat Etelä-Suomen kuuden pohjoisimman piirimetsälautakunnan alueella 1979-1982 ja niiden kehittyminen koko Etelä-Suomessa 1977-1982. Folia Forestalia 568. 79 p.	Finnish/ English	Printed
6th NFI	1974	Etelä-karjalan, Pohjois-Savon, Keski-Suomen ja Itä-Savon metsävarat vuonna 1973. Folia Forestalia 207. 35 p.	Finnish/ English	Printed
6th NFI	1974	Ahvenanmaan maakunnan, Helsingin, Lounais-Suomen, Satakunnan, Uudenmaan, Hämeen, Pirkka-Hämeen, Etelä-Savon ja Etelä-Karjalan piirimetsälautakunnan metsävarat vuosina 1971-1972. Folia Forestalia 191. 64 p.	Finnish/ English	Printed
6th NFI	1978	Suomen metsävarat ja metsien omistus 1971-1976. Communicationes Instituti Forestalis Fenniae 93.6. 107 p.	Finnish/ English	Printed
5th NFI	1972	Suomen metsävarat ja metsien omistus 1964-70 sekä niiden kehittyminen 1920-70. Communicationes Instituti Forestalis Fenniae 76.5 126 p.	Finnish/ English	Printed

5th NFI	1968	Etelä-Savon, Etelä-Karjalan, Itä-Savon, Pohjois-Karjalan, Pohjois-Savon ja Keski-Suomen metsävarat vuosina 1966-1967. Folia Forestalia 42. 54 p.	Finnish/ English	Printed
5th NFI	1967	Helsingin, Lounais-Suomen, Satakunnan, Uudenmaan-Hämeen, Pohjois-Hämeen ja Itä-Hämeen metsävarat vuosina 1964-1965. Folia Forestalia 27. 56 p.	Finnish/ English	Printed
	1966	Ålands skogar 1963-1964. Folia Forestalia 21. 18 p.	Swedish	Printed
4th NFI	1968	IV valtakunnan metsien inventointi. 4. Suomen metsävarat vuosina 1960-63. Communicationes Instituti Forestalis Fenniae.		
4th NFI	1963	IV valtakunnan metsien inventointi. 1. Maan eteläpuoliskon metsänhoitolautakuntien alueryhmät. Communicationes Instituti Forestalis Fenniae 57(4) 99 p.	Finnish/ English	Printed
4th NFI	1962	IV valtakunnan metsien inventointi. 1. Maan eteläpuoliskon vesistöalueryhmät. Communicationes Instituti Forestalis Fenniae 56(1). 112 p.	Finnish/ English	Printed
3rd NFI	1958	Suomen metsät päävesistöalueittain. Valtakunnan metsien inventoinnin tuloksia. The forests of Finland by the main water system areas. Results of the national forest inventory. Communicationes Instituti Forestalis Fenniae 47.4. Valtioneuvoston kirjapaino. Helsinki. 88 p.	Finnish/ English	Printed
3rd NFI	1957	Suomen metsät metsänhoitolautakuntien toiminta-alueittain. Communicationes Instituti Forestalis Fenniae 47(3) 128 p.	Finnish/ English	Printed
3rd NFI	1956	Suomen metsät vuosista 1921-1924 vuosiin 1951-1956. Kolmeen valtakunnan metsien inventointiin perustuva tutkimus. Communicationes Instituti Forestalis Fenniae 47(1). 227 p.	Finnish/ English	Printed
2nd NFI	1943	Metsänhoitolautakuntain toimintapiirien metsät. II valtakunan metsien arvioinnin tuloksia. keskusmetsäseura Tapio, Helsinki 130 p.	Finnish	Printed

2nd NFI	1942	Suomen metsävarat ja metsien tila. II valtakunnan metsien arviointi. Communicationes Instituti Forestalis Fenniae 30. 446 p.	Finnish/ English	Printed
1st NFI	1927	Suomen metsät. Tulokset vuosina 1921-1924 suoritetusta valtakunnan metsien arvioimisesta. Communicationes Ex Instituto Quaestionum forestalieun Finlandiea Editae 11. 421+199 p.	Finnish/ English	Printed

2.7.2 The list of contents of the latest report and update level

The latest results of the inventory are published as a Forest Statistics Bulletin (Tomppo and Henttonen 1996), including results for South Finland, for North Finland and for the whole country, such as:

- land use classification
- volume, increment and pole size of growing stock
- age distribution of forests.

The content of inventory report 'Forest resources of Southernmost Finland 1986–1988' (Salminen 1993), which includes results by forestry board districts, is:

Land use classifications
Site classes on forest land
Areal classifications based on tree species proportions, age and dimensions
Volume, increment and pole size in the growing stock
Stand quality
Completed and proposed cutting, silvicultural and soil preparation actions
Reliability of the results

2.7.3. Users of the results

The traditional role of the National Forest Inventory has been to provide unbiased, reliable large area forest resource information. The information has been utilized in large area forest management planning, such as determining the level of cuttings and other treatments needed, as the information basis of the official forest policy, in the strategic planning of the forest industries and, more recently, to calibrate the level of estimates of standlevel inventories. Also the compilation of national and international statistics utilize the information produced by the NFI.

Examples of typical users of the inventory information include:

1) Forestry Department of Ministry of Agriculture of Forestry
2) Ministry of Environment
3) Ministry of Finance (for forest income taxation)
4) Ministry of Trade and Industry

5) other decision makers of forestry policy
6) forestry organizations, of which the organizations of private forestry have been the most important ones
7) forest industry companies and the association of forest industries
8) forest research organizations and individual researchers
9) national and international forest statistics.

2.8. FUTURE DEVELOPMENT AND IMPROVEMENT PLANS

2.8.1. Next inventory period

The 9th successive inventory was started in 1996 and it is expected to be completed in 1999. The analysis of the results will be finished in 2000 and the results will be published in 2001.

2.8.2. Expected changes

The 9th inventory will utilize field measurements, remote sensing data and digital map data.

Changes in the field data: temporary and permanent field plots will be located in separate tracts. New field variables will be added. Examples are those describing forest biodiversity, more detailed information of soil type, site fertility, forest health, etc.

Changes in remote sensing data: the most accurate satellite sensors will be used. An airborne imaging spectrometer (AISA) may be in operation by the latter period of the inventory.

Digital map data: new digital maps are e.g. digital soil map.

2.8.2.1. Nomenclature

Additional attributes will be included. These include e.g. attributes describing forest biodiversity, detailed soil information, detailed site fertility information and forest protection.

2.8.2.2. Data sources

Field data, digital map data, satellite image and possibly airborne imaging spectrometer data will be used.

2.8.2.3. Assessment techniques

The final estimates both for large and small areas will be computed by means of the multi-source technique. The GPS will be used for locating the plot centres.

2.8.2.4. Data storage and analysis

The analysis of data will be based on our own models and image analysis utilities and on some commercial software packages like SAS, ARC/INFO.

2.8.2.5. Reliability of data

The arrangements of the field crew training and field data checks are maintained. Assessment of the reliability of the estimates given by the multi-source technique will be further developed.

2.8.2.6. Models

Relevant models will be revised.

3. PERMANENT FIELD PLOTS IN THE NATIONAL FOREST INVENTORY OF FINLAND

3.1. OVERVIEW

The first Permanent field plot /National Forest Inventory of Finland (Valtakunnan metsien inventointi, pysyvät koealat) covers the whole Finland. The National Forest Inventory of Finland was extended by a permanent field plot survey mainly for forest health monitoring purposes. The permanent field plot extension has been under development since its establishment and the analysis has followed the analysis of the normal National Forest Inventory presented in Chapter 2. The first permanent field plots were establishes in the National Forest Inventory of Finland in 1984–1985. The plots have been remeasured in 1990– 1991 and again in 1995. The responsible organization for the National Forest Inventories is the Finnish Forest Research Institution (Metla) (Pysyvien koealojen... 1995).

3.1 NOMENCLATURE

3.1.1 List of the attributes directly assessed

A) Geographic regions

Attribute	Data source	Object	Measurement unit
The Finnish Uniform Coordinates of the tract *(traktin koordinaatit)*	1:20000 map	tract	m
Land use class *(maaluokka)*	field assessment	stand	categorical
Land use class specification *(maaluokan tarkennus)*	field assessment	stand	categorical
Bearing to the next location point or field plot centre *(suunta)*	previous inventory	location point/ field plot centre	degree
Location description *(kuvaus)*	previous inventory	location point/ field plot centre	literal description
Deviation of the track line *(linjan siirtymä)*	field assessment /geographic map	field plot	m/categorical
Bearing of the slope *(kaltevuuden suunta)*	field assessment	fixed radius plot r=20m	categorical
Slope *(kaltevuus)*	field assessment	fixed radius plot r=20m	categorical
Topographic location *(topografinen asema)*	field assessment	fixed radius plot r=20m	categorical

Percentage of the plot covered by the stand *(kuvion koko koealalla)*	field assessment	field plot	categorical
Percentage of the small sample plot covered by the stand *(kuvion koko pienkoealalla)*	field assessment	field plot	categorical
Distance to the amenity boundary *(maisemarajan etäisyys)*	field assessment	amenity boundary	categorical

B) Ownership

Attribute	Data source	Object	Measurement unit
Owner *(omistaja)*	cadaster map	field plot	categorical

C) Wood production

Attribute	Data source	Object	Measurement unit
Stand basal area observation *(kuvion pohjapinta-alahavainto)*	field assessment	basal area field plot	m^2/hectare
Location of the stand basal area observation *(pohjapinta-alahavainnon sijainti)*	field assessment	stand	categorical
Stand basal area *(kuvion pohjapinta-ala)*	field assessment	stand	m^2/hectare
Pole size of the growing stock *(järeys)*	field assessment	stand	cm
Proportion of saw-timber in final cutting *(tukkikokoisten puiden osuus päätehakkuussa)*	field assessment	dominant tree storey	categorical
Quality distribution of the saw-timber in final cutting *(tukkikokoisten puiden laatujakauma päätehakkuussa)*	field assessment	dominant tree storey	categorical
Condition of the growing stock *(laatu)*	field assessment	dominant tree storey	categorical
Cause for revision of the condition of the growing stock	field assessment	dominant tree storey	categorical

Size of the tally tree plot *(lukupuukoealan koko)*	field assessment	radius of the tally tree plot	categorical
Distance from the plot centre *(etäisyys)*	field assessment	tally tree	cm
Bearing *(suunta)*	field assessment	tally tree	angle
Breast height diameter *(rinnankorkeusläpimitta)*	field assessment	tally tree	mm
Timber assortment class *(puuluokka)*	field assessment	tally tree	categorical
Timber assortment class specification *(puuluokan tarkennus)*	field assessment	tally tree	categorical
Crown storey *(latvuskerros)*	field assessment	tally tree	categorical
Size of the sample tree plot *(lukupuukoealan koko)*	field assessment	radius of the sample tree plot	categorical
Length of the broken part *(katkenneen osan pituus)*	field assessment	sample tree	m
Height at the base of dead crown *(kuolleen latvuksen alaraja)*	field assessment	sample tree	dm
Height at the base of green crown *(elävän latvuksen alaraja)*	field assessment	sample tree	dm
Height to the lowest dead branch *(alimman kuolleen oksan korkeus)*	field assessment	sample tree	dm
Height to the highest dead branch *(ylimmän kuolleen oksan korkeus)*	field assessment	sample tree	dm
Diameter of the thickest living branch *(paksuimman elävän oksan läpimitta)*	field assessment	sample tree	mm
Diameter of the thickest dead branch *(paksuimman kuolleen oksan läpimitta)*	field assessment	sample tree	mm
Height to the thickest dead branch *(alimman kuolleen oksan korkeus)*	field assessment	sample tree	dm
Width of the crown *(latvuksen leveys)*	field assessment	sample tree	dm
Sweepness *(lenkous)*	field assessment	sample tree	cm

Lenght of saw-timber *(tukkikelpoisen osan pituus)*	field assessment	sample tree	dm
Timber quality *(laatu)*	field assessment	sample tree	categorical
Lenght of quality timber *(laatuosan pituus)*	field assessment	sample tree	dm
Cause for quality revision *(laadun alenemisen tai pakkokatkaisun syy)*	field assessment	sample tree	categorical

D) Site and soil

Attribute	Data source	Object	Measurement unit
Land use class effect *(maaluokkien vaikutus)*	field assessment	stand	categorical
Stand history *(kuvion historia)*	field assessments	stand	categorical
Main site class *(kasvupaikan päätyyppi)*	field assessment	stand	categorical
Main site class specification for peatland *(suon sekatyyppi)*	field assessment	stand	categorical
Site fertility class *(kasvupaikkatyyppi)*	field assessment	stand	categorical
Site fertility class specification for peatland *(kasvupaikkatyypin lisämääre)*	field assessment	stand	categorical
Thickness of the organic layer *(orgaanisen kerroksen paksuus)*	field assessment	stand	cm
Type of the organic layer *(orgaanisen kerroksen laatu)*	field assessment	stand	categorical
Thickness of soil mantle *(maapeitteen paksuus)*	field assessment	stand	categorical
Soil type *(maalaji)*	field assessment	stand	categorical
Drainage stage *(ojitustilanne)*	field assessment	stand	categorical
Draining proposal *(ojitusehdotus)*	field assessment	stand	categorical
Applied soil preparation *(tehdyt maanparannustoimenpiteet)*	field assessment	stand	categorical

Time of the completed soil preparation *(maanparannustoimenpiteen aika)*	field assessment	stand	categorical
Taxation class *(veroluokka)*	field assessment	stand	categorical
Taxation class specification *(veroluokan tarkennus)*	field assessment	stand	categorical

E) Forest structure

Attribute	Data source	Object	Measurement unit
Tree storey position *(jakson asema)*	field assessment	stand	categorical
Origin of the stand *(perustamistapa)*	field assessment	stand	categorical
Development class *(kehitysluokka)*	field assessment	tree storey	categorical
Dominant tree species *(pääpuulaji)*	field assessment	tree storey	categorical
Number of stems *(runkoluku)*	field assessment	field plot	number/ hectare
Total number of seedlings *(taimien kokonaismäärä)*	field assessment	stand	1000/hectare
Number of seedlings capable for further development *(kehityskelpoisten taimien lukumäärä)*	field assessment	stand	100/hectare
Mean age of the growing stock *(puuston keski-ikä)*	field assessment	stand	a
Tree species *(puulaji)*	field assessment	tally tree	categorical
Birth type *(syntytapa)*	field assessment	tally tree	categorical
Time of the cutting or mortality *(hakkuun tai luonnonpoistuman aika)*	field assessment	tally tree	categorical
Height *(pituus)*	field assessment	young tally tree	dm
Height *(pituus)*	field assessment	dead sample tree	dm
Upper diameter *(yläläpimitta)*	field assessment	dead sample tree	cm
Height *(pituus)*	field assessment	sample tree	dm
Stump diameter *(kantoläpimitta)*	field assessment	sample tree	cm

Upper diameter *(yläläpimitta)*	field assessment	sample tree	cm
Breast height age *(rinnankorkeusikä)*	field assessment	sample tree	a

F) Regeneration

Attribute	Data source	Object	Measurement unit
Tree species *(puulaji)*	field assessment	seedlings by strata	categorical
Mean height *(keskipituus)*	field assessment	seedlings by strata	dm
Number of seedlings	field assessment	seedlings by strata	categorical

G) Forest condition

Attribute	Data source	Object	Measurement unit
Description of the damage *(tuhon ilmiasu)*	field assessment	stand	categorical
Causing agent of the damage *(tuhon aiheuttaja)*	field assessment	stand	categorical
Degree of the damage *(tuhon aste)*	field assessment	tree storey	categorical
Abundance of beard lichens *(naavamaiset jäkälät)*	field assessment	tree storey	categorical
Abundance of leaf lichens *(lehtimäiset jäkälät)*	field assessment	tree storey	categorical
Abundance of *Scoliciosporum clorococum* and *Desmococcus olivaeus (vihersukkula-jäkälät)*	field assessment	tree storey	categorical
Defoliation *(harsuuntuminen)*	field assessment	stand	categorical
Defoliation *(harsuuntuminen)*	field assessment	sample tree	categorical
Description of the damage *(tuhon ilmiasu)*	field assessment	dead sample tree	categorical
Causing agent of the damage *(tuhon aiheuttaja)*	field assessment	dead sample tree	categorical
Description of the damage *(tuhon ilmiasu)*	field assessment	sample tree	categorical
Causing agent of the damage *(tuhon aiheuttaja)*	field assessment	sample tree	categorical

Damage degree *(tuhon aste)*	field assessment	sample tree	categorical
Number of the needle storeys *(neulasvuosikerrat)*	field assessment	sample tree	categorical
Moss sample *(sammalnäyte)* *(Hylocomium splendens/ Pleurozium schreberi)*	field extraction	tract plot	varying volume
Humus layer sample *(humusnäyte)*	field ectraction	tract plot	varying volume

H) Accessibility and harvesting

Attribute	Data source	Object	Measurement unit
Applied cutting methods *(tehdyt hakkuut)*	field assessment	stand	categorical
Time of applied cuttings *(tehtyjen hakkuden ajankohta)*	field assessment	stand	categorical
Applied silvicultural treatments *(tehdyt metsänhoitotyöt)*	field assessment	stand	categorical
Time of the applied silvicultural treatments *(tehtyjen metsänhoitotöiden ajankohta)*	field assessment	stand	categorical
Proposed cuttings *(hakkuuehdotus)*	field assessment	stand	categorical
Urgency of the proposed cuttings *(ehdotetun hakkuun ajankohta)*	field assessment	stand	categorical
Proposed silvicultural treatments *(ehdotetut metsänhoitotyöt)*	field assessment	stand	categorical
Proposed soil preparation *(maapinnan käsittelyehdotus)*	field assessment	stand	categorical

I) Attributes describing forest ecosystem

Attribute	Data source	Object	Measurement unit
Tree species *(puulaji)*	field assessment	dead tally tree	categorical
Cause of death *(kuolinsyy)*	field assessment	dead tally tree	categorical
Stage of the decay *(lahon aste)*	field assessment	dead tally tree	categorical

Breast height diameter *(rinnankorkeusläpimitta)*	field assessment	dead tally tree	cm
Height *(pituus)*	field assessment	dead tally tree	dm
Decaying fallen tree class *(maapuuluokka)*	field assessment	dead fallen tally tree	categorical
Stage of the decay *(lahon aste)*	field assessment	dead fallen tally tree	categorical
Stump diameter *(kantoläpimitta)*	field assessment	dead fallen tally tree	cm
Top diameter of the fallen part *(koealalla olevan osan latvaläpimitta*	field assessment	dead fallen tally tree	cm
Lenght *(pituus)*	field assessment	dead fallen tally tree	cm

J) Non-wood goods and services

Attribute	Data source	Object	Measurement unit
Fringing effect *(reuna-alue)*	field assessment	stand	categorical
Micro habitat *(pienhabitaatti)*	field assessment	stand	categorical
Timber production restrictions based on nature conditions *(luonnonoloista johtuvat puuntuotannon rajoitukset)*	map/field assessment	stand	categorical
Timber production restrictions based on multiple-use *(monikäytöstä johtuvat puuntuotannon rajoitukset)*	map/field assessment	stand	categorical
Multiple use specification *(monikäytön tarkennus)*	cadaster map/field assessment	stand	categorical

3.1.2. List of derived attributes

As in section 2.1.2.

3.1.3. Measurement rules for measurable attributes

Note: only variables which differ from those presented in section 2.1.3. are presented.

C) Wood production

Distance from the plot centre
a) Distance from the plot centre to the middle of the tally tree stump
b) height over 1.35m
c) 1 cm classes
d) rounded towards the closest class centre
e) measuring tape
f) field assessment

Bearing
a) bearing from the plot centre to the middle of the tally tree stump
b) height over 1.35m
c) 1/400 classes
d) rounded towards the closest class centre
e) compass
f) field assessment

Height of the base of death crown
a) height of death branches or branch stubs over 5 mm diameter without bark
b) coniferous species and *Betula* spp. over 1.35 height
c) 1 dm classes
d) rounded towards the closest class centre
e) Suunto altimeter
f) field assessment

Diameter of the thickest living branch
a) diameter of the thickest living branch
b) measured only if previous measurement exists
c) 1 mm classes
d) rounded towards the closest class centre
e) callipers or ocular estimation
f) field assessment

Width of the crown
a) the largest width of the crown is measured based on estimated ground projection
b)
c) 1 dm classes
d) rounded towards the closest class centre
e) measuring tape
f) field assessment

Sweep of the tree
a) sweep of the tree is measured from the stem to the height of 4m
b) coniferous species and *Betula pendula*, as well as *Betula pubescens* with diameter over 16.5 cm
c) 1 cm classes
d) rounded towards the closest class centre
e) measuring bar
f) field assessment

Height
a) length of the dead sample tree
b) length of the dead sample or tally tree from the stump to the top or the plot border
c) 1 dm classes
d) rounded towards the closest class centre
e) measuring tape
f) field assessment

I) Attributes describing the forest ecosystem

Top diameter of the fallen part

a) the top diameter of the dead fallen tally tree
b) measured when the top is within the plot
c) 1 cm classes
d) rounded towards the closest class centre
e) caliper
f) field assessment

Length

a) length of the fallen tally tree
b) length from the stump to the top or to the plot border
c) 1 dm classes
d) rounded towards the closest class centre
e) measuring tape
f) field assessment

3.1.4. Definitions for attributes in nominal or ordinal scales

Note: only variables which differ from those presented in section 2.1.3. Definitions of some of the variables may have minor differences between the sections.

A) Geographic regions

Bearing of the slope

a) Points out the downwards bearing of the slope
b) 0) flat
   1) North-East
   2) East
   3) South-East
   4) South
   5) South-West
   6) West
   7) North-West
   8) North
c) field assessment

Topographic location

a) General land class defining risk for the wind damages
b) 0) flat or slight slope < 4:20
   1) slope > 4:20
   2) summit or the upper part of the slope < 4:20, part of the plot at least 10m higher compared with the surroundings
   3) valley or an area otherwise protected from wind
c) field assessment

Percentage of the plot covered by the stand

a) when the plot extends to more than one stand, the plot coverages of the stands are recorded
b) 10 % classes
c) field assessment

Distance to the amenity boundary

a) Distance to the closest amenity boundary
b) 0) plot centre further than 50 m from the amenity boundary
   1) A plot centre 30–50m from the amenity boundary
   Plot centre 10–30 m from the amenity boundary
   2/b) plot centre stand differs from the amenity boundary forest edge stand
   3) plot centre stand belongs to the amenity boundary stand, the boundary has appeared less than 5 years ago
   4) plot centre stand belongs to the amenity boundary stand, the boundary has appeared more than 5 years ago
   Plot centre 10m or closer form the amenity boundary
   5/E) part of the plot does not belong to the amenity boundary
   6) plot forms part of the amenity, the boundary appeared less than 5 years ago

7) plot forms part of the amenity, the boundary appeared less than 5 years ago
c) field assessment
d) Letter codes stand for smooth amenity boundary and numeric codes for sharp amenity boundary

C) Wood production

Location of the stand basal area observations
a) Defines the location of the basal area observations from the plot centre
b) P) 20m North of the plot centre
I) 20m East of the plot centre
E) 20m South of the plot centre
L) 20m West of the plot centre
K) observed at the plot centre
M) other observation point
c) field assessment

Size of the tally tree plot
a) Defines the size of the tally tree plot according to the diameter of the tree
b) d< 4.5 cm r = 5.64 m
4.5< d < 10.5 cm r = 5.64 m
d > 10.5 cm r = 9.77 m
c) field assessment

Size of the sample tree plot
a) Defines the size of the sample tree plot according to the diameter of the tree
b) d< 4.5 cm r = 2.82 m
4.5< d < 10.5 cm r = 2.82 m
d > 10.5 cm r = 4.89 m
c) field assessment

D) Site and soil

Effect of the land use class
a) defines the possible effect of the surrounding land use class, especially on small or narrow stands
b) 0) no effect of the land use classes 4–9
4, D effect of the class 'other forestry land'
5, E effect of the land use class 5
6, F effect of the land use class 6
7, G effect of the land use class 7
8, H effect of the land use class 8
9, I effect of the land use class 9
c) field assessment

Stand history
a) Variable describing the history of the near surroundings of the field plot
b) 0) no history information
1) plot has belonged to meadow or pasture land
2) plot has belonged to agricultural land
3) plot has belonged to build up land
4) plot has been burned during the existing tree generation
5) plot has been burned during the previous tree generations
c) field assessments

Thickness of the soil mantle
a) Estimates based on the existence of rock and metallic peak
b) 1) more than 70 cm
2) 20–70 cm
3) less than 20 cm
4) varying thickness, fracture plains of the rocks visible
c) field assessment

E) Forest structure

Time of the cut or mortality
a) time of the cut or mortality
b) 0) inventory year
1) one year ago
2) two years ago
.
6) six or more years ago
c) field assessment

F) Forest condition

Number of the needle storeys
a) Number of the needle storeys counted from the pines taller than 1.35m preferably from the crown top
b) 1) 1 storey
.
9) no visibility to the top
c) field assessment

I) Attributes describing the forest ecosystem

Cause of death

a) cause of death
b) 0) unknown
1) cutting (not transported log or unmerchantable top) or other manmade activity
2) burned
3) damages
4) competition between trees
c) field assessment

Severity of decay

a) severity of decay
b) A) most of the bark and branches still remain in the tree
B) coniferous tree which has lost bark, most of the branches have fallen, deciduous trees have not lost the bark but the stem starts to rot from the inside
C) Stem has become rotten and deciduous trees have lost the branches
D) dry standing tree; of deciduous trees only aspen normally reaches this stage
c) field assessment

Decaying fallen tree class

a) The class separates trees completely lying on the ground and trees partly in the air
b) 0) mostly not lying on the ground
1) mostly lying on the ground but dried instead of rotting
2) mostly lying on the ground
3) lying on the ground in several pieces, length or bearing of the felling hard to determine
c) field assessment

Severity of the decay

a) the severity of decay in the largest part of the fallen tree
b) 1) recently fallen, timber still hard, bark remains on the stem
2) timber still rather hard, often covered by bark. Knife penetrates the timber 1–2 cm
3) rather soft lying tree, bark often lost in large areas. Knife easily penetrates the timber 3–5m
4) timber soft and rotten, knife penetrates the timber completely
5) timber very rotten and soft, material dispersed when handled
c) field assessment

K) Forest functions

Fringing effect

a) If the following fringes exist closer than 30m to plot centre they are recorded
b) 0) no fringe
1) border between open bog and mineral soil
2) brook, width < 5m
3) river, width > 5m
4) bond or lake
5) sea
6) road
7) build up land
8) wire line
9) agricultural land
M) other
c) field assessment
d) for the class M literal description is given

Micro habitat

a) Following micro habitats are recorded when they are located closer than 30m from the plot centre
b) 0) no micro habitat
1) spring
2) small moist ground spot
3) small grove
4) precipice
5) open rock
6) rocky area larger than 0.05 ha
7) group of large aspens or other exceptionally large stems or rare tree species
8) ruins (stone)
9) remains of buildings
A) remains of human activities such as tar-burning or charcoal pit, trenches,

peat or gravel digging sites
B) remarks form the animal activity such as nest trees, nests of bear, visible beaver dwelling site. etc.
C) notable human transportation such as footpaths, ski tracks, snowmobile tracks
M) other
c) field assessment
d) for the class M literal description is given

Timber production restrictions based on the condition of the environment
a) Timber production restrictions marked on map and inspected on field
b) 0) no restrictions
1) protection forest outside the protected areas
2) an area where regeneration is difficult because of low day degree sum (700–800 d.d.)
3) summit forest under 700 d.d.
4) protection forest where d.d. exceeds 700
5) protection forest where d.d. does not exceed 700
c) map information and field assessment

Timber production restrictions based on multiple-use
a) Timber production restrictions based on multiple-use
b) 0) no restrictions
1) strict nature reserve or national park
2) wilderness forest
3) peatland reserve
4) peatland reserve under designing
5) drained area reserved by the Forest and Park Service
6) hiking areas, park forest and other multiple-use areas
7) build up areas where felling of trees is not allowed without permission
8) areas belonging to shore plan
c) map information and field assessment

3.2 DATA SOURCES

3.3. ASSESSMENT TECHNIQUES

3.3.1 Sampling frame

The sample has been located according to a systematic grid covering the whole Finland including water areas (lakes, rivers and seas). Plots which,according to the base maps, are located on lakes or at sea are not measured in the field. In total 3009 plots have been established on land.

3.3.2. Sampling units

Field assessment

Tract: Sampling unit is a tract consisting of 4 fixed radius circular field plots in Southern Finland and 3 similar plots in Northern Finland. In Southern Finland the distance between tracts is 16 km in North-South and East-West directions. In Northern

Finland the distance between the tracts is 24 km in North-South direction and 32 km in East-West direction.

Field plot: In Southern Finland the field plots are located along a line running in North-South direction, and the distance between plots is 400m. In Northern Finland the plots are located on the North-South line and the distance between plots is 600m. The coordinates of the trees on permanent plots are registered (Figure 10). The location of the field plots is described by marking up enough location points around the plot and by describing the path from a known map point to the plot in a way that it can easily be found . The radius of a field plot varies according to the tree size (Table 2). The same centre point is used for all the field plots.

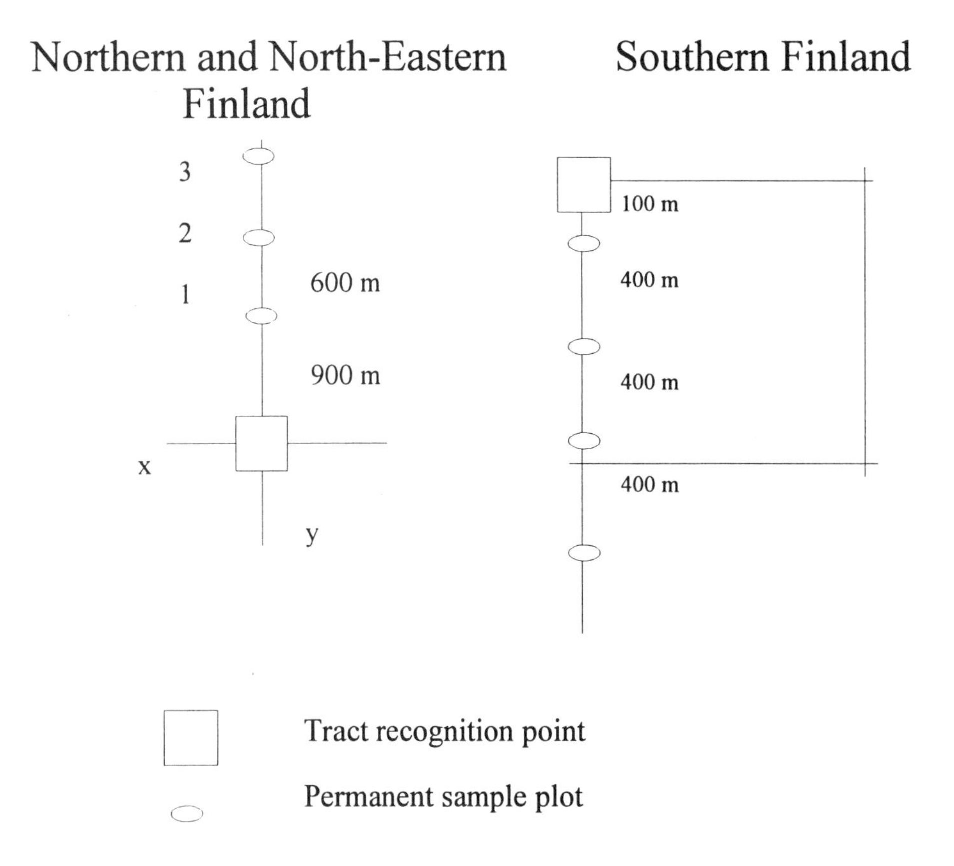

Figure 10. Tracts and location of the permanent field plots.

Table 2. Determination of the radius of the field plots

Measurement object	Breast height diameter cm	Radius m
Tally tree	4.5< d <10.5	5.64
Tally tree	d ≥ 10.5	9.77
Sample tree South Finland	4.5< d <10.5 d ≥ 10.5	2.82
Sample tree South Finland	d ≥ 10.5	4.89
Sample tree North Finland	4.5< d <10.5 d ≥ 10.5	3.26
Sample tree North Finland	d ≥ 10.5	6.91
Young tree	4.5 ≥ d	5.64
Dead tree standing tree	d ≥ 10.5	9.77
Dead fallen tree	d (largest) ≥ 10.5	9,77

Stand: The forest stand which includes a plot or a part of a plot. The plot may fall on two or several stands. In these cases all the stands are described in the stand description sheet.

The collected laboratory samples: Moss samples are collected from two plots in each tract and from a third plot if the first two samples are too small due to lack of moss. If the moss sample can not be collected from any of the plots, one sample is collected from any stand outside the plot on a tract. Each sample is collected from 8–12 points within a plot, about four liters in total. The humus layer sample is collected from one plot in each tract. The plot must be one of the moss sample plots. The sample is collected from 8–12 points on the plot with a sample drill.

3.3.3 Sampling designs

Tract: In Southern Finland (Finland excluding four northernmost Forestry Board Districts) the distance between tracts is 16 km in the North-South and East-West directions. In Northern Finland (four northernmost Forestry Board Districts) the distance between tracts is 24 km in the North-South direction and 32 km in the East-West direction.

3.3.4. Techniques and methods for combination of data sources

Field information

The field information is collected from the calculation areas using map coordinates of the field plots. The elevation of the field plot is detected from a digital elevation model and combined with the plot information also with the help of map coordinates. The increment information cores of the sample trees are measured in a laboratory, and the time needed for reaching the measuring height threshold of 1.35 m is combined with the sample tree information based on the tract field plot and tree numbering.

3.3.5. Sampling fraction

Data source and sampling unit	Proportion of forested area covered by sample (0.03 ha sample plots)	represented mean area per sampling unit
Field sample plots	0.000004	7644 ha

3.3.6. Temporal aspects

Inventory cycle	Time interval of data assessment	Publication of results	Time interval between assessments	Reference date
1st	1985–86	–	4	
2nd	1990–91	–	4	
3rd	1995	.		

3.3.7. Data capturing techniques in the field

Data is mainly recorded with hand computers which perform some plausibility checks during the recording. The information is transferred to the main frame of the Inventory Department of the Finnish Forest Research Institute in Helsinki via cellular phone every second or third work day.

If the hand computers are malfunctioning or not available, manual entry to inventory forms is mainly used. The forms are mailed to the Inventory Department of The Finnish Forest Research Institute in Helsinki for recording.

3.4. DATA STORAGE AND ANALYSIS

The inventory data are stored in The Finnish Forest Research Institute. The person responsible for the information is Dr. Erkki Tomppo, Professor of Forest Inventory (Dr. Erkki Tomppo, Helsinki Research Centre, The Finnish Forest Research Institute, Tel: +350 0 857051, Fax: +358 0 85705717, email:erkki.tomppo@metla.fi)

Following sections as in chapter 2

REFERENCES

Ilvessalo, Y. 1927. The Forests of Suomi (Finland). Results of the general survey of the forests of the country carried out during the years 1921-1924. Communicationes Ex Instituto Quaestionum Forestalium Finlandiae. Editae 11.

Kujala, M. Runkopuun kuorellisen tilavuuskasvun laskentamenetelmä. Summary: A calculation method for measuring the volume growth over bark of stemwood. Folia Forestalia 441:1-8.

Kuusela, K. & Salminen, S. 1969. The 5th National Forest Inventory in Finland. General design, instructions for field work and data processing. Commun. Inst. For. Fenn. 69.4.1-72.

Laasasenaho, J. 1982. Taper curve and volume functions for pine Spruce and Birch. Communicationes Instituti Forestalis Fenniae 108:1-74.

Lappi, J. 1992. JLP A linear programming package for management planning. The Finnish Forest Research Institute. Research Papers 414:1-134.

Matérn, B. 1960 Spatial variation. Medd. statens skogsforskn. inst. Bd. 49(5).

Pysyvien koealojen 3. mittaus 1995 maastotyön ohjeet, kuvio- ja puutiedot, näyteiden keruu. 1995. Metsäntutkimulaitos, Helsingin tutkimuskeskus, Valtakunnan metsien inventointi. pp. 101.

Ojansuu, R., Hynynen, J., Koivunen, J. & Luoma, P. Luonnonprosessit metsälaskelmassa (MELA) -Metsä 2000 -versio. Metsäntutkimuslaitoksen tiedonantoja 385:1-42.

Ranneby, B. 1981. Medelsformer till skattningar baserade på material från den 5:e riksskogstaxering. Abstract: Estimation of standard error in systematic sampling. Swedish university of Agricultural Sciences, section of Forest Biometry. Report 21 19 s.

Salminen, S. 1985. Metsien inventointimenetelmien tilastomatemaattinen perusta. Summary: Mathematic statistical foundation of the forest inventories. Silva Fennica 3:226-232.

Siitonen, M. A long term forestry planning system based on data from the Finnish national forest inventory. In publication: Forest inventory for improved management. Helsingin yliopiston metsänarvioimistieteen tiedonantoja 17:195-207.

Statistical Yearbook of Forestry. 1995. Aarne, M. (edit). Metsäntutkimuslaitos.

Tomppo, E, 1992. Multi-Source national forest inventory of Finland. In: Proceedings of Ilvessalo Symposium on National Forest Inventories; 1992 August 17-21; Helsinki, Finland; The Finnish Forest Research Institute Research Papers 444:52-50.

Tomppo, E. 1993. Multi-Source National Forest Inventory of Finland. In Publication: Nyyssönen, A. Poso, S. and Rautala, J (Edit.), 1993: Proceedings of Ilvessalo Symposium on National Forest Inventories, Organized by IUFRO S4.02, The Finnish Forest Research Institute and Department of Forest Resource Management at the University of Helsinki. The Finnish Forest Research Institute, Research Papers 444:52-60.

Tomppo, E. 1996. Multi-Source national forest inventory in Finland. In Proceedings of the Subject Group S4.02-00 'Forest Resource Inventory and Monitoring and Subject Group' S4.12-00 'Remote Sensing Technology', IUFRO XX World Congress, 6-12 August 1995, Tampere, Finland. New Trusts in Forest Inventory, EFI proceedings 7:27-41.

Tomppo, E. & Henttonen, H. 1995. Suomen metsävarat 1989-1995 ja niiden muutokset vuodesta 1951 lähtien. Metsätilastotiedote 354. Metsäntutkimuslaitos, Helsinki. 18 s.

Tomppo, E. & Korhonen, K.T. 1996. Manuscript 'Result computation of the 8th National Forest Inventory in Finland'. Finnish Forest Research Institute, Helsinki, Finland.

Päivinen, R., Pussinen, A. & Tomppo, E. 1993. Assessment of Boreal Forest stands using field assessment and remote sensing. Paper presented at the International symposium' operationalizing of Remote Sensing', 19-23 April 1993, ITC Enshedene, The Netherlands.

Poso, S. 1972. A method for combining photo and field samples in forest inventory. Commun. Inst. For. Fenn. 76(1):1-133.

Vähäsaari, H. 1989. Yläläpimitan estimointi kahden tunnuksen runkokäyrän ja metsikkötunnusten avulla. Research report of The Finnish Forest Research Institute.

Vuokila, Y. 1986. Puuntuotoksen tutkimussuunnan kestokokeiden periaatteita ja suunnitelmia. Metsäntutkimuslaitoksen tiedonantoja 239: 1-229.

COUNTRY REPORT FOR FRANCE

Isabelle Lagarde
Agence MTDA
France

Summary

2.1 Nomenclature

Assessments are made at department level. Departments are subdivided according to land use, forest ownership, ecological regions with the help of existing maps (topography, geology, ...) and aerial photographs. Photo-interpretation of wooded lands leads to the identification of vegetation types (defined from forest management and forest composition). Two other main criteria are interpreted from photographs: rough classes of volume per hectare and the size of the forested area. These attributes are assessed on stratification purposes.

Field assessments consist of two steps: control of photo-interpretation and measurements of other attributes for forest production estimations in production wooded lands. The main measured attributes regarding wood production are: circumference at breast height, diameters and heights at different levels of the trunk (for volume calculation), radius increase in the past 5 and 10 years, bark depth, age of the stand. Wood quality assessment and accessibility and hauling distance are also assessed on the field.

Beside tree measurements, many parameters regarding forest ecosystems are measured (soil description, floristic description).

2.2 Data sources

Apart from field assessments, the main data sources are aerial photographs. They are interpreted department-wise (about 1,200 to 2,000 photographs for one department - 96 departments). Photos are shot especially for the National Forest Survey. Results from photo-interpretation are drawn on topographic maps. Attempts were made to use space borne digital remote sensing data: they proved less accurate than aerial data and are not used in the assessments

2.3 Assessment techniques

Sampling design:

The entire country is covered by the assessment. Production forests, tree lines, hedges, poplar plantings, sparse trees are sampled on photos and on the field. Other forests (non-production forests) are identified on aerial photos but not sampled on the field.

The sampling design is a **3 steps double sampling for stratification**.

First, aerial **photographs are interpreted** to divide the department according to three criteria: land cover type, localisation, ownership category. The limits of these subdivisions are drawn on maps to calculate their size. These subdivisions are called “mapped domains”.

Then, systematic photo-plots are distributed on photographs with a transparent grid (about 1 point for every 30 to 40 ha). The photo-plots represent a 25 m radius plot on the field. A **plot is assigned to a strata** according to: the vegetation type, the estimated volume per ha (estimated in a few rough classes), the local composition of the stand, the size of the forest, the function (production or other function). The number of plots with the same values for all parameters are calculated. When multiplied by the extension factor of a plot, they provide an estimation of the size of the strata, according to the previously calculated size of the mapped domains.

The second step is the **checking of photo-interpretation** data on the field. The plots to be checked are chosen randomly among the first plots, they belong to the various strata. After field checking, a plot that proved not properly classified is moved to its proper category.

In the third step, **field sample plots** are chosen randomly among the plots from the second phase, in the various strata. Field plots are temporary plots.

The **sampling fraction may vary from one department to another**, and from one strata to another. Estimated mean values for proportions of forested area covered by samples are about 2 % for photo-interpretation and 0.07 % for field sampling.

The values of the attributes assessed on field plots are weighted with the strata sizes derived from the photo-interpretation to get total values.

Control of photo-interpretation is made in 25 m radius field plots. Field plots for wood production attributes consist in three concentric circle plots: according to their Circumference at breast height class, trees are measured on plot with a different radius (6 m, 9 m and 15 m). Regeneration is assessed on nine plots 2.26 m radius. Only the stems considered as viable trees (able to be part of a stand) are taken into account. The development potential is an ocular estimation based on forestry knowledge.

Data are recorded on **tally sheet and edited by hand into the computer**.

The current cycle is the third one; it started in 1987. because assessments are made at department level, **cycles overlap** (the second cycle ended in 1994).

2.4 Data storage and analysis

Data are stored in a national Forest survey database. The data in the data bank are not updated to a common point in time. National results are the sum of departmental results obtained at different points in time. Updating to a specific point in time can be done with the help of specific software on special requests. National Forest Inventory provides the service.

For data analysis, the principle of the calculation is:

1. calculating the volume and growth of each measured tree, convert these values per hectare; the numbers of trees are also calculated per ha
2. these average values are then multiplied by the "extension factor" of the sample plot

Strictly, the method only enables assessments for the sizes of the "studied domains" and of the strata, and for the volumes and growth in theses areas.

When sample plots are numerous enough in a strata, the error is estimated as the variance of the data. Generally, relative error is published only for areas, volumes, and growth for all the forests in the department and per ownership category.

For other data, error estimation can only be approximate.

The total volume of the strata is calculated as the multiplication of the average volume per hectare calculated on the sample plots times the size of the strata.

The precision of these part results *(résultats partiels)* can only be calculated approximately. The calculation consists in calculating the variances of these results from the variance of the inputs weighted by the proportion of the part area sizes or volumes.

Results are obtained per strata. A strata includes the criteria of forest ownership, forest region and vegetation type. Total results are the mean of the values of all the strata with this characteristic.

Calculation of the results are made with especially developed WINDOWS software.

Raw data and derived attributes can be made available for outside persons as long as they pay for it.

The contents of the database are basically tree data, subunit by species data and plot data, from which forexample the following items may be deduced:
1) For each forest region: size, volume, growth (by ownership category, vegetation type)
2) Distribution of one species in the department: global results for volume, growth, in-growth, area covered.
3) Global results for a department: by ownership category, vegetation type, species or groups of species.

The term subunit refers to a population of trees distinguished within the plot by their origin (high forest or coppice), state (alive, dead, cut, broken..),dimension class and storey.

2.5 Reliability of data

Photo-interpretation is checked by another photo-interpreter. The results of photo-interpretation are also checked on the field (13 % of photo plots). Some field plots measured twice, either by the same team (time interval: three to four weeks) or by another team. If an important mistake is noticed, the first value is replaced by the correct value. Only sampling error is assessed. Other errors (models, measurements, ...) are not assessed. For consistency check of data, field measurements are controlled on the tally sheet. The data are edited into the computer

2.6 Models

Volume is calculated using the values of attributes measured on the field. Total volume is the sum of the volume of basal log and the volume of upper log.

From these results, volume tariffs are derived, when enough measurements are available. Different volume functions are calculated for specific species groups, vegetation regions, and social position of trees in forest structures. This tariff is sued in the next inventory cycle to calculate the volume of a single tree: then, all trees are not thoroughly measured. Only input parameters to the tariff are measured: circumference at breast height and total height of the tree and metric decrease.

Models are applied to calculate in-growth and mortality (see also §2.4.5). A model is applied to calculate the volume increment per year and the average volume increment for planting (calculatedas volume/age of the planting) .

2.7 Inventory reports

Results are published by department, in three volumes.

Vegetation types (forest composition and structure) are presented as paper maps and as numerical maps. Special request are achieved mainly for international organisations, research institutions, private consulting firms, and private forest organisations.

1. OVERVIEW

1.1 GENERAL FORESTRY AND FOREST INVENTORY DATA

The data in this chapter were published in "*Les indicateurs de gestion durable des forêts françaises*" (*Ministère de l'Agriculture et de la pêche, 1995.)*

In 1993 (latest SCEEC/Teruti inquiry results from the Central Service for statistics- see §1.2.1), total wood and forest lands (including poplar plantings) area reaches **14.810 millions ha**. This represents **27 %** of the country.

Thickets and non-forest areas (moor *(landes)*, garrigues), hedges, sparse trees are not included in this number : they cover 3.598 millions hectares (6.5 % of total area of the country), distributed as follows :

	Area in hectares (1993)
Thickets	643,000
Hedges and sparse trees	1,021,000
Moor and garrigue	1,934,000

Deciduous trees cover $2/3$ of total wooded lands. The main tree species are:

1) *Quercus pedunculata*
2) *Quercus robur,*
3) *Pinus pinaster*
4) *Pinus sylvestris*
5) *Fagus sylvatica*

The average volume per hectare is 138 m^3 (National Forest Survey, 1994). The average increment per hectare is about 5.42 $m^3ha^{-1}yr$-1 (National Forest Survey, 1994).

Forest ownership is distributed as follows:

Forest ownership	Area (millions hectares)	% of total forest area (1)	Average size of forest (hectares) (2)
Private (*forêt privée*)	10.91	73.7 %	2.6
Community (*forêt communale*)	2.38	16.1 %	177
State (*forêt publique*)	1.51	10.2%	1 162

(1) Data from National Forest Survey

(2) For State forests and Community forests : data from National Forest Office, 1994.
For Private forests : data from Central Service for Statistic Inquiries and Studies (SCEES), 1983.

The number of inhabitants per ha forested land is 3.8 (from the latest census of population from 1990 - population of France: 56.6 million).

1.2 OTHER IMPORTANT FOREST STATISTICS

1.2.1 Other forest data and statistics on the national level

Forest area at national level :

It is calculated from 555,845 plots distributed on the national territory. This inquiry is called SCEEC/Teruti; it is achieved by a Central Service for Statistic Inquiries and Studies (Service Central d'enquêtes et d'études statistiques - SCEES). 99 categories are distinguished for land-use, and six categories describe wooded lands (lands where trees represent more than 10 % of total coverage and size over 0.5 ha). Wooded lands include poplar plantings in this survey. The land-use categories are identified by photo-interpretation. This inventory allows the presentation of results obtained on the same year for the entire country.

Protective forests

Size of protective forests are obtained from the National Forest Office (Office National des Forêts- O.N.F.) for State Forests.

Game

The abundance of game in a forest is assessed from inventories of live game and of dead animals killed by hunters. A national level organism is in charge with game management the National Hunting Office (Office National de la Chasse- O.N.C.). Hunters federations also play an important part in decision making.

Forest frequenting

No precise data are available about forest frequenting by visitors, except for one State forest (Fontainebleau forest).

Some wooded lands are particularly devoted to recreation. The size of forests under concern are provided by the National Forest Office for State Forests and by the Department of Agriculture (Central Service for Statistic Inquiries and Studies- Service Central d'enquêtes et d'études statistiques - SCEES).

The attractive role played by forests around cities is assessed by the ratio:

$$\frac{\text{Total forest surface in a 200 km diameter area around the city}}{\text{Total number of inhabitants of the city}}$$

Removals

Removals are assessed by the Department of Agriculture (Central Service for Statistic Inquiries and Studies- *Service Central d'enquêtes et d'études statistiques - SCEES*). Every year, the total volume of wood produced for the various uses (woodwork, industry, ...) is assessed, through inquiries at wood processing industries level. These data are not included in the National Forest Inventory results reports.

1.2.2 Delivery of statistics to UN and community institutions

The information in this chapter have been provided by Mr. VALDENAIRE (National forest Survey)

1.2.2.1 Responsibilities for international assessments

Person responsible for providing forest resource data :

a) to FAO/ECE for the 1990 Forest Resource Assessment (FRA 1990)

Mr CINOTTI
Present address :
Office National de la Chasse
85 bis avenue de Wagram
BP 236 75 822 PARIS Cedex 17
Tel : 44 15 17 17

At the moment, his function at the National Forest Survey is held by M. VALDENAIRE, responsible for international assessments - see §1.2.2.d).

b) To EUROSTAT

Mr CASAGRANDE
Service Central des Enquêtes et des études statistiques
Complexe d'enseignement agricole
BP 88
31 250 CASTANET TOLOSAN
Tel : 61 28 82 00
Fax : 61 28 83 83

c) to OECD

In 1980-1988 and in 1993, the ministry of Environment gathered the data provided by he NFS, the SCEES (Central Service for statistics) and the ministry of Agriculture to answer OECD request.

Mrs QUINCY / Mr PELISSIE
Ministry of Environment
Service des Affaires Internationales
20, avenue de Ségur
75 302 PARIS 07 SP
Tel : 42 19 17 68 / 42 19 17 55
Fax : 40 81 18 18

d) Other intergovernmental enquiries

To apply Helsinki conference decisions, data regarding French Forest have been analysed and presented by the Ministry of Agriculture, by the person responsible for the forest service.

Mr Christian BARTHOD
Ministère de l'agriculture
Direction de l'Espace rural et de la Forêt
Sous-direction de la Forêt
19, avenue du Maine
75 732 PARIS
Tel : 49 55 51 19

Generally speaking, the person responsible for providing forest resource data on international purposes is now:

Mr VALDENAIRE
Inventaire Forestier National
Château des Barres
45 290 NOGENT-sur-VERNISSON
Tel : 38 28 18 00
Fax : 38 28 18 28

1.2.2.2 Data compilation

The way of compiling the data depends on the request. But, generally speaking, no special analysis is conducted: existing data is compiled. They are not updated to a common point in time. Data at national level are the compilation of the latest available data for each department.

1.2.3 National assessment described

The survey that is described further in this report is the **National Forest Survey** *(Inventaire Forestier National - IFN).*

It is continuous. Field samplings are achieved by department. France consists of 99 departments - the average size of a department is 500,000 to 600,000 ha. The survey covers the entire France. The time interval between two surveys in the same department is about 10 years, except for the first cycle. The first cycle started in 1960 and lasted nearly 20 years. The second cycle ended in 1994. The current cycle is the third one: field measurements started in 1987 and are still going on in some departments. The cycle described in this report is the **third cycle**.

The National Forest Inventory Service is part of the **Department of Agriculture**. It was created in 1960. It aims to help a forest policy in rural management. It has to locate wooded lands in the territory. Therefore, it must first study all types of land-use in a department: (from "*Buts et méthodes de l'Inventaire Forestier National*" - *Inventaire Forestier National,* 1985) :

- forest and woodlands, and also "other wooded lands" (poplar planting, lines of trees, wooded hedges, sparse trees)
- others areas: waste lands, cultivation, water areas, non-productive lands from forest and agriculture points of view.

The second step is getting the information concerning the productive forests and woodlands. Only the survey of productive forests is described further. Sampling methods for "other wooded lands" are briefly described in § 2.9.

Two kinds of results are obtained :

- statistical results (descriptive and synthetic results) regarding present resource and potential resources (described by volume and volume increase) ;
- analytical and explicate results: correlation between the production of a forest stand and its qualitative and quantitative characteristics, and environment conditions. These results should help in making decisions in the frame of forest management for one precise forest.

Today, National Forest Inventory consists in:

- one central service in Loiret Department (in Nogent-sur-Vernisson)
- five inter-regional services in charge with the inventory of one part of French territory
- one service in charge with data analysis in Nancy
- one research service in Montpellier.

The staff (information provided by Mr WOLSACK - National Forest Survey) consists in **180 persons** including :

- 90 technicians achieving field measurements,
- about 20 persons for administration
- about 25 persons for survey planning, methods adjusting, and data analysis.

Part of the work is achieved by private firms paid by the NFS (photo-interpretation, field measurements, and edition of data in computers).

The financial means granted to NFS are now **60 millions francs**, including the state staff charges. The money comes from the State (Department of Forest and agriculture), from some taxes, and from the sale of information by NFS.

The decision of starting a national forest inventory was made by the French government. In 1958, two new articles were defined in the **forestry code**:

- "the Civil Service will inventory continuously the national forest resources, whoever their owner may be" (article L. 521.1, 521.2)
- this inventory will be ruled by the same law as cadastral and geodesic works (laws from July 6 1943 and March 28 1957).

Therefore, a Prefect decree *(arrêté préfectoral)* allows the NFS field team to enter any private property to achieve measurements.

2. NATIONAL FOREST SURVEY (*INVENTAIRE FORESTIER NATIONAL)*

2.1 NOMENCLATURE

2.1.1 List of attributes directly assessed

A) Geographic region

Attribute	Data source	Object	Measurement unit
Forest region (*Région forestière*)	existing maps: topography, soil, climate, vegetation	stand	categorical
Administrative unit: department (*département*)	existing maps	stand	categorical

B) Ownership

Attribute	Data source	Object	Measurement unit
Ownership (*catégorie de propriété*)	public forests maps	stand	category: public/private forest

C) Wood production (see figure 1)

Attribute	Data source	Object	Measurement unit
Circumference at breast height (1.30): CBH	field assessment	tree	cm
Height at 7 cm diameter of the trunk: Hd (*Hauteur à la découpe 7 cm*)	field assessment	tree	decimetre
Diameter at height 2.60 m: D2.6	field assessment	tree	cm
Diameter at height:1.30 + Hd/2: Dm:	field assessment	tree	cm
(Hd + 2.6)/2	field assessment	tree	dm
Total Height: H (*hauteur totale*)	field assessment	tree	dm
Bark thickness at breast height (*épaisseur d'écorce à 1.30 m*)	field assessment	tree	mm
Age	field assessment	tree	years
Radius increase in the past 10 years: Ir10	field assessment	tree	mm
Radius increase in the past 5 years: Ir5	field assessment	tree	mm

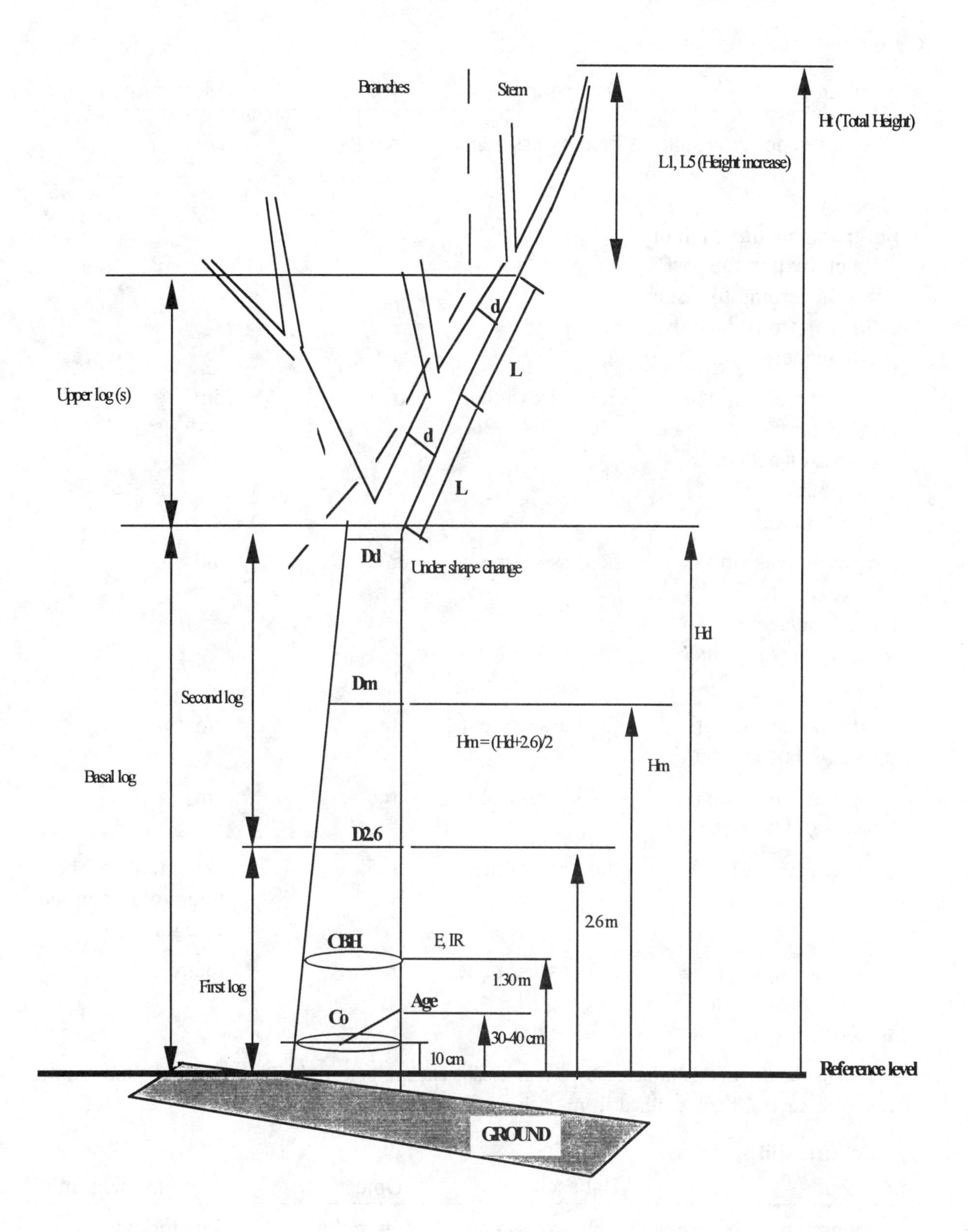

Figure 1. Wood production.

(C9 Wood Production continues)

Attribute	Data source	Object	Measurement unit
Circumference at ground level: Co (*circonférence à la base*) For trees that died, fell or were cut within the past 5 years (according to ocular estimate), Co is the only attribute measured.	field assessment	tree	cm
Height increase in the past 5 years: L5 (*accroissement en hauteur sur les 5 dernières années)*	field assessment	tree	dm
Height increase in the past year: L1 *(accroissement en hauteur sur la dernière année*)	field assessment	tree	dm
Medium diameter of upper log(s)* (*surbille)*	field assessment	tree	cm
Length of upper log(s) (*surbille)*: L	field assessment	tree	cm
Quality (*qualité*)	field assessment	tree	percentage of the total volume in the category
Flawless length of the trunk (*Longueur de fût sans défaut)*	field assessment	tree	metres

*If the shape of the tree is affected by a significant change before the 7 cm diameter, the stem is measured in several parts defined from each shape change.

D) Site and soil

Attribute	Data source	Object	Measurement unit
Topography (*topographie locale*)	field assessment	6m radius sample plot	categorical
Aspect (*exposition*)	field assessment	25m radius sample plot	categorical
Slope (*pente*)	field assessment	25m radius sample plot	categorical
Elevation (*altitude*)	field assessment	Plot centre	meters

Soil texture (*texture*), compactness *(compacité)*, acidity *(acidité)*	field assessment	Soil description plot (as near as possible plot centre, and in an area representative of the area used for forest ecosystem description)	categorical
Limestone (*calcaire*)	field assessment	Soil description plot	categorical
Soil depth (*profondeur de sol)*	field assessment	Soil description plot	centimetres
Stones (*cailloux*)	field assessment	6m radius sample plot	categorical
Outcrops (*affleurements rocheux*)	field assessment	6m radius sample plot	categorical
Bedrock (*roche mère*)	field assessment	Soil description plot	categorical

E) Forest structure

Attribute	Data source	Object	Measurement unit
Vegetation type	photo-interpretation	stand (minimum 15-20 ha)	Categorical
Regularity/irregularity (for timber wood, or coppice, or mixed timber and coppice stands) *(Régularité/irrégularité - futaies, taillis, taillis avec réserves)*	field assessment	1 ha sample plot (including the 20 acres sample plot for tree measurements)	categorical
Composition of the stand (deciduous/coniferous)	field assessment	20 acres sample plot	categorical, according to crown cover percentage
Absolute crown cover percentage of the stand	field assessment	20 acres sample plot	crown cover percentage
Tree species	field assessment,	20 acres sample plot	crown cover percentage
Age of the prevailing species	field assessment	20 acres sample plot	age (5 year classes before 100, 1 year classes from 100)

Relative importance of the prevailing species, by species	field assessment	20 acres sample plot	categorical (according to relative crown cover percentage)
Circumference class of prevailing species (by species)	field assessment	20 acres sample plot	categorical (according to diameter classes)
Absolute crown cover percentage of timber	field assessment	20 acres sample plot	percentage
Absolute crown cover percentage of prevailing stems	field assessment	20 acres sample plot	categorical, according to crown cover percentage
Average height of prevailing stems (H)	field assessment	20 acres sample plot	dm
Relative crown cover percentage of dominating stems by species	field assessment	20 acres sample plot	percentage
Absolute crown cover percentage of dominated stems (trees with CBH > 24.5 cm and height <2/3H)	field assessment	20 acres sample plot	categorical, according to crown cover percentage
Relative crown cover percentage of dominated stems (measurable trees under 2/3H) by species	field assessment	20 acres sample plot	percentage
Evolution of the stand	field assessment	20 acres sample plot	categorical

F) Regeneration

Attribute	Data source	Object	Measurement unit
Density	field assessment	200 or 400 m² sub- plots	Number of stems
Tree Species	field assessment	200 or 400 m² sub- plots	tenths of global number of stems in the regeneration
Absolute crown cover percentage of regeneration	field assessment	20 acres sample plot	categorical, according to crown cover percentage
Average height of regeneration	field assessment	20 acres sample plot	dm
Relative crown cover percentage of regeneration by species	field assessment	20 acres sample plot	percentage

Regeneration of the species	field assessment	20 acres sample plot	proportion of stems of the species in the regeneration (in tens)
Date of the clear cut or age of the planting	field assessment	20 acres sample plot	date

G) Forest condition

Attribute	Data source	Object	Measurement unit
Nature of the tree cutting	field assessment	20 acres sample plot	categorical, number of felled trees

H) Accessibility and harvesting

Attribute	Data source	Object	Measurement unit
Hauling distance (*distance à une piste de débardage*)	field assessment	20 acres sample plot	meter
Slope	field assessment	20 acres sample plot	percent
Ground ability for timber unloading (*débardage*)	field assessment	20 acres sample plot	category
Possibility to create a new unloading road (*piste de débardage*)	field assessment	20 acres sample plot	metres (distance from the plot)
Possibility to use cables for unloading (*Possibilité de débardage au câble*)	field assessment	20 acres sample plot	meters (distance)

I) Attributes describing forest ecosystems

Attribute	Data source	Object	Measurement unit
List of unmeasured species over 1.50 m high (with CBH < 24.5 cm)	field assessment	6 m radius sample plot	categorical
Total crown cover percentage of species over 1.50 m high (with CBH < 24.5 cm)	field assessment	6 m radius sample plot	percent
Relative crown cover percentage of species under 1.50 m high (with CBH < 24.5 cm)	field assessment	6 m radius sample plot	percent
List of unmeasured species over 1.50 m high (with CBH < 24.5 cm)	field assessment	6 m radius sample plot	categorical

Total crown cover percentage of species under 1.50 m high (with CBH < 24.5 cm)	field assessment	6 m radius sample plot	percent
Relative crown cover percentage of species under 1.50 m high (with CBH < 24.5 cm)	field assessment	6 m radius sample plot	percent

J) Non-wood goods and services

Attribute	Data source	Object	Measurement unit
Forest Function	existing maps, local organisms	stand	categorical

Poplar plantings are described with a special method.

Trees along roads, hedges and sparse trees are also described.

K) Miscellaneous

Attribute	Data source	Object	Measurement unit
Size of the wooded land including the photo-plot	Aerial-photos	stand	categorical (forest, small wood, thicket)

2.1.2 List of derived attributes

A) Geographic region : none

B) Ownership : none

C) Wood production

Attribute	Measurement unit	input attributes
Volume of basal log (*bille de pied*), including bark - For trees under 2.60 m height (Vb)	m^3	Hd: height at 7 cm diameter D ½: diameter at Hd level
Volume of basal log (*bille de pied*), including bark - For trees beyond 2.60 m height (Vb)	m^3	Volume of two logs (under 2.60 m and beyond 2.60 m) -DBH (D1.30) - D2.60: diameter at height 2.60 - Dm: diameter at 1.30 + Hd/2 - Hd (height at diameter 7 cm)
Volume of upper log(s) (*surbilles*) of the stem (Vs)	m^3	- l: length of the log (*surbille*) - d: medium diameter of the upper log
Total volume of tree stem	m^3	Vb + Vs

Current increase of volume including bark	m^3	- Previously measured total volume (5 years ago) - D1.30 : DBH - E: Bark thickness - Ir5: thickness of the last 5 age rings - Ht: total height
Volume per hectare	m^3ha^{-1}	- single tree volume - sample plot size - weighting factor of the area represented by each sample plot

D) Site and soil: none

E) Forest structure

Circumference class of the prevailing species (*catégorie de dimension de l'espèce dominante)*

a) Circumference class of the prevailing species is calculated as the mean of the circumferences of trees measured on the 15 m radius sample plot. If this plot is empty, all trees beyond this radius and included in the 25 m radius sample plots are measured and the mean value of their circumferences is calculated.

b) Classes :

Circumference at breast height = 24.5 to 72.4 *(Petits bois)*
Circumference at breast height = 72.5 to 120.4 *(Bois moyens)*
Circumference at breast height = >120.5 *(Gros bois)*
c) field assessment

F) Regeneration: none

G) Forest condition: none

H) Accessibility and harvesting

Hauling distance

a) The hauling distance is the distance between the plot centre and a road suitable for rough timber transport. On flat areas, it is measured along the slope. On slopes, it is calculated from the slope measured on the field (between plot-centre and existing road) and the horizontal distance measured on a map or on the aerial photo.
b) No threshold value
c) metres
d) three kinds of classes:
Case 1: no road creation need: hauling distance: 0 - 200 m, 200 - 500 m, 500 - 1,000 m, 1,000-2,000 m, > 2,000 m).
Case 2: a road must be created: length of the required road: < 500 m, 500 - 1,000 m, > 1,000 m
Case 3: impossibility to create a road but a cable can be set up: length of cable system required: < 500 m, > 500 m.

e) Slope is measured with a slope meter (*clisimètre*)
f) field assessment and topographic maps

I) Attributes describing forest ecosystems

Abundance of unmeasured species over 1.50 m and under 1.50 m
a) Absolute percentage of cover of species
b) No threshold value
c) percent
d) Classes
   1 = < 5 %
   2 = 5 to 25 %
   3 = 25 to 50 %
   4 = 50 to 75 %
   5 = > 75 %
e) ocular estimation
f) field assessment

J) Non-wood goods and service: none

K) Miscellaneou: none

2.1.3 Measurement rules for measurable attributes

For each of the measurable attributes the following information is given:

a) measurement rule
b) threshold values
c) measurement scale
d) rounding rules
e) instrument used for measurement
f) data source

A) Geographic region

-

B) Ownership

-

C) Wood production

Circumference at breast height (1.30): CBH
a) Measured at 1.30 m height above ground
   - On slopes, measured from uphill side. One reading, the measuring tape being perpendicular to the axis of the stem.
b) Minimum CBH: 24.5 cm
c) centimetre
d) rounded to the closest cm (for example, 24.6 to 25.4 rounded to 25, 25.5 to 25.9 rounded to 26)
e) Forestry measuring tape *(ruban forestier)*
f) field assessment

Height of 7 cm diameter of the trunk: Hd (Hauteur à la découpe 7 cm)
a) The location of the 7 cm diameter on the trunk is an ocular estimation
b) No threshold
c) dm
d) rounded to the closest decimetre.
e) Measured with poles if Hd is less than 15 m, with Christen Height meter (*dendromètre Christen*) if Hd > 15 m (when poles can no longer be used).
f) field assessment

D2.6: diameter at height 2.60,

a) Measured at 2.60 m height above ground (the 2.6 m height is measured with poles) - On slopes, measured from uphill side
b) No threshold value
c) centimetre, in classes of one centimetre (19.5 to 20.4 cm are recorded as 20 cm)
d) rounded to closest cm
e) finish calliper (*compas finlandais*) graduated with the centimetre classes (if D2.6 < 38 cm) or diameter ruler (*règle à diamètre*) if D2.6 >38 cm
f) field assessment

Dd: diameter at the top of trunk (if Dd is not 7 cm)

a) Measured at the top of the stem (under starting point of crown)
b) 7 cm
c) centimetre
d) rounded to closest cm
e) finish calliper (*compas finlandais*) or diameter ruler (*règle à diamètre)*
f) field assessment

Dm: diameter at height Hd + 1.30

a) On slopes, measured from uphill side
b) No threshold value
c) centimetre, in classes of one centimetre (19.5 to 20.4 cm are recorded as 20 cm)
d) rounded to closest cm
e) finish calliper (compas finlandais) graduated with the centimetre classes (if Dm<38 cm) or diametre ruler (règle à diamètre) if Dm>38 cm

Hm

a) On slopes, mesured from uphill side
b) No threshold
c) dm
d) rounded to the closest decimetre
e) Measured with poles if Hm is less than 15 m, with Christen Heightmeter (dendromètre Christen) if hm>15 m (when poles can no longer be used)
f) field assessment

Total Height (Hauteur totale): H

a) On slopes, measured from uphill side
b) No threshold
c) dm
d) rounded to the closest decimetre.
e) Measured with poles if H is less than 15 m, with Christen Height meter (*dendromètre Christen*) if H > 15 m (when poles can no longer be used).
f) field assessment

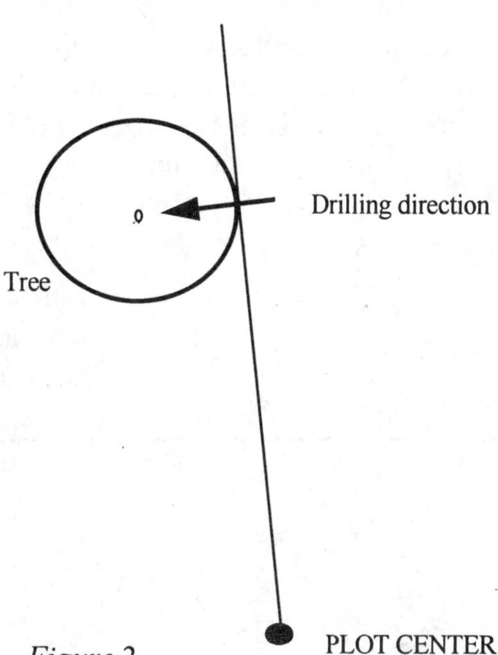

Figure 2

Bark thickness at breast height (épaisseur d'écorce)

a) Measured at 1.30 m, on the contact point between the tree and a tangent to the tree pointing to the plot-centre. Contact point always chosen on the right or on the left of the tree seen from the plot-centre. If there are more than one contact point, the point at mid-distance between the two is the measurement point. The drilling is made on the perpendicular to this tangent. (see figure 2).
b) No threshold value
c) mm (exact value)
d) rounded to the closest mm
e) bark gauge (*jauge à écorce*)
f) field assessment

Age of the prevailing species (age de l'espèce dominante)

a) Age of the prevailing species in stands of structure 1 (Regular timber wood), 3 (Timber wood mixed with coppice), 4 (Coppice) or 5 (Timber wood mixed with coppice) and of species planted since 40 years (see § 2.1.4 A) Forest structure definition). A species is prevailing when it has the highest relative crown coverage. Age is measured as near as possible the bottom of the tree. Two trees belonging to the upper diameter class are measured, and one tree from the other two classes. The trees on which age is measured are the first ones in the numbering process. (They are chosen objectively). Only, if the objective procedure leads to measure trees that are not representative at all of the rest of the plot, more trees are measured. The age of these extra trees is recorded separately. Measurements are not conducted on Poplars and *Juglans sp.* for these species seemed to be seriously damaged by such measurements.

b) No threshold value

c) years

d) Classes:

e) *Pressler* drill (*tarière de Pressler)*

f) field assessment

Age (years)		Code for species with stems of the same age in the stand (measured age)	For species with stem of different ages (estimated ages)
From	to		
0	19	0,01, ...19	91
20	24	20	91
25	29	25	91
30	34	30	92
35	39	35	92
40	49	40	92
50	59	50	92
60	69	60	93
70	79	70	93
80	99	80	93
100	119	81	94
120	139	82	94
140	159	83	94
160	179	84	95
180	199	85	95
200	239	86	95
240	more	87	96

Circumference at 10 cm above ground (Co) (Circonférence à la base)

a) Measured at 10 cm with a measuring tape. The measuring tape is placed in a plan perpendicular to the axis of the stem. (after removing moss, ivy, etc ...)
b) No threshold value
c) cm (exact value)
d) rounded to the closest cm
e) Forestry measuring tape *(ruban forestier)*
f) field assessment

Radius increase in the past 5 years (Ir5) and radius increase in the past 10 years (Ir10) (Epaisseur des 5 derniers cernes et des 10 dernier cernes)

a) Measured at 1.30 m, on the contact point between the tree and a tangent to the tree pointing to the plot-centre. Contact point always chosen on the right or on the left of the tree seen from the plot-centre. If there are more than one contact point, the point at mid-distance between the two is the measurement point. The drilling is made on the perpendicular to this tangent. The measured value is the minimum distance between the two age rings under consideration.
b) 5 and 10 years
c) mm (exact value)
d) rounded to the closest mm.
e) Pressler drill (*tarière de Pressler*)
f) field assessment

Length of upper log(s) (longueur des surbilles)

a) An upper log starts at the point where the shape of the tree changes
b) No threshold value. (The threshold value for the diameter of the upper end of the upper log is 7 cm)
c) dm
d) rounded to the closest decimetre.
e) Measured with poles if the height of the upper end of the log is less than 15 m, with Christen Height meter (*dendromètre Christen*) if it is over 15 m (when poles can no longer be used). The poles are set vertically. If the axis of the log to be measured is not vertical, and ocular estimate provides the correction to be applied to the measured length.
f) field assessment

Medium diameter of upper log(s) (Diamètre médian des surbilles): d

a) Measured at the middle of the upper log. The location of the middle is measured with poles with the same procedure as described above.
b) No threshold value (But it is necessarily more than 7 cm)
c) centimetre, in classes of one centimetre (19.5 to 20.4 cm are recorded as 20 cm)
d) rounded to closest cm
e) finish calliper (*compas finlandais*) graduated with the centimetre classes (if d < 38 cm) or diameter ruler (*règle à diamètre*) if d >38 cm
f) field assessment

Height increase in the past 5 years (L5) (Accroissement en hauteur sur les 5 dernières années)

a) Length of the latest five annual shoots. The location of limits of the various shoots is an ocular estimation. If some branches or some cut trees are on the plot, their shoots are examined to adjust the likely value of the shoots. When poles are used, the poles are set vertically. If the axis of the shoots to be measured is not vertical, and ocular estimate provides the correction to be applied to the measured length.
b) No threshold
c) dm
d) rounded to the closest decimetre
e) Measured with poles if total height less than 15 m, with Christen Height meter (*dendromètre Christen*) if it is over 15 m (when poles can no longer be used).
f) field assessment

Height increase in the past year (L) (Accroissement en hauteur sur la dernière année)

a) Length of the latest annual shoot. The location of limits of the various shoots is an ocular estimation. If some branches or some cut trees are on the plot, their shoots are examined to adjust the likely value of the shoots. When poles are used, the poles are set vertically. If the axis of the shoots to be measured is not vertical, and ocular estimate provides the correction to be applied to the measured length.
b) No threshold
c) dm
d) rounded to the closest decimetre
e) Measured with poles if total height less than 15 m, with Christen Height meter (*dendromètre Christen*) if it is over 15 m (when poles can no longer be used).
f) field assessment

Flawless length of trunk (longueur de fût sans défaut)

a) Flawless length of trunk under Hd. Only for trees belonging to the upper circumference class and for the part of the trunk belonging to the best quality (1). (see also Quality § 2.1.4.-C)
b) No threshold value
c) Metres, grouped into classes. First class from 0 to 1.49, other classes: 1.50 m
d) rounded to the closest metre
e) Measured with poles
f) field assessment

D) Site and soil

Slope

a) Slope is measured along a 50 m line centred on the centre of the plot (see figure 3). One reading in the upper slope direction. One reading in the lower slope direction. The mean value is calculated. (The calculated value is the tangent).
b) No threshold value
c) percent
d) rounded to the closest percent
e) "Slope meter" (*Clisimètre*)
f) field assessment

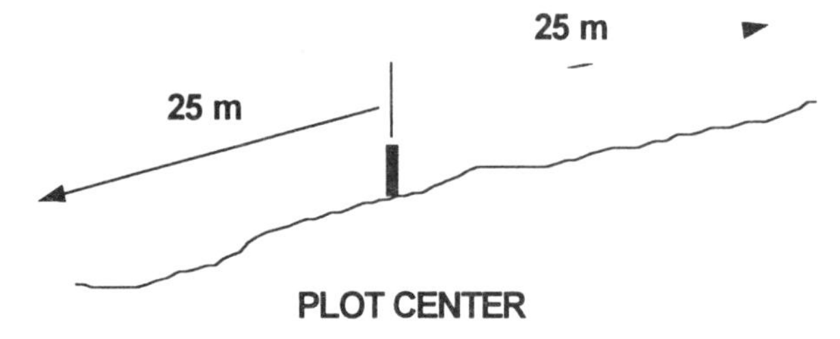

Figure 3

Aspect

a) Aspect is measured as the aspect of the ground along the main slope direction. The operator is located at the centre if the 25 m radius plot.
b) No aspect for flat areas
c) grade
d) rounded to the closest grade
e) compass
f) field assessment

Elevation

a) Elevation is read from a contour map
b) No threshold value
c) metres
d) rounded to the elevation of the closest contour line. Contour line provide elevation in classes of 10 metres generally, of 5 metres in flat areas.
e) contour map
f) plot location on a contour map

Soil thickness (profondeur de sol) and thickness of particular layers

a) Soil description is made as near as possible to the centre of the plot in an area that is representative of the area used for the floristic description (estimation of the coverage of unmeasured species). Soil depth is the distance between the surface of ground (excluding humus) and the bedrock. The thickness of other particular layers or particular characteristics of soil is also

measured: for pseudogley, gley and the presence of oxydation or carbonatation.
b) No threshold value
c) ten classes 9 cm wide (< 5 cm, 5-14 cm, ..., 75-84 cm, > 85 cm)
d) rounded to the closest centimetre.
e) ruler
f) field assessment

Outcrops

a) Percentage of cover of outcrops is assessed on the 25 m radius sample plot
b) No threshold value
c) Ten classes 9 % wide (< 5 %, 5-14 %, ..., 75-84 %, > 85 %)
d) -
e) ocular estimate
f) field assessment

Stones

a) Percentage of cover of stones is assessed on the 25 m radius sample plot
b) No threshold value
c) Ten classes 9 % wide (< 5 %, 5-14 %, ..., 75-84 %, > 85 %)
d) -
e) ocular estimate
f) field assessment

Litter

a) Litter thickness is assessed on the soil description plot (this plot being chosen near as possible the centre of the plot in an area that is representative of the area used for the ecological description).
b) No threshold value
c) 0= absence
1= presence
2 = one continuous layer
3 = two layers
4 = litter thickness > 1 cm
5 = litter thickness > 5 cm
d) rounded to the closest centimetre
e) ocular estimate
f) field assessment

Acidity of soil (acidité)

a) The pH of the first layer (*A1 horizon*) is measured
b) No threshold value
c) Value of the pH
d) -
e) pH meter
f) field assessment

Abundance of stains and origin of satins

a) The presence of stains is assessed in each layer . the limits of layers are rounded to the closest decimetre.
b) No threshold value
c) Abundance is assessed for oxydation, concretion, ...
1 = < 5 %
2 = 5-25 %
3 = 25-50 %
4 = 50-75 %
5 = > 75 %
d) 25 % wide classes
e) ocular estimation
f) field assessment

E) Forest structure

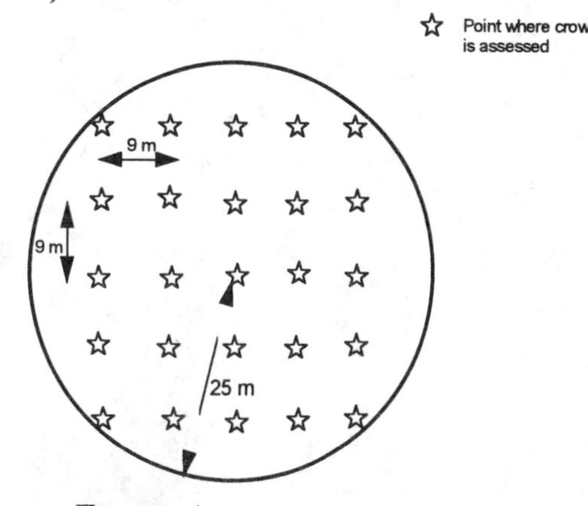

Figure 4

Absolute crown cover percentage of the stand (pourcentage de couvert absolu)

a) Values are assessed on 20 acres plots. The assessed parameters must be homogeneous on the 20 acres. Some simplified methods can be used: placing 25 points (5 lines with 5 points - see figure 4), distance between points measured by steps, and assessing crown cover percentage at the 25 points (each point represents 4 % of the cover).
b) A 20 acres plot is measured only if the

crown cover percentage is over 10 %

c) The percentages are grouped into classes :

For stands where absolute cover of measured trees (DBH > 7.5 cm) is under 10 %

0 = lower than 10 %
1 = 10 to 24 %
2 = 25 to 49 %
3 =≥50 %

For stands where absolute cover of measured trees is over 10 %

5 = 10 to 19 %
6 = 20 to 24 %
7 = 25 to 49 %
8 = 50 to 74 %
9 =≥75 %

d) see the classes

e) Ocular estimation.

f) field assessment and photo-interpretation

F) Regeneration

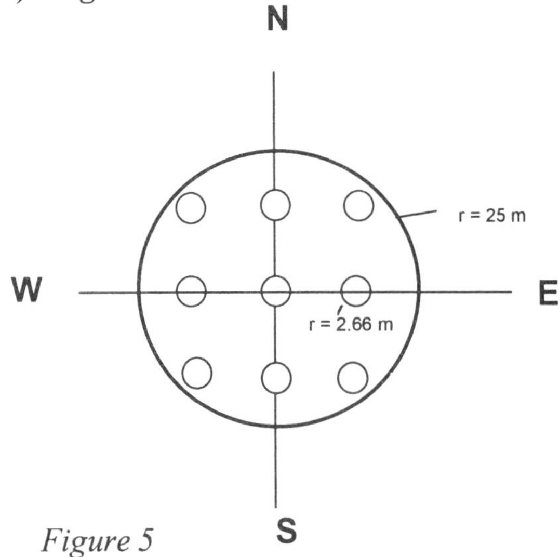

Figure 5

Density

a) The stems taken into account as regeneration are the stems considered as viable trees, able to be part of a stand (*arbres d'avenir*). Regeneration is assessed on nine 2.66 m radius plots (see figure 5)

b) The maximum is one stem for 2 m².

c) - = Less than 100 stems per hectare (all species together)

0 = 100 to 500
1 = 500 to 1,000
2 = 1,000 to 1,500
3 = 1,500 to 2,000
4 = 2,000 to 2,500
6 = 2,500 to 3,000
7 = 3,000 to 3,500
8 = 4,000 to 5,000
9 = > 5,000

d) see classes

e) counting and ocular estimation of development potential

f) field assessment

G) Forest condition

-

2.1.4 Definition for attributes on nominal or ordinal scale

The following information is provided for each attribute:

a) the definition
b) the categories (classes) and
c) the data sources
d) remarks

A) Geographic region

Forest region (Région forestière)

a) 309 "forest regions" defined by NFS
b) Name of the region
c) existing maps (climate, soil, ...), field assessments, contacts with specialised services

B) Ownership

a) Legal status of the forest regarding its management
b) 1) State forest
2) public forest submitted to a state management *(soumise au régime forestier)*
3) private forest (= not submitted to state management)
c) existing maps

C) Wood production

Quality

a) Wood quality of the various parts of the tree
b) 1 = best quality, for veneering *(tranchage)* and Woodwork *(menuiserie)*
2 = all other uses *(autres utilisations)*
3 = fire wood (*bois de chauffage*)
c) field assessment: ocular estimate
d) The values are given in tenths of the total volume belonging to the various qualities. Total volume and volume per quality classes are assessed with the help of a table providing volume values according to circumferences and height.

D) Site and soil

Topography (Topographie)

a) local configuration of ground on the 25 m radius sample plot
b) 0) Flat ground (slope < 5 %)
1) Sharp summit, round summit, escarpment *(escarpement)*
3) round summit
2) top of slope *(haut de versant)*
3) Mid slope of a concave side (*mi-versant concave)*
4) Mid slope of a rectilinear side *(mi-versant rectiligne)*
5) Mid slope of a convex side*(mi-versant convexe)*
6) Shelf (*replat*)
7) bottom of slope *(bas de versant)*
8) Valley
9) Dip, coomb, narrow valley
c) field assessment

Soil texture

a) Soil texture is assessed in the first layer. If there is a noticeable change between layers, texture is assessed in each layer.
b) 0 - absence of soil
1 = Sand (sand prevailing)
2 = Sand and silt (sand prevailing)
3 = Sand and gley (sand prevailing)
4 = Silt and sand (silt prevailing)
5 = Silt and gley or silt gley and sand (silt prevailing)
6 = Silt
7 = gley and silt (gley prevailing)
8 = Gley and sand and silt or gley and sand (gley prevailing)
9 = Gley
c) field assessment

Soil structure
a) Structure of upper layer *(Horizon A1)*
b) seven classes
c) field assessment

Compactness of layers
a) assessed on the soil description plot
b) four classes
c) field assessment

Bedrock (Roche mère)
a) Bedrock is estimated with the help of existing geological maps and from field observations on the soil description plot.
b) 12 classes divided into subclasses
c) field assessment and existing geological maps

Humus type
a) assessed on the soil description plot
b) 17 humus types are distinguished (from Mor to Turf)
c) field assessment

E) Forest structure

Vegetation type (type de peuplement)
a) Photo-interpreted, it is the combination of forest structure, and stand composition.
b) Structure: see below: Forest structure
Composition: A stand is "pure" if the relative crown cover percentage of one species is over 75 %. A stand is a mixture of two species if the relative crown cover percentage of each species is over 25 % (and < 75 %) . (The trees species are described below)
c) photo-interpretation

Forest structure (Regularity/irregularity of the stand)
a) Definition of the stand structure from forestry point of view
b) Classes
1 = Regular timber wood *(Futaie régulière)*
2 = Irregular timber wood *(Futaie irrégulière)*
3 = Timber wood mixed with coppice, deciduous trees prevailing
4 = Coppice *(Taillis simple)*
5 = Timber wood mixed with coppice, coniferous trees prevailing
0 = Area temporarily deforested (for less than 5 years)
c) field assessment

Composition of the stand (deciduous/coniferous)
a) Relative proportion of coniferous and deciduous species in the stand, in crown cover percentage
b) Classes
0 = Area temporarily deforested (for less than 5 years)
1 = Pure deciduous stand (relative crown cover percentage of deciduous trees > 75 %)
2 = Pure coniferous stand (relative crown cover percentage of coniferous trees > 75 %)
3 = deciduous stand mixed with coniferous species : relative crown cover percentage of coniferous trees between 50 and 75 %)
4 = coniferous stand mixed with coniferous species : relative crown cover percentage of coniferous trees between 50 and 75 %)
c) field assessment

Evolution of the stand
a) Origin and evolution of the stand, assessed on the 25 metres radius plot: existence of regeneration, change in management regime, natural extension of forested area, ...
b) Classes:
1 = Natural or artificial regeneration, without change in the species
2 = Natural or artificial regeneration, with change in the species
3 = Artificial stand less than 40-year-old, on previously unwooded land
4,5,6 = Artificial stand, less than 40-year-old, on clear-cut land (4), after "shed cut" (5) *(coupe d'abri)*, plantings in strips (6)
7 = Transformation into timber forest (*conversion feuillue*)

8 = Development of a natural wood on a bare ground, occurred after the latest inventory
9 = Destroyed stand *(peuplement ruiné)*
0 = Other cases (no evolution)

c) field assessment and photo-interpretation of the previous cycle.

Tree species

a) List of 38 broadleaf trees and 22 coniferous trees and two additional classes (other exotic coniferous trees and other exotic broadleaf trees).
b) 60 tree species and two superior classes
c) field assessment

Absolute crown cover percentage of timber (couvert absolu de la futaie)

a) Absolute crown cover percentage of timber in timber forest or in mixed timber and coppice
b) Classes
   0 = less than 10 %
   1 = 10 to 24 %
   2 = 25 to 49
   3 = 50 to 74
   4 = 75 % and more
c) ocular estimate

Absolute crown cover percentage of coppice in vegetation stands consisting in timber and coppice (couvert absolu du taillis)

a) Absolute crown cover percentage of coppice in mixed vegetation stands timber and coppice
b) Classes
   0 = less than 10 %
   1 = 10 to 24 %
   2 = 25 to 49
   3 = 50 to 74
   4 = 75 % and more
c) ocular estimate

Stand composition: Relative importance of species, by species (composition du peuplement)

a) Importance of a species is defined from its relative crown cover percentage on the 25 metres radius sample plot.
b) Classes
   1 = Pure: relative crown cover percentage > 75 %
   2 = Prevailing: highest relative crown cover percentage
   3 = Important: relative crown cover percentage between 25 and 74 %
   4 = Minor: relative crown cover percentage < 25 %
   9 = Species disappeared since 5 years (cut or accident)
c) field assessment

F) Regeneration

Tree species

a) List of scientific names of 38 broadleaf trees and 22 coniferous trees and two additional classes (other exotic coniferous trees and other exotic broadleaf trees).
b) 60 tree species and two superior classes
c) field assessment

G) Forest condition

Nature of tree cuts (nature des coupes)

a) Tree cuts occurred on the 25 m radius plot. The cut is assessed only if it has affected the area including the plot-centre.
b) Nature of the cut: thinning, clear cut, accident (fire, storm, ...)
c) field assessment

H) Accessibility and harvesting

Ground ability for timber removal

a) Slope and soil quality on the plot
b) Two kinds of classes:
Classes of slope in % (0-15, 16-30, 31-70, > 70)
Soil type: load-bearing, uneven, wet, damp.
c) field assessment

I) Forest ecosystem

Unmeasured species over 1.50 m

a) An indicative list of species is provided, but some species can be added.
b) All species have to be recorded
c) field assessment

Unmeasured species under 1.50 m
a) An indicative list of species is provided, but some species can be added.
b) All species have to be recorded
c) field assessment

J) Non-wood goods and services

Forest function
a) Forest function
b) green spaces, natural preserves, protection forests, no-admittance areas
c) existing maps, photo-interpretation

K) Miscellaneous: size of the wooded land including the photo-plot

a) size of the wooded land including the photo-plot
b) Classes
- forest: included in a larger forested area covering at least 4 ha, and the average width of canopy being at least 25 m,
- small woods: include in smaller forested areas covering 50 acres to 4 ha, and the average width of canopy being at least 25 m,
- thickets: included in small forested areas covering 5 to 50 acres, and the average width of canopy being at least 15 m,

c) photo-interpretation

2.1.5 Forest area definition and definition of "Other Wooded Land"

The aerial photo-interpretation is made at the scale of a whole department. The department area is divided into land-use categories.

Wooded land consist of forest and "other wooded lands"

A) Forests

They are identified from aerial photos, from ocular estimates. They must have the following characteristics:
- either measured trees (diameter > 7.5 cm) have a crown cover percentage reaching at least 10 % (ground projection of crowns)
- or there are more than 500 stems per ha that are viable trees (able to make a stand) : seedlings, plants or shoots, vigorous, well shaped and regularly distributed.
- These characteristics, identified by photo-interpretation, are then checked up on the field.
- cover at least 5 acres, the average width of canopy being at least 15 m,

Among orchards, only Castanea sativa are classified as Production forest. Others are considered as cultivation.

Among forests, only production forests are sampled on the field. A forest is a production forest if its main function neither protection nor recreation.

Production forests are divided into :
- forest: larger forest area covering at least 4 ha, and the average width of canopy being at least 25 m,

- small woods: small forested areas covering 50 acres to 4 ha, and the average width of canopy being at least 25 m,
- thickets: smaller forested areas covering 5 to 50 acres, and the average width of canopy being at least 15 m,
- and all wooded areas with an average width of canopy between 15 and 25 m (without any maximum area threshold).

B) Other wooded lands

They are defined according to the same criteria as production forests. The only difference is that their main function is not production. They are not sampled on the field. They mainly consist in :

- unmanaged forests: inaccessible forest or forest located on too steep slopes
- protective forests (where cuttings are forbidden) and recreation forest.
- green spaces
- no-admittance areas (military grounds for example).

C) Other land cover types described by the survey:

1) Poplar plantings
   - planted poplars regularly spaced
   - poplars are the only species or are clearly prevailing
   - density of stems over 100 per hectare (more than 50 live stems per hectare)
   - minimum size: 5 acres and average width of canopy: 15 metres.

2) Hedges
   - wooded lines , the average maximum canopy width being 15 m
   - minimum length: 25 m
   - including at least three forest species trees with DBH over 7.5 cm,
   - the minimum average density of trees with DBH over 7.5 cm being one tree per 10 m.

3) Tree lines
   - Tree line consisting in forest species regularly spaced
   - the average maximum canopy width being 15 m
   - minimum length: 25 m
   - including at least three forest species trees with DBH over 7.5 cm,
   - the minimum average density of trees with DBH over 7.5 cm being one tree per 25 m.

4) Sparse trees

   forest species trees that are either isolated or forming groves under 5 acres area in “moors”, the crown coverage of these trees must be lower than 10 %.

2) Moors

   This category covers: moors, waste lands, empty lands that are not cultivated and not regularly kept by grazing.

   A “moor “ may contain trees as long as their crown cover percentage or their density does not reach the values for “wooded lands”.

3) Cultivation and non-productive lands (roads, rivers, ...).

2.2 DATA SOURCES

A) Field data

Field samplings are described in § 2.3

B) Questionnaire

No standardised questionnaire

C) Aerial photography

- Photographs are taken especially for the NFS, so that they have the required characteristics.
- The average scale is 1:17000, photographs can be more accurate in more difficult region (mountainous, diversified forests - scale 1:15000) or less accurate in "easier" regions (scale 1:20000).
- Two types of photos are taken at the same time: panchromatic and infrared (either black and white infrared or colour infrared in regions where the photo-interpretation of vegetation is very difficult, such as Mediterranean regions).
- Pictures are shot between June and mid-September (optimum conditions: absence of clouds, reduction of shadows.
- The photos cover an entire department
- They are interpreted with the help of a stereoscope
- 1,200 to 2,000 photographs are used for a department-wise assessment.
- The cost of aerial photographs is 4 million francs per year (for 8 to 9 departments).

D) Spaceborne or Airborne digital remote sensing

Such data is not used. All attempts made with them have been considered as providing less accurate results than aerial photographs.

E) Maps

Topographic map

- *Institut Géographique National* (IGN°: 107, rue La Boetie - 75 008 PARIS (Tel: 1 - 42 56 06 68)
- latest update level: from about 1975 to 1995
- scale: 1:25000 (also 1:100 000, 1:50000)
- printed maps, also exist as digital maps in ARCINFO format (DEM spatial resolution: 75 m)

Geological maps

- *Bureau de Recherche Géologique et Minière* - Direction du service Géologique et des laboratoires - BP 818 - 45 000 ORLEANS LA SOURCE
- date: depending on areas
- scale: 1:50 000

F) Other geo-referenced data

Not relevant at the moment.

2.3 ASSESSMENT TECHNIQUES

2.3.1 Sampling frame

The entire country is covered by the assessment. Production forests, tree lines, hedges, poplar plantings, sparse trees are sampled on photographs and on the field.

Other forests (non-production forests) are identified on aerial photographs but not sampled on the field (no description of species, no measurements for volume calculations).

Cultivation and non-productive areas (roads, towns, ...) are not sampled on the field.

2.3.2 Sampling units

A) Field assessment:

- Type of sampling unit: concentric circle plots
- Size and shape of sampling units:

Measurements / assessments	Shape	Size of sample plot (acres)
Live stems: CBH = 24.5 to 72.4 cm [1]	circle: 6 m radius	1.13 acres
Live stems: CBH = 72.5 to 120.4 cm	circle: 9 m radius	2.54 acres
Live stems: CBH = 120.5 cm and over	circle: 15 m radius	7.07 acres
Dead trees, felled trees (in the past 5 years): CBH = 24.5 cm and over	circle: 15 m radius	7.07 acres
Forest composition (crown cover percentage by species)	circle: 25 m radius [2]	20 acres
Forest structure	The shape of this sample plot is not defined but it must, as far as possible, include the 25 m sample plot and have rectilinear limits.	1 ha
Forest Ecosystem	circle: 6 m radius	1.13 acres
Regeneration: randomly distributed seedlings	nine plots 2.26 m radius forming a square around plot centre	total: 200 m^2
Regeneration: stripes distributed plants or seedlings	four rectangular plots, perpendicular to planting lines: 4 meters wide and 16 meters spaced	total: 400 m^2

[1] The CBH is measured for all trees beyond the CBH threshold in that circle. If the total number of trees in that circle is N, and if N>15, only one tree out of n is thoroughly measured The value of n depend on N as follows:

N	n
15	1 (all trees are measured)
16 to 25	2
26 to 35	3
36	4
etc ...	

The measured trees are chosen according to a precise procedure: all trees are numbered according to a pre-established order. For the calculation of total values, the values of the attributes from the measured trees are assigned to the next n-1 unmeasured trees.

[2] this plot should be centred on the sample point. If this is impossible, the 20 acres area must still include the sample plots used for wood production attributes measurements. The assessed criteria must be homogeneous on the sample plot, as far as possible.

- Method for slope correction: The distance between the plot-centre and a tree is measured horizontally. The slope is measured on the ground line between a tree and the plot-centre. A table is used to provide the maximum distance on this line that allows the tree to be part of the circle with the radius under concern.
- There are no permanent plots on the field. (In the 1980s, some permanent plots were laid out, but there were always some temporary plots because French forest is changing. Permanent plots have been abandoned now for some methodological and some practical reasons).
- Spatial distribution of sampling units: sampling units are systematically distributed (except for the forest composition sometimes) around the sample point. The sample point is located from an aerial photo-plot.
- System of plot-centre location in the field: an easily locatable starting point (D) is identified on the photograph (cross-roads, isolated tree, ...), located as near as possible the photo-sample point (A). Azimuth and distance between A and D are calculated from the photo (according to the local scale of the photograph). The distance is measured on the photograph with a "parallax ruler" (*règle de parallaxe*). The accuracy is 0.25 mm. A mark is set on the field at the location of the point A (plot-centre). Then the location of A on the field checked by comparison with the photo-plot with a stereoscope.

B) Aerial photography sample plots :

- Type of sampling unit: a fixed area plot
- a circle: 25 m radius (on the field) sample plot.
- For the survey cycle n, the department is covered with photographs on which new photo-plots are photo-interpreted and checked on the field. Besides, the second step photo-plots of the survey cycle n-1 are interpreted again on these new photos, to make a survey comparison.

- Spatial distribution of sample plots: Sample plots are systematically distributed according to a grid.

2.3.3 Sampling designs

The sampling design is a three-steps double-sampling for stratification.

1.First step: photo-interpretation

The preliminary phase is a photo-interpretation aiming at identifying the studied domains (*domaines d'études*). Studied domains are the domains for which results are to be obtained. They combine three criteria :

1. the land cover type (divided into: production wooded lands, other wooded lands, poplar plantings, moors, waste lands, agriculture, non-productive lands - see definitions above),
2. the localisation (the forest region, that is an ecological criteria)
3. the ownership category (private, state, community forest).

From this interpretation, the limits of land cover types are mapped, on photographs and on paper maps. "Mapped domains" (*domaines cartographiés*) are obtained. They are made of the studied domain, of other wooded lands and also contain other elements (such as roads, rivers, ...). In fact, not all mapped domains are sampled further. For example, a domain must cover at least 3 000 to 5 000 ha in a department to be a studied domain (domains sampled on the field).

A studied domain is a population that is inventoried independently of the other studied domains. Studied domains are divided into strata. A strata is homogenous regarding the parameter to be assessed. Strata assignment is made by photo-interpretation of plots systematically distributed on a transparent grid. Practically, theoretical nadir of photos are located on a map (scale 1:100000) and then located on photographs. The transparent grid is placed upon the photograph parallel to the nadir lines axes. The average density of photo-plots is 1 point for every 30 to 40 ha, depending on the scale of the photograph. If a studied domain is very little represented in the department (< 10,000 ha), the density of plots can be raised to 1 point for every 8 to 13 ha. This is only achieved for public forests or community forest submitted to forest "regime" (*forêts soumises au régime forestier*) (state management).

Strata are defined according to:

- the vegetation type (forest management type and prevailing species),
- the estimated volume per ha estimated in a few rough classes. (The partition in volume classes aims at creating homogeneous strata. Therefore, the limits of volume rough classes must be adapted for each studied domain and each department.)
- the local composition of stands (in the 25 m radius photo-plot)
- the size of the forest
- occasionally site (*aspect*) (site is a particular ecological condition)

- the function (production or other functions: protection, natural preserves, green spaces, no-admittance areas).

These parameters are identified on systematically distributed 25 m diameter photo-plots. The number of plots with the same values for all parameters are calculated. When multiplied by the extension factor of a plot, they provide an estimation of the size of the strata.

For inventory results comparisons, some photo plots are “permanent plots”: they are interpreted in two consecutive inventory cycles.

2. Second step: field checking of some of the photo-interpreted plots (see also §2.1.5).

The second step is the checking of photo-interpretation results on the field. The plots to be checked are chosen randomly among the first plots, they belong to the various strata from the various studied domains: forest (production and non-production forests), non-wooded lands. Plots in a strata are chosen independently from the plots in other strata. Two kinds of methods are used to chose the plots: either the plots are chosen with the same probability, or the probability is weighted by the represented area per sample point.

After field checking, a plot that proved not properly classified is moved to its proper category.

The total number of plots for the 2nd and 3rd step is chosen according to the homogeneity of the strata (which is known or assumed), according to the estimated economic value of the strata (and also according to the available financial means for the sampling!).

3.Third step: Field measurements

Field sample plots are chosen randomly among the plots from the second phase, in the various strata of the various studied domains. The number of field sample plots is defined on economical basis (given budget). This number of plots is distributed into the various strata according to the economical value of the forests, the size of each strata and to the admitted value for the variance. The required accuracy level and the economical value of stands are also taken into account in this distribution of plots.

2.3.4 Techniques and methods for combination of data sources

The size of the mapped wooded lands and poplar plantings, tree lines, hedges can be measured on maps.

The strata are defined from the photo-interpretation. The size of the strata are estimated from the proportion of photo-plots in the strata, after the 2nd step checking. This proportion is used as a weighting factor for the size of the mapped domain. The values of the attributes assessed on field-plots are weighted with the strata sizes derived from the photo-interpretation to get total values.

2.3.5 Sampling fraction

As mentioned above, the sampling fraction may vary from one department to another, and from one strata to another. Therefore, only mean values can be presented here (percentages of the whole wooded area).

Data source and sampling unit	Proportion of forested area covered by sample(estimated mean value)	Represented forested area per sampling unit(estimated mean value)
aerial photography, fixed area sample plots	0.018 to 0.02 (1.8 to 2 %)	8 to 11 ha
field assessment, fixed area sample plots	0.0007 (0.07 %)	100 ha

2.3.6 Temporal aspects

The French survey is continuous: departments are sampled one after the other and data are regularly published for each department. The time interval between two surveys is 10 to 15 years. The coverage of the whole country is not obtained at the same date.

Inventory cycle	Time period of data assessment	Publication of results	Time period between assessments	Reference date
1st NFS	1962-1980	According to department		According to department
2nd NFS	1977-1993	According to department	about 10 years	According to department
3rd NFS (current inventory)	1987 ...	According to department	about 10 years	According to department

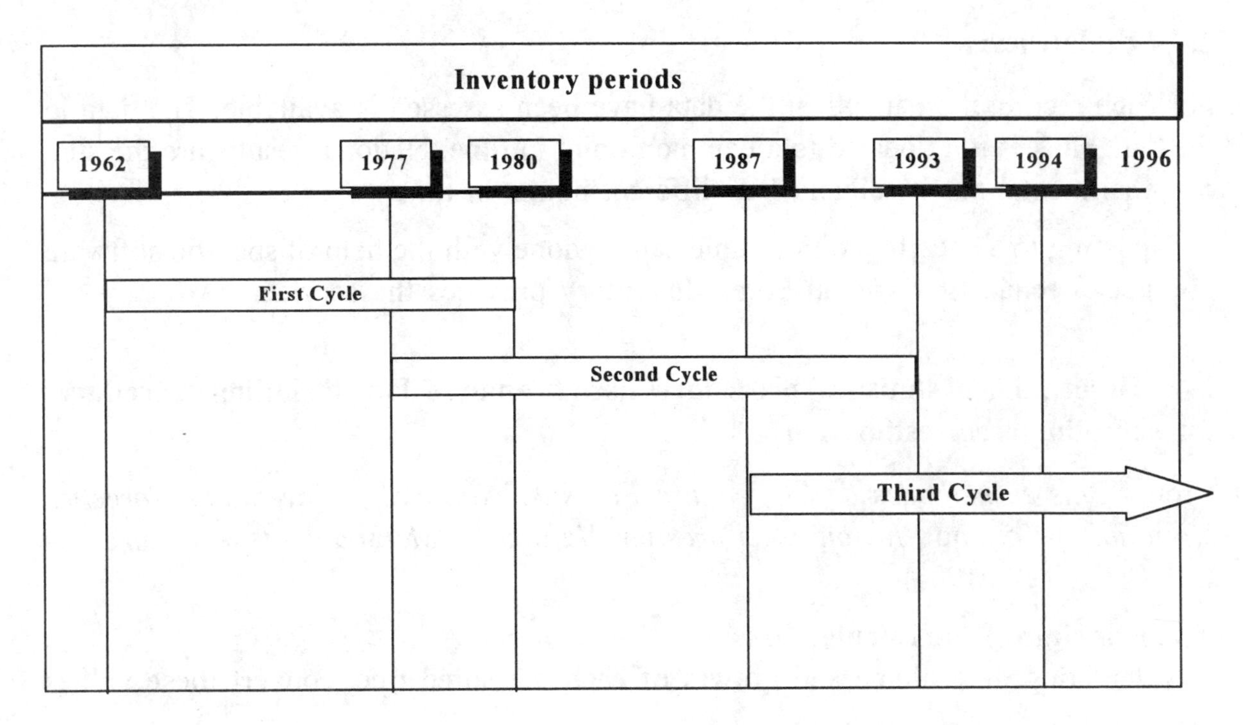

2.3.7 Data capturing techniques in the field

Data are recorded on tally sheet and edited by hand into the computer.

2.4 DATA STORAGE AND ANALYSIS

2.4.1 Data storage and ownership

The data of the assessment are stored in a data base system at the National Forest survey service located in Nancy. The person responsible for the data is the director of the service in Nancy: Mrs Carlin - Inventaire Forestier National - 14, rue Girardet - 54 000 NANCY (Tel: 83 35 02 97 - Fax: 83 32 46 99). A partial copy of the data base is located in Montpellier at the "Calculation National University Center" (Center National Universitaire Sud de Calcul- C.N.U.S.C.).

2.4.2 Database system used

The data are stored in a SQL database.

2.4.3 Data bank design (if applicable)

The database consists of:

1. raw data:
   - the general description of the field sample plots,
   - the global analysis of each forest species on each sample plot,
   - the field measurements achieved on the tree from the sample plots,
   - (about 200,000 sample plots and 1.5 million trees are described in this database).
2. documentary data: the history of procedure and codification used throughout the years (since 1962).

2.4.4 Update level

For each record, the date when the data have been assessed is available. The data in the data bank is not updated to a common point in time. National results are the sum of departmental results obtained at different points in time.

Updating to a specific point in time can be done with the help of specific software on special requests. National Forest Inventory provides the service.

2.4.5 Description of statistical procedures used to analyse data, including procedures for sampling error estimation

(from "*Buts et méthodes de l'Inventaire Forestier National" - Inventaire Forestier National,* 1985 and "*Inventaire Forestier National - Méthodes et procédures" -* R.B. CHEVROU, 1985)

The principle of the calculation is:

- calculating the volume and growth of each measured tree, convert these values per ha; the numbers of trees are also calculated per ha

- these average values are then multiplied by the "extension factor" of the sample plot.

General procedure for error estimation:

Strictly, the method only enables assessments for the surfaces of the "studied domains" and of the strata, and for the volumes and growth in theses areas.

When sample plots are numerous enough in a strata, the error is estimated as the variance of the data. The fact that some sample plots are placed into another strata can be taken into account in this calculation:

$$Var(Yj) = \frac{D^2}{\bar{b}^2} * \left[\sum (Pi * \bar{bi} * \bar{yij})^2 * \frac{pij(1-pij)}{n} + \sum (Pi * pij * \bar{yij})^2 * \frac{S^2bi}{N} + \sum (Pi * pij * \bar{bi})^2 * \frac{S^2yij}{nij} + \frac{1}{N} * \left\{ \sum_i Pi * (pij * \bar{bi} * \bar{yij})^2 - \left(\sum Pi - pij * \bar{bi} * \bar{yij} \right)^2 \right\} \right]$$

Where :

D = total size of the area under concern (department or forest region)
yij = value per ha of the measured parameter on a sample plot in the strata i, belonging to the domain j (after field checking)
s^2yij = variance of yij (calculated from all the measured values yij)
N = total number of photo-plots of the area under concern
Ni = number of photo-plots of the strata i
ni = number of sample-plots randomly chosen among the Ni
nij = number of the ni plots actually belonging to the strata i after field checking

$$Pi = \frac{Ni}{N}$$

$$pij = \frac{nij}{ni}$$

The bi factors take into account the distribution of the nadirs of photos on an horizontal plan on the ground: B is the distance between two nadir along the flight line of the plane, I is the distance between two flight lines.

$\bar{bi}$ = mean value of B*I for the strata i

$\bar{b}$ = mean value of B*I for the area under concern

S^2*bi = variance of B*I

NB: In the latest inventories, a new method was adjusted: the nadirs are regularly distributed, their location on photos being calculated from their location on maps. Thus the B and I values are constant. The weighting with B and I in the above equation becomes more simple.

Generally, relative error is published only for areas, volumes, and growth for all the forests in the department and per ownership category.

For other data, error estimation can only be approximate.

a) Area estimation:

The size S of "mapped domains" is measured on maps (scale 1:100000) after drawing their limits from photo-interpretation. It is either measured on paper maps or calculated from numeric maps, if they exist. It is assumed that this size is perfectly known. This can only be accepted if the limits drawn on photos take into account the parallax. Otherwise, the error may reach several percents.

The size of strata is assessed from the proportion of photo-plots in the strata.

It is calculated as follows:

$$Si = \frac{SNi}{N}$$

where :

Si = size of the strata i
S = size of the mapped domain
N = total number of photo plots in the mapped domain
Ni = number of photo plots in the strata

The relative error in the estimation of $\frac{Ni}{N}$ is

$$Er = \frac{1}{\sqrt{Ni}}$$

After the 2nd step (photo-plots checking), the size of the strata can be corrected as follows :

$$Si = \sum_{j} \left(\frac{SNj}{N}\right)\left(\frac{nji}{nj}\right)$$

where :

nj = number of field sample plots in the strata j
nji = number of field sample plots that have been moved from the j strata to the i strata;

b) Aggregation of tree and plot data

Tree data are transformed into data per ha by weighing them according to the size of the plot.

Circumference class	weighting factor
CBH = 24.5 to 72.4 cm	88.42
CBH = 72.5 to 120.4 cm	39.29
CBH =≥120.5 cm	14.15

Volume, gross growth, number of trees per ha are obtained by multiplying tree data by these factors.

After the 3rd phase, the total volume for the strata i (Vi) is calculated as follows :

$$Vi = \sum_{j} \left(\frac{SNj}{N}\right)\left(\frac{nji}{nj}\right) vji$$

where :

vji = mean volume per ha calculated on the nji plots.
nj = number of field sample plots in the strata j
nji = number of field sample plots that have been moved from the j strata to the i strata;

c) Estimation of total values

The total volume of the strata is calculated as the multiplication of the average volume per hectare calculated on the sample plots times the size of the strata.

The precision of these part results *(résultats partiels)* can only be calculated approximately. The calculation consists in calculating the variances of these results from the variance of the inputs weighted by the proportion of the part sizes or volumes.

Total values are the sums of previous values.

d) Estimation of ratios

The total volume per hectare is the mean of ratios calculated on the sample plots.

e) Sampling at the forest edge

A special procedure is used when the sampling plot is crossed by the limit of the forest, of the studied domain, of the strata, or of the type of vegetation (described on the 20 acres sample plot). The centre of the plot is moved so that the plot is completely included into the strata, the studied domain, or the forest. The circles are moved as little as possible, to be tangent to the limit. Therefore, the center of the 3 concentric circles are different. (see figure 5 below).

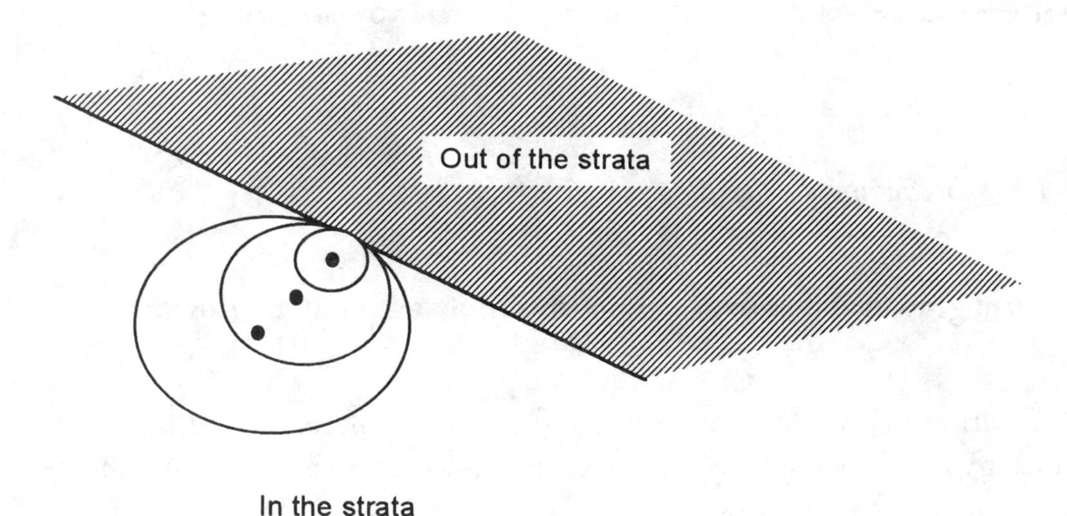

Figure 5. Sampling at forest edge

This method creates a bias *(biais)* that is known as being usually small. It is due to the fact that the trees at the limit are slightly different from the trees inside the strata, the forest, ...

The distance between the tree and the centre of the plot is measured. A tree belongs to the plot if the circle of the limit of the plot crosses the tree at 10 cm from the ground. The tree is taken into account (as part of the plot) if it is located on the West side of the circle. it is not taken into account if it is on the East side of the circle. This creates a bias by replacing the circle by two half-circles and the total diameter of the circle is increased by the radius of the tree at 10 cm. The bias is positive (overestimation) but it is neglected.

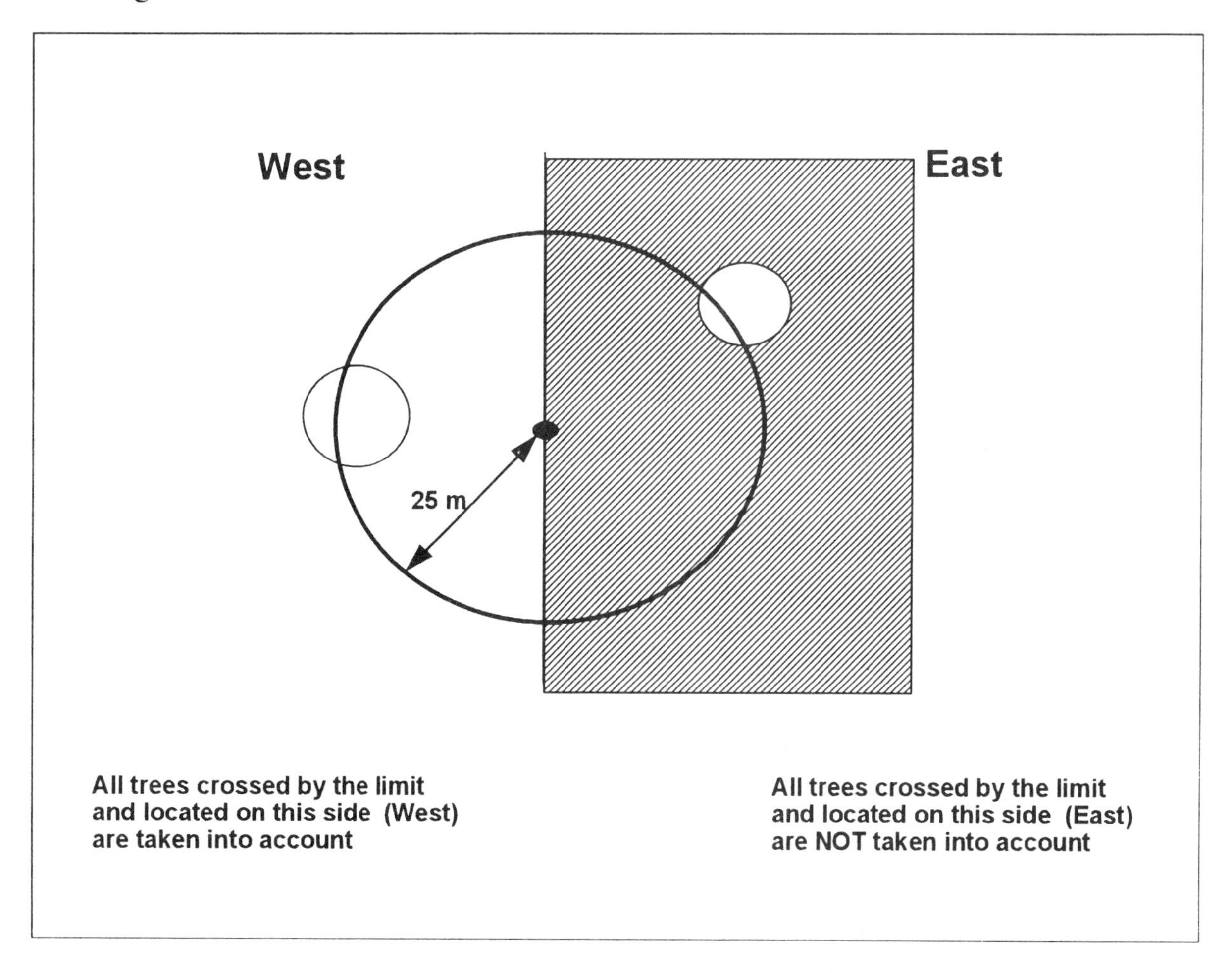

Figure 6. Trees belonging to the plot.

f) Estimation of growth and growth components (mortality, cut, in-growth)

Mortality:

Trees felled during the past 5 years are measured (circumference at ground level: Co). The mean volume, per species, per year, that has been removed during the past five years is calculated as follows:

$$Vc = \frac{1}{5} * V * \frac{s}{S} * K'$$

where:

s = sum of the squares of Co of dead trees of the species measured on the plot
S = sum of the squares of Co of live trees of the same species measured on the plot
V = total volume of live trees of the species of the plot
K' = weighting factor between the area where felled trees are measured (a 15 m radius plot) and the extension factor of the plot. (see § 2.3.2)

If no live trees of the same species have been measured on the plot, the volume of felled trees is calculated with a model (volume=f(Co)) calculated from all live trees from this species in the whole department.

The volume of dead trees and fallen trees is calculated with the same method.

Growth from trees that died, fell or were felled during the past 5 years:

It is assumed that they have been providing some growth in the stand during one half of the five years period. This growth is calculated as:

$$dVc = \frac{1}{2} * \frac{1}{5} * dV * \frac{s}{S} * K'$$

where:

dV is the growth of live trees from the same species measured on the plot
K' is a weighting factor between the area where felled trees are measured (a 15 m radius plot) and the extension factor of the plot.

<u>In-growth :</u>

Measured trees that were not measured five years ago (CBH < 24.5 cm five years ago) are identified from the calculation when: DBH-2Ir5 is inferior to 7.5 cm.

It is assumed that they reached the CBH limit 2½ years ago. Therefore, only half of the growth they produced during the last five years must be taken into account.

If V is the present volume of these trees, dV their growth, the in-growth (in volume) is:

$$R = V - \frac{5\ dV}{2}$$

the in-growth per year is:

$$r = \frac{V}{5} - \frac{dV}{2}$$

This in-growth per year is added to the annual growth (*accroissement courant annuel*) calculated for the other trees on the plot (with DBH that was already equal to the DBH limit five years ago): the result is gross production (*production brute*).

g) aggregation of stand and area related data to single sample plots/sample points

The values estimated or assessed on the sample plot are averaged at plot level. Only one value is assigned to the plot: the mean value of the attributes on the plot. (Each class of stand or area related data are not assigned to a proportion of the plot area.)

h) Hierarchy of analysis: how are sub-units treated?

The method provides results per strata. A strata is defined from the forest ownership category, the forest region and the vegetation type. Data are also produced for sub-units: results by ownership category, by vegetation type, by species or groups of species. Results for a sub-unit are the mean of the values from all the plots belonging to this sub-unit.

The error for the results presented by sub-units is calculated as follows:

Relative error for the size of the sub-unit is proportional to $\frac{1}{\sqrt{N}}$, where N is the number of plots.

N (the number of plots) depends on the size of the sub-unit, therefore:

- if S is the total size of the strata and e its relative error
- if S1 is the size of the sub-unit and e1 its relative error, e1 can be calculated as:

$$\frac{e1}{e} = \sqrt{\frac{S}{S1}}$$

In fact, the error made by NFS is inferior to this calculated value for the most interesting vegetation types because more sample plots are laid out in them. On the contrary, the error is greater than this value in less interesting vegetation types.

The error for volumes is calculated in the same way:

$$\frac{e1}{e} = \sqrt{\frac{V}{V1}}$$

2.4.6 Software applied

Calculation of the results are made with especially developed WINDOWS software.

The software used to extract data from the database is “QMF”. It enables the extraction and the presentation of results in a tabular form.

From basic analysis of forest resource in a precise place with QMF, complementary software allow:

- the extraction of the sample plots or sampled trees included in the considered place
- the definition of a silviculture scenario
- the edition of resulting assessments

These software are:
TARIF: to estimate tariff functions
SIMUL: to simulate the evolution of regular stands, by age classes
DISFO: to estimate wood resource at mid-term

Maps are created with ARC INFO. About 10 software products have been developed especially for the calculation and presentation of the results in map format.

2.4.7 Hardware applied

(Information provided by Mr WOLSACK - National Forest Survey)

Photo-interpretation results are edited into PC computers, as well as field data.

Maps are created on SUN Stations.

2.4.8 Availability of data (raw and aggregated data)

a) raw data and derived attributes can be made available for outside persons. They have to pay for it.

c) data compiled due to special requests:

The Resource evaluation Cell *(Cellule d'évaluation de la Ressource-CER)* can achieve specific studies. It works as a private consulting firm.

This service also adjust the new software to users needs and teaches the use of software and the statistics methods of NFS.

2.4.9 Sub-units (strata) available

The contents of the database are basically tree data, subunit by species data and plot data, from which forexample the following items may be deduced:

1) For each forest region: size, volume, growth

- by ownership category
- by vegetation type
- by species or groups of species

2) Distribution of one species in the department: global results for volume, growth, in-growth, area covered.

3) Global results for a department

- by ownership category
- by vegetation type
- by species or groups of species

The term subunit refers to a population of trees distinguished within the plot by their origin (high forest or coppice), state (alive, dead, cut, broken..),dimension class and storey.

2.4.10 Links to other information sources

The maps used for results publication are the topographic maps from National Geographic Institute (IGN).

2.5 RELIABILITY OF DATA

2.5.1 Check assessment

A) Photo-interpretation

1 % to 5 % of plots are checked. More checking are made when operators are less experienced and less when they have more experience. Experience is evaluated from the results of previous checking.

Photo-interpretation is checked by making two different persons achieve photo-interpretation on the same area.

The results of photo-interpretation are also checked on the field. The proportion of photo-plots that are checked is about 13 %. (see also §2.3.3)

B) Field plots

A technical supervisor makes sure that all teams are following exactly the same procedures. He also decides to have some field plots measured twice, either by the same team (time interval: three to four weeks) or by another team (the time interval depends on the opportunity for the team to be the neighbourhood of the plot to be checked).

The controls are made for measurements and for localisation of sampling plots. If differences appear, they try to find out where the problem comes from.

If an important mistake is noticed, the first value is replaced by the correct value.

The values of the two measurements are compared: an assessment of the error can thus be provided on special request, but it is not usually taken into account in the publication of results.

2.5.2 Error budget

Only sampling error is assessed. Other errors (models, measurements, ...) are not assessed.

2.5.3 Procedures for consistency check of data

Field measurements are controlled on the tally sheet: a person looks for omissions and mistakes, and checks up consistency between measurements and the description of the plot.

The data is edited into the computer twice. They are checked by automatic processes. A last control is made when the database for the public is created.

2.6 MODELS

2.6.1 Volume (functions, tariffs)

a) Outline of the model

Two methods are used to calculate volumes:

1. Volume is calculated using the values of attributes measured on the field:

Total volume is the sum of the volume of basal log and the volume of upper log(s). (See also figure 1).

Volume of basal log Vb is calculated as follows:

- in two parts for trees with Hd > 2.60 m (Hd: height of the 7 cm diameter or height at the point where the tree shape changes).

  the volume of the first 2.60 m log is calculated like the volume of a cylinder:

  $$V1 = \frac{CBH^2}{4 * \Pi} * 2.60$$

  the volume of the second log is calculated with Newton equation:

  $$V2 = \Pi * [(D2.6)^2 + 4 * (Dm)^2 + (Dd)^2 * (Hd - 2.60)] / 24$$

  Then Vb = V1 + V2

- For trees with Hd < 2.60:

  $$Vb = \Pi * (d1/2)^2 * Hd / 4$$

  Where :
  Hd = height of the 7 cm diameter
  D½ = diameter at height $\frac{Hd}{2}$

  Volume of upper log(s) Vb is calculated like the volume of a cylinder

  $$Vs = (\Pi * d^2 * L) / 4$$

  where :
  L = length of upper log
  d = medium diameter of the upper log

  Total volume is V = Vb + Vs

2. From these results, volume tariffs are derived, when enough measurements are available (at least 50 trees distributed in the diameter classes and the height classes to be used in the tariff). Different volume functions are calculated for specific species groups, vegetation regions, and social position of trees in forest structures (for example dominated or prevailing stems in timber vegetation stand).

This tariff is sued in the next inventory cycle to calculate the volume of a single tree: then, all trees are not thoroughly measured. Only input parameters to the tariff are measured: circumference at breast height and total height of the tree and metric decrease.

Tariffs are also derived for an entire stand. For a stand, the input is the prevailing height *(hauteur dominante),* defined as the average height of the highest trees in the stand.

b) Overview of error prediction

Thorough field measurements for a single tree provide a volume with a very small variance. Volume are calculated for trees with CBH > 24.5 cm.

Tariffs functions are calculated from previous inventories. The input data to use the tariff are the circumference at 10 cm from ground level. and the total height.

The error due to the tariff is neglected in common presentation of the results.

d) Data material for derivation of the model

Tariff functions are calculated for a studied domain or for a group of studied domain. Tariffs are calculated by species and by studied domain every time it is possible.

(from "*Buts et méthodes de l'Inventaire Forestier National*" - *Inventaire Forestier National,* 1985)

2.6.2 Assortments

Quality assessments are made on the filed.

2.6.3 Growth components

Annual volume increase

The inputs are: the volume, bark thickness, circumference at breast height, radial increase (IR), height, height increase. The formula which is used (proposed by L. BRENAC) has not been studied to know its reliability but "it is commonly assumed in France being the best one" (in "*Inventaire Forestier National - Méthodes et procédures*" - R.B. CHEVROU, 1985).

2.6.4 Potential yield

-

2.6.5 Forest functions

Not relevant

2.6.6 Other models applied

Models are applied to calculate in-growth and mortality (see also §2.4.5).

A model is applied to calculate the volume increment per year and the average volume increment for planting (calculated as volume/age of the planting) .

A model has been developed to find a relationship between the DBH and the CBH. in the past, DBH was measured. Now, CBH is measured. The CBH classes correspond to the previous DBH classes according to this model.

2.7 INVENTORY REPORTS

2.7.1 List of published reports and media for dissemination of inventory results

Interactive access to data base : the data base at CNUSC can be accessed through Transpac, Réseau téléphonique commuté, Minitel, EARN (European Academic Research Network).

The data are available for department.

- Printed reports consist of two volumes: the first one presents general forestry data of the department, the second one consists in tables regarding wooded lands.

Inventory	Year of publication	Citation	Language	Dissemi nation
First French NFI	between 1962 and 1980	Department of ... - Results from the first forest inventory (19...-19...) (*Département de ... - Résultats du premier inventaire forestier (19...-19...)- Ministère de l'Agriculture et de la pêche- Direction de l'espace rural et de la Forêt- Inventaire Forestier National.*)	French	printed
Second French NFI	between 1977 and 1993	Department of ... - Results from the second forest inventory (19... - 19...) (*Département de ... - Résultats du premier inventaire forestier*)	French	printed
Third French NFI	between 1987 and 199..	Department of ... - Results from the third forest inventory (19... - 19...) (*Département de ... - Résultats du premier inventaire forestier*)	French	printed

- forest atlas: four volumes, by administrative region and by department including results from the survey and maps of wooded lands.
- forest maps can be obtained from the NFS (Montpellier Service)

Type	Scale	Date
forest regions (ecosystems)	- 1:1000000 for the whole country	- 1994
	- 1:500000 for the 22 administrative regions	- 1994
	- 1:500000 for each department	- 1994 for the departments recently sampled
Forest maps	- 1:200000 for each department - and 1:25000 for each department	the date depends on the date of the latest survey in the department

- the numerical maps can be obtained from the NFS
- they are available to anybody who is ready to pay for them (international/ charge)
- theme of data: forest maps (vegetation types)
- spatial resolution: three resolutions for raster maps: 25, 50 and 100 metres,
- area covered: in 1996, each department of France has been covered by a numerical map.
- date: the date is the date of the latest sample (survey) in the department
- format: for raster maps: EPPL7, or ERDAS
  vector maps: ARC/INFO format.

2.7.2 List of contents of latest report and update level

After field assessment, results are published by department in three volumes. No publication is available at national level. Therefore, the number of pages can not be indicated here in the list of contents of publication. they vary from one department to another.

Volume 1:

I- General presentation of department
Description of forest regions and of vegetation types
II - Results of the inventory: size, volume, growth
III - Analysis of results
Present state of the forest
Comparison with past inventories

Volume 2: Tables

Tables A - Number of trees, volume, growth, total height, presented in the following strata:

- vegetation type
- owner ship
- forest region
- species
- diameter classes

Tables B: relative to coppice: structure and age of coppice stands. Size, volume, and growth are presented by prevailing species and age classes.

Tables C: relative to coniferous stands: structure and age classes of stands: size, volume, growth by prevailing species, forest region, age.

2.7.3 Users of the results

The following data are percentages of the total amount of money from data sale and services provided by the National Forest service in 1995. These information have been provide by the NFS.

Percentage of the various issues and services :

Reports	4%
Paper maps	4%
Forest atlas	1%
Aerial photographs	15%
Numerical maps (files)	9%
Wood production attributes	10%
Special requests:	
Special analysis	30%
Standing volume	13%
Remote sensing	11%
Teaching	2%
Miscellaneous	1%

Special requests by institutions :

National Forest Office	1%
Private forest organisations	14%
Research institutions	18%
Wood industry	4%
Consulting firms	19%
Environmental institutions	< 1 %
Communities, departments and regions	< 1 %
International organisations	38%
Miscellaneous	5%

2.8 FUTURE DEVELOPMENTS AND IMPROVEMENT PLANS

2.8.1 Next inventory period

The current cycle started in 1987 and is still going on.

2.8.2 Expected or planned changes

2.8.2.1 Nomenclature

-

2.8.2.2 Data sources

Tests will be made to get directly photo-interpretation results in numerical format.

2.8.2.3 Assessment techniques

-

2.8.2.4 Data storage and analysis

The main planned changes are the creation of GIS. This should enable the assessment of accessibility and harvesting. data should also be available at commune level.

Studies are on to create an ecological database to be added to the present forest database.

2.8.2.5 Reliability of data

-

2.8.2.6 Models

-

2.8.2.7 Inventory reports

-

2.8.2.8 Other forestry data

-

2.9 MISCELLANEOUS

-

COUNTRY REPORT FOR GERMANY

Christoph Kleinn
Matthias Dees
Heino Polley
Germany

Summary

2.1. Nomenclature

Altogether about 50 attributes were directly assessed and about 10 derived. Focus in terms of number of attributes assessed is in the groups wood production and forest structure. About two third of the attributes are on a ordinal or nominal scale, one third is on metric scale.

Forest according to NFI definition are areas of minimum size 1,000 m^2 which are stocked with woody plants. Exceptions to be included (like clearcut areas) or to be excluded (like christmas tree plantations) are defined. In general, there have been only few problems with respect to the forest/ non-forest decision.

2.2. Data sources

Principal data source for NFI I were field assessments. In the pre-clearing phase when the forest/non-forest decision was made tract-wise, other data sources were used, where available. These sources included aerial photography, forest management maps, cadastrial maps and documents and other documents.

2.3. Assessment techniques

NFI I refered to all forest in the Federal Republic of Germany in the territory before unification (11 federal states). The inventory was carried out between 1986 and 1989. The inventory used square tracts (clusters) which were systematically distributed over the country. Base grid was 4km x 4km in some regions a smaller grid size (2.83km or 2km) was adopted. Altogether

28,978 tracts were pre-cleared out of which 12,580 turned out to be forest tracts. On the tracts four different sampling units were assessed: Point sample (Bitterlich sample) and concentric circular plots at the tract corners to assess tree related information, and line intercept and line intersect sampling on the tract line to assess area information and data on road lengths. Tract corners are permanently marked to provide possibility for relocation during next inventory cycle.

Field assessment was up to the federal states and lead by the state inventory supervisor. The design had been defined by a committee of experts (*Gutachterkommission*) from several forest research institutions and BML.

2.4. Data storage and analysis

Ownership of data is with BML and the federal states. Data are generally available to researchers and other users, if no significant reason stands against the release. Main concern is that with the goegraphic coordinates a forest owner could be tracked, which possibly is against the principles of data security.

Data were stored and analysed in BFH, Hamburg. After reunification the BFH institute in Eberswalde is in charge of all NFI affairs. Data were analysed with custom-tailored Pascal programs, the data were organized in a Pascal file structure, all on VAX computers under VMS. Today the data are in BFH Eberswalde rearranged and stored in data base system Informix under Unix. Reference date for all results is October 1, 1987.

Data analysis was done looking at the cluster design as a two-stage sampling procedure. The procedure is described in detail in the inventory report (BML 1992a). Information is available for the following strata: Entire inventory region, federal states, systems of management ownership types, age classes, and species groups.

Federal states have their own data sets. These data are stored in dBase format and used on PCs. A regionalization program developed by FVA Baden-Württemberg allows the whole range of possible evaluations for states and subunits of states.

2.5. Reliability of data

Check assessments were required for 5% of tracts and carried out for about 10% of tracts. Check assessments were used for tracing of errors and for training purposes. Data were pre-checked by the state inventory supervisors and entered into computers through BEL Frankfurt. Subsequently plausibility checks were automatically carried out and the necessary corrections were done.

Sampling error is given in result reporting mainly for area and volume.

2.6. Models

Several models were used in data analysis. Volume was calculated with set of taper form functions with dbh, height and d_7 as input variables. The corresponding computer program BDAT is also used for assortment purposes and to calculate dbh if it hadd been measured in a height different from 1.3m. For model development data of about 36,000 trees were utilized, precision of the model is assumed very high, above all for the common species and height/diameter ranges.

Other models were used to estimate bark thickness, to calculate the area of ideal pure stands from mixed stands and to derive dbh if it was not possible to measure it in a height of in 1.3m.

Potential yield modelling was not part of the inventory exercise. But on the basis of the inventory data a yield prediction model is now available (*Holzaufkommensprognose*).

2.7. Inventory reports

Official publications were launched by BML (Federal Ministry of Agriculture): Two volumes for the whole of Germany and one volume for each state. Structure of all these reports is the same. For EFICS purposes the results for the entire inventory region are of highest interest:

BML 1992a: German National Forest Inventory, Volume I: Inventory report and summary tables (*Bundeswaldinventur, Band I: Inventurbericht und Übersichtstabellen*), 118 p.

BML 1992b: German National Forest Inventory, Volume II: General tables (*Bundeswaldinventur, Band II: Grundtabellen)* . 366 p.

There are many journal publications interpreting and analysing the NFI results. They are, however, not formal part of the NFI itself.

Users of the results are the forest administrations and the scientific community. No systematic assessment has been made as to what extent the NFI results were used or processed in whatever field.

2.8. Future development and improvement plans

NFI II is currently under discusssion but no decision has been made yet (April 1996). An expert committee *(Gutachterkommission*) is working on possible adaptation of the design, out of the experiences made during NFI I and out different needs.

Principal characteristics and changes: The same grid of sample locations is used. Line sampling is dropped. Radii of circular plots are changed. Several ecological attributes are added for assessment (like assessment of dead wood). Time consuming measurement of height and upper diameter is dropped and replaced by models. Minimum dbh in point samples will be 7cm (instead of 10 cm before). And - of course - in data analysis it will be possible to derive growth data. Required models will be developed.

Abbreviations used:

AFZ	Allgemeine Forst Zeitung (Journal)
BEF	Bundesamt für Ernährung und Forstwirtschaft (Federal Institute for Food and Forestry), FrankfurtNew name today: Federal Institute for Food and Agriculture (Bundesamt für Ernährung und Landwirtschaft)
BFH	Federal Research Centre for Forestry and Forest Products, Hamburg (Bundesforschungsanstalt für Forst- und Holzwirtschaft).
BML	Federal Ministry of Food, Agriculture and Forestry, Bonn (Bundesministerium für Ernährung, Landwirtschaft und Forsten).
BW	Federal state of Baden-Württemberg
BWI	Bundeswaldinventur = German National Forest Inventory
BY	Federal state of Bayern (Bavaria)
BZE	Survey of soil condition (Bodenzustandserhebung)
FVA	Forest Research Station Baden-Württemberg. (Forstliche Versuchs- und Forschungsanstalt Baden-Württemberg).
NFI	National Forest Inventory

Brand names used:

- Access
- DEC
- Excel
- IBM
- Informix
- SAS
- SUUNTO

1. OVERVIEW

1.1 GENERAL FORESTRY AND FOREST INVENTORY DATA

Key data on forests in Germany are given in Table 1.

Table 1. Key data of German forests (Sources as given in the header if not stated otherwise in the cells).

Attribute	Old states Source: NFI	New states Source: BML 1994	Total Source: Polley 1994
Total land area	249,000 km^2 (Source: Anonymous 1993)	108,000 km^2 (Source: Anonymous 1993)	357,000 km^2
Total forested area	7.75 million ha	2.98 million ha	10,7 mio. ha
Proportion of forest land	31%	28%	30%
Main tree species (reference area: managed forest area)	Conifers: 63% Deciduous:37% Spruce: 38% Pine: 18% Beech: 17% Oak: 10%	Conifers: 75% Deciduous:25% Pine: 52% Spruce: 20% Beech: 7% Oak: 6%	Conifers: 67% Deciduous:33% Spruce: 33% Pine: 28% Beech: 14% Oak: 9%
Average volume per hectare	302 m^3 over bark	212 m^3 over bark	277 m^3 over bark
Average increment per hectare	7.0 m^3/ha (Source: Kuusela 1995)	6.75 m^3/ha (Source: Kuusela 1995)	6.9 m^3/ha
Ownership proportions	Private: 46% State: 28% Communities:24% Federal: 2%	Preliminary data: Private: 26% State: 33% Communities:9% Federal: 10% Trust-Company (temporary): 23%	Private: 40% Non-priv.:60%
Number of inhabitants per hectare forested land	8.32 persons/ha forest or 0.12 ha/person	3.99 persons/ha forest or 0.25 ha/person	7.13 pers./ha forest or 0.14 ha/person
Average size of a stand	Minimum size 0.1ha according to NFI guidelines	-	

Germany is politically subdivided in federal states. They are refered to as 'states' in the following. This political subdivision has also some implications for forestry issues as forestry is state-wise organized. General information about total area and forest area of the states is given in Table 2 and Figure 1 and Figure 2.

Table 2. General information about the federal states of Germany. The five new states are bold framed; they were not covered by the German NFI. See also Figure 1 and Figure 2.

State	Abbreviation	Total area (km^2)	Forest area (1000 ha)	Forest cover (%)
Baden-Württemberg	BW	35 741	1 353	38
Bayern	BY	70 553	2 526	36
Berlin	BE	889	17	19
Brandenburg	BB	29 479	993	34
Hamburg	HH	755	4	5
Hessen	HE	21 114	870	41
Mecklenburg-Vorpommern	MV	23 415	532	23
Niedersachsen	NI	47 438	1 068	23
Nordrhein-Westfalen	NW	34 067	873	26
Rheinland-Pfalz	RP	19 847	812	41
Saarland	SL	2 568	90	35
Sachsen	SN	18 407	502	27
Sachsen-Anhalt	ST	20 444	424	21
Schleswig-Holstein	SH	15 727	155	10
Thüringen	TH	16 181	510	33
Total		356 625	10 741	30

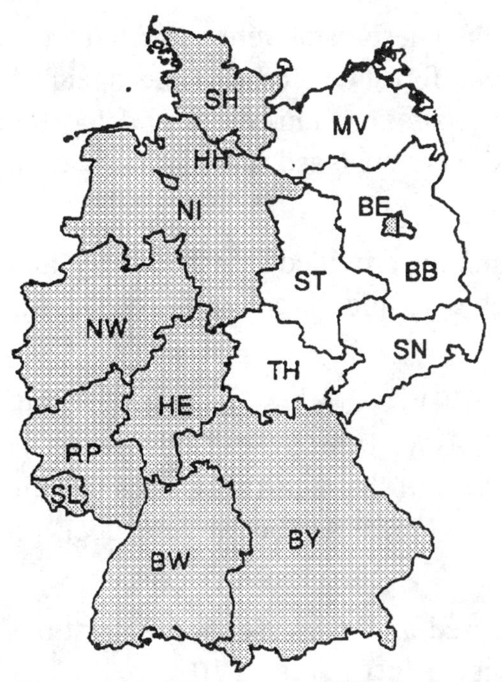

Figure 1. Germany. Shaded area: Old states (Acronyms see Table 2).

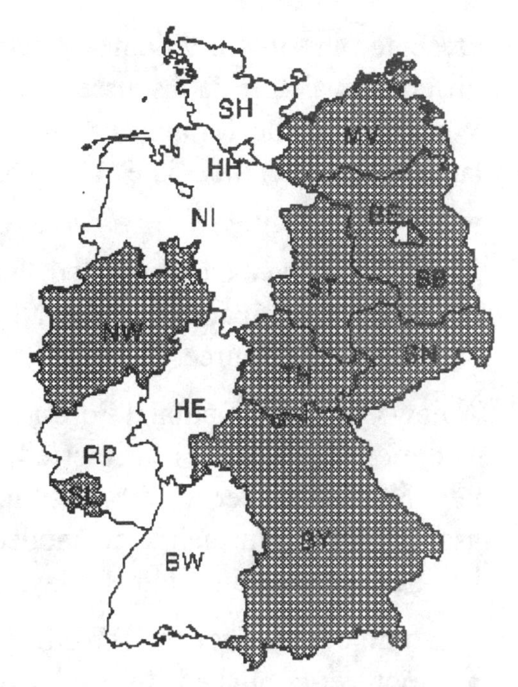

Figure 2. Germany. Shaded area: States where regional inventories had been carried out (Former German Democratic Republic 1961-1974, Bayern 1970, Saarland 1990) or are currently (1996) under discussion (Nordrhein-Westfalen). (Acronyms see Table 2).

National and regional assessments:

1. The first and most up-to-date National Forest Inventory (Bundeswaldinventur, BWI) in Germany was carried out between 1986 and 1990. Reference date is 1 October 1987. As this inventory was before re-unification of East and West Germany, the inventory and its results refer to the western states (Bundesländer)only.

   The NFI is regarded as the first and only statistically sound source of forest data of the whole of Germany at the stated reference date. It will form therefore the principal reference in the present paper.

   <u>Legal basis of the NFI:</u>

   On 27 June 1984, a new article was included in the Federal Forest Act (Act Governing the Preservation of Forest and the Promotion of Forestry) of the Federal Republic of Germany: §41 a of this act demanded the implementation of a forest inventory for the whole of the Federal Republic. The general principles of the inventory are defined here. The Ministry of Agriculture (BML) was commissioned to work out the corresponding legal regulation (Verordnung) and administrative guidelines (Allgemeine Verwaltungsvorschrift) which were published on 10 March 1986 and 28 May 1986, respectively.

2. For the five new states and the eastern part of Berlin, the Federal Ministry of Agriculture (Bundesministerium für Ernährung Landwirtschaft und Forsten)

presented in 1994 a volume of actual forest data. These data were extracted from different available data sources (mainly the general forest data base =Datenspeicher Waldfond of the former centralised forest administration, uniformly updated the last time to reference date 1 January 1993) and were analysed in a similar way to the analysis in the NFI.

It is, however clearly stated that the information provided is preliminary and not directly comparable to the NFI results. Table 1, middle column, gives some data from this source.

3. A new cycle of a National Forest Inventory (BWI II) is currently under discussion. Therefore the last NFI is referred to as NFI I or BWI I throughout this text. The new NFI will cover for the first time the entire area of the reunified Germany. The assessment is in principle demanded by the Federal Forest Act but no final decision has been met so far (state 1 May 1996).

   Some of ideas for the BWI II, that are discussed and expressed in publications and not-yet-published discussion papers, are listed in Chapter 2.10.

4. Forest Health Monitoring (Waldschadenserhebungen) is carried out every year according to UN/ECE regulations as described in the Manual on methodologies and criteria for harmonised sampling, assessment, monitoring and analysis of the effects of air pollution on forests (UN/ECE 1994, 3rd ed.). Federal states (Bundesländer) are responsible for the Forest Health Monitoring. The survey designs of all federal states meet the norm set by the ICP standards. Still in detail, all of the states differ in their sampling designs with respect to plot design, number and nomenclature of attributes, data storage and analysis. It is beyond the scope of this study to cover all of the 15 somewhat different variations, but since they all cover ICP- Standards, the general features will be outlined. This survey is conducted yearly, the last survey being in 1995.

5. Inventory of the Forests Soil condition (Bodenzustandserhebung im Wald (BZE): Federal states (Bundesländer) are responsible for the Implementation of BZE. The survey designs of the federal states generally meet the norm set by the ICP-standards which are defined in Commission Regulation (EEC) No 926/93 of 1 April 1993 and in International cooperative programme on assessment and monitoring of air pollution effects on forests. Manual on methods and criteria for harmonised sampling, assessment, monitoring and analysis of the effects of air pollution on forests (N.N., 1994).

   The surveys in the federal states differ to a small extent. There are two reasons for this:

   1. In Germany an inventory design, that sets standards for the federal states, was established in 1990. This standard is the basis for the 'Nation-wide inventory of the forests soil condition (bundesweite Bodenzustandserhebung im Wald). This inventory design was established before the ICP standards were fixed.

   2. Some federal states have designed their own BZE, before the nation-wide standard for Germany had been fixed (e.g. BW and BY). Still, design and

nomenclature do not differ widely. Since the European standard is well known, mainly the differences to this standard are given in the chapter on this survey.

The differences are presented in detail by a comparative study comprised at the BFH by Riek and Wolff (1995).

6. Forest Enterprise Survey (*Testbetriebsnetz*): This national inventory is conducted by BML. Sampling frame consist of all forest enterprises (Forstbetriebe) with at least 200 ha of forest area under management (Forstliche Betriebsfäche).

   The only actual data collected in this survey on natural resources is the attribute 'commercial volume of cut per species group'. Main interest of this survey is the economic situation of the forest enterprises in Germany and not the forest resource on which the focus of EFICS is. Most data on natural attributes are taken from the surveys of forest management planning. They date back up to 10 years and are often of minor data quality. This was one motivation to establish the National Forest Inventory BWI (see chapter 2) to get more precise data on these attributes.

7. Only a few regional forest inventories took place or are under preparation. In 1970/71 a regional forest inventory was carried out in the state of Bayern (Bavaria). No second cycle was implemented.

   In 1990 a regional inventory in the state of Saarland was carried out. A base grid of 1 km by 1 km was used, altogether 825 sample points were on forest area. The inventory used elements of the Swiss *Kontrollstichprobe* (sort of continuous forest inventory). Details can be found in Heupel (1994).

   In the federal state Nordrhein-Westfalen, the forest department now (1995/1996) prepares a state forest inventory (Landeswaldinventur LWI). The intention of this inventory is to provide information on forest at a state level down to forest districts (Forstämter), i.e. down to a unit of reference size of 10.000 ha forest area. It is designed as a multi-source inventory. It is planned to include all possible and relevant sources like digital elevation models, digital maps of protected areas and others. Field assessment is also included.

   The inventory is still in the planning phase. A pilot inventory in the public forests was only made in 1995. A final decision in favour or against implementation has not yet been made.

   In case the NFI II will not materialise within the next years, some federal states are considering to carry out regional inventories on their territory, following the NFI scheme.

   The one-time Bavarian inventory is too old to be referred to in this context, the Nordrhein-Westfalen inventory is not yet implemented.

Historical remarks: The first statistical and sample based forest inventory took place from 1986 to 1990 (NFI I). In former times information on the forest resource in Germany were collected through forest surveys (*Forsterhebungen*) based upon questionnaires sent out to forest owners. The first survey of this kind was in 1878, the last one for the whole of Germany in 1937 and the last for the old states of the

Federal Republic of Germany in 1961 (BML 1992a). In the years from 1961 to 1974 a permanently installed large area forest inventory was implemented in the German Democratic Republic. It was a statistically sound sample based inventory andd is described in more detail in GROSSMANN (1972).

1.2 OTHER IMPORTANT FOREST STATISTICS

1.2.1 Other forest data and statistics on the national level

There are several other forest statistics which, however, do either not cover the entire region or do not treat the whole of the forests and forest products. It is therefore not possible to derive statements about the whole of German forests from these statistics alone. Yet they can serve as complementary information in connection to other information sources.

General forest data base (*Datenspeicher Waldfond*):

In the new federal states there is a general forest data base (*Datenspeicher Waldfond*) which has been set up during GDR (German Democratic Republic) times. These data are brought to an actual state every year through propagation models and they are periodically updated. The last propagation had been done with a reference date of January 1, 1993. These data were used for a publication which analysed the *Datenspeicher Waldfond* data according to the analysis procedures of NFI I. The *Datenspeicher Waldfond* was a centralized institution. Today all the five new federal states have their own data bases with different structure, update procedures and development plans. This makes a common analysis virtually impossible.

Holzeinschlagsstatistik (wood harvest statistic):

There is a *Holzeinschlagsstatistik* (wood harvest statistic) which is published quarter yearly by BML. But this statistic, though refering to the forest resource, does not give an idea of state, potential or development of the natural resource forest.

1.2.2. Delivery of statistics to UN and EU institutions

1.2.2.1. Responsibilities for international assessments

Delivery of forestry and wood statistics by BML to international organizations can be seen from Table 3. The information was made available by BML and reflects the status by July, 1995.

Table 3. Delivery of forestry and wood statistics to international organizations.

Organisation collecting data	Type of data	Responsible BML section (**bold**) and other concerned sections (*Referate am BML*)
ECE/FAO timber section	forestry and wood market data	**611**, 612,613
Eurostat	forestry and wood market data	**212**, 611, 612, 613
EU-Comm. DGVI	forestry and wood market data (EFICS) data on forest condition and permanent observation plots	**611**, 612, 613 **615**
EU-Comm. DGIII	forestry and wood market data (planned)	**611**, 612, 613
ITTO (UN)	forestry and wood market data	**614**, 611, 612, 613
ICP forests (UN)	data on forest condition	**615** Delivery of the statistics to EC 'Level I-Program' by BFH, Hamburg.
Ministerial conference for the protection of forests in Europe (Helsinki process)	forestry data	**614**, 611
CSD Rio process	forestry data	**614**, 611
OECD	forestry data	**611**, 613

1.2.2.2. Data compilation

Data provision to international institutions goes through BML as can be seen from Table 3. Data on forest resources generally originate from the NFI results for the old federal states, and from the *Datenspeicher Waldfond* (forestry data base) for the new federal states. Special algorithms to adapt data are not used. If no appropriate information is available, expert guesses are used in some cases.

2. NATIONAL FOREST INVENTORY, NFI (*BUNDESWALDINVENTUR, BWI*)

Pre-remarks:

In Germany general regulations on forestry are given in the Federal Forest Act in its latest version dated from July 27, 1984. In this act the implementation of a National Forest Inventory is demanded. Forest management, however, is under state forest legislation. Each of the federal states has its own state forest act.

NFI was organized in a labour-sharing manner: The federal institutions assumed the tasks of general organization and coordination (in detail listed in BML 1986a I.2), the states were responsible for the data collection (in detail listed in BML 1986a I.3).

The results were published in 1992 for the whole of the inventory region, which is the area of the Federal Republic of Germany before October 3, 1990, including Berlin (West).

The current report refers to the NFI and its methodology as applied in a uniform way to the entire inventory area.

Some federal states utilized the opportunity and collected additional information in the course of the NFI. This lead to regionally more detailed information. These enhancements are not refered to in detail. The data set which is stored in BFH andd comprises data of all of the surveyed federal states contains only the information that has been uniformly assessed over the entire inventory region.

The inventory bases upon systematically distributed square tracts in a base grid of 4km by 4km size. Implementation consisted of two principal components:

1. Pre-clearing of tracts: Multi-stage procedure to find out whether a sample location is to be classified forest or non-forest. To find out about that, all available sources of information (avoiding expensive field visit) were processed. In BML 1986a the following sources of information are explicitly mentioned:
   - Maps
   - Documents
   - Consultation of local forest officers
   - Consultation of cadastrial survey
   - Up-to-date air photos
2. Field measurements. Four different sample elements are utilized on a tract to assess different attributes:
   - Line intercept sample (*Taxationslinie*)
   - Point sample (*Winkelzählprobe*)
   - Sample plot (*Probefläche*)
     - Sample plot 1m radius
     - Sample plot 2m radius
     - Sample plot 4m radius
   - Line intersect sample (*Linienschnittpunkte*).

2.1 NOMENCLATURE

2.1.1 List of attributes directly assessed

A) Geographic regions

Attribute	Data source	Object	Measurement unit
Political/administrative unit (*Kreis*) BML 1986 III.5.2	Pre-celaring: Maps and general administrative guidelines	Political unit (*Kreis*)	List of *Kreise* (list names and of codes = *Kreiskennziffern)*
Responsible forest office (*Forstamt*) BML 1986a III.5.2	Pre-clearing: Maps and forest administrative guidelines	Forest office (*Forstamt*)	List of *Forstämter* (list of names and of codes=*Forst-amtskennziffern)*
Growth region (*Wuchsgebiet*) BML 1986a III.5.2	Pre-clearing: Maps and administrative guidelines	Growth region (*Wuchsgebiet)*	List of *Wuchsgebiete* (list of names and of codes = *Wuchs-gebietskennziffern)*
Site unit (*Standortseinheit*) (optionally assessed by federal states, but not in data set of BFH) BML 1986a III.5.2	Pre-clearing: Site maps and documents	Site unit (*Standorts-einheit*)	Statewise different

B) Ownership

Attribute	Data source	Object	Measurement unit
Ownership type (*Eigentumsart*) BML 1986a III.5.4.2	Pre-clearing: Maps, cadastrial and other documents, interviews.	Stand which intersects the taxation line	four ownership classes
Ownership size class (*Eigentumsgrößenklasse)* BML 1986a III.5.4.3	Pre-clearing: Maps, cadastrial and other documents, interviews.	Property of the owner of a stand which intersects the taxation line	seven or eight size classes (unit: ha)
Trails, roads and closest truck motorable road (*Wege, Straßen und nächste LKW-befahrbare Straße*) BML 1986a III.5.3	Pre-clearing: Maps and cadastrial documents	Roads	four ownership classes; dawn in tract sketch: Attribute 'closest truck motorable road' cannot be further processed (only in sketch)

C) Wood production

Attribute	Data source	Object	Measurement unit
Species (*Baumart*) BML 1986a IV.6.4.2	Field: Point sample	Tree	Species code list
Diameter at breast height dbh *(Brusthöhen-durchmesser BHD)* BML 1986a IV.6.4.6	Field: Point sample	Tree	Metre
Height *(Baumhöhe)* BML 1986a IV.6.4.9	Field: Point sample	Tree	Metre
Height code *(Höhenkennziffer)* BML 1986a IV.6.4.10	Field: Point sample	Tree	Code 1 or 2
Stem height for hardwood *(Stammhöhe, Laubbäume)* BML 1986a IV.6.4.11	Field: Point sample	Tree	Metre
Stem code *(Stammkennnziffer)* BML 1986a IV.6.4.12	Field: Point sample	Tree	Code 1 to 5
Upper stem diameter *(Oberer Stammdurch-messer)* BML 1986a IV.6.4.13	Field: Point sample	Tree	Metre
Code for upper stem diameter *(Kennziffer des oberen Stammdurch-messers)* BML 1986a IV.6.4.14	Field: Point sample	Tree	Code 1 or none
Code for secondary stand *(Nebenbestandskennziffer)* BML 1986a IV.6.4.15	Field: Point sample	Tree	Code 1 or none
Stem damages *(Stammschäden)* BML 1986a IV.6.4.16	Field: Point sample	Tree	Code 1 to 6

D) Site and soil

Attribute	Data source	Object	Measurement unit
Elevation BML 1986a IV.1	Topographic Map 1:25000	Tract centre	Rounded off to full metres
Slope BML 1986a IV.5.12	Field: Relascope	Relief, along gradient, not taxation line	Percent
Exposition BML 1986a IV.5.13	Field: Compass Not measured for slope less than 5%	Relief, measu-red in direction of slope	gon (400 gon scale)

E) Forest structure

Attribute	Data source	Object	Measurement unit
Species and age BML 1986a III.5.6	Pre-clearing: Forest management plans and similar documents (Only assessed during pre-clearing if these documents are available)/Field	Stand	Species codes and age in years (reference date 1 October 1987)
System of management (*Betriebsart*) BML 1986a IV.5.6	Field: Line intercept sample	Stand	six classes of management systems
Vertical structure BML 1986a IV.5.8	Field: Line intercept sample	Stand	six classes of layer composition
Growth class BML 1986a IV.5.9	Field: Line intercept sample	Stand	six classes of age/growth status
Age BML 1986a IV.5.10 and 5.11	Different alternatives: 1 = From pre-clearing phase (documents) 2 = From stumps 3 = Counting of branch whorls *(Astquirlzählung)* 4 = Borings 5 = Estimation	Stand Age of principal stand/age of trees measured in point sample	Years
Crown cover (*Schlußgrad*) BML 1986a IV.5.15	Field: Line intercept sample	Stand	five classes
Stand type (*Bestandestyp*) BML 1986a IV.5.16	Field: Line intercept sample	Stand	Classification with species composition with up to four components

F) Regeneration

Attribute	Data source	Object	Measurement unit
Pre-regenerated area (*vorausverjüngte Fläche*) BML 1986a IV.5.18	Field: Line intercept sample	Stand (only stand part around taxation line considered, up to 10m)	Estimation of pre-regenerated proportion of area. Proportional line length recorded in metres
Species (*Baumart*) BML 1986a IV 7	Field: Plot sample 1m, 2m, 4m	Tree	Species list
Number of trees - broken down by protection status and damage BML 1986a IV 7	Field: Plot sample 1m, 2m, 4m	Plot	Frequency
Tree size (*Größe)* BML 1986a IV.7.2.3	Field: Plot sample 1m, 2m, 4m	Tree	seven size classes
Damages (*Schäden*) BML 1986a IV.7.2.4	Field: Plot sample 1m, 2m, 4m	Tree	three classes
Protection measures (*Schutzmaßnahmen*) BML 1986a IV.7.2.5	Field: Plot sample 1m, 2m, 4m	Tree	three classes

G) Forest condition

Attribute	Data source	Object	Measurement. unit
Deficiencies in tending (*Pflegerückstände*) BML 1986a IV.5.14	Field, line intercept sample Not assessed in regeneration stands	Stand	Yes or no (Example: Missing thinnings or thinnings with negative impact on stand development)

H) Accessibility and harvesting

Attribute	Data source	Object	Measurement unit
Trails, roads and closest truck metalled road BML 1986a III.5.3	Pre-clearing: Maps	Roads	Drawn in tract sketch
Road ownership class (Eigentumsart) BML 1986a IV.8.3	Pre-clearing	Roads	four classes
Valence of road (*Wertigkeit*) BML 1986a IV.8.4	Field: Line intersect sample	Roads	two classes

Width of roadway *(Fahrbahnbreite)* BML 1986a IV.8.5	Field: Line intersect sample	Roads	five classes
Road surface *(Fahrbahndecke)* BML 1986a IV.8.6	Field: Line intersect sample	Roads	three classes
Slope of road (*Gefälle des Weges*) BML 1986a IV.8.7	Field: Line intersect sample	Roads	Per cent
Slope of terrain (*Gefälle des Geländes*) BML 1986a IV.8.8	Field: Line intersect sample	Terrain around roads	Per cent

I) Attributes describing forest ecosystems

None

J) Non-wood goods and services

None

K) Miscellaneous

Attribute	Data source	Object	Measurement Unit
Utilisation restrictions (*Feststellung von Nutzungsbeschränkungen*) BML 1986a III.5.5	Pre-clearing	Forest stands that intersect taxation line	Dichotom (yes or no) Examples: Nature conservation area, National Park, Military training area, protection forest
Fenced area *(gezäunte Fläche)* BML 1986a IV.5.17	Field: Line intercept sample	Fenced area	Length in metres of taxation line inside fenced area
Mirroring code *(Spiegelungskennziffer)* BML 1986a IV.6.4.3	Field: Point sample	Tree	Code 1 or 2 (only rarely applied)
Azimuth BML 1986a IV.6.4.4	Field: Point sample	Tree	Compass: 400gon scale

The following ′miscellaneous′attributes are only listed without further description and without giving more details in the following sections, because they refer to organizational aspects of the inventory and not to relevant technical ones. They appear on the form sheets used for field assessment but they do not appear in legal regulations about implementation of the NFI.′

Attribute	Data source	Object	Measurement unit
Gauss-Krüger geodetical coordinates of South-West corner of the tract	Form sheets 1-7		
Crew leader (name)	Form sheets 1-7		
Control crew leader (name)	Form sheets 1-7		
Name of topographic map used for locating the tract	Form sheet 2		
Scale of topographic map used for locating the tract	Form sheet 2		
Number of form sheets 4 used for the tract	Form sheet 4		
Number of form sheets 5 used for the tract	Form sheet 4		
Number of form sheets 6 used for the tract	Form sheet 4		
Number of form sheets 7 used for the tract	Form sheet 4		
Tract corner number	Form sheets 5 and 6		

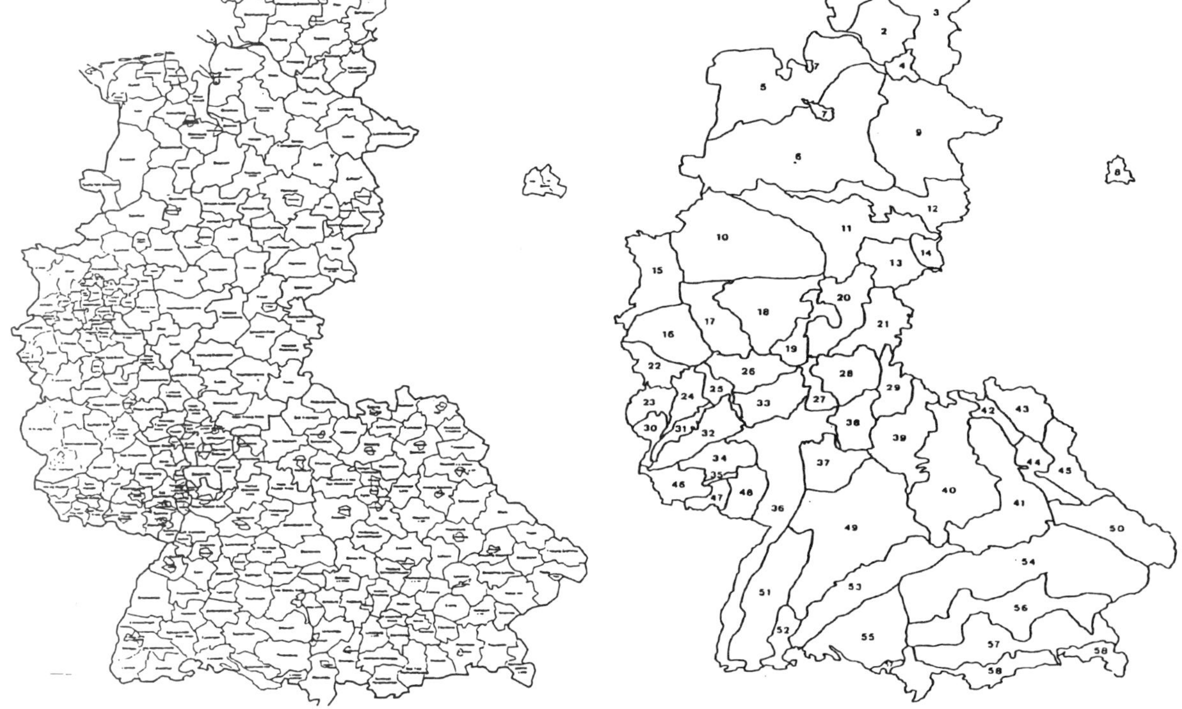

Figure 3. Map of Kreis-boundaries: administrative units used in NFI, old federal states (From BML 1986a).

Figure 4. Map of growth regions (Wuchsgebiete) as optionally assessed by the states, old federal states (from BML 1986a)

2.1.2 List of derived attributes

The following list refers to the two-volume NFI results published by the Federal Ministry of Agriculture (BML 1992a, 1992b). In the meantime the inventory data have also been used to carry out additional analyses. These are published in scientific journals, not being part of the inventory procedure. The contents of these papers was therefore not refered to in this listing.

A) Geographic regions

None

B) Ownership

None

C) Wood production

Attribute	Measurement unit	Input attributes
Tree volume (*Baumvolumen*)	m^3 over bark (*Vorratsfestmeter mit Rinde*)	Species, dbh, d_{7m}, total height: Model for taper curve and single tree volume
Assortment (*Sortierung*)	Assortment classes	Species, dbh, d_{7m}, total height: Model for taper curve
Diameter class (*Durchmesserklasse*)	18 classes: 1 = 7-9.9cm 2 = 10-14.9cm .... 5 cm classes up to 18 = > 89.9cm	dbh
Basal area	m^2	dbh, number of stems
Mean diameter	cm, calculated as diameter of the tree of mean basal area	dbh, number of stems
Mean height	m, calculated as height according to Lorey	total height, number of stems

D) Site and soil

None

E) Forest structure

Attribute	Measurement unit	Input attributes
Species groups (*Baumartengruppen*) BML 1992a p.45 ...	List of nine species groups in BML 1992a p. 109	Species
Pure and mixed stands *(Rein- und Mischbestände)* BML 1992a p.64	six classes of different mixtures	Tree species, stand type (*Bestandestyp*)
Age class (*Altersklasse*)	nine age classes in 20 years intervals from 1-20 years up to > 160 years	Age

F) Regeneration

None

G) Forest condition

None

H) Accessibility and harvesting

None

I) Attributes describing forest ecosystem

None

J) Non-wood goods and services

None

K) Miscellaneous

None

2.1.3 Measurement rules for measurable attributes

For each of the measureable attributes the following information is given:

a) measurement rule
b) threshold values
c) measurement scale
d) rounding rules
e) instrument used for measurement
f) data source

A) Geographic regions

—

B) Ownership

—

C) Wood production

Diameter at breast height dbh (Brusthöhendurchmesser) BML 1986a IV.6.4.6

a) Stem diameter measured in 1.3m height perpendicular to stem axis. Tape has to be tightened, loose bark particles, lichens or moss has to be removed first. Figure 5 shows the measurement procedure. If diameter has to be taken in a height different from 1.3m this must be recorded in a separate field. Trees bifurcated below 1.3m are recorded as two trees.
b) 10cm for sample trees in point sample
c) Meter
d) Measured to millimeter accuracy
e) Diameter-tape (steel tape) for dbh, ruler (*Meßstock*) of 1.3m length for measuring breast height
f) Field: Point sample

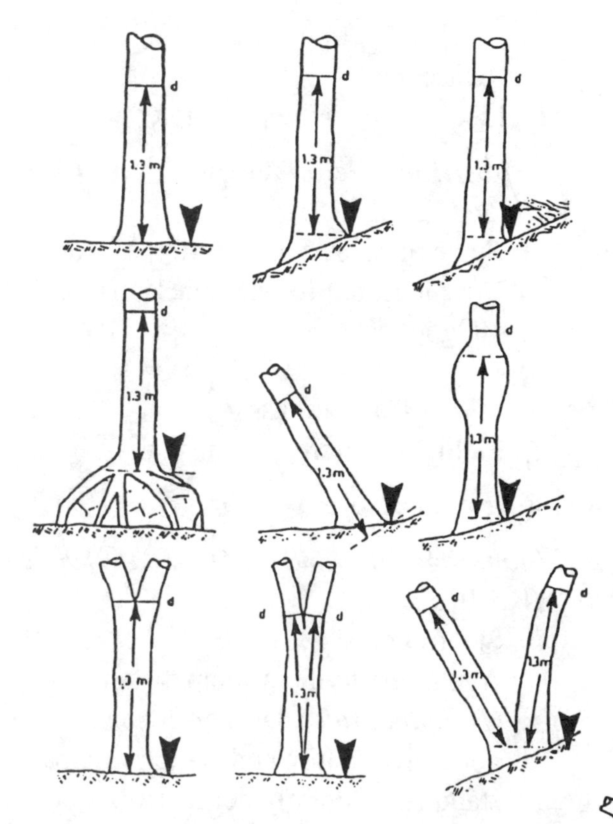

Figure 5. Guidelines how to measure dbh (BML 1986b)

Height (Baumhöhe) BML 1986a IV.6.4.9

a) Distance from stem base to top. Only measured for living trees. Standard rules for trigonometrical height measuremenets: For upright trees in plane area: any measurement direction. For oblique trees: Relascope must be adjusted according to the slant stem. At slopes to be measured from mountain side or at least from the same elevation. For broadleaf trees: top is difficult to aim at. Standard measurement distance: 20m measured with Ultra-sonic distance meter (for tall trees: 30m, for small trees 10m).
b) -
c) Meter
d) Decimeter accuracy
e) Relascope (percent scale, 2nd scale from right hand side)
f) Field: Point sample

Stem height for hardwood (Stammhöhe, Laubbäume) BML 1986a IV.6.4.11

a) Distance from base to the point at the stem where the crown begins and the stem ends. See also Figure 6. Measurement is done from the same location as measurement of total height. Same general rules like for measurement of total height.
b) Only for broadleaf trees with more than 20cm dbh and stem code (*Stammkennziffer*) not 2 to 5.
c) Meter
d) Measured to decimeter accuracy
e) Relascope
f) Field: Point sample

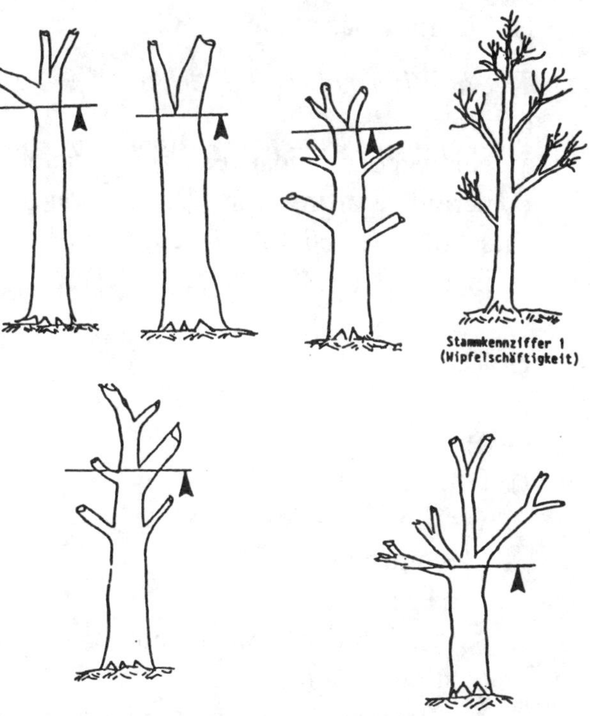

Figure 6. Stem height for hardwood (BML 1986b)

Upper stem diameter (Oberer Stammdurchmesser) BML 1986a IV.6.4.13

a) Stem diameter in 7m height. Measurement direction: from centre of point sample
b) Only measured for trees with minimum dbh 20cm and - in case of broadleaf trees - stem length more than 7m. In case of stem codes (*Stammkennziffern*) 2 to 5 no upper diameter is taken.
c) Meter
d) Measured to half-centimeter accuracy
e) Finnish caliper. If caliper not applicable (branches etc., or diameter too big): Measurement with relascope, the height of 7m is to be determined with the 7m stick of the caliper.
f) Field: Point sample

Tree age (Baumalter) BML 1986a IV.6.4.7/ 6.4.8

a) Number of calendar years or vegetation periods, starting from germination of seed; referred to reference date of inventory (BML 1992a, p. 108).
b) -
c) Years
d) -
e) (see f) data source)
f) Different alternatives:
   1 = From pre-clearing phase (documents)
   2 = From stumps
   3 = Counting of branch knots (Astquirlzählung)
   4 = Borings: Only with permission of forest owner. Not admitted in 1.3m height.
   5 = Estimation

D) Site and soil

Slope (Geländeneigung) BML 1986a IV.5.12

a) To be measured in direction of main relief gradient, not along taxation line
b) -
c) Percent
d) -
e) Relascope
f) Field: Line intercept sample

Exposition (Geländeexposition) BML 1986a IV.5.13

a) To be measured in direction of slope
b) Not measured for slope less than 5%
c) 400gon scale
d) -
e) SUUNTO compass
f) Field, line intercept sampling

E) Forest structure

Stand age (Bestandesalter) BML 1986a IV.5.10 and 5.11

a) Stand age.
   Age refers to the principal stand (*Hauptbestand*). Depending on vertical structure codes the principal stand is differently defined:
   for vertical structure code 2: age of upper layer
   for 3: age of stand, reserved standards disregarded
   for 4 and 5: age of old stand
   for 6: age of upper layer. No age for plenter forest.
   According to BML 1992a, p. 108, for stands where inside the principal stand there are considerable differences in age a mean age is to be calculated, weighted by area (*das mit der Fläche gewogene Durchschnittsalter*).
b) -
c) Years
d) -
e) (see f) data source)
f) Different alternatives:
   1 = From pre-clearing phase (documents)
   2 = From stumps
   3 = Counting of branch knots (Astquirlzählung)
   4 = Borings: Only with permission of forest owner. Not admitted in 1.3m height.
   5 = Estimation

Line length in line intercept sampling (Linientaxation) BML 1986a IV.5.2

a) Length measurement
b) -
c) Metric
d) Rounded to whole meters
e) Ultrasonic distance meter or tape
f) Line intercept sample (*Linientaxation*)
e) Length measurements used for all area attributes assessed during line intercept sample

F) Regeneration

Pre-regenerated area (vorausverjüngte Fläche) BML 1986a IV.5.18

a) Estimate area proportion of pre-regenerated area and record the corresponding proportion of line length. Only the immediate surrounding of the taxation line is assessed (up to 10m). Example: If 100m of the taxation line are in a stand the pre-regeneration proportion of which is estimated to be 15%, a line length of 15m is recorded.
b) -
c) Meters
d) -
e) Estimation
f) Field: Line intercept sample
g) Does apply only for vertical structure 4.

Number of trees - broken down by protection status and damage BML 1986a IV 7

a) Counting
b) Sample plot 1m: Trees from 20cm to 50cm height
Sample plot 2m: Trees from 50cm to 130cm height
Sample plot 4m: Trees from 130cm height and with less than 10cm dbh.
c) Frequency
d) -
e) -
f) Field: sample plots 1m, 2m, 4m

G) Forest condition

-

H) Accessibility and harvesting

Slope of road (Gefälle des Weges) BML 1986a IV.8.7

a) Slope measured at intersection of taxation line and road
b) -
c) Percent
d) -
e) Relascope
f) Field: Line intersect sample

Slope of terrain (Gefälle des Geländes) BML 1986a IV.8.8

a) Mean slope of terrain through measurements below and above the road.
b) -
c) Percent
d) -
e) Relascope
f) Field: Line intersect sample

I) Attributes describing forest ecosystems

-

J) Non-wood goods and services

-

K) Miscellaneous

Fenced area (gezäunte Fläche) BML 1986a IV.5.17

a) Length of line segement within fenced area.
b) -
c) Meters
d) -
e) Ultrasonic distance meter or tape
f) Field: Line intercept sample

Azimuth BML 1986a IV.6.4.4

a) Measured from centre of point sample: Angle between north and the (imagined) centre of the sample tree. No correction for magnetic declination.
b) -
c) 400 gon scale
d) -
e) SUUNTO compass
f) Field: Point sample

Horizontal distance (Horizontal-entfernung) BML 1986a IV.6.4.5

a) Measured from centre of point sample: Horizontal distance to imagined centre of sample tree
b) -
c) Meters, recorded to centimeters
d) -
e) Ultra-sonic distance meter or distance tape
f) Field: Point sample

2.1.4 Definitions for attributes on nominal or ordinal scale

The following information is provided for each attribute:

a) the definition
b) the categories (classes)
c) the data sources
d) remarks

A) Geographic regions

Federal state (Bundesland)

a) Federal state to which the tract centre belongs to. In case of boundary tracts, line segments of the are assigned to the federal state they fall into.
b) Map
c) In pre-clearing phase, see b)
d) -

Political/administrative unit (Kreis) (BML 1986a, III.5.2)

a) Political unit to which the tract centre belongs to
b) Map with *Kreis*-boundaries and list of *Kreise (*list of codes = *Kreiskennziffern)*
c) In pre-clearing phase, see b)
d) —

Responsible forest office (Forstamt) (BML 1986a, III.5.2)

a) Forest office to which the tract centre belongs to
b) Map with *Forstamt*-boundaries and list of *Forstämter (*list of codes = *Forstamtskennziffern)*
c) In pre-clearing phase, see b)

Growth region (Wuchsgebiet) (BML 1986a, III.5.2)

a) Growth region to which the tract centre belongs to
b) Map and list of *Wuchsgebiete (*list of codes = *Wuchsgebietskennziffern)*
c) In pre-clearing phase see b)
d) Attribute optionally assessed by federal states

Site unit (Standortseinheit) (BML 1986a, III.5.2)

a) Site unit to which the tract centre belongs to
b) Site maps and documents, statewise different
c) In pre-clearing phase
d) Attribute optionally assessed by federal states. This attribute is not contained in the data set stored BFH.

B) Ownership

Ownership type (Eigentumsart) (BML 1986a, III.5.4.2)

a) Land owner of the forest stands which intersect the taxation line

b) 10 = Federal property (*Staatswald (Bund))*

20 = State property (*Staatswald (Land))*

30 = Community forest (*Körperschaftswald*)

40 = Private (*Privatwald*)

c) In pre-clearing phase, forest ownerhip map, cadastrial maps

d) In unclear cases state legislation is decisive (*Zuordnung entsprechend den Vorschriften des Landes*).

Second digit of code can be used by states for further breakdown of ownership.

Ownership size class (Eigentumsgrößenklassen) (BML 1986a, III.5.4.3)

a) Size in ha of the forest area which belongs to the owner of a stand which intersects the taxation line, and which is managed by the same unit. This refers to the ownership types private and community forest.

b) 3 = up to 20 ha

4 = over 20 ha to 50 ha

5 = over 50 ha to 100 ha

6 = over 100 ha to 200 ha

7 = over 200 ha to 500 ha

8 = over 500 ha to 1000 ha

9 = over 1000 ha.

Alternatively class 3 may be subdivided into two classes:

1 = up to 5 ha

2 = over 5 ha to 20 ha.

c) Pre-clearing: Maps, cadastrial and other documents, interviews with forest officers.

d) Whether to use class 3, or1 and 2, must be decided uniformly within the single states. Owners were not directly contacted. There is the possibility that the owner of a particular forest stand owns more forests beyond the present forest district and beyond the knowledge of the contacted sources. This might lead to an underestimation of ownership size class.

Trails, roads, closest truck-motorable road (1986a, III.5.3) (Wege, Straßen, nächste LKW-befahrbare Straße)

Ownership of roads: See section H) Accessibility and Harvesting

C) Wood production

Species (Baumart) BML 1986a IV.6.4.2

a) Species or species group

b) see attribute ´species and age´ under E) FOREST STRUCTURE.

The following codes are also possible (BML 1986b S.38):

Sfi = *Picea sitchensis (Sitca-Fichte)*

Kta = *Abies grandis (Küstentanne)*

Elä = *Larix decidua (Europäische Lärche)*

Jlä = *Larix japonica (Japanische Lärche)*

Thu = *Thuja sp.*

Tsu = *Tsuga sp.*

Wey = *Pinus strobus (Weymouthkiefer)*

Bah = *Acer pseudoplatanus (Bergahorn)*

Sah = *Acer platanoides (Spitzahorn)*

Fah = *Acer campestre (Feldahorn)*

Ul = *Ulmus sp. (Ulme)*

Hbu = *Carpinus betulus (Hainbuche)*

Rob = *Robinia pseudoacacia (Robinie)*

El = *Sorbus trominalis (Elsbeere)*

Ka = *Castanea sativa (Kastanie)*

Bpa = *Populus sp. (Balsampappel)*

Wei = *Salix sp. (Weiden)*

c) Field: Point sample

Height code (Höhenkennziffer) BML 1986a IV.6.4.10

a) Describes damages of the tree top. Necessary to be recorded in order not to derive mistaken diameter/height relationships
b) 1 = broken top (*Wipfelbruch*) (estimated length of broken part less than 3m)
   2 = broken crown (*Kronenbruch*) (estimated length of broken part more than 3m)
c) Field: Point sample
d) Analysis together with height measurement

Stem code (Stammkennnziffer) BML 1986a IV.6.4.12

a) Important stem characteristics.
b) 1 = Stem up into the crown (broadleaf species) (*Wipfelschäftigkeit*)
   2 = Bifurcation between 1.3m and 7m (for conifers and broadleaf species)
   3 = No clear single stem. Stem base to begin of crown less than 3m (for conifers and broadleaf species)
   4 = Dead or broken stem. Wood still utilizable, not rotten. Length of broken stems minimum 2/3 of total tree height.
   5 = Dead. Wood not any more utilizable.
c) Field: Point sample

Code for upper stem diameter (Kennziffer des oberen Stammdurchmessers) BML 1986a IV.6.4.14

a) Characterisation of upper diameter measurement.
b) no record = Measurement with Finnish caliper
   1 = Measurement with Relascope
c) Field: Point sample

Code for secondary stand (Nebenbestandskennziffer) BML 1986a IV.6.4.15

a) All sample trees of point sample are classified to be members of the main stand or of the secondary stand. Secondary stand comprises all trees with no contact to the crown space of the main stand.
b) no record = Main stand
   1 = Upper stand
   2 = Lower stand
c) Field: Point sample

Damages (Schäden) BML 1986a IV.6.4.16

a) Only assessed for living trees. Up to 3 damages can be recorded in the order of their significance.
b) 1 = Bark peeling damage, new (within last 12 month) (*Schälschaden*)
   2 = Bark peeling damage, old
   3 = Skidding damage, new (within last 12 month) (*Rückeschaden*)
   4 = Skidding damage, old
   5 = Other stem damages
   6 = Dying
c) Field: Point sample
d) Category 6 characterises the general impression of the tree. If this applies, it has to be listed at first position. Old peeling damages were sometimes difficult to identify.

D) Site and soil

—

E) Forest structure

Species and age (Baumart und Alter) BML 1986a, III.5.6

a) Species and age of stands that intersect the taxation line
b) Species codes:
c) Pre-clearing phase: Forest management plans or similar documents
d) Assessed in pre-clearing phase as ancillary attribute only. Not further processed. Tree species assessed in attribute stand type (*Bestandestyp*).

System of Management (Betriebsart) BML 1986a IV.5.6

a) Type of forest management
b) 1 = Age class high forest *(Schlagweiser Hochwald)*
   2 = Plenter forest *(Plenterwald)*
   3 = Composite forest *(Mittelwald)*
   4 = Coppice shoot forest *(Stockausschlagwald)*
   5 = Unproductive forest area *(unproduktive Waldflächen)* like dwarf pines, green alder fields shrub areas and other low production areas with average annual increment less tan $1m^3$/a/ha.
   6 = Non-stocked areas in forest *(Nichtholzboden)* like forest roads broader than 5m, forest aisles, log storage sites, nursery, meadows for wildlife, meadows for recreation, water areas, rocky areas etc.
c) Field: Line intercept sample

Vertical structure / layer composition (Aufbau) BML 1986a IV.5.8

a) Vertical structure of forest
b) 1 = one layer *(einschichtig)*
   2 = two layers *(zweischichtig)*
   3 = two layers (upper layer reserved standards or harvest rest) *(zweischichtig: Überhälter oder Nachhiebsrest)*
   4 = two layers (lower layer: pre-regeneration) *(zweischichtig: Unter-schicht Vorausverjüngung)*
   5 = two layers (lower layer: Undercrop) *(zweischichtig: Unter-schicht: Unterbau)*
   6 = multi layer or plenter type *(mehrschichtig oder planterartig)*
c) Field: Line intercept sample

Growth class (Wuchsklasse) BML 1986a IV.5.9

a) Rough classification of growth/stand age
b) 1 = Clear area / glade *(Blöße)*
   2 = Young plantation *(Kultur)*
   3 = Sapling (up to 10cm mean dbh) *(Dickung)*
   4 = Pole timber (up to 20cm mean dbh) *(Stangenholz)*
   5 = Timber (more than 20cm mean dbh) *(Baumholz)*
   6 = Old grown stand (final harvest) *(Altholz)*
c) Field: Line intercept sample

Crown cover (Schlußgrad) BML 1986a IV.5.15

a) Degree of coverage of stand area by tree crowns
b) 1 = Aggregate *(gedrängt):* Crowns overlap intensively, or dense natural regeneration.
   2 = Closed *(geschlossen):* Crowns touch, or closed natural or artificial regeneration with up to 10% gaps.
   3 = Loose *(locker):* Crowns with small distances to each other, or natural or artificial regeneration with 10% to 30% gaps.
   4 = Clear *(licht):* Crowns hava a distance of about one crown diameter, or natural or artificial regeneration with 30% to 50% gaps.
   5 = Gappy *(räumdig):* Crowns have a distance of several crown diameters, or natural or artificial regeneration with more than 50% gaps.
c) Field, line intercept sample
d) Cover percentages are estimated by field crews.

Stand type (Bestandestyp) BML 1986a IV.5.16

a) Description of stand type along taxation line with species composition with up to 4 components (species codes)
b) 1st component: Leading species, decisive for length of rotation period (share generally more than 50%)
   2nd component: Next important species with silvicultural significance, share more than 10%.
   3rd component: Further leading, stand-characterising species; ecologically important admixture even with less than 10%.
   4th component: Most important species of secondary stand (*Nebenbestand*)
   Following species codes were used:
   Fi = Spruce *(Fichte)*
   Ta = Fir *(Tanne)*
   Ki = Pine *(Kiefer)*
   Ski = *Pinus nigra (Schwarzkiefer)*
   Dgl= Douglas fir *(Douglasie)*
   Lä = Larch *(Lärche)*
   sNb = Other Conifers *(sonstige Nadelbäume)*
   Bu = Beech *(Buche)*
   Ei = Oak *(Eiche)*
   Rei = Red Oak *(Roteiche)*
   Pa = Poplar *(Pappel)*
   Es = Ash *(Esche)*
   Ah = Maple *(Ahorn)*
   Bi = Birch *(Birke)*
   Erl = Alder *(Erle)*
   Kir= Cherry *(Kirsche)*
   sLb = Other deciduous *(Sonstiges Laubholz)*
c) Field: Line intercept sample
d) For management system 4 (coppice shoot forest): 4th component always empty.
   - For vertical structure 3 (two layers with reserved standards): 4th component is for the leading species of reserved standards, first three components for the stand.
   - For vertical structure 4 and 5 (pre-regeneration and undercrop): 4th component for leading species of pre-regeneration or undercrop, first three components for old stand.

F) Regeneration

Species (Baumart) BML 1986a IV 7

a) see C) WOOD PRODUCTION
b) see C) WOOD PRODUCTION
c) Field: Sample plots 1m, 2m and 4m
d) -

Tree size (Größe) BML 1986a IV.7.2.3

a) Size classes of regeneration according to height and diameter
b) 1: 50cm to 130cm (sample plot 2m)
   2: Height more than 130cm, up to 4.9cm dbh (sample plot 4m)
   5: 5.0 to 5.9cm dbh (sample plot 4m)
   6: 6.0 to 6.9cm dbh (sample plot 4m)
   7: 7.0 to 7.9cm dbh (sample plot 4m)
   8: 8.0 to 8.9cm dbh (sample plot 4m)
   9: 9.0 to 9.9cm dbh (sample plot 4m)
c) Field: Sample plots 2m and 4m

Damages (Schäden) BML 1986a IV.7.2.4

a) Damages to single trees of regeneration
b) 1 = Browsing damage, new; within last 12 month (*Verbißschaden*)
   2 = Browsing damage, old
   3 = Peeling or fraying damage (*Schäl- oder Fegeschaden*)
c) Field: Sample plots 1m, 2m, 4m
d) Browsing damage only when terminal bud is affected.

Protection measures (Schutzmaßnahmen) BML 1986a IV.7.2.5

a) Artificial measures to protect regeneration trees from damages
b) 0 = Unprotected
   1 = Fence
   2 = Individal protective measures (wire hose, chemicals)
c) Field: Sample plots 1m, 2m, 4m
d) Fence as protection does also apply if fence is damaged and game can enter into fenced area.

G) Forest condition

-

H) Accessibility and harvesting

Trails, roads, closest truck-motorable road (1986a, III.5.3) (Wege, Straßen, nächste LKW-befahrbare Straße)

a) Roads crossing the taxation line and closest truck motorable road
b) Ownership classes
   1 = Public roads (öffentliche Wege und Straßen)
   2= Common roads (Gemeinschaftswege)
   3 = Private roads (Privatwege)
   4 = Ownership status unknown
c) In pre-clearing, maps and documents
d) Closest truck-motorable road only recorded on tract sketch; thus not processable.

Valence of road (Wertigkeit von Straßen und Wegen) BML 1986a IV.8.4

a) Presence of forest around a road
b) 1 = Forest on both sides of the taxation line/road.
   2 = Forest on one side of the taxation line/road.
c) Field: Line intersect sample
d) Attribute has not been evaluated in the analysis phase.

Width of roadway (Fahrbahnbreite) BML 1986a IV.8.5

a) Self explaining
b) 1 = Foot trail and horse trail. (*Fuß- und Reitwege*)
   2 = Skidding roads (*Rückewege*)
   3 = Drive roads 2-3m (*Fahrwege*)
   4 = Drive roads >3-5m (*Fahrwege*)
   5 = Drive roads > 5m (*Fahrwege*)
c) Field: Line intersect sample

Road surface (Fahrbahndecke) BML 1986a IV.8.6

a) Self explaining
b) 1 = No artificial fortified surface (*unbefestigt*)
   2 = Fortified (Compressed ground or added stones) (*befestigt, verdichtetes anstehendes oder zugeführtes Steinmaterial*)
   3 = Special surface (Concrete, Asphalte)
c) Field: Line intersect sample

I) Attributes describing forest ecosystems

-

J) Non-wood goods and services

-

K) Miscellaneous

Mirroring code (Spiegelungskennziffer) BML 1986a IV.6.4.3

a) If a sample tree is recorded from sample point that has been mirrored beyond the stand boundary, a mirror code is recorded
b) Code 1: If tree is recorded from one mirrored sample point
   Code 2: If tree is recorded from two different mirrored sample points (in case of broken boundary line)
c) Field: Point Sample
d) Only rarely applied. In most cases the stand boundary line was recorded relative to the plot centre and mirroring done automatically.

2.1.5 Forest area definition and definition of "other wooded land"

BML 1986a: Forest within the meaning of the National Forest Inventory is, regardless of the information in the cadastral or similar records any area stocked with woody plants. Forests include clear-felled or cleared areas, forest roads, forest meadows, game pasture, timber yards, pipe routes located in the forest, further recreation facilities connected with the forest, overgrown heathens and moors, overgrown former meadows, alpine areas and rough grazings as well as dwarf pine and green alder areas. Heathens, moors, meadows, alpine areas and rough grazings are considered overgrown when the naturally occurring stocking has an average age of 5 years and at least 50% of the area is stocked. Stocked areas in the field or in built-up areas less than 1,000 sq. m., strips of woody plants less than 10 m wide and Christmas tree and ornamental branch crops as well as parks in residential areas are not forests according to the NFI.

Forest area is the sum total of all areas defined as forests, consisting of productive wooded area and non-wooded areas.

A definition of forest boundary is not present in the German NFI system.

Forest and non-forest distinction is generally no problem under German forest conditions as land use is very intensive and forests on marginal stands which display a low density are rare (dry forests, swamp forests, forests near in the mountains near the tree line). In some cases there were doubts on relatively young succession areas, stocked e.g. with hazel.

2.2 DATA SOURCES

A) Field data

Field data used in the German NFI are collected in the course of the sampling procedure. The procedure is described in Chapter 2.3

B) Questionnaire

No formalized questionnaires were used.

In the pre-clearing phase state forest offices were contacted to find out about forest / non-forest on the sample points. For this they received a list of geodetic coordinates of the sample points.

Not clear points were in some cases discussed personally between forest officers and field crews. Thus, interviews were in some cases another source of information.

C) Aerial photography

Not used systematically as a standard tool in the German NFI.

There are two mentionings of air photos in the NFI regulations:

1. BML 1986a III.5.1: In the course of pre-clearing, when the forest/non-forest decision is tried to be done from available documents, actual air photos are admitted as source of information.

2. BML 1986a IV.3: Air photos can be used to locate the sample tract in the field. If it can be stated clearly from the air photo that no tract corner falls into forest, then the line intercept sample can be done in the air photo.

If air photos are used they are required to be up-to-date (*aktuelle Luftbilder*). Orthophotos are to be used if possible. Otherwise distortions have to be taken into respect.

There is no formalized regulation about purchase and technical characteristics of air photos used, and no extra flights were commissioned in the framework of the German NFI. Air photos were used to facilitate work when they were available. In many cases they were made available by the forest offices responsible for the management of the forests considered.

There is no estimation available to what extent air photos were actually used.

The same statements as made before for the use of air photos hold also for the use of orthophotos: They were used where available. No orthophotos have been produced specifically for the German NFI.

D) Spaceborne or airborne digital remote sensing

—

E) Maps

In pre-clearing and for field work maps were used to a large extent. Land survey responsibility is with the federal states (*Bundesländer*). Topographic maps are distributed by the state survey (*Landesvermessungsamt*) of each state. A complete list of addresses of all state surveys, federal survey institutes and of the geological survey can be obtained from every of the state surveys. A copy of a list of these addresses is attached to this study in the Appendix.

Map: Topographic map 1:25.000 (*TK25*)
Scale: 1:25,000
Date: Different from region to region. Generally very actual.
Use: 1. All tracts (forest and non-forest) are drawn in TK25
2. Orientation in the field
Format: Analog
Source:Updated, edited and published statewise by state survey (*Landesvermessungsämter*).

Map: German base map (*Deutsche Grundkarte*)
Scale: 1:5,000
Date: Different from region to region. Generally very actual, update status generally not older than about 10 years.
Use: In pre-clearing phase position of all tracts that are clearly not non-forest is drawn in the large scale maps
Format:Analog
Source:Updated, edited and published statewise by state survey (*Landesvermessungsämter*). Not for all regions available.

Map: Forest management map (*Forstbetriebskarte*)
used where accessible and available.
Scale. 1:10,000
Date: Depending on last forest managament planning (< 10 years).
Use: In pre-clearing phase and for measurement field work.
Format: Analog
Source: Forest offices

F) Other geo-referenced data

In the pre-clearing phase (BML 1986a III.5) maps and documents of the corresponding forest offices and land registries were consulted to find out whether a tract is a forest tract or not. In the legal regulation maps and documents are not further specified.

The following geo-referenced data had been used with varying frequency:

- Forest management maps
- Forest management plans
- Sketches on forest ownership, produced in the forest office
- Forest site maps (to identify site type). These maps have not been used in the field. The site type information was complemented afterwards.
- Cadastrial maps
- Land register

Pre-clearing of forest/non-forest status of tracts took place through several sources of information utilized subsequently. The procedure is described in BML 1986a and depicted in Figure 7.

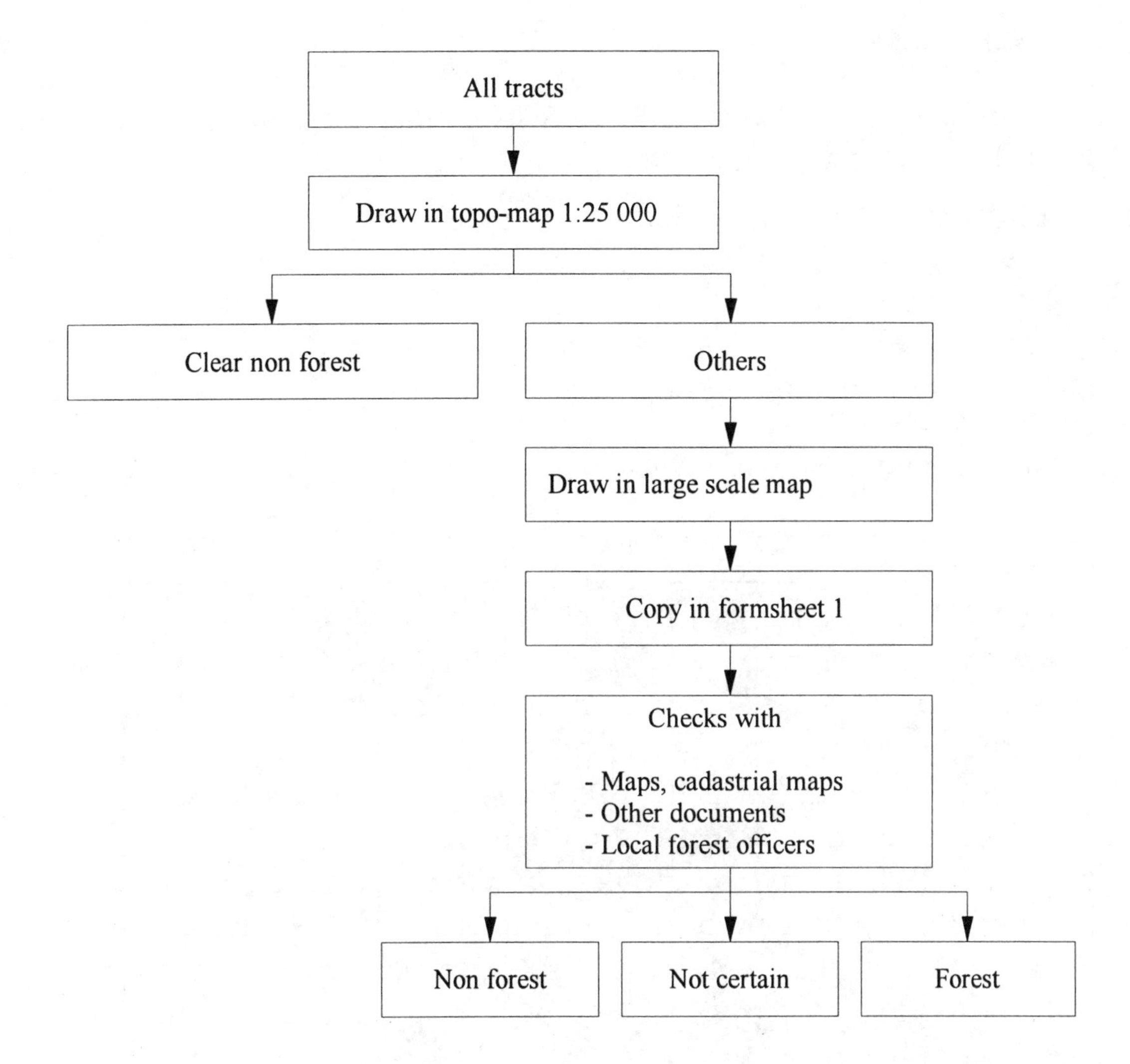

Figure 7. Pre-clearing procedure to determine forest/non-forest status (following section III.4 - 5 in BML 1986a).

2.3 ASSESSMENT TECHNIQUES

2.3.1 Sampling frame

Object of the NFI was forest in Germany (before October 3, 1990). The area comprises the old federal states including Berlin (West).

Definition of forest is given in Chapter 2.1.5. No land areas falling under the definition of forest were explicitly excluded from the survey.

There were only few non-accessible points. None of these fell into managed forest, they were part of the 'unproductive forest' class. There are no statistics available on these points.

2.3.2 Sampling units

The sample design of the German NFI consists of four different sampling units simultaneously assessed in a square tract:

Line intercept sample (*Taxationslinie*)
Point sample (*Winkelzählprobe*)
Sample plot (*Probefläche*)
  Sample plot 1m radius
  Sample plot 2m radius
  Sample plot 4m radius
Line intersect sample (*Linienschnittpunkt*).

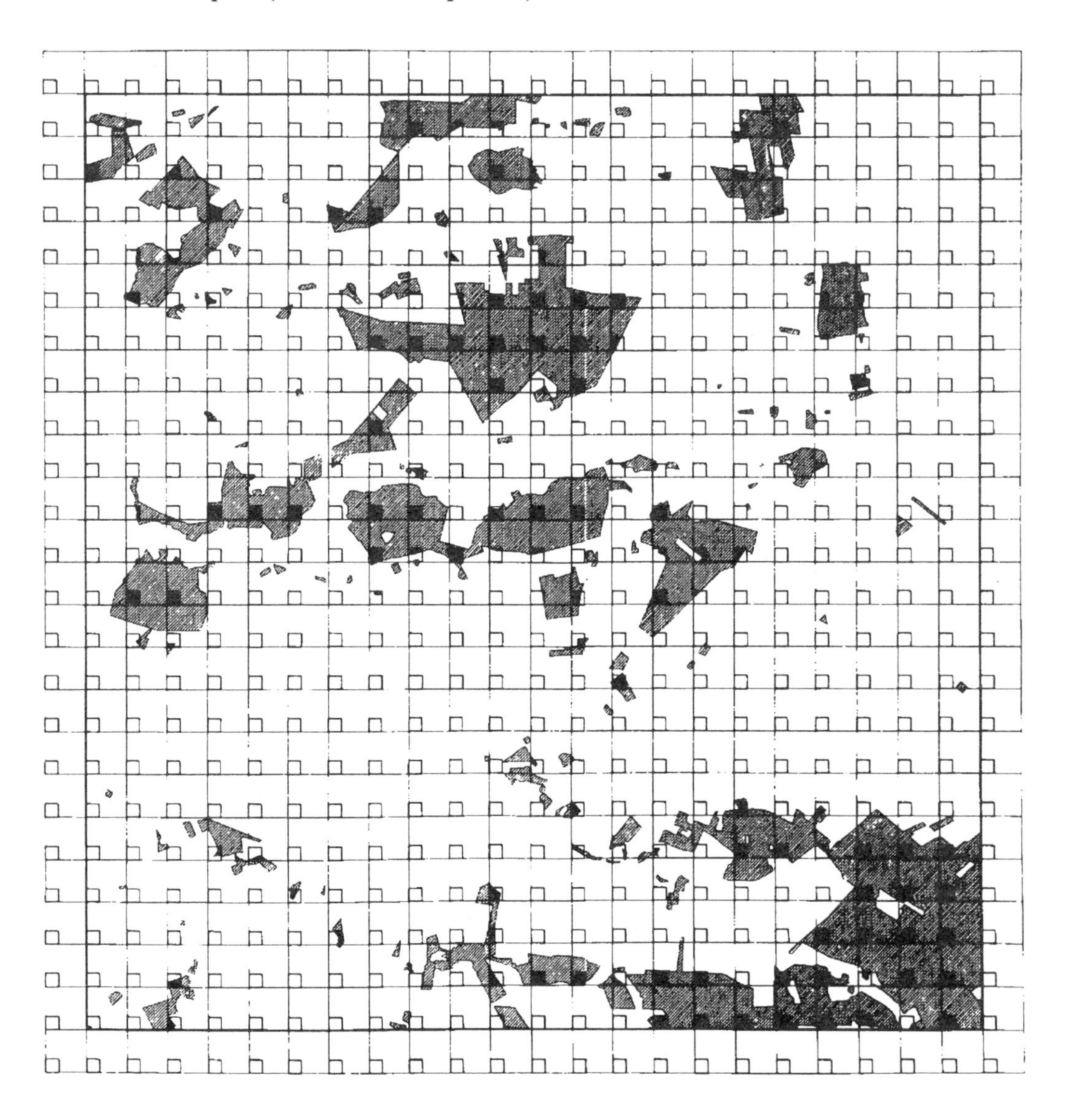

Figure 8. Example for tract distribution. Only where a tract line cuts forest area the tract is assessed. All other tracts are non-forest tracts (BML 1992a).

The tract design is schematically depicted in Figure 9. Tracts are systematically distributed in a base grid of 4km by 4km over the entire inventory region. Position refers to the Gauß-Krüger coordinate-system. Starting point of the grid is R 3556.2 and H 5566.2 (BML 1986a III.1).

Tract distribution is shown schematically in Figure 8. In the regulations on the implementation of the NFI (in BML 1986b) several regions are listed, where a denser grid was employed within the 4km x 4km network (see Figure 10).

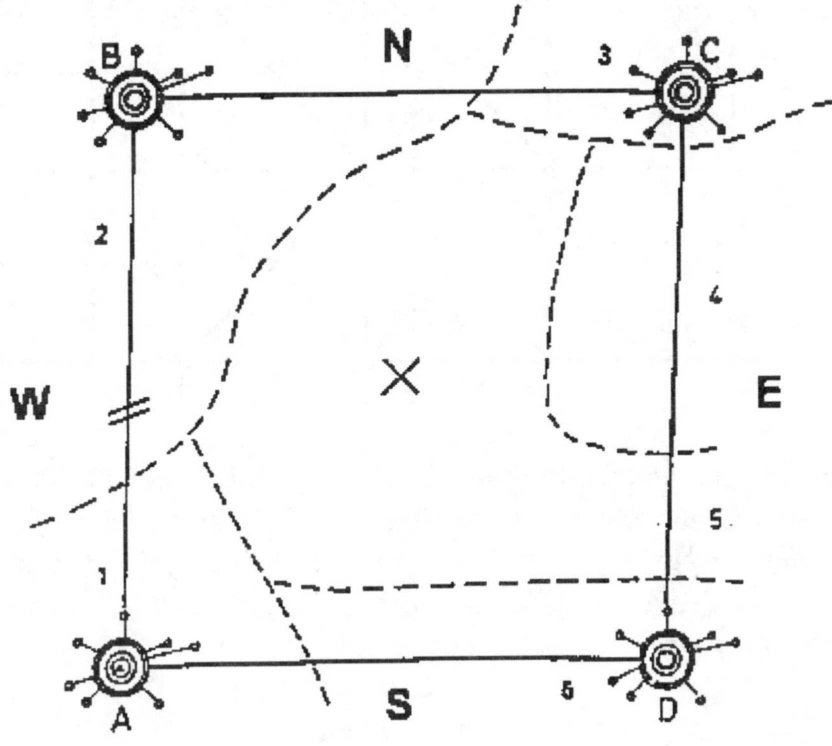

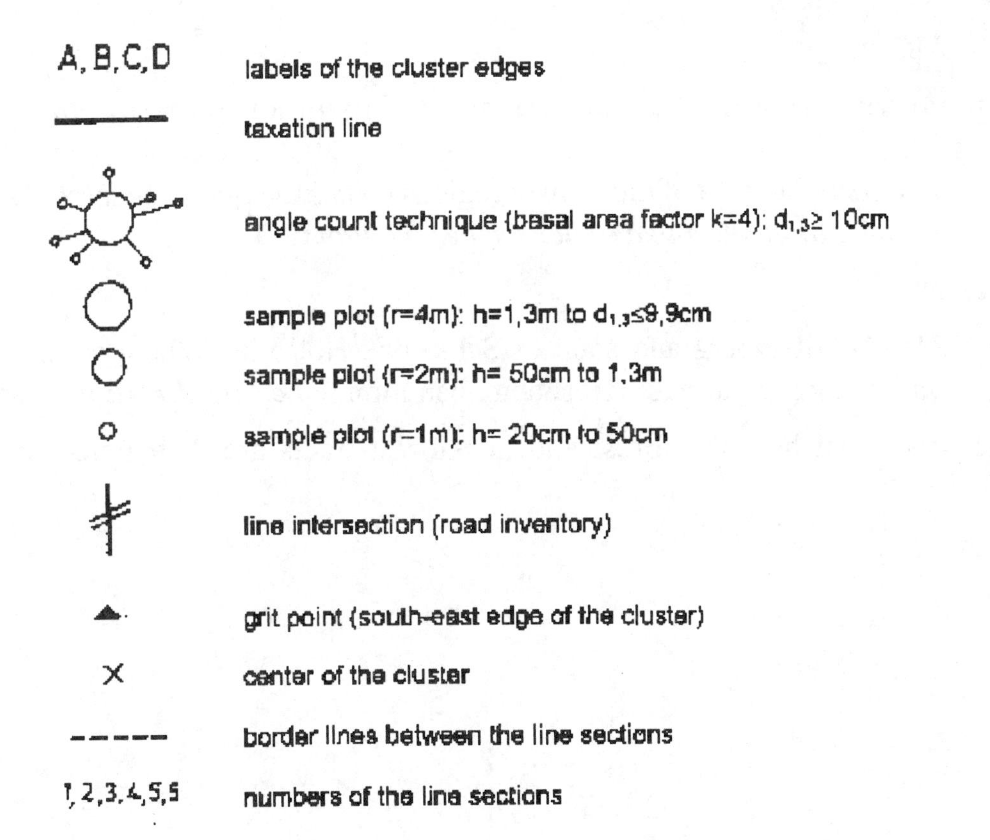

Figure 9. Tract desgin of German NFI I (BML 1986a, p. 19).

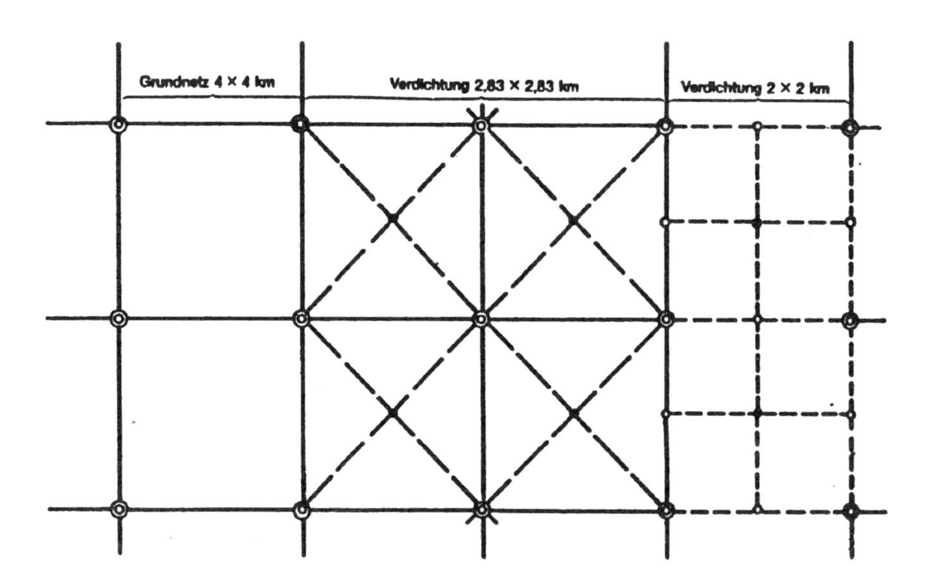

Figure 10. Grid for allocation of tracts (BML 1986b, Annex to §2). Left 4km x 4 km grid Grundnetz = *base grid), centre 2.83km x 2.83 km grid (i.e. doubling the number of grid points compared to base grid of 4km size), right 2km x 2km (i.e.quadrupling the number of grid points compared to base grid of 4km size).*

2.83km by 2.83km:

State of Bavaria: Area of the state forest administration (*Oberforstdirektion*) Augsburg.
State of Lower Saxonia: Area of the growth regions (*Wuchsgebiete*) *Niedersächsischer Küstenraum* and *Mittel-Westniedersächsisches Tiefland.*

2km by 2km:

State of Baden-Württemberg and state of Schleswig-Holstein: Whole area.
State of Bavaria: Area of the state forest administration (*Oberforstdirektion*) Ansbach.

In Table 4 is listed how the forest and non-forest tracts are distributed over the country

Table 4. Number of forest and non-forest tracts in the 11 federal states covered by German NFI 1 (BML 1992a).

Federal State	Forest tracts	Non-forest tracts	Total
Baden-Württemberg	4,685	4,285	8,970
Bayern	3,279	3,182	6,461
Berlin (West)	5	25	30
Bremen	0	26	26
Hamburg	4	50	54
Hessen	702	623	1,325
Niedersachsen	1,424	3,162	4,586
Nordrhein-Westfalen	905	1,239	2,144
Rheinland-Pfalz	687	559	1,246
Saarland	82	81	163
Schleswig-Holstein	807	3,166	3 ,973
Total	12,580	16,398	28,978

Locating of the tract starts with any tract corner. Normally this is the most easy accessible tract corner. Measurement starts from a unambiguously identifyable reference point nearby. This reference point must be clearly identifyable in the field and in the map. Measurement is done with SUUNTO-compass and ultra-sonic distance meter or distance tape. Magnetic declination has to be considered. Measurement procedure is comprehensively recorded in the form sheets (BML 1986a IV.1).

Side length of the square tract is 150m. The perimeter line of 600m is used for the line intercept sample and for the line intersect sample. At the corner points concentric sample plots and point samples are located. Lines are measured with ultra-sonic distance meter or tape and SUUNTO compass. In very steep terrain short segments are measured to avoid oblique aiming with the SUUNTO compass. Slope correction has to be applied. Correction table is provided in 5% steps. In some states (like Baden-Württemberg) pocket calculators were used to do slope correction.

After having completed measurement of a tract the following dislocations between starting point and end point are tolerated (BML 1986b p. 25):

- in normal terrain: 5m,
- in difficult terrain: 10m,
- in very difficult terrain: 20m.

Very difficult terrain means extremely steep terrain.

If measurement result does not comply with the tolerance range, the tract has to be re-measured.

Tract corners inside forest are permanently marked with iron tubes at plot centre (BML 1986a IV.2). If it is not possible to fix an iron stick there a nearby reference point is marked, measured and refered to in the tract map. In case that no one tract corner lies in forest but parts of the taxation line do, the segments in forest are permanently marked with iron tubes at the intersections forest boundary/taxation line.

General description of the four sample concepts:

Line intercept sample (*Taxationslinie*):

Perimeter of the square tract is used as taxation line for estimation of area strata. Stand type, forest structure attributes, ownership status etc. are registered for each change in stand type, ownership or ownership size class (segment of the sample line). Line segments outside forest are only registered as 'non-forest'. Does a stand boundary line have multiple intersections with the taxation line, the corresponding stand(s) are only described once in the form sheets.

Measurement procedure is described above in the context of tract measurement.

As tracts are permanently marked this holds also for the taxation line.

Line intersect sample (*Linienschnittpunkt*):

Perimeter of the square tract is also used as taxation line for road length estimation by line intersect sampling. Number of intersections are counted. If a road has multiple intersections each intersection is recorded. Roads at which loading of log trucks is not permitted (highways etc. (*Bundesautobahnen, Bundesstraßen*) are not recorded. Intersections are also drawn in the tract sketch.

Technical aspects of line measurement see Line intercept sample the paragraph before.

Point sample (*Winkelzählprobe*) (BML 1988a IV.6)

At each tract corner that comes to lie in forest a point sample is established. Basal area factor is 4. Minimum dbh for sample trees is 10cm. If not clear, in/out decision has to be made by distance measurement. Counting is done with Relascope so that no explicit slope correction is required. Centres of point samples are marked with iron tube. Near stand boundaries mirror technique is applied. Either the measurements are repeated from the new, mirrored plot centres (thus tallying some trees again), or the course of the stand boundary is registered so that mirror technique can be carried out automatically by a computer program. Location of trees is recorded in the point sample with measurement of horizontal distance and azimuth from plot centre.

Measurement procedure how to locate the plot centre is given in the context of the general tract measurement procedure.

Sample plots (*Probeflächen*):

Three concentric sample plots are located around the tract corners which come to lie in forest. Plot centre is the same like for point sample:

Sample plot: 1m radius: Trees with minimum height 20cm up to 50cm.

Sample plot: 2m radius: Trees from 50cm up to 1.3 m height.
Sample plot: 4m radius: Trees with more than 1.3m height and dbh less than 10cm.

Minimum distance to stand boundary is 1m, 2m or 4m. If this is not true, plot centre is shifted so that the whole plot lies in forest (BML 1986a IV.7). In the concentric circular plots tree regeneration is tallied. Damages to the regeneration (peeling, browsing, faying) and protective measures (fences, chemicals etc.) are recorded.

Measurement procedure how to locate the plot centre is given in the context of the general tract measurement procedure.

2.3.3 Sampling design

Basic sampling design of the first German NFI is cluster sampling with a systematic arrangement of clusters.

2.3.4 Tecniques and methods for combination of data sources

Different data sources to be combined in data processing and analysis of first German NFI were the line intercept sample on the one hand and the other sample types (point sample, concentric circular plots) on the other hand.

In order to combine the line sample and the cluster sample approach, analysis of the inventory data was done following a two-stage -sampling concept (BML 1992a, p. 34ff): Tract line is regarded as a 600m strip. The strip is the primary sampling unit. The strip is partitioned in smaller units, the secondary units, out of which (up to) four are selected. Selection, however, is not random but always the sample at the four corners of the tract are selected.

2.3.5 Sampling fraction

Four different sampling techniques are incorporated in the German NFI, as mentioned before.

For point sampling the sampling fraction can not directly be given in terms or proportion of area assessed. Therefore only the represented area per plot is given in the listing below.

For fixed area plots sampling intensity is given either as percentage area or as size of the represented area per plot.

For line sampling (line intercept sample and line intersect sample) sampling intensity is given in m/km^2.

Table 5. Sampling fraction and sampling intensity.

Sample unit	4km grid		2.83 km grid		2km grid	
	Sample fraction	Represented area per plot	Sample fraction	Represented area per plot	Sample fraction	Represented area per plot
Point sample		400 ha		200 ha		100 ha
Plot 1m radius	7.86 E-07	400 ha	1.57 E-06	200 ha	3.14 E-06	100 ha
Plot 2m radius	3.14 E-06	400ha	6.28 E-06	200 ha	1.26 E-05	100 ha
Plot 4m radius	1.26 E-05	400ha	2.51 E-05	200 ha	5.03 E-05	100 ha
Line sample	37.5m per km^2		75m per km^2		150m per km^2	

2.3.6 Temporal aspects

There is to date only one German national forest inventory completed.

Inventory cycle	Time period of data assessment	Publication of results	Time period between assessments	Reference date
1st NFI	1986-1989	1992	-	1 October, 1987

2.3.7 Data capturing techniques in the field

Data are recorded on tally sheets. Seven different form sheets were developed:

Sheet 1: Pre-clearing of tract area.
Sheet 2: Localisation and measurement of tract.
Sheet 3: Sketch of tract.
Sheet 4: Line taxation (line intercept sample)
Sheet 5: Point sample.
Sheet 6: Concentric circular plots.
Sheet 7: Road inventory (line intersect sample).

Data were manually entered into the computer.

2.4. DATA STORAGE AND ANALYSIS

2.4.1. Data storage and ownership

Ownership of the data lies with BML (Federal Government, address on last pages of country report) and the state governments. Requests for submission of raw data have to be directed to BML. From there confirmation is requested from the concerned state governments. If no objections are expressed from their side the data can be released to the interested party. For scientific purposes this procedure is free of cost, for private companies the real cost for making the data available are billed.

Data are stored in the computers of BFH (Federal Research Centre for Forestry and Forest Products). This data set is the reference data set for all analysis. All federal states have the data set of their region.

2.4.2. Data basement used

Currently the complete data set of the first German NFI is managed under the data base system Informix on DEC-station in BFH, Eberswalde.

The data sets of the single federal states whoch were sent to the state forestry administration are stored there mainly on Personal Computers in dBase format.

2.4.3. Data bank design (if applicaple)

No formal description of data base design and structure available for German NFI I.

2.4.4. Update level

Field work has been done from 1986 to 1989, in most states between 1986 and 1987. For each field datum it is recorded on which day it was recorded. Given the field assessment period, the maximum deviation between specific measurements and the reference date of October 1, 1987, was about 1.5 years. Given this, it was not deemed necessary to use models to more exactly update the data to the common reference date.

The only updated datum was age (*Alter*). It was generally refered to the references date of October 1, 1987, and not the the date when measurements were taken.

2.4.5. Description of statistical procedures used to analyse data; including procedures for sampling error estimation

Description of the statistical data analysis is done refering to BML 1992a, Chapters 3 and 5.

A) Area estimation

Area calculations are done for federal states separately. Areas are calculated from the line intercept sample. Total line length (forest and non-forest tracts) in a state divided by the (known) total area of this state yield a state-specific expansion factor for area estimates (BML 1992a p.22).

Calculation: If in tract i the line length in forest is W_i and the total line length L_i then the estimate p_i is an estimate for the forest area proportion in the inventory region: $p_i=W_i/L_i$. The mean forest proportion $\bar{p}$ of the entire inventory region is

$$\bar{p}=\frac{\sum_{i=1}^{n} L_i p_i}{\sum_{i=1}^{n} L_i}=\frac{\sum_{i=1}^{n} W_i}{\sum_{i=1}^{n} L_i}$$

with n: Number of tracts (forest and non-forest) in a state

Total forest area F_w of a federal state with total area F is then $F_w=\bar{p}F$.

For error calculation two assumptions are made:

1. Tracts are either completely in forest or completely outside forest.
2. Forest follows spatially a random distribution so that the statistical distribution of the forest/non-forest tracts is binomial.

The relative error of estimation (*relativer Fehler der Schätzung*) eF_{Wa} is then

$$eF_{Wa}=\sqrt{\frac{1-\bar{p}}{n\bar{p}}}$$

For estimation of the error of area proportions for species groups this calculation turned out to be too rough. In this case $\mathrm{var}(\bar{p})$ is calculated directly from the observations p_i:

$$\mathrm{var}(\bar{p})=\frac{1}{n(n-1)}\sum_{i=1}^{n}(p_i-\bar{p})^2$$

B) Aggregation of tree and plot data

See under D) Estimation of ratios.

Road length estimation is done with the concept of line intercept sampling. Road length W_L in kilometer is calculated according to $W_L=\frac{\pi\ n\ 5}{L}F$, where F is the area in hectare of the unit of reference, L is the total length (in meter) of all taxation lines and n is the number of interesections of roads with the taxation line. Road length W_D in meter per hectare in forest is then $W_D=\frac{W_L 1000}{W_F}$, where W_F is the forest area in hectare of the unit of reference.

C) Estimation of total values

As pointed out under D) and H) totals are calculated by aggregating results of the inferior strata. This is generally done using the mean values and weighting them by strata area.

Error for total volume eV_{Ges} is calculated from error of volume eV_{Ha} and error of area eF: $eV_{Ges} = \sqrt{eV_{Ha}^2 + eF^2}$

D) Estimation of ratios (proportions, estimates related to unit area)

For the concentric circular plots calculations are done using expansion factors according to plot size. These factors are 3183.1 for the 1m circle, 795.8 for the 2m circle and 198.9 for the 4m circle.

For the point sample and the estimates of area related stand characteristics: These are derived with the concept of a two-stage sample: Primary unit is a tract, regarded as a 600m strip of 20m width. This is about the width of observation for stand description for the line intercept sample. The strip is segmented in several strata according to species composition, age class etc. Mixed stands are converted in a set of ideal pure stands. The algorithm for that is given in BML 1992a p.38ff.

The inventory region is covered with a population of N strips. From this population a sample of size n is drawn. It is assumed the principles of random sampling apply.

A strip i of length l_i is now segmented in $M_i = l_i/a$ squares of side length a, where M_i is assumed to be a whole number. These squares are the secondary sampling units. A sample of m_i is drawn from them with m_i 4, as each selected secondary unit corresponds to a tract corner. Systematic arrangement of tract corners within the set of secondary units is disregarded.

In each selected secondary unit sample measurements have been taken for stand characteristics. In the following Y stands for volume growing stock per hectare. The secondary unit j in strip i then yields the hectare volume $Yha_{i,j}$ or volume $Y_{i,j}$ in secondary unit j of size a^2.

Is the entire strip in single a pure stand the mean $\bar{Y}_i$ is the most precise estimator of the mean volume in one strip:

$$\bar{Y}_i = \frac{1}{m_i} \sum_{i=1}^{m_i} Y_{i,j}$$

Does a strip intersect more than one stand, for each segment the mean volume of the corresponding secondary units is calculated.

The situation occurs that the strip intersects stands but no tract corner (with a sample plot or point sample) falls into this stand. No tree information is available there, these stands do not enter the volume calculation.

An estimate $\bar{Y}$ for the mean volume $\bar{y}$ of the secondary units within one stratum of analysis is then

$$\overline{Y} = \frac{\sum_{i=1}^{n} M_i \overline{Y}_i}{\sum_{i=1}^{n} M_i} = \frac{\sum_{i=1}^{n} l_i \overline{Y}_i}{\sum_{i=1}^{n} l_i}$$

with n: Number of tracts in a stratum.
Mi: Number of secondary units in line segments with tree measurements
l_i: Sum of lengths of line segments with tree measurements in same stratum of analysis in tract i.

From this mean volume $\overline{Y}$ for the area a^2 one can now calculate the mean volume per hectare

$$\overline{Y}_{\text{Ha}} = \frac{10000}{a^2} \overline{Y} .$$

The variance of $\overline{Y}_{\text{Ha}}$ is estimated as

$$\hat{\text{var}}(\overline{Y}_{\text{Ha}}) = \frac{1}{(\sum_{i=1}^{n} M_i)^2} \frac{n}{n-1} \sum_{i=1}^{n} (M_i^2 (\overline{Y}_i - \overline{Y})^2)$$

An estimate for the standard error of the mean volume per hectare eV_{Ha} for one species group within a superior stratum is

$$eV_{\text{Ha}} = \frac{1}{\sum_{i=1}^{n} l_i} \sqrt{\frac{n}{n-1} \sum_{i=1}^{n} l_i^2 (\overline{Y}_{\text{Ha}\,i} - \overline{Y}_{\text{Ha}})^2} .$$

Results for superior strata are calculated by weighting with the strata areas.

E) Sampling at the forest boundary

Treatment of sample plots near forest boundary is different for point samples and for concentric circular plots.

For the second, plots are shifted into the stand so that for data analysis the situation of boundary plots does not occur.

For point samples the mirror method has been selected and applied. In the initial phase of the inventory the classical mirror method according to Schmid was applied, mirroring the plot centre out of the forest and redoing the measurements. This means that some trees are counted twice. The procedure is critical if the outside point is not accessible (e.g. Highway). So in the further development of the inventory the procedure was changed in that only the course of the stand boundary was measured relative to the plot centre with polar coordinates. For each tallied tree an individual basal area factor was then calculated according the area of the ideal plot area (*Grenzkreis*) of the tree.

F) Estimation of growth and growth components (mortality, cut, ingrowth)

Does not apply (first NFI).

G) Allocation of stand and area related data to single sample plots/sample points.

Descriptive attributes of stands are assessed along the taxation line, not only at the tract corners. The samples at tract corners are assigned to the stratum which the line belongs to at the tract corner points.

For species-wise evaluations mixed stands have to be split up into ideal pure stands (*ideelle Reinbestände*) of only one species each. This splitting up is done according to the crown projection area (*Standfläche*) the individual trees exhibit. This is different for different species. Therefore crown-area-equations (*Standflächengleichungen*) have been developed giving the crown projection area s as a function of dbh. Regression model is

$$s = b_0 + b_1 \frac{\pi}{4} dbh^2$$

The coefficients were determined for 13 species groups as given in Table 6. Each tree in the point sample represents N_i trees per hectare. Stand area for a specific species group is calculated from all trees of this group as

$$S'_{\mathrm{Ba}} = \sum_i N_i s_i$$

The sum of all species group stand areas must sum up to one hectare, so that finally

$$S_{\mathrm{Ba}} = S'_{\mathrm{Ba}} \frac{10000}{\sum S'_{\mathrm{Ba}}}$$

Area is estimated from the line length within a stratum (see A)). Therefore line length l is subdiveded and assigned segmentwise to the ideal pure stands proportionally to the stand area.

Line segments without sample plots cannot be treated in the same way as no information is available on species composition. These line segments are partitioned and assigned to the stand areas proportional to the species shares in the corresponding strata.

Table 6. Coefficients for the regression model for calculating crown projection area as a function of dbh.

Species	b_0	b_1
Spruce	2.85	195
Fir	2.85	200
Douglas Fir	5.00	200
Pine	1.00	300
European Larch	5.00	285
Japanese Larch	5.00	260
Beech	1.33	300
Oak	1.11	395
Red Oak	2.50	350
Ash	2.50	330
Alder	2.50	435
Birch	2.50	525
Poplar	23.00	320

H) Hierarchy of analysis: How are subunits treated?

During data analysis the inventory region is stratified according to federal states, systems of management, ownership, age classes and species groups. Smallest logical unit of analysis is therefore a species group in a age class in an ownership type in a system of management in a federal state.

All area related stand characteristics (volume/ha, number of stems/ha ...) are calculated first for this smallest unit. The values for the superior classes are with strata areas weighted mean values.

2.4.6. Software applied

Data of the first German NFI are currently managed under the data base system Informix on DEC-station. The data of the federal states are stored there in the responsible state forest administrations on Personal Computers in dBase format.

Data processing and reporting of the first NFI took place in BFH (Hamburg, at that time), under self-written, custom tailored Pascal programmes. No commercial data base software was used at that time. No further description of the program system or of single modules is available.

Parallel to data processing in BFH, Hamburg, FVA Freiburg developed a PC based regionalization software with which results can be produced for (almost) any subregion of the inventory region. This software uses dBase database logic and Clipper for program development. It is not part of the actual NFI but an additional development. The software was made available to the single states and enabled them to make their own data analysis with the data sets of the state.

2.4.7. Hardware applied

Data analysis took place on VAX computers of DEC.

2.4.8. Availability of data (raw and aggregated data)

Results are publicly available for the whole of the country and for the federal states.

Raw data and aggregated data are not publicly available. As mentioned above, requests for submission of raw data, of aggregated data or requests for special evaluations have to be directed to BML. From there confirmation is requested from the concerned state governments. If no objections are expressed data can be released to the interested party.

There is no legal regulation on cost. Normally, for scientific purposes delivery of data is free of cost. For private companies the actual cost for making the data available are invoiced.

Main concern is that data could allow to make evaluations about large private forest owners. This might be in conflict with legal regulations about the protection of private data. No final decision has been made in this respect.

2.4.9. Subunits (strata available)

Analysis is done for six strata:

1. Entire inventory region (*Bundesgebiet*)
2. Federal states (*Bundesstaaten*)
3. Systems of management (*Betriebsarten*)
4. Ownership types (*Eigentumsarten*)
5. Age classes (*Altersklassen*)
6. Species groups (*Baumartengruppen*)

For some states additional analysis has been done. For Baden-Württemberg (FVA Baden-Württemberg 1993) for example results are presented for Forest Directorates (*Forstdirektionen*), for growth regions (*Wuchsgebiete*) and for elevation classes (*Höhenstufen*).

2.4.10. Links to other information sources

—

2.5. RELIABILITY OF DATA

2.5.1. Check assessments

a) How check assessments are organised and carried out:

According to the inventory guidelines (BML 1986a I.9.) control measurements have to be carried out on a minimum of 5% of the forest tracts. The state inventory supervisor (*Landesinventurleiter*) selects the tracts to be controled. On these tracts all measurements are re-done by an independent check crew.

Errors and deviations, particularly those with a systematic character, were to be cleared with the field crew leader. For each controlled tract a protocol was written stating errors and deviations and the means taken for correction. These protocols were sent to the organisation which was responsible for data management.

The actual form and procedure of control measurements was up to the state inventory supervisors (*Landesinventurleiter*). Selection of tracts to be re-measured was generally a random selection with some more weigth on those crews where problems were expected.

Some federal states made controls exceeding the defined 5% of samples.In the state of Baden-Württemberg for example (FVA Baden-Württemberg 1993, p.173) control measurements were effected at about 10% of the samples. In BML 1992a p.20 it is stated that in the average 10% of the tracts were controlled in the single states.

b) How the results of check assessments are used in the inventory system

In BML 1986a I.9 no actual means are listed, what the possible alternatives of reaction are in case of errors. This was up to the state inventory supervisor (*Landesinventurleiter*). Generally, in case of clear errors the crews had to re-measure the plot. If the deviations were small no means were taken at all.

No further analysis of the results of the check assessments has been carried out. They were neither used to model or quantify measurement errors.

2.5.2. Error budgets

Statistical error is estimated and reported for the most important attributes (area, stock volume). For total volume the errors of area and volume/ha are combined.

Formal error budget are not part of the inventory results. Sources of error other than statistical error are not formally documented (like measurement errors, model errors).

2.5.3. Procedure for consistency checks of data

(BML 1992a p. 20) Data were gathered state-wise on form sheets. They were pre-viewed by the state inventory supervisors and checked for completeness.Then they were forwarded to the inventory coordination centre at BFH.

Data were then entered in digital form in the Federal Office of Food and Agriculture (Frankfurt). There formal plausibility checks were carried out, checking mainly for completeness and formal correctness.

More checks were carried out after data were sent in digital form to BFH: Attributes which exhibit relations were cross-checked for plausibility. For individual characteristics like age, height, diameter, threshold values were defined. Over- or undershooting of these values were individually checked. Errors were corrected and re-checked for plausibility. Most of the checks and corrections, however, were done automatically.

There is no publication where procedures and algorithms for plausibility checks are listed in more detail.

Most of the errors found in the first round of plausibility tests were formal errors. Most of the errors could be corrected in the coordination centre at the BFH (Hamburg). In about 10%-15% of the cases the errors had to be corrected by the state inventory supervisors.

As several institutions were involved in different steps of collecting, entering, checking etc. the data. Table 7 gives a idea of the data flow. In BML 1992a, the data flow is more detailed given as shown in Figure 11.

Table 7. Data flow and plausibility checks of data in German NFI I.

1	Data capturing in the field	States: Field crews supervised by state inventory supervisor
2	Check for completeness of form sheets	State inventory supervisor
3	Data entering into computer and simple formal checks	Federal Institute for Food and Agriculture, Frankfurt and private service companies
4	Logical plausibility checks	BFH, Hamburg station with support by FVA Baden-Württemberg
5	Data storage	BFH, Hamburg station
6	Data analysis, completing with derived attributes Data analysis and reporting of results	BFH, Hamburg station
7	Submitting statewise data-sets to the state forest administrations	

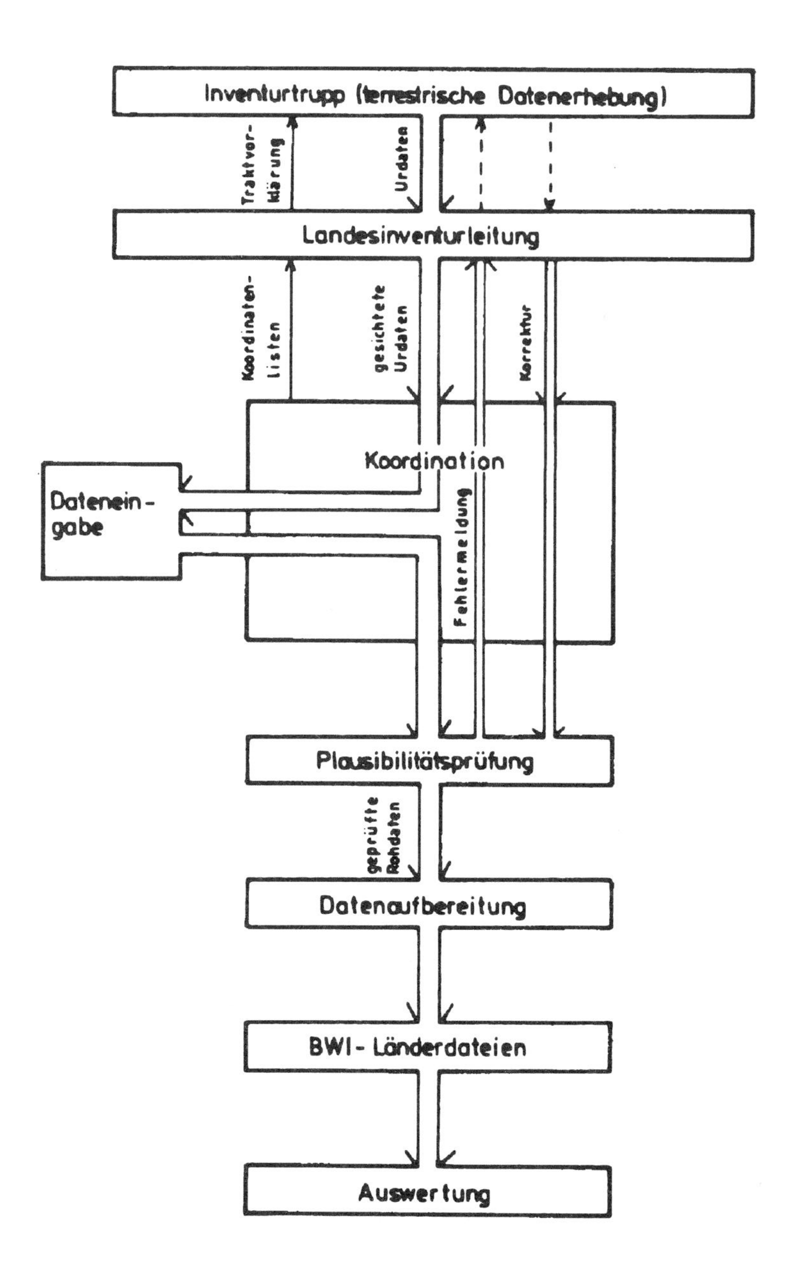

Figure 11. Flow of data processing in German NFI I (BML 1992a).

Translation of German terms:

Main field of activities:
Inventurtrupp (terrestrische Datenerhebung) Field crew
Landesinventurleitung State inventory supervisor
Dateneingabe Data input
Koordination Coordination
Plausibilitätsprüfung Data preparation and organisation
BWI-Länderdateien Data sets for federal states
Auswertung Data analysis

Flows:
Trakvorkärung Pre-clearing phase
Urdaten Original field data
Koordinatenlisten List with geographical coordinates
gesichtete Urdaten Pre-checked original data
Korrektur Correction
Geprüfte Rohdaten Checked original field data

2.6. MODELS

2.6.1. Volume (functions, tariffs)

Modelling of volume and assortment is done with the computer program BDAT which is described in full detail in a publication (Kublin and Scharnagl 1986).

a) Brief outline of the model: Input attributes for volume calculation are dbh, height and an additional upper diameter which was measured at absolute height of 7m. These three attributes are used to identify species-specific taper equations. These equations can be used to calculate the volume of any stem section by simple integration.

Derivation of species-specific taper equations was subject of a comprehensive research project carried out at FVA, Freiburg. In a first step a model was derived which used diameters D in relative heights of 5% ($D_{0.95}$) and 30% ($D_{0.70}$) of total height. This model could be fitted very well to the data sets of the sample trees. As the measurement heights of $D_{0.95}$ and $D_{0.70}$ are - for mean stem length - close to the absolute heights where diameters were measured in the inventory, it was possible to establish good regressions of $D_{0.95} = f(D_{1.3m}, \text{height})$ and $D_{0.70} = f(q_{7m}, \text{height})$, where $q_{7m} = D_{7m} / D_{1.3m}$. The model of taper equation used is a composite equation model based upon spline regression with four additional knots at 30%, 50%, 70% and 90% of total height.

Part of the volume modelling is also modelling of losses due to harvesting. Part of it are bark thickness equations described briefly in 2.6.6. Other losses were also modelled but are not described here (see Kublin and Scharnagl 1988, p. 67ff)

b) Overview over prediction errors: For the prediction error the condition was pre-formulated that it must be less than the one for the standard volume tables used in forest management planning. More information about actual precision is given below under ´model validation´.

As is to be expected precision is less for broadleaf trees.

c) Data material used: Main point in model development was the derivation of species-specific taper equations. For the eight main species the form of between 1,200 and 5,900 sample trees was examined, altogether about 36,000 sample trees.

d) Methods applied to validate the model: Thorough check of model validity was carried out: Original form series of the sample trees were compared with the results of the model. Objective was to check the general validity of the model and to find out whether there are systematic deviations for any ranges of stem forms.

For all species it could be stated that for the range of the data sets which is sufficiently covered with sample tree data the model fit is very good. For more extreme taper forms, higher deviations were observed, in some cases for stem tips these deviations even exceeded the 5% mark. As the stem top is of minor interest in calculation of volumes and for assortment purposes, this drawback is of minor importance for the model.

2.6.2. Assortments

Modelling of volume and assortment is done with a computer program BDAT which is described in detail in a publication (Kublin and Scharnagl 1986). Main elements of the models used are given in Chapter 2.6.1 already, mainly with respect to precision and model validation.

Volume calculation uses species-specific taper equations. It is clear that presence of a taper equation allows to do assortments of single stems and of whole stands. Assortment therefore is a commercial grading according to dimensions (length and middle or top diameter). There were not sufficient stem quality attributes assessed to perform a quality grading, too.

Program BDAT follows the official guidelines on round wood assortments *(Handelsklassensortierung für Rohholz).* Wood volume (under bark) was calculated with the taper equations and specific bark thickness equations, which were also derived in the context of developing BDAT.

For softwood up to 4 different round wood products were distinguished on a single stem: (1) low quality, damaged parts at stem base (*X-Holz)*, (2) timber (*Stammholz)*, (3) additional log at upper stem position (*Abschnitt HL im oberen Schaftbereich*), (4) pulpwood (*Industrieholz*).

For broadleaf trees assortments algorithm uses the attribute stem length assessed on the sample trees.

2.6.3. Growth components

Does not apply.

2.6.4. Potential yield

In the framework of German NFI I no calculation of potential yield has been done.

In follow up studies a yield prediction model (*Holzaufkommensprognose*) has been built, on the basis of NFI I data. The model was developed in FVA Baden-Württemberg (Bösch 1995).

With this computer program which does access the NFI I data base the potential yield for any sufficiently large region can be predicted. Detailed reports are in press and will be published in 1996.

2.6.5. Forest functions

No modelling done.

2.6.6. Other models applied

Bark thickness equations:

(Kublin and Scharnagl 1988, p. 61f) Information about bark thickness was required for deriving the wood volume from the taper curves which yielded volume over bark only. Wood volume was required for assortment purposes. Altogether 26 bark thickness equations groups were calculated for 23 species. Other species were assigned to a specific species group. For Norway spruce, white fir and Douglas fir two equations each were derived, one for the middle diamter assortment and one for the *Heilbronner Sortierung*, a special assortment scheme for softwood.

Basis for the bark thickness equations was a data set of about 20,000 stems of different species. For the middle diameter assortment scheme three equations per species were derived, one for the butt log, one for the middle part of the stem and one for the upper part. With this partitioning a very good model fit and high precision could be reached.

Modelling areas of pure stands from measurements of mixed stands:

For species-wise evaluations mixed stands had to be split up into ideal pure stands (*ideelle Reinbestände*) consisting of only one species. This splitting up is done according to the crown projection area (*Standfläche*) the individual trees exhibit. Crown projection area (*Standfläche*) is different for different species and depends on dbh. Crown area equations (*Standflächengleichungen*) have been developed giving the crown projection area as a function of dbh. The corresponding coefficients of the regression model are given in Table 6, a graphical presentation in Figure 12.

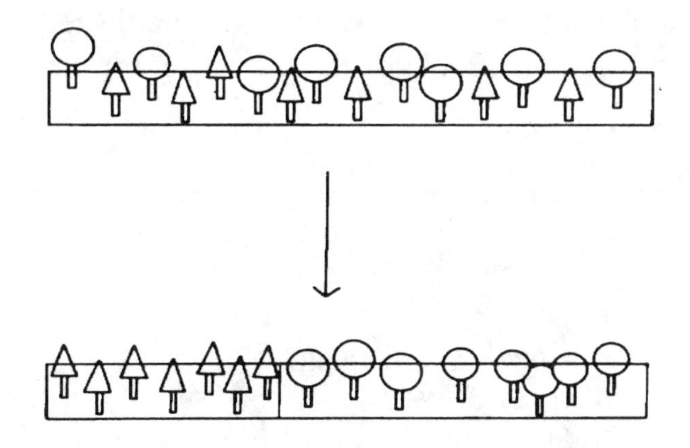

Figure 12. Modelling ideal pure stands from mixed stands (BML 1992a).

Dbh measurement on other heights than 1.3m:

If it was not possible to measure dbh in 1.3m height it was measured at another height. This height was recorded and used afterwards to make an estimate of dbh.

2.7. INVENTORY REPORTS

2.7.1. List of published reports and media for dissemination of inventory results

A two volume inventory results report has been published through BML:

Inventory	Year of publication	Citation	Language	Dissemina tion
First German NFI	1990a	BML: *Bundeswaldinventur, Band I: Inventurbericht und Übersichtstabellen* (German National Forest Inventory, Volume I: Inventory report and summary tables). 118 pp.	German	printed
First German NFI	1990a	BML: *Bundeswaldinventur, Band II: Grundtabellen* (German National Forest Inventory, Volume II: General tables. 366 pp.	German	printed

For each federal state BML published one volume of state results. These reports have exatcly the same structure as the report for the whole of Germany. However for the states of Saarland, Berlin (West) and Hamburg not all tables are given due to insufficient sample size.

The two volumes for the whole of Germany and the statewise reports are the only official inventory results which are formal part of the NFI.

Several other papers in the context of NFI results were published either from BML of from state inventory supervisors. Some of them are listed in the following:

Inventory	Year of publication	Citation	Language	Dissemina tion
First German NFI. Results for state Baden-Württemberg	1993	Forstliche Versuchs- und Forschungsanstalt Baden-Württemberg, 1993: *Der Wald in Baden-Württemberg im Spiegel der Bundeswaldinventur 1986-1990.* 180 pp. (Forest in the state of Baden-Württemberg according to the German NFI)	German	Printed
A similar publication as before has been published for the state of Bavaria, too.				
Saarland, Regional Inventora	1990	Heupel 1990	German	Printed
First German NFI	1994	BML: *Bundeswaldinventur 1986-1990: Eine Wertung.* 29 pp. (Valuation of German NFI)	German	Printed

Available forest data and first German NFI	1994	BML: *Der Wald in den neuen Bundesländern. Eine Auswertung vorhandener Daten nach dem Muster der Bundeswaldinventur.* 20 p. + tables. (Forest in the new federal states. An analysis from available data according to the methods of the German NFI)	German	Printed

In the meantime there are many journal publications interpreting and analysing the NFI results. These were published mainly in the German forestry journals *Forst und Holz* and *Allgemeine Forstzeitschrift*. They are written in German. These publications are - of course - not formal part of the NFI itself, they are more on the scientific and forest policy side.

2.7.2. List of contents of latest report and update level

Reference of latest report (2 volumes):

BML 1992a: *Bundeswaldinventur, Band I: Inventurbericht und Übersichtstabellen* (German National Forest Inventory, Volume I: Inventory report and summary tables), 118 p.

BML 1992b: *Bundeswaldinventur, Band II: Grundtabellen* (German National Forest Inventory, Volume II: General tables. 366 p.

Update level of inventory data used in the report: Field data 1986-1990, reference date October 1, 1987.

Volume I:

Chapter	Title	Number of pages
I.	Inventory report	
1	Introduction	2
2	Inventory concept and data assessment	6
3	Data analysis	3
4	Inventory results	7
5	Methods of data analysis	10
II.	Summary tables	
Section 1	Areas	4
Section 2	Growing stock (Volume)	8
Section 3	Forest structure	4
Section 4	Age classes	3
Section 5	Assortment	2
Section 6	Tree damages	13
Section 7	Roads	5
Appendix	Glossary	9
	Acronyms	2
	References	1

Volume II:

Chapter	Title	Number of pages
Section 1	Areas and growing stock per ownership	14
Section 2	Dendrometric characteristics, number of stems, distributions of diameter classes and assortment classes	203
Section 3	Forest structure, species composition, stand types	30
Section 4	Tree damages	56
Section 5	Roads	10

2.7.3. Users of the results

The inventory results for the whole of Germany (state of Oct. 3, 1990) have been published and are available for everyone. Through BML they were distributed to the forest administrations of the federal states, to Universities and other forestry related institutions.

No formal statistics have been established of who ordered copies of the results volumes.

About 5-10 requests were made for raw data sets. They came from universities and other research institutes.

2.8. FUTURE DEVELOPMENT AND IMPROVEMENT PLANS

Expert commission (*Gutachterkommission*) and BFH have developed a concept for the second German NFI. Not all technical details are yet decided upon, nor is there a comprehensive publication.

No decision has been made yet about whether and when the second NFI will be carried out. A group of experts is currently (beginning of 1996) working on it. Main point of discussion are costs and reliability of results for single federal states.

2.8.1. Next inventory period

Not decided yet.

2.8.2. Expected or planned changes

The following description of expected or planned changes bases on a paper presented during an inventory meeting in March 1996 (Polley 1996) and on communication with Dr. Polley, Eberswalde.

2.8.2.1. Nomenclature

Dbh:

Minimum dbh for the point samples is now 7cm (not 10cm like in the first NFI) and corresponds with the general threshold value for solid volume (*Derbholz*) as applied in German forestry. In NFI I the tree information in the diameter range between 7cm and 10cm had to be taken from the concentric circular plots.

Dead wood (*Totholz*):

For ecological reasons dead wood is assessed.

A differentiation is made according to dimension (3 classes).

- Thick stems (more than 20cm at thick end)
- Small (less than 20cm at thick end)

For *big stems* (more than 20cm at the big end) the following attrributes are recorded

- Species / Species group - if identifiable
- Middle diameter
- Length / height for stumps
- Degree of decomposition:
  1 = recently died *(frisch abgestorben)*
  2 = starting decomposition *(beginnende Zersetzung)*
  3 = proceeded decomposition *(fortgeschrittene Zersetzung)*
  4 = heavily decomposed *(stark vermodert)*
- Type
  1 = standing *(stehend)*
  2 = lying *(liegend)*
  3 = root stumps (from 50cm height or 60cm diameter at cut) (*Wurzelstöcke*)

Prevailing species group and mean degree decomposition are also recorded.

Small wood (less than 10cm at the thick end): Area coverage is estimated in 10%-steps.

Height and upper diameter:

For sample trees measured at first occasion (NFI I), height and upper diameter measurement are skipped and replaced by models. This, of course, does apply only for the inventory region of the first German NFI.

Stand type:

It showed that the attributes assessed for stand description in NFI I were not sufficient. It will be enhanced through a detailed assessment of vertical and horicontal structure. Also elements of shrub layer and ground vegetation will be assessed.

Damages:

Damages will be assessed in a more detailed manner.

Forest boundary:

Length and structure of the forest boundary will be assessed for ecological reasons.

Potential natural forest formation:

Potential natural forest formation will be assessed.

2.8.2.2. Data sources

Line intercept sample dropped:

One of the most important technical changes will be that the line intercept sample is dropped. The separation of assessment of stand description attributes (line intercept sample) and of dendrometric chracteristics (tract corners, point sample and concentric circular sample plots) is therefore given up. It had caused considerable problems in data analysis and created the situation that stand types were identified through the line sample without having dendrometric data for it, because no tract corner fell into this stand type.

Stand description attributes will now be made on reference areas at the tract corners where also the point sample and the circular concentric plots are located. In this reference area area attributes are assessed which can not be measured on a dimensionless point (plot centre).

Diameters of concentric circular plots changed:

A 10m radius circle is for assessment of dead wood. This information is assessed for ecological reasons (see 2.10.3.1 Nomenclature)

The 4m and 2m circles are replaced by a 1.75m circle. 1.75m is in point samples the radius of the imaginary boundary circle of a tree of 7cm diameter. Thus the concentric circles join gapless to the point sample.

Other information sources:

It is planned to combine the results with other inventories, information sources and models.

It is planned to utilize other maps on a regular basis where they are available. This refers particularly to maps of the site mapping campaign and to maps of the biotop mapping campaign.

2.8.2.3. Assessment techniques

While in German NFI I all data were recorded on form sheets in the field, handheld computers for data capturing are under discussion for the second cycle.

2.8.2.4. Data storage and analysis

——

2.8.2.5. Reliability of data

——

2.8.2.6. Models

For reasons of rationalisation of field work height measurements and measurement of upper diameter are skipped for sample trees measured at first occasion (NFI I). Measurements are replaced by models which are to be developed. In the new federal states height and upper diameter are measured.

Other inventories and information sources will be utilized in combination with the NFI II results. To implement this connection specific models have to be developed.

Growth models will have to be developed to derive growth data from the remeasured trees.

2.8.2.7. Inventory reports

Does not yet apply.

2.9. MISCELLANEOUS

Main change in the second German NFI will be the reference area. While the first inventory took place before German unification the second will cover the whole unified Germany (about 350,000 km^2 instead of about 250,000 km^2).

Rationale for planning the second NFI was that consistency and compatibility to the first inventory cycle is guaranteed to allow a statistically sound change assessment by comparing the new with the old data.

3. FOREST HEALTH MONITORING (*WALDSCHADENSERHEBUNG*)

3.1 NOMENCLATURE

Attributes covering ICP Standards and attributes to geographic regions are listed. Additional attributes collected by single federal states are only given for one example, for the federal state Baden-Württemberg (signed with *BW* -). Since the nation-wide surveyed attributes are defined according to ICP Standards, only the main attributes are described in detail.

3.1.1 List of attributes directly assessed

A) Geographical regions

Attribute	Data source	Object	Measurement unit
Forest district *(Forstamt)*	thematic map	plot	categorical
Growth zone (*Wuchsgebiet)*	thematic map	plot	categorical

C) Wood production

Attribute	Data source	Object	Measurement unit
Dbh (not measured every year)	field assessment	tree	cm
BW - Cuttings during the last five years	local forest service and field assessment	plot	yes/no
BW - Tree location	field assessment	tree	distance and azimuth to plot centre

D) Site and soil

Attribute	Data source	Object	Measurement unit
Altitude (*Höhe*)	topographic map	plot	m
Aspect (*Exposition*)	topographic map	plot	categorical
Soil type (*Bodenart*)	thematic map and field assessment	plot	categorical
Mean age of predominant species	field assessment	plot	years
Humus cover	field assessment	plot	categorical
Availability of water to principal species	field assessment	plot	categorical
BW - Slope	field assessment	plot	per cent

BW - Climatic zone	thematic map	plot	categorical
BW - Geology	thematic map	plot	categorical
BW - Percentage of stones in the soil (*Bodenskelett*)	field assessment	plot	categorical
BW - Fertilisation (other types)	local forest service	plot	year
BW - Soil depth	field assessment	plot	categorical
BW - Terrain condition (*Geländeform*)	field assessment	plot	categorical
BW - Silvicultural system	field assessment	plot	categorical
BW - Dominating species	field assessment	plot	species and per cent (per cents sum up to 100)
BW - Crown closure (*Kronenschluß*)	field assessment	plot	categorical
BW - Stand structure (*Bestandesaufbau)*	field assessment	plot	categorical
BW - Mixture distribution (*Mischungsform*)	field assessment	plot	categorical

E) Forest structure

Attribute	Data source	Object	Measurement unit
Tree species (*Baumart*)	field assessment	tree	categorical

G) Forest condition

Attribute	Data source	Object	Measurement unit
Defoliation (*Nadel-/Blattverlust*)	field assessment	tree	5% steps (0%, 5%, ...,95%, 100%
Discoloration (*Vergilbung der Nadeln und Blätter*)	field assessment	tree	categorical
Easily identifiable damage - biotic damage and desease (above all insects and mycosis) (*Schädlingsbefall durch biotische Schaderreger, vor allem Insekten und Pilze*)	field assessment	tree	name of insect or desease

Easily identifiable damage - biotic damage and disease (above all insects and mycosis) (*Schädlingsbefall durch biotische Schaderreger, vor allem Insekten und Pilze*)	field assessment	tree	categorical

K) Miscellaneous

Attribute	Data source	Object	Measurement unit
Longitude		plot	
Latitude		plot	
Date of observation		plot	
Plot number		plot	1 to 9999

3.1.2 List of derived attributes

Attribute	Object	Measurement unit	Input attributes
Damage Class Type 1 (*Schadstufen*)	tree	ordinal categories	defoliation
Damage Class Type 2 (*kombinierte Schadstufen*)	tree	ordinal categories	defoliation, colour change

3.1.3 Measurement rules for measurable attributes

G) Forest condition

Defoliation

a) according to the International Co-operative -Programme on Assessment and Monitoring of Air Pollution Effects on Forests (ICP)
b) -
c) in 5% steps (0%, 5%, ...,95%, 100%)
d) see measurement scale
e) leaf loss in contrast to a reference tree
f) expert estimation

Discoloration

a) according to the International Co-operative -Programme on Assessment and Monitoring of Air Pollution Effects on Forests (ICP)
b) -
c) in 5% steps (0%, 5%, ...,95%, 100%) percentage of discoloured (*vergilbt*) leaves
d) see measurement scale
e) -
f) expert estimation

3.1.4 Definitions for attributes on nominal or ordinal scale

G) Forest condition

Damage class type 1

a) Cassification according to ECE and EU

b)

Defoliation class	Defoliation %
0	≤10
1	11-25
2	26-60
3	61-99
4	100

c) Field

d) -

Damage class type 2

a) Cassification according to ECE and EU

b)

Damage class type 1	Colour change, % of leaves with changed colour		
	0-25	26-60	61-100
0	0	1	2
1	1	2	2
2	2	3	3
3	3	3	3

c) Field

d) -

3.1.5 Forest area definition and definition of "other wooded land"

The forest definition of the plots of the 16 km * 16 km plots follows the ICP-Standard. This standard follows a definition of FAO. The plots of the grids of higher density follow forest definitions of the federal states, according to the definitions for forest management surveys.

3.2 DATA SOURCES

A) Field assessment.

E) Several mapsets, different in the 15 federal states.

3.3 ASSESSMENT TECHNIQUES

3.3.1 Sampling frame

Sampling frame is forest area of Germany, including all federal states (*Bundesländer*). Only forest with trees higher than 60 cm and trees from the top stand layer (*Oberschicht*) (trees of Kraft's Classes 1-3) are included in the survey.

3.3.2 Sampling units

Different cluster designs are applied in the federal states.

Generally:

Clusters of 24 to 30 trees
Permanent plot designs

Example: The state Baden-Württemberg the following design is used:

Four satellite plots (north, south, east, west) with 6 trees per plot and a distance to the cluster centre of 25 m.

The plot centre is located by rapid measurement by tape and compass or equivalent equipment; sample trees are numbered with persistent colour.

3.3.3 Sampling designs

One cluster per grid point is sampled. Grid size varies from year to year and from state to state. Principal grid sizes are:

Grid size	Cycle
16km * 16km	Yearly
4km * 4km	Every about 3 years (1984, 1986, 1991, 1994)

In some years in some states and regions other square grid sizes were used like 2km x 2km, 1km x 1km or even 0.5km x 0.5km. Also rectangular grids are used in some years in some states and regions, like 2km x 4km, 4km x 8km and others.

A complete sample (*Vollstichprobe*) is assessed every about 3 years, in the other years a partial sample (*Teilstichprobe*) is measured. Only in rare cases only the European base grid of 16km x 16km is measured.

3.3.4 Techniques and methods for combination of data sources

In regions with critical forest health conditions aerial photographs were taken.

3.3.5 Sampling fraction

Since distance methods are used to select a fixed number of trees, no fixed plot size can be calculated. Therefore no sampling fraction is calculated.

Grid size	Represented area per cluster
16 * 16 km	25 600 ha
4*4 km	1 600 ha
2*2 km	400 ha
1*1 km	100 ha
0.5*0.5 km	25 ha

3.3.6 Temporal aspects

In the 'old' federal states yearly assessments are conducted since 1984, with exception of 1990, where some federal states did not assess the forest condition because of a hurricane.

In the 'new' federal states, the former GDR (*DDR*), an assessment with this design is conducted since 1990.

3.3.7 Data capturing techniques in the field

Data recorded on tally sheets and edited by hand into computer.

3.4 DATA STORAGE AND ANALYSIS

3.4.1 Data storage and ownership

The Level I data of the 16 * 16 km grid are stored at the BFH , Hamburg.

Responsible: Dr. Lorenz, BFH Hamburg 040-73962 119.
Data base management: Dr. Becker, BFH Hamburg 040-73962 121

The complete raw data are stored and owned by the State Forestry Organisations. A list of these organisations is included in the appendix.

3.4.2 Data base system used

At BFH, Hamburg: SAS.

3.4.3 Data bank design (if applicable)

The data management follows ICP Standards. A brief prescription of the attributes on the entities plot and tree is given in BFH (1996).

3.4.4 Update level

Reference date is the field date of data collection. No updates to other common reference dates are made.

3.4.5 Description of statistical procedures used to analyse data, including procedures for sampling error estimation

Since there is no general design, there is no general algorithm for calculation of the results. The federal states have different procedures. The results for Germany are calculated on the base of federal state results: Totals are totals of state totals and so on. There exists no general procedure to calculate the sampling error. A general statement is, that, to achieve an error less than 5%,at least 300 trees have to be in the sample of the unit of reference (BML 1993).

3.4.6 Software applied

The federal states use different software.

3.4.7 Hardware applied

The federal states use different hardware.

3.4.8 Availability of data (raw and aggregated data)

Raw data are theoretically available. Aggregated date are available by states and for the whole of Germany.

3.4.9 Subunits (Strata) available

Available strata are the states. Several State Forestry Organisations provide aggregated data for subregions.

3.4.10 Links to other information sources

In all states, except for BW, the sampling grid is identical to other parts of the grid of the Level I program.

3.5 RELIABILITY OF DATA

3.5.1 Check assessments

Central training of inventory leaders of all states ensures a minimum of measurement error (BML 1995). Check assessments are standard, but differ within the states.

3.5.2 Error budgets

Attribute	Error source
Damage Class Type 1 (*Schadstufen*)	sampling error, measurement error of the input variable
Damage Class Type 2 (*kombinierte Schadstufen*)	sampling error, measurement error of the input variables

3.5.3 Procedures for consistency checks of data

Checks by inventory leaders of federal states. No generalized procedure.

3.6 MODELS

-

3.7 INVENTORY REPORTS

3.7.1 List of published reports and media for dissemination of inventory results

- Governmental Forest Condition Report (*Waldzustandsbericht der Bundesregierung*)- yearly national reports since 1984.
- Yearly reports on state level since 1983 in some and since 1991 by all states.
- Reports on special topics (examples):

Schöpfer, W., Bösch, B., Hannak, Ch. & Gross, C.P. 1986. Terrestrische Waldschadensinventur Baden-Württemberg 1986. Mitteilungen der FVA Baden-Württemberg Nr. 125.

Schöpfer, W. & Hradetzky, J. 1984. Analyse der Bestockungs- und Standortsmerkmale der terrestrischen Waldschadensinventur Baden-Württemberg. Mitteilungen der FVA Baden-Württemberg Nr. 110.

Schöpfer, W. & Hradetzky, J. 1984. Ausmaß und räumliche Verteilung der Walderkrankungen in Baden-Württemberg. Mitteilungen der FVA Baden-Württemberg Nr. 109.

Schumacher, W. 1993. Dokumentation neuartiger Waldschäden in Baden-Württemberg. Schriftenreihe der Landesforstverwaltung Baden-Württemberg.

3.7.2 List of contents of latest report and update level

The Governmental Forest Condition Report 1995 (*Waldzustandsbericht der Bundesregierung 1995*) is the latest national report of the survey of 1995.

Chapters concerning the forest health monitoring:	Number of pages
1. Forest condition in the Federal Republic of Germany	3
2. Forest condition in the federal states	10
3. Forest condition of main species	5
4. Discoloration	2
5. Replacing sampling trees	2

3.7.2 Users of the results

No published information available. It is general understanding and the experience of the organisations providing the data, that the users are: Scientists, politicians, press and public.

3.8 FUTURE DEVELOPMENT AND IMPROVEMENT PLANS

3.8.1 Next inventory period

1996. For interval details see the chapter on sampling design.

3.8.2 Expected or planned changes

No changes are planned.

3.9 MISCELLANEOUS

-

4. SURVEY OF FOREST SOIL CONDITION (*BODENZUSTANDSERHEBUNG IM WALD, BZE)*

4.1 NOMENCLATURE

The nomenclature of the state inventories are close to the ICP standards. In field assessment and laboratory analysis some slight differences exist. They are precisely listed by Riek and Wolff (1995).

4.2 DATA SOURCES

A) Field assessment.

E) Several mapsets, different in the 15 federal states.

4.3 ASSESSMENT TECHNIQUES

A detailed manual is published has been by BML (1994, 2nd ed.) which states:

- Objectives
- Survey method
- Field work on sample point
- Soil sample taking
- Laboratory analysis

DATA PROCESSING AND ANALYSIS

4.3.1 Sampling frame

The sampling frame is the forest of Germany, including all federal states (*Bundesländer*). The forest definition follows the definition of the Forest Health Monitoring.

4.3.2 Sampling units

Data are collected in a soil profile in the plot centre and in 8 satellite places.

- Soil: Sample taken at soil profile and at 8 satellite places. Mixed sample in different depth-classes down to 2m.
- Humus: Mainly mixed samples from 8 satellite places (L and Of-horizon, Oh-horizon).
- Needle/leaf: For each leading species from 3 individuals of Kraft class 1 and 2 (upper layer trees, *vorherrschend und herrschend*) samples are taken. For conifers also older than only the most recent needle generation are included in the sample (spruce last 3 years, pine, last two years).
- Assessment according to Forest Health Monitoring: Normally this assessment is made and analysis of needle and leaf samples carried out together.

4.3.3 Sampling designs

One cluster per grid point of a systematic grid.

16 km x 16 km grid: EU -norm, Baden-Württemberg (BW).
8 km x 8 km grid: Basic grid in all states.

In some states and regions denser square or rectangular grids are used (4km x 4km, 4km x 8km, 8km x 16km and others).

In BW only those stands, that are dominated by coniferous trees, are part of the sample.

In all states the plots of the Survey of Forest Soil Condition and the Forest Health Monitoring are identical, except for BW. They use different locations, that are identical with a special survey on air pollution impacts in the forest *(Immissionsökologische Waldzustandserfassung IWE).*

If a stand is qualified as a disturbed area the plot centre is moved. The details of this procedure differ within the states.

4.3.4 Techniques and methods for combination of data sources

-

4.3.5 Sampling fraction

Since the different attributes are collected by techniques, that cannot be referred to an exact area, no fixed plot size can be calculated. Therefore no sampling fraction is calculated.

Grid size	Represented area per cluster
16 * 16 km	25 600 ha
8*8 km	6 400 ha
4*4 km	1 600 ha

4.3.6 Temporal aspects

It is not decided yet, whether the survey is repeated. Since the plots are marked, a repetition at the identical grid points is possible.

4.3.7 Data capturing techniques during in the field

Data recorded on tally sheets and edited by hand into computer.

4.4 DATA STORAGE AND ANALYSIS

All raw data are collected and stored by the federal states.

For the most relevant attributes the data were also sent to BFH to prepare a report for the whole of the country.

Responsible is

Dr. Barbara Wolff
Bundesforschungsanstalt für Forst- und Holzwirtschaft
Institut für Forstökologie und Walderfassung
Eberswalde
Tel. 03334/65304

The complete raw data are stored and owned by the State ForestryAdministrations. A list of these organisations is included in the appendix.

All national data are already available at the BFH. A national report is planned for the end of 1996. An analysis of the representativity (*Repräsentativität*) of the results will be made, comparing the results of different sub-grids, but no sampling error will be calculated. It is planned by the BFH to calculate sampling errors in a later stage. The data will be analysed on a national level only. Several strata types will be analysed: soil types, humus types, altitude-strata, stand types.

Aggregated data will then be available on a national level only. An analysis with growth zone strata is planned by BFH for a late stage. There exist aggregated data on state level, provided by the forest administrations of the federal states. They differ with respect to analysis. It would be an easy procedure for BFH to produce state estimates as well. To get such harmonised data for all states or other subregions, the political approval of the federal states is necessary.

A complete documentation of the data management is in preparation, only a preliminary version is now available (BFH Institut für Forstökologie und Walderfassung, 1996).

Software used: SAS, Excel, Access.
Hardware used: DEC, x86 based Pcs.

4.5 RELIABILITY OF DATA

Responsible for data collection and data checks are the federal states. In BFH completenessof data sets is checked and internal plausibility checks are carried out.

In the manual a detailed description of permitted techniques for extraction of samples and for laboratory analysis is given. Proven standard techniques are employed. Additionally techniques are employed to cross-check variation between different laboratory analysis techniques (König and Wolff 1993).

4.6 MODELS

In the analysis of the data pedo-chemical models are used.

4.7 INVENTORY REPORTS

4.7.1 List of published reports and media for dissemination of inventory results

- An unpublished national report based on the data of the ICP-16 km * 16 km grid is already available (RIEK, W. and B. WOLFF 1995).
- A national report based on all data will be published by the end of 1996 by BFH, *Institut für Forstökologie und Walderfassung*, Eberswalde.
- Reports on state Level have been published by several states:

Backes, J., 1993. Aufbau eines Waldbodeninformationssystemes und Ergebnisse der saarländischen Waldbodeninventur. Saarbrücken.

Benecke, P., Eberl, C. & Schulte-Bisping, H. 1994. Die Waldböden Schleswig-Holsteins. Eine Auswertung der Bodenzustandserhebung im Wald 1990-1992. Ministerium für Ernährung, Landwirtschaft Forsten und Fischerei Schleswig-Holstein.

Buberl, H. G., v. Wilbert, K., Trefz-Malcher & Hildebrand, E.E. 1994. Der chemische Zustand von Waldböden in Baden-Württemberg. Ergebnisse der Bodenzustandserhebung im Wald 1989-92. Mitteilungen der FVA Baden-Württemberg 182/1994.

Gehrmann, J., 1993. Zwischenbericht über Ergebnisse der landesweiten Bodenzustandserhebung im Wald - BZE. Stand September 1993. Landesanstalt für Ökologie, Landesentwicklung und Forstplanung Nordrhein-Westfalen. Recklinghausen.

Gulder, H. J. & Kolbel, M. 1993. Waldbodeninventur in Bayern. Forstwissenschaftliche Fakultät der Universität München und FVA Bayern (Hrsg.) Forstliche Forschungsberichte München Nr.: 132.

Hocke, R. 1995. Waldbodenzustand in Hessen. Hessische Landesanstalt für Forsteinrichtung, Waldforschung und Waldökologie. Hessisches Ministerium des Innern und für Landwirtschaft, Forsten und Naturschutz (Hrsg.). Wald in Hessen Forschungsbericht Bd. 19.

Sachsische Landesanstalt für Forsten 1994. Waldschadensbericht 1994. Sächsisches Staatsministerium für Landwirtschaft, Ernährung und Forsten (Hrsg.), Graupa.

4.7.2 List of contents of latest report on national level

The latest national report is RIEK, W. and B. WOLFF 1995: German Contribution to the European Forest Health Monitoring (Level I) (*Deutscher Beitrag zur europäischen Waldbodenzustandserhebung (Level I*)). Its an unpublished internal report of the BFH.

Chapters:	No of pages
1 Introduction	
1.1 National BZE (Bundesweite Bodenzustandserhebung im Wald)	2
1.2 European BZE (Europaweite Bodenzustandserfassung)	1
2 Material and methods (Material und Methoden)	13
3 Results	
3.1 General site data	4
3.2 Chemical soil condition (Bodenchemischer Zustand)	18
4 Discussion	
5 Summary	3

4.7.3 Users of the results

No published information available. It is general understanding and the experience of the organisations providing the data, that the users are: Scientists, politicians, press and public.

4.8 FUTURE DEVELOPMENT AND IMPROVEMENT PLANS

A repetition of the inventory on a national level is in discussion. No concept is developed yet. Several states plan to establish grids of higher density on their territory and to repeat the BZE. No decision has been made yet.

4.9 MISCELLANEOUS

-

5. FOREST ENTERPRISE SURVEY (*TESTBETRIEBSNETZ*)

5.1 NOMENCLATURE

5.1.1 List of attributes directly assessed

A) Geographical regions

Attribute	Data source	Object	Measurement unit
Federal state (*Bundesland*)		enterprise	

B) Ownership

Attribute	Data source	Object	Measurement unit
Owner type (*Waldbesitzart*)		enterprise	

C) Wood production

Attribute	Data source	Object	Measurement unit
Commercial volume of cut per species group		enterprise	m**3 commercial volume (*Erntefestmeter*)

5.1.2 List of derived attributes

5.1.3 Measurement rules for measurable attributes

All standard techniques allowed.

5.1.4 Definitions for attributes on nominal or ordinal scale

Species groups:
1. Quercus sp. (*Eiche*)
2. Fagus sylvatica and other broad leaf species (*Buche und andere Laubbaumarten*)
3. Picea abies, abies sp. and pseudotsuga sp. (*Fichte, Tanne und Douglasie*)
4. Pinus sp., larix sp. and other coniferous species (*Kiefer, Lärche und anderes Nadelholz*)

5.1.5 Forest area definition and definition of "other wooded land"

The forest definition follows the definitions of forest management surveys of the federal states.

5.2 DATA SOURCES

Questionnaire.

5.3 ASSESSMENT TECHNIQUES

Questionnaire.

5.4.1 Sampling frame

The sampling frame are 'all' forest enterprises with more than 200 ha. Since it is a voluntary decision for a private forest enterprise to be a part of the survey, nonresponse might influence the results.

5.4.2 Sampling units

Sampling unit is a single forest enterprise.

5.4.3 Sampling designs

The forest enterprises owned by the nation and by the federal states (*Bundes- und Staatswald*) are covered completely.

The private forest enterprises and those owned by public owners, that are not the nation and by the federal states ('corporation forest') (*Körperschaftswald*) are collected randomly within each state, stratified into 3 size groups.

Size group	forest area under management
1	200 - ≤ 500 ha
2	500 - ≤ 1000 ha
3	>1000 ha

5.4.4 Sampling fraction

Size group	forest area under management	sampling fraction
1	200 - ≤ 500 ha	6 %
2	500 - ≤ 1000 ha	11 %
3	>1000 ha	25 %

5.4.5 Temporal aspects

No update to a fixed date. Data are collected per year. The date of report varies from forest enterprise to forest enterprise.

5.4.6 Data capturing techniques in the field

Data are recorded on tally sheets and edited by hand into computer.

5.4 DATA STORAGE AND ANALYSIS

The data are stored and analysed on a national level at the BML. The data are also analysed on state level , but these data are not available for reasons of data security.

5.4.1 Description of statistical procedures used to analyse data including procedures for sampling error estimation

An expert report on this topic is available at the BML. It states that the sampling error is generally under 5 % for national estimates.

5.4.2 Availability of data (raw and aggregated data)

Only data on a national level are available for reasons of data security.

5.5 RELIABILITY OF DATA

The raw data are checked for consistency. No check assessments.

5.6 MODELS

-

5.7 INVENTORY REPORTS

5.7.1 List of published reports and media for dissemination of inventory results

The results are published in the Governmental Report on the Agrar Sector (*Agrarbericht der Bundesregierung)* (BML 1996) and in the Governmental Statistical Report on the Agrar Sector *(Statistisches Jahrbuch über Ernährung, Landwirtschaft und Forsten der Bundesrepublik Deutschland) (*BML 1995).

5.7.2 List of contents of latest report on national level

The Governmental Forest on the Agrar Sector 1996 is the latest national report and is covering the results of the survey of 1995 (BML 1996). The results of Forest Enterprise Survey are only a part of this report.

Chapters concerning the forestry	Number of pages
Chapter III.1 Forestry	
1. Structure (Struktur)	1
2. National results (Gesamtrechnung)	2
3. Results of enterprises (Betriebsergebnisse)	4

5.7.3 Users of the results

No published information available. It is general understanding and the experience of the organisations providing the data, that the users are: Scientists, politicians, press and public.

5.8 FUTURE DEVELOPMENT AND IMPROVEMENT PLANS

No change planned.

5.9 MISCELLANEOUS

——

6. FOREST INVENTORY OF THE FEDERAL STATE NORDRHEIN-WESTFALEN (*LANDESWALDINVENTUR NORDRHEIN-WESTFALEN*)

In the federal state Nordrhein-Westfalen, the forest department now (1995/1996) prepares a state forest inventory *(Landeswaldinventur LWI)*. A pilot inventory comprising the public forests only has been made in 1995. First results, that are not published yet, are available by Landesamt für Ökologie, Bodenordnung und Forsten ofNRW.

The design is:
Systematic Grid: 1000m * 1000m ; in public forest 500m * 500m

Field measurements:

5(6) different sample elements are utilised on a tract to assess different attributes:
1-5 concentric sample plots (*Probefläche*)

1 Sample plot 1.5m radius : regeneration
2 Sample plot 2m radius: single trees < 10 cm bhd
3 Sample plot 3m radius: single trees >= 10 cm bhd
4 Sample plot 10m radius: single trees >= 10 cm bhd
5 Sample plot 12m radius: ground vegetation (*Bodenvegetation*) assessment
6 Point sample (*Winkelzählprobe*), if a plot meets the BWI grid.

The south corner 'A' of the BWI cluster (*Ecke A eines Bundeswaldinventur-Trakts)* is assessed by a point sample with count factor (Zählfaktor) 4.

The attribute list and the definition of attributes follows mainly German National Forest Inventory. The definition of forest and 'other wooded land' is identical to the definition of the national Inventory BWI: (LÖBF, 1996). Beside field assessment a wide range of other data is used. Satellite data and aerial photos are integrated in a research part on the design. This part is still in the planing stage. Inventory reports are not available yet.

REFERENCES

Chapter 2: NFI I (*Bundeswaldinventur*):

Anonymous 1993. Facts about Germany, Societäts-Verlag, Frankfurt.

Bitterlich, W. The Relascope Idea. Commonwealth Agricultural Bureaux (sic), Norfolk 1980.

BML 1984. *Gesetz zur Erhaltung des Waldes und zur Förderung der Forstwirtschaft—Bundeswaldgesetz* [Act Governing the Preservation of the Forest and the Promotion of Forestry (Federal Forest Act)]. Promulgated on 2 May 1975 (Fed. Law Gazette I p. 1037), amended by the First Act Amending the Federal Forest Act of 27 July 1984 (Fed. Law Gazette I p. 1034).

BML 1986a. *Allgemeine Verwaltungsvorschrift zur Durchführung der Bundeswaldinventur 1986-1989 (BWIVwV)*—General Administrative Guideline Implementing the National Forest Inventory 1986-1989. Federal Ministry of Food, Agriculture and Forestry, Bonn.

BML 1986b. *Instruktion für die Traktaufnahme* (Guideline for Recording Tracts). Federal Ministry of Food, Agriculture and Forestry, Bonn.

BML 1992a. *Bundeswaldinventur, Band I: Inventurbericht und Übersichtstabellen* (German National Forest Inventory, Volume I: Inventory report and summary tables), 118 p.

BML 1992b. *Bundeswaldinventur, Band II: Grundtabellen* (German National Forest Inventory, Volume II: General tables. 366 p.

Bösch, B. 1995. Ein Informationssystem zur Prognose des künftigen Nutzungspotentials (An information system on prediction of potential yield). Forst und Holz 50(19):587-593.

Cochran, W.G. Sampling Techniques. Wiley, New York 1977.

Grossmann, H. 1972 Zehn Jahre permanente Großrauminventur in der DDR (*Ten years permanent large area forest inventory in GDR*). Die sozialistische Forstwirtschaft 22(3):74-76.

Heupel, G. M. 1994. Zur Entwicklung einer Forstinventur auf Landesebene und auf der Basis von permanenten Probekreisen am Beispiel der Landesfforstinventur des Saarlandes. Bericht des Forschungszentrums Waldökosysteme, Reihe A, Heft 120, Göttingen.

Hradetzky, J. 1990. Stichproben an Bestandesrändern (*Random Sampling at Stand Margins*). Mitteilungen der Forstlichen Versuchs- und Forschungsanstalt Baden-Württemberg, No. 152. Freiburg.

Kennel, E. 1973. Bayerische Großrauminventur 1970/71; Inventurabschnitt I: Großrauminventur Aufnahme- und Auswertungsverfahren, Forschungsberichte der FVA München 11.

Kramer, H. 1980. Begriffe der Forsteinrichtung (*Forestry Management Planning Terms*). Schriftenreihe der Forstlichen Fakultät der Universität Göttingen, Volume 48. Göttingen.

Kuusela, K. 1995. Forest Resources in Europe. Cambridge University Press.

Kublin, E. & Scharnagl, G. 1988. Verfahrens- und Programmbeschreibung zum BWI-Unterprogramm BDAT (*Description of the Procedure and Programme of the NFI Subprogram BDAT*). Forstliche Versuchs- und Forschungsanstalt Baden-Württemberg, Freiburg

Polley, H. 1994. Ein Überblick aus BWI und Datenspeicher - Der Wald in den Bundesländern. (*Overview derived from NFI and data base of eastern federal states - Forests in the federal states*). AFZ 49(6):318-321.

Schmid, P. 1986. Stichproben am Waldrand (*Random Sampling at the Forest Margin*). Mitteilungen der Schweizerischen Anstalt für das Forstliche Versuchswesen, Volume 45, No. 3. Birmensdorf.

Verordnung über gesetzliche Handelsklassen für Rohholz (Regulations for the Assortment of Marketable Classes of Timber and Timber Varieties). Promulgated on 31 July 1969, Fed. Law Gazette I pp1075.

Chapter 3: Forest Health Monitoring (*Waldschadenserhebung*)

BML (eds.) 1993. Terrestrische Waldschadenserhebung. Aufgaben, Methoden und Stellenwert. BML, Bonn.

BML (eds.) 1995. Waldzustandsbericht der Bundesregierung 1995 - Ergebnisse der Waldschadenserhebung.

BFH (eds.) 1995. Dauerbeobachtungsflächen zur Umweltkontrolle im Wald. Bundesministerium für Ernährung, Landwirtschaft und Forsten.

König, N. & Wolff, B. 1993. Abschlußbericht über die Ergebnisse und Konsequenzen der im Rahmen der bundesweiten Bodnezustandserhebung im Wald (BZE) durchgeführten Ringanalysen. Berichte des Forschungszentrums Waldökosysteme, Reihe B, Band 33.

Schöpfer, W., Bosch, B., Hannak, Ch. & Gross, C.-P. 1986. Terrestrische Waldschadensinventur Baden-Württemberg 1986. Mitteilungen der FVA Baden-Württemberg Nr. 125.

Schöpfer, W. & Hradetzky, J. 1984. Analyse der Bestockungs- und Standortsmerkmale der terrestrischen Waldschadensinventur Baden-Württemberg. Mitteilungen der FVA Baden-Württemberg Nr. 110.

Schöpfer, W. & Hradetzky, J. 1984. Ausmaß und räumliche Verteilung der Walderkrankungen in Baden-Württemberg. Mitteilungen der FVA Baden-Württemberg Nr. 109.

Schumacher, W. 1993. Dokumentation neuartiger Waldschäden in Baden-

Württemberg. Schriftenreihe der Landesforstverwaltung Baden-Württemberg.

BFH (eds.) 1996. Transnational Forest Damage Data. Formats and files prescribed for the data submission of PCC WEST. Unpublished internal report.

Chapter 4: Inventory of the Forests Soil condition (FSCI) (*Bodenzustandserhebung im Wald (BZE)*)

BFH Institut für Forstökologie und Walderfassung (eds.) 1996. Datenmanagement. Unpublished internal report.

BML (eds.) 1994. Bundesweite Bodenzustandserhebung im Wald (BZE) Arbeitsanleitung. 2. Auflage (1. Auflage 1990). Bundesministerium für Ernährung, Landwirtschaft und Forsten, Bonn.

König, N. & Wolff, B. 1993. Abschlußbericht über Ergebnisse und Konsequenzen der im Rahmen der bundesweit durchgeführten Bodenzustandserhebung im Wald (BZE) durchgeführten Ringanalysen. Berichte des Forschungszentrums Waldökosysteme. Reihe B, Bd. 33.

N.N. 1994: International cooperative programme on assessment and monitoring of air pollution on forests. Manual on methods and criteria for harmonized sampling, assessment, monitoring and analysis of the effects of air pollution on forests. Hamburg und Prag. (reference in RIEK, W., WOLFF, B. 1995).

Riek, W. & Wolff, B. 1995. Deutscher Beitrag zur europäischen Waldbodenzustandserhebung (Level I). Unpublished internal report, BFH.

Chapter 5: Forest enterprise survey (*Testbetriebsnetz*)

BML (eds.) 1993. Buchführung der Testbetriebe. Ausführungsanweisung zum Erhebungsbogen für Betriebe der Forstwirtschaft. Unpublished internal paper.

BML (eds.) 1994. Erhebungsbogen für Betriebe der Forstwirtschaft. Unpublished internal paper.

BML (eds.) 1995. Statistisches Jahrbuch über Ernährung, Landwirtschaft und Forsten der Bundesrepublik Deutschland. Landwirtschafts-Verlag Münster-Hiltrup.

BML (eds.) 1996. Agrarbericht der Bundesregierung.

Chapter 6: Forest Inventory of the federal state Nordrhein-Westfalen (*Landeswaldinventur Nordrhein-Westfalen*)

LÖBF; Landesamt für Ökologie, Bodenordnung, und Forsten (eds.) 1994/95. Vorerhebung zur Landeswaldinventur - Aufnahmeanweisung. Unpublished internal report.

Spelsberg, G. & Wessels, W. 1994. Landeswaldinventur Nordrhein Westfalen. AFZ, pp.1292-1293.

Appendix I: Contact addresses For German NFI

Bundesministerium für Ernährung, Landwirtschaft und Forsten
(Federal Ministry of Food, Agriculture and Forestry)
Rochusstraße 1, 53123 Bonn, Germany
Tel.: +49-228-529-0
Fax: +49-228-529-4262

Bundesforschungsanstalt für Forst- und Holzwirtschaft
(Federal Research Centre for Forestry and Forest Products)
responsible for NFI I until 1992:
Leuschnerstraße 91
21031 Hamburg, Germany
Tel.: +49-40-73962-0
Fax: +49-40-73962-480

Responsible for data management and planning of NFI II:
Dr. Heino Polley
Institut für Forstökologie und Walderfassung
Postfach 10 01 47
16201 Eberswalde, Germany
Tel: +49 3334-65 318
Fax: +49 3334-65 354

COUNTRY REPORT FOR GREECE

Ioannis M. Meliadis
NAGREF - Forest Research Institute
Thessaloniki, Greece

ABSTRACT

The first attempt for the National Forest Inventory in Greece started in 1963, in a cooperative program between the Greek Forest Service and FAO. This effort continued for a period of 20 years, with many gaps.

The inventory is the only one existing this time in Greece. It was based on aerials photographs, at different scales, orthophotographs and field assessments.

The sampling frame of the National Forest Inventory is given by the definition of industrial and non industrial forests. (Industrial forests areas characterized by high trees and produced merchantable wood products. Non-industrial forests areas with multibrushed dwarf trees and bushes, which for the time being can not produce merchantable wood products but they have value for grazing and for protection). 86,2% of the whole country has been covered by the inventory, the rest belongs to agricultural or crop lands. For the inventory 2,744 ground-plots have been used by the responsible service. The ground plots have been determined by the azimuth and the distance of characteristics points that can be well recognized on aerial photographs. For the sampling area, 10 sampling points have been used and arranged in a spiracle way. Point No 1 was 10 m south of the central of the ground plot and the remaining 9 were determined according to the No 1. These ten points of the sample were in a distance of 20 m and covered an in area of 0.5 ha. The tree selection in each sampling point has been done by the use of a metric angle, equal with 10 m^2 / ha.

In the aerial photographs a number of photo-plots has been determined. The photointerpretation of these photo-plots (95,220 totally) was used for the stratification. The ground-plots were determined (2,744 totally) by the use of random number table. The strata were: non-forest, forest without volume and forest with volume. In each region the number of ground-plots have been selected as follows: for the non-forest 1 photo-plot per 35, for the forest without volume 1 per 50 and for the forest with volume 1 per 15.

The data of the inventory are stored as printed material, and some of them in magnetic tapes in different formats.

Forests covers 6.513.068 há (49.3% of the whole country). From this, 3.359.186 Ha are industrial forests (25,4%) and the rest are non-industrial forests (23,9%). From the industrial forests 736.978 Ha (21,94%) are composed with trees having diameter more than 30 cm, 616.785 Ha (18,36%) with diameter between 10-30 cm and 618.435 Ha (18,41%) with diameter less than 5 cm.

The average volume per hectare is 41 m^3/ha

The average increment per hectare is 1,14 m^3/ha

The public forests occupy 1,644,005 (65.5%), while the private forests 199,870 ha (8%). The remaining 26.5% belongs to the state by under a special legislation (e.g. co-operative forests, charity institution forests, monasteries, etc.).

The publication of the results appears in 1992.

There is an indication for a new Forest Inventory.

1. OVERVIEW

Greece is a mountainous area situated in the eastern part of Europe. The total area is 131,957,40 ha. Forests and partial forest lands cover 5,754,558 ha (2,512,418 and 3,242,140 respectively), which is 43.6% of the total area of the country.

The main tree species are the following:

Coniferous	%	Broadleaf	%
Abies cephalonica, A. alba, A. cephalonica X alba	13.1	*Quercus*	22.6
Pinus halepensis and P. brutia	8.72	*Fagus*	5.15
Pinus nigra	4.33	*Platanus*	1.33
Pinus sylvestris	0.32	*Castanea*	0.51
Pinus leucodermis	0.13	*Betula*	0.02
Picea excelsa	0.04		

The average volume per hectare is 41 m^3/ha. The average increment per hectare is 1,14 m^3/ha.

Public forests cover 1,644,005 ha (65.5%), while private forests cover 199,870 ha (8%). The remaining 26.5% of the area belongs to the state by special legislation (e.g. cooperatives forests, charity institution forests, monastery, etc.).

The number of inhabitants per hectare in the forested land is 941,973 (9.67%) (this figure, according to the National Statistical Service of Greece, represents the inhabitants that live in altitudes more than 800 m).

The average size of the stands is unknown, because this inventory did not use stands as measurable data.

1.2. OTHER FORESTRY DATA

1.2.1 Other forest data and statistics on the national level

For each Forest Office (109 totally) the following data exist:
Industrial wood (cm^3) (Coniferous-Broadleaf)
Firewood for market (tn) (Coniferous-Broadleaf)
Firewood imperfectly selected (tn) (Coniferous-Broadleaf)
Resin (tn)

No data has been collected for non-wood goods and services.

1.2.2 Delivery of statistics to UN and EU institutions

Ministry of Agriculture. Department of Development - Plans for Forestry. Aharnon 3-5 Athens. Contact person Mr. Fragkos.

2. NATIONAL FOREST INVENTORY (*ΕΘΝΙΚΗ ΑΠΟΓΡΑΦΗ ΔΑΣΩΝ*)

2.1 NOMENCLATURE

2.1.1 List of attributes directly assessed

In this inventory 86.2% of the whole country has been covered.

A) The inventory units (*Εργα*) were the followings ten (Figure 1)

1. Central Greece (or "Work 81") (*Κεντρική Ελλάδα* ή *Εργο 81*)
2. Mornos (*Μόρνος*)
3. Evinos (Ευήνος)
4. Peloponnisos (*Πελοπόννησος*)
5. Western Greece (*Δυτική Ελλάδα*)
6. Eastern Macedonia - Thraki (*Ανατολική Μακεδονία - Θράκη*)
7. Western Macedonia (*Δυτική Μακεδονία*)
8. Eastern Central Greece (*Ανατολική Στερεά Ελλάδα*)
9. Euboea (*Εύβοια*)
10. Aegean - Ionian islands - Crete (*Νησιά Ιονίου - Αιγαίου - Κρήτη*)

Forest regions or eco-regions (Figure 2)

1. Thremo-mediterranean formation of Eastern Mediterranean (*Θερμομεσογειακές διαπλάσεις (Oleo-Ceratonion) Ανατολικής Μεσογείου*).
2. Mediterranean formation of Quercus ilicis Balkanian and Eastern Mediterranean type (*Μεσογειακή διάπλαση Αριάς τύπος Βαλκανικός και Ανατολικής Μεσογείου*)
3. Sub-mediterranean formation of Ostryo-Carpinion (*Υπομεσογειακή διάπλαση*)
4. Thermophilic subcontinental deciduous oaks formations (*Διαπλάσεις θερμοφίλων υποηπειρωτικών φυλλοβόλων δρυών*)
5. Mountainous-mediterranean formations of Southern Greece (*Ορομεσογειακή διάπλαση Ν. Ελλάδας*)
6. Mountainous-mediterranean formations of Northern Greece (*Ορομεσογειακή διάπλαση Β. Ελλάδας*)
7. Mountainous-mediterranean subalpic formations (*Ορομεσογειακές, υπαλπικές διαπλάσεις*)

B) In this inventory no data on the ownership situation in Greece has been collected (the National Cadastral is working on it at the moment).

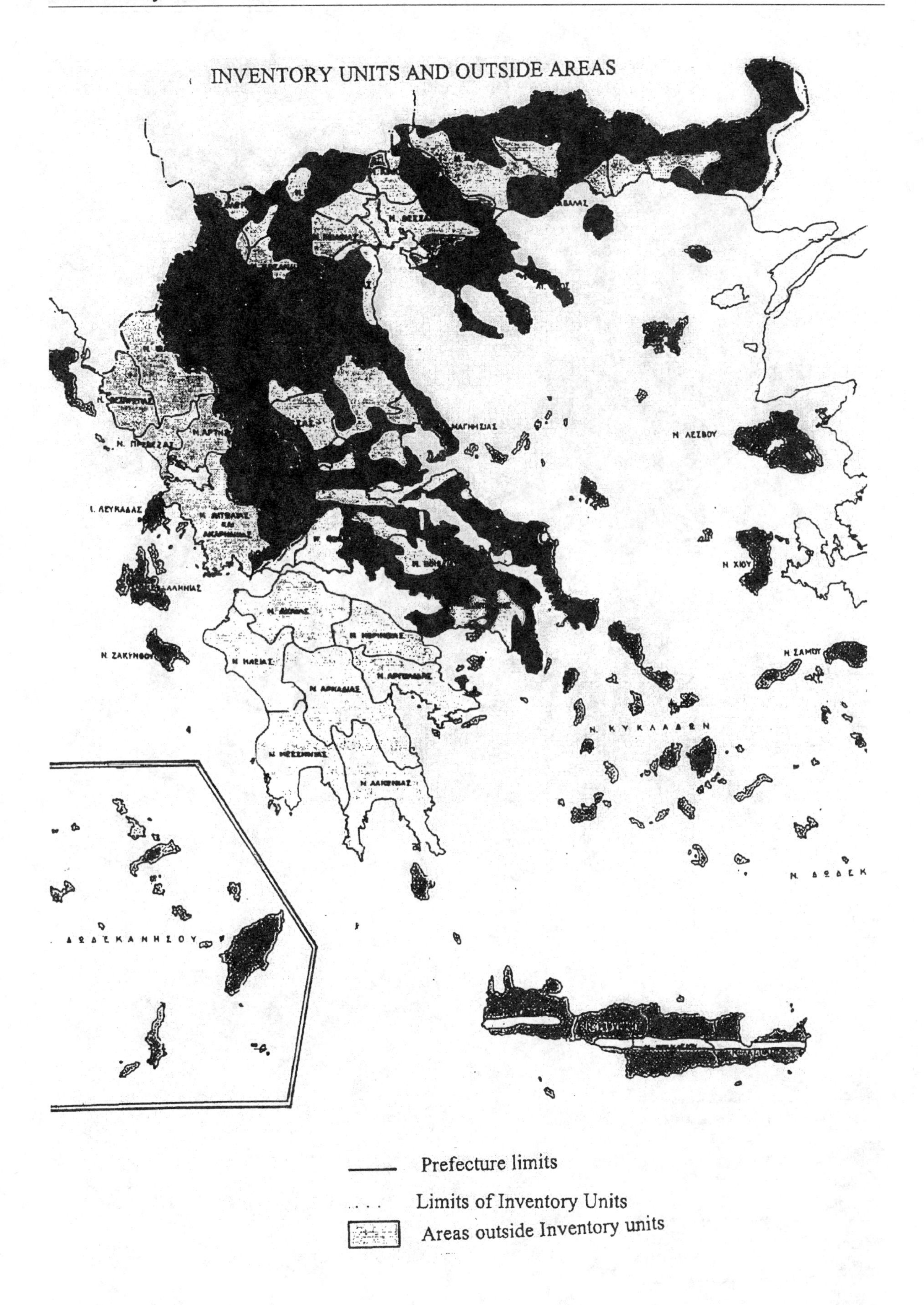

Figure 1. Inventory units and outside areas.

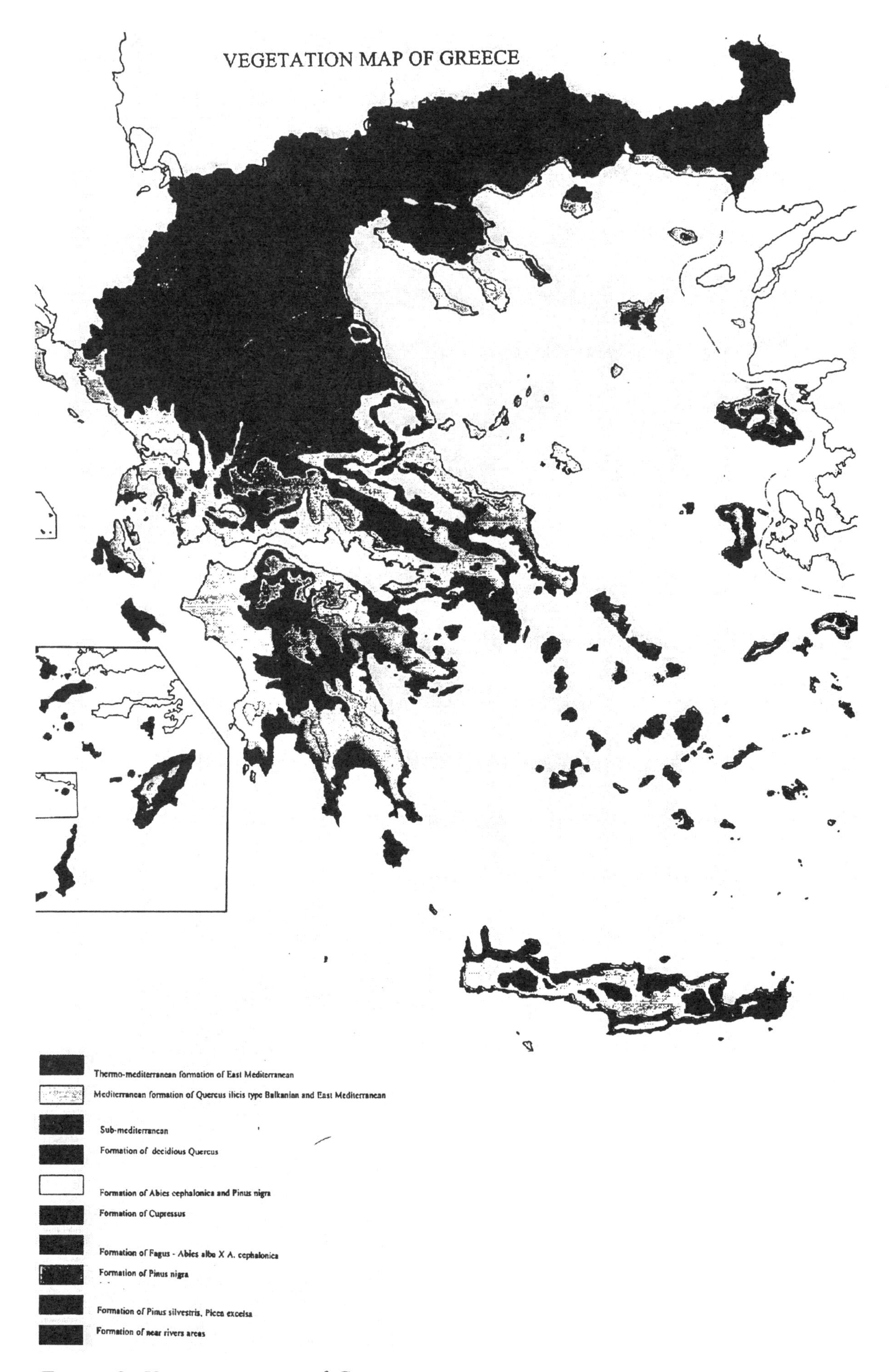

Figure 2. Vegetation map of Greece.

C) Wood production (Παραγωγή ξύλου)

Attribute	Data source	Object	Measurement unit
Diameter at breast height (*στηθιαία διάμετρος*)	field assessment	tree	cm
Total height (*ολικό ύψος*)	field assessment	tree	cm
Merchantable height (*εμπορεύσιμο ύψος*)	field assessment	tree	cm
Sawtimber height (*πριστό ύψος*)	field assessment	tree	cm
Pressler height (*ύψος Pressler*)	field assessment	tree	cm
Total Volume (*ολικός όγκος*)	field assessment	tree	m^3
Merchantable volume (*εμπορεύσιμος όγκος*)	field assessment	tree	m^3
Crown closure (*βαθμός συγκόμωσης*)	field assessment	tree	categorical
Type of damages (*κλάση ζημιών*)	field assessment	tree	categorical
Bark thickness (*πάχος φλοιού*)	field assessment	tree	cm
Tree species (*δασοπονικό είδος*)	field assessment	tree	

D) Site and soil

In the inventory no data for the site and soil has been collected.

E) Forest structure

Attribute	Data source	Object	Measurement unit
Age (ηλικία)	field assessment	tree	categorical
Mean tree height (μέσο ύψος δένδρου)	photointerpretation	tree	cm
Crownclosure (Συγκόμωση)	photointerpretation	tree	categorical
Crown density (Πυκνότητα κόμης)	photointerpretation	tree	categorical
Forest species (Δασοπονικά είδη)	field assessment - pho-tointerpretation	tree	categorical

F) Regeneration

Attribute	Data source	Object	Measurement Unit
regeneration	field assessment	tree with d.b.h. < 5 cm	cm

G) Forest condition

Attribute	Data source	Object	Measurement Unit
damage	field assessment	tree	categorical

2.1.2 List of derived attributes

Attribute	Measurement unit	Input attributes
% of healthy merchantable wood volume (*εκατοστιαία αναλογία για τον υγιή όγκο της εμπορεύσιμης ξυλείας*)	%	dbh:5-30cm, tree height, tree species
% of healthy volume of sawtimber volume (*εκατοστιαία αναλογία για τον υγιή όγκο της ξυλείας για πρίση*)	%	dbh>10cm, tree height, tree species
Single tree volume (*συνολικός όγκος δένδρου*)	m^3	d.b.h., diameter in 10 m height, tree height, tree species
Merchantable volume (*εμπορεύσιμος όγκος*)	m^3	dbh, db in 30 cm height, top diameter, tree species
Wood volume for sawing (*όγκος ξυλείας για πρίση*)	m^3	dbh, db in 30 cm height, top diameter = 20cm tree species

2.1.3 Measurement rules for measurable attributes

For each of the measurable attributes the following information is given:

a) measurement rule
b) threshold values
c) measurement scale
d) rounding rules
e) instrument used for measurement
f) data source

C) Wood production

Diameter at breast height (στηθιαία διάμετρος):
a) the diameter of the tree measured at 1,3 m above the ground
b) minimum dbh: 10cm
c) cm
d) rounded to the class limits
e) calliper, Bitterlich instrument
f) field assessment

Total Height (ολικό ύψος)
a) length of the tree from the ground level to the top of tree
b) no threshold
c) m
d) rounded to the closest m
e) Blume-Leiss
f) field assessment

Merchantable height (εμπορεύσιμο ύψος)
a) length of the tree from the ground up to the top where the diameter is no less than 5 cm
b) 5 cm
c) m
d) rounded to the closest m
e) Bitterlich instrument
f) field assessment

Sawtimber height (πριστό ύψος)
a) length of the tree from the ground up to the top where the diameter is no less than 20 cm
b) 20 cm
c) m
d) rounded to the closest m
e) Blume-Leiss
f) field assessment

Tree quality (ποιότητα δένδρου)
a) optical observation of the tree
b) no threshold
c) categorical
e) ocular estimate
f) field assessment

Thickness of the bark (πάχος φλοιού)
a) thickness of the tree bark
b) .> 5 cm
c) cm
d) rounded to the closest m
e) Bark counter
f) field assessment

Crown closure (βαθμός συγκόμωσης)
a) ratio of the area covered by tree crowns and the forested area
b) no threshold
c) in 3 classes
d) categorical
e) photo interpretation
f) aerial photographs

Regeneration (αναγέννηση)
a) all the above mentioned trees
b) 1cm
c) cm recorded classes: < 5, 5.1-10, 10.1-30, >30.1
d) rounded to the class limits
e) calliper
f) field assessment

Age (ηλικία)
a) all trees
b) no threshold
c)
d) categorical
e) Pressler instrument
f) field assessment

2.1.4 Definitions for attributes on nominal or ordinal scale

The following information is provided for each attribute:

a) the definition
b) the categories (classes) and
c) the data sources
d) remarks

Forest species (δασοπονικά είδη)

a) -

b) *Acer tribolium*

Acer creticum
Fraxinus oxyphylla
Fraxinus excelsior
Fraxinus ornus
Ulmus campesrtis
Ulmus montana
Ulmus effusa
Tilia tomentosa
Tilia parvifolia
Tilia gradifolia
Tilia vulgaris
Juglans regia
Sorbus torminalis
Sorbus domestica
Forest species
Sorbus aucuparia
Sorbus aria
Celtis australis
Ceratonica siligua
Salix alba
Salix fragilis
Salix purpurea
Salix carpea
Salix viminalis
Salix pentandra
Salix triandra
Salix cinerea
Salix incana
Salix autria
Coryllus avelana
Coryllus colurna
Populus alba
Populus tremula
Populus nigra
Alnus glutinosa
Betula verrucosa
Arbutus unedo
Arbutus adrache
Cercis siliquastrum
Cornus mas
Cornus sanguinea
Ilex aquifolium
Ryracantha coccinera
Crataegus oxyacantha
Crataegus monogyna
Erica arborea
Erica verticillata
Erica carnea
Pistacia lentiscus
Myrtus communis
Olea europaea
Lauris nobilis
Prunus spinosa
Prunus pseudoamenica
Prunus mahaleb
Prunus amygdaliformis
Prunus malus
Prunus communis
Aesculus hippocastaneum
Buxus sempervirens
Phus cotinus
Rhus coriaria
Spartium jucenium
Cystius laburnum
Colutea arborescens
Coronila emeroides
Calysotone villosa
Forest species
Paliurus australis
Berberis vulgaris
Berberis cretica
Nerium oleander
Vitex agnus castus
Tamarxi sp.
Eucalyptus rostrata
Eucalyptus globulus
Eucalyptus gomphoceplala
Clematis flammula
Clematis vitalba
Evonimus latifolius
Evonimus europaeus
Evonimus verrucosus
Rhamnus cathartica
Rhamnus graeca
Rhamnus oleoides
Rhamnus alaternus
Rhamnus fallax
Rhamnus prunifolia
Rhamnus frangula
Rhamnus rupestris
Lonicera periclymenum
Lonicera implexa
Lonicera etrusca

Viburium lantana
Viburium opulus
Viburium tinus
Sambucus nigra
Rosa canina
c) ocular estimate
f) field assessment and aerial photographs

Type of damage (κλάση ζημιών)
a)-
b) -
c) field assessment

Land use (μορφές χρήσεις γης)
a)-
b) Categories
- Industrial forests (*βιομηχανικά δάση*): areas characterized by high trees and produced merchatable wood products. (Another definition is the following: areas which are capable of producing 1 m3 of wood per hectare every year.
-Non-industrial forests (μη-βιομηχανικά δάση): areas with multibrushed dwarf trees and bushes, which for the time being can not produce merchantable wood products but are used for grazing and for protection.
- Non-forest land (μη δασική γη): areas not classified as forest lands.
- Range land (βοσκότοποι) : non forest land used for grazing.
- Crop land (γεωργική γη): non forest land which is used for crops.
- Bare land (γυμνή γη): non forest land without any vegetation in more than 50% of the area.
- Urban, Residential and Industrial land (αστική και βιομηχανική γη): non forest land which is used for industry, residential, etc.
- Water (υδάτινες επιφάνειες) : rivers, lakes, swamps.
c) aerial photographs

Tree size classes (κλάσεις διαστάσεων δένδρων):
b) - Reproduction (αναγέννηση) well developed trees of desirable species with dbh less than 5 cm.
- Saplings (κορμίδια): desirable tree species with dbh 5-10cm
- Poletimber (δένδρα με διαστάσεις στύλων): trees with dbh 10-30cm
- Sawtimber (δένδρα με διαστάσεις ξυλείας για πρίση) : trees with dbh >30cm
c) field assessment

Volume (ογκος): (figure 4)
b) - Merchantable volume (εμπορεύσιμος όγκος): the net volume of the tree with a diameter of more than 5cm between the stump and the point, where the top is 5cm, or the point if there is distortion
- Sawtimber volume (όγκος ξυλείας για πρίση): the net volume of the tree with a diameter of more than 30cm between the stump and the point, where the top is 20cm or the point if there is distortion.
- Top (κορυφή): the net volume of the tree with a diameter of more than 5cm between the top of the merchantable wood and the upper edge of the tree.
c) field assessment

Growth (προσαύξηση):
b)- the gross annual growth (μικτή ετήσια προσαύξηση) : the annual volume growth that has been measured in trees with dbh more than 5 cm, without the subtraction of the annual mortality.
- the net annual growth (καθαρή ετήσια προσαύξηση): the result derived by the subtraction of the gross annual growth and annual mortality
-Annual mortality (ετήσια θνησιμότητα): the annual loss of volume due to competition, fires, diseases, animals, weather.
c) field assessment

Tree quality for commercial purposes (ποιότητα δένδρων για εμπορικούς σκοπούς):

b) - Desirable (επιθυμητά): trees with merchantable value. They adjust to the forest environment well, they do not have distortion or other damages, their shape and health are good.
- Acceptable (παραδεκτά): trees with merchantable value. They adjust to the forest environment well, their merchantable wood has not been distorted more than 50%, and their shape and health are quite good.
- Rejected (απορριπτέα): trees with no merchantable value
- Dead (νεκρά): trees that have been dead the last 5 years.

c) field assessment

Tree damage classes (κλάσεις ζημιών δένδρου):

b) - None (καμμία)
- Insects (εντομα)
- Disease (ασθένειες)
- Fires (πυρκαγιές)
- Logging (υλοτομικές εργασίες)
- Illegal cutting (παράνομες υλοτομίες)
- Animals (ζώα)
- Weather (καιρός)
- Erosion (διάβρωση)
- Other damages (άλλες ζημιές)

c) field assessment

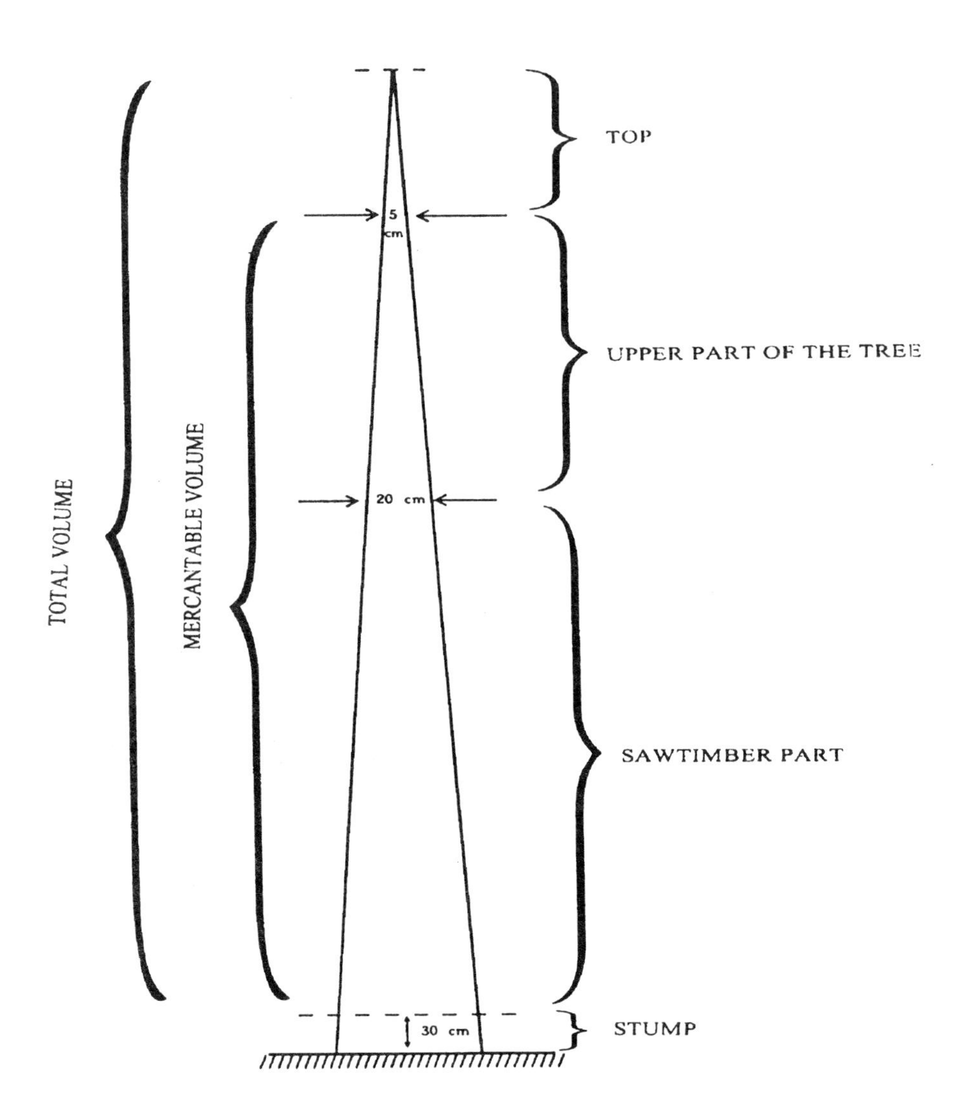

Figure 4.

2.1.5 Forest area definition and definition of "Other Wooded Land"

Forest lands (δασική γη) includes: 1) areas with 0.5 ha or 30cm strips with a tree crown closure that covers 10% of the area, or areas with 250 trees of a regeneration size, not used for other purposes, and which are able to produce forest products or/ and other services.

2) areas where trees have been removed with a crown closure less than 10% and they have not been used for other purposes.
3) areas with regeneration
4) brush lands (areas covered by evergreen broad-leaved trees)

2.2 DATA SOURCES

A) field data

The field data came from mensurations, estimations and ocular observations. The data was written on a tabulate sheet.

B) questionnaire

No questionnaire has been used in the inventory.

C) aerial photographs

- Geographic Military Service
- 1:42,000, 1:30,000, 1:20,000, orthophotos
- 1945-1980
- the whole country
- pocket mirrors, mirror stereoscope
- unknown
- 1,300 drx each (approximately 3.5ECU) (today's price)

D) Spaceborne or Airborne Digital Remote Sensing

No digital data was used.

E) Maps

- Geographic Military Service
- 1971 -1972
- 1:50,000
- topographic in an analog format

F)- Map of forests in different Greek prefectures:

Forest Service
International / no charge
forest species
1:200,000
Different Prefectures of Greece
1971
printed material

Figure 3. General soil map of Greece.

- Soil map of Greece (Figure 3)
  Forest Institute of Athens
  International / no charge
  soil types
  1:20,000
  Different areas in Greece
  1977
  printed material
- Vegetation map of Greece (Figure 2)
  Forest Institute of Athens
  International / no charge
  forest regions
  1:20,000
  Greece
  1978
  printed material
- Hydrological atlas for Greece
  Ministry of Agriculture
  International/no charge (restrictions)
  hydrological basins
  1:200,000
  Greece
  1972
  printed material

2.3 ASSESSMENT TECHNIQUES

2.3.1 Sampling frame

The sampling frame of the National Forest Inventory is given according to the definition of industrial and non industrial forests. 86,2% of the whole country has been covered by the inventory, the rest consists of agricultural or crop land. 2,744 ground-plots have been used for the inventory.

2.3.2 Sampling units

Field Assessments:

The ground plots have been determined by the azimuth and the distance of the characteristics points can be well recognized in aerial photographs. For the sampling area, 10 sampling points were used and arranged in the form of a spiracle. Point No. 1 was 10 m south of the center of the ground plot and the remaining 9 were determined according to the point no. 1. These ten points of the sample were at a distance of 20 m and covered an area of 0.5 ha (figure 5). The tree selection in each sampling point has been done by the use of a metric angle, equal to10 m2 / ha .

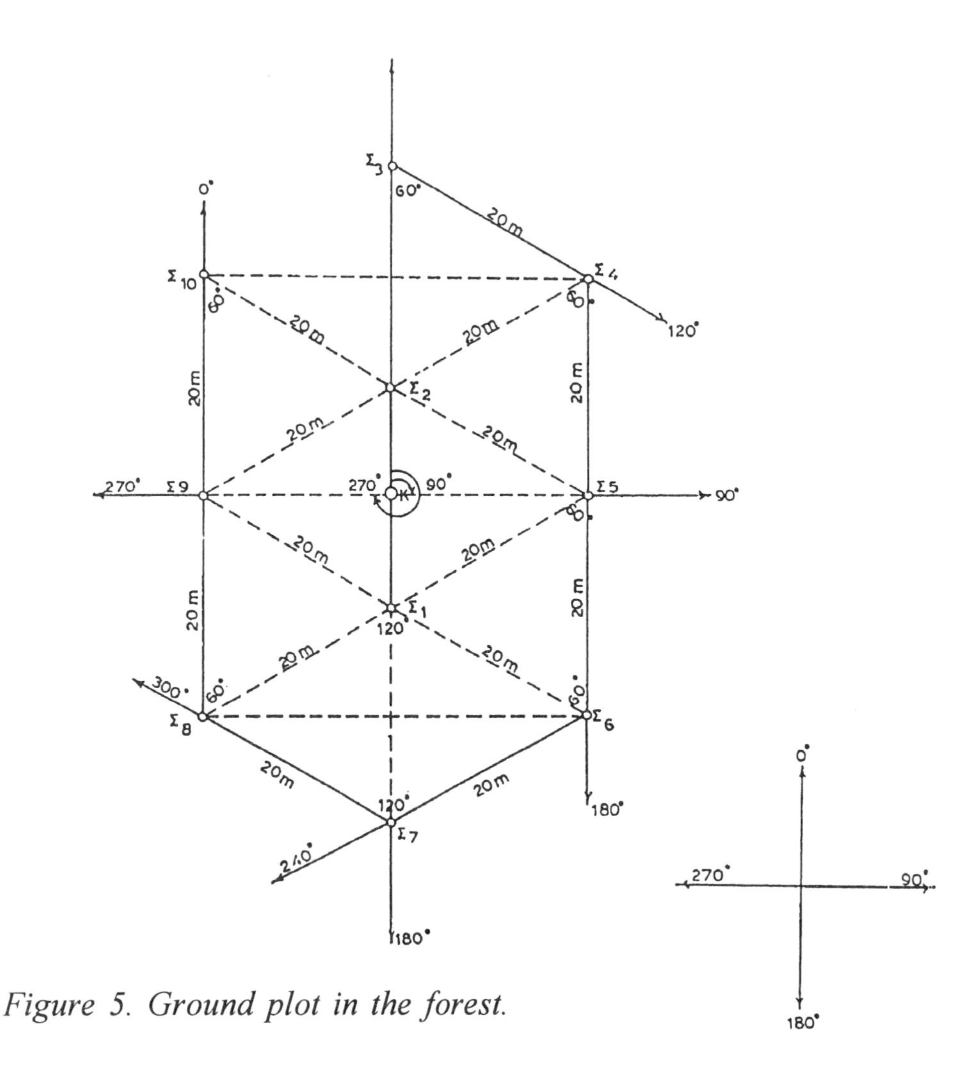

Figure 5. Ground plot in the forest.

Aerial Photography Sample Plots:

In the aerial photographs a number of photo-plots has been determined. The photointerpretation of these photo-plots (95,220 in total) was used for the stratification. The ground-plots were determined (2,744 in total) by the use of a random number table and they were divided into three groups: non-forest, forest without volume and forest with volume. In each region, the number of the ground-plots has been selected as follows: for the non-forest 1 photo-plot per 35, for the forest without volume 1 per 50 and for the forest with volume 1 per 15.

2.3.3 Sampling Designs

On the aerial photographs the photo - plots were determined. After a photointerpretation some information on each of them was ascertained. The information was used for the stratification. Using a table of random numbers the ground plots were divided into three groups: non- forest, forest with volume and forest without volume. In the field assessment the ground plots were visited and different variables were measured. These ground - plots were not permanent, but all the necessary information for each of them was recorded (for possible future inventories).

2.3.4 Techniques and methods for the combination of the data sources.

The attributes were assessed in the field and on the photo-plots. With the statistical method of optimum sample distribution, a selection was made for the classified photo-plots. Then, using a table of random numbers, the ground-plots were determined.

2.3.5 Sampling fraction

Aerial photography, random area sample

field assessment, each tree represents an area of $10m^2$/ha

2.3.6 Temporal aspects

Inventory cycle	time period of data assessment	publication	time period between assessments	reference data
1st Inventory	1962-1982	1992	20 years	

2.3.7 Data collecting techniques in the field

The data was recorded on paper and later by computer.

2.4 DATA STORAGE AND ANALYSIS

2.4.1 Data storage and ownership

The data of the inventory is stored as printed material, and some of it on magnetic disks in different formats. The person responsible for the data is the Director of the Department of Forest Cadastral, Mapping, Inventory and Classification of Forests and Forestlands, Mr. V. Giotakis.

2.4.2 Data-base system used

The data is stored in different formats in different systems such as DBASE III, LOTUS, WP51.

2.4.3 Data-bank design

There is no set design for the stored data. There are different ways to present the same results.

2.4.4 Update level

There is no updated data. When the data is stored, the results that are derived from it can later be used with other data that has been collected at another time (years later).

2.4.5 Description of statistical procedures used to analyze data including procedures for the sampling error estimation

a) Area estimation: a network of dots was used on maps (scale 1:50,000) and the areas were estimated according to the number of dots that were inside them.

Area (ha) $A_i = \Sigma\ AM_i\ /M$

M_i is the number of photo-plots in the survey area by volume strata (A_i), M is the number of photo-plots in the survey area.

b) Volume: the General Volume Formula was the following:

$$V = \Sigma\ (U \times \Pi\ (.5d.h.h.)^2\ HF\ A)\ /\ \Pi\ R^2\ N$$

where U area for the land unit (m^2), Π=3.14, d.b.h. = diameter at breast height (m), H = height of the tree, F form factor of the tree, R = the maximum distance from the point to the tree (m), N the total number of ground plots in the area surveyed. Based on the general formula the following were estimated: the net volume per hectare including limbs , the net volume per hectare for main stem inside the bark, the net volume per hectare for merchantable stem and the net volume per hectare for the sawtimber volume.

c) Growth: the General Growth Formula was the following:

$$G = \Sigma\ (HFA/N)\ P_v$$

where G = the volume of the annual growth for the survey area (m^3), P_v is the annual growth.

No data has been collected for non-wood goods and services.

2.4.6 Software applied

As was mentioned before in 2.4.2. the software used is not compatible with other systems.

2.4.7 Hardware applied

PC IBM compatible.

2.4.8 Availability of data

All data in printed or magnetic format is available to everyone without any restriction. Most of the data is not updated and it can primarily be used by people who have been involved in the inventory.

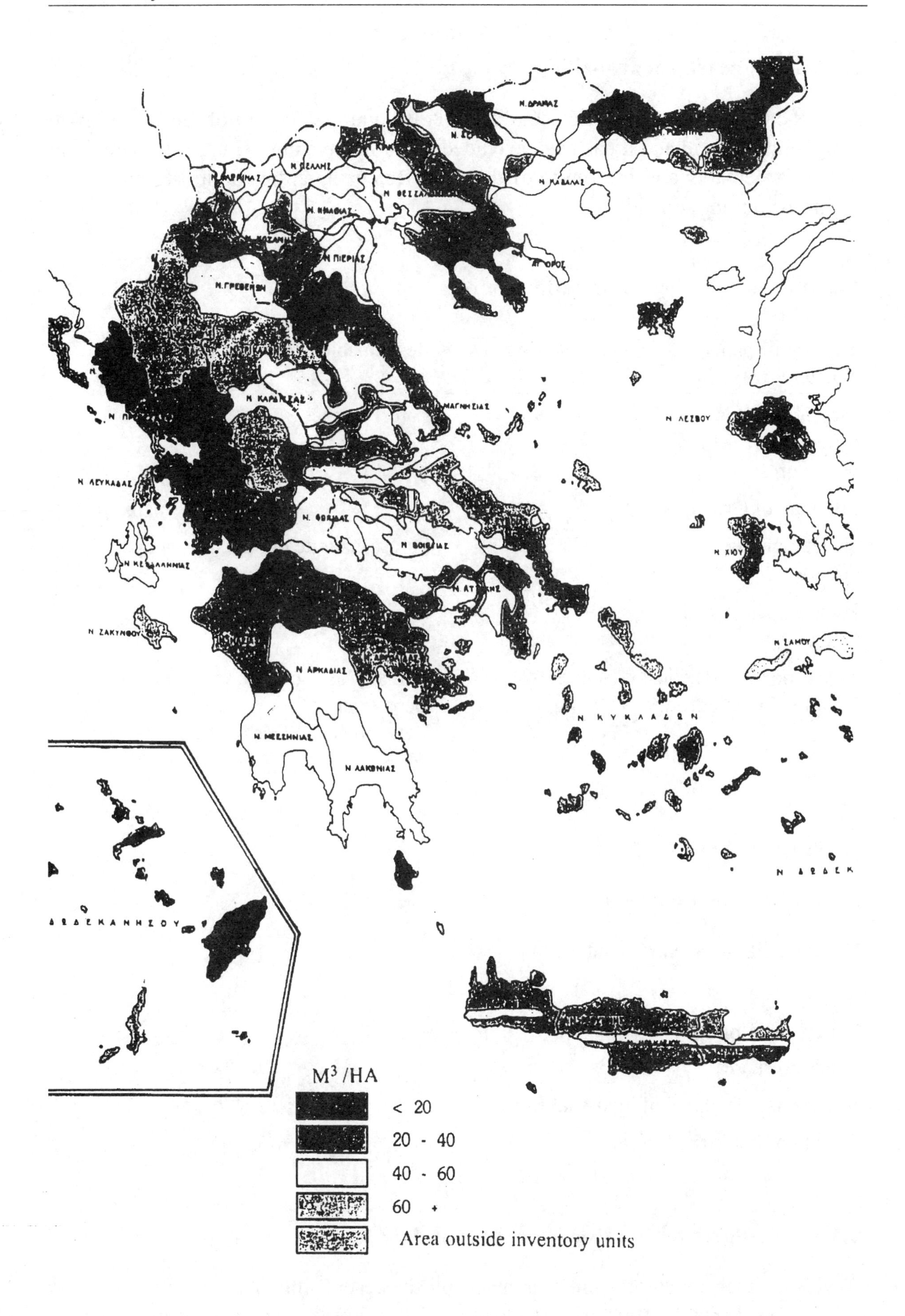

Figure 6. Mean mercantable volume per ha in the industrial forests per prefecture.

2.4.9 Subunits (strata available)

Results can be obtained for 86,2% of the whole country. The rest of the area consists of agricultural land and it has been excluded from the inventory (Figure 1). The limits of the Prefectures and the limits of the 10 different inventory units (Figure 1 and Figure 6) do not coincide.

2.4.10 Links to other information sources

Topic and spatial structure	Age/period availability	Responsible agency	Kind of data-set
Map of forest in different Greek prefectures	since 1962	Forest Service	maps
Soil map of Greece			maps
Vegetation map of Greece	1978	Forest Institute of Athens	map
Hydrological atlas for Greece	1972	Ministry of Agriculture.	maps
Year-Book of Statistical Data for Greek Forestry	since 1984 available yearly, limited	Ministry of Agriculture. Department of Development-Plans for Forestry	tables

2.5 RELIABILITY OF DATA

2.5.1 Check assessments

No check has been carried out

2.5.2 Error budgets

Total forest area	± 0,2%
Merchantable volume of industrial forests	±2,6%
Growth of industrial forests	±3,1%

2.5.3 Procedures for consistency checks of data

A visit to each inventory unit was made by the responsible staff, before any work was done for a better classification of the area. The next step was the photointerpretation of the aerial photographs and the identification of the photo - plots. On each of them the staff ascertained some information (such as forest species, crown closure, etc.) which helped the stratification. The ground - plots were examined on the aerial photographs and in the field. The staff which was responsible for the field assessment made a draft check of the information derived by the photointerpretation. The different variables in the field assessment were recorded in a tabulate sheet of paper. In the beginning of the inventory, all the data was collected by hand and only much later by the use of computers.

2.6 MODELS

2.6.1 Volume

$$V = \sum (U \times \Pi (.5d.h.h.)^2 HF\ A) / \Pi R^2 N$$

where U is the area of the land unit (m^2), Π=3.14, d.b.h. = diameter at breast height (m), H = height of the tree, F is the form factor of the tree, R = the maximum distance from the point to the tree (m), N is the total number of the ground plots in the area surveyed. Based on the general formula the following was estimated: the net volume per hectare including limbs , the net volume per hectare for the main stem inside the bark, the net volume per hectare for merchantable stem and the net volume per hectare for the sawtimber volume.

2.6.2 Assortments

2.6.3 Growth components

c) Growth: the General Growth Formula was the following:

$$G = \Sigma (HFA/N) P_v$$

where G = the volume of annual growth for the survey area (m^3), P_v is the annual growth.

2.7 INVENTORY REPORTS

2.7.1 List of publishes reports and media for dissemination of inventory results

Inventory	Year of publication	Citation	Language	Dissemination
First National Forest Inventory	1991	Ministry of Agriculture General Secretariat of Forests and Natural Environment. General Department of Forests and Results of Natural Environment, 1991.	Greek	printed

2.7.2 List of contents of the latest report and the updated level

There is only the present report and no any updated version exists.

2.7.3 Users of the results

The results are used by the Greek Forest Service. The value of the results cannot be estimated due to the big time gap between the assessment and the publications. This report can only give some general information on the condition of the forests in Greece.

2.8 FUTURE DEVELOPMENT AND IMPROVEMENT PLANS.

The staff and the whole Service that carried out the existing inventory have been transferred to another department recently. The future of a new National Forest Inventory is therefore unknown and no development plan exists at the moment.

2.9 MISCELLANEOUS

There are some problems related to the data from the First Forest Inventory in Greece. The most important one is the fact that the service responsible for the delivery of the data has now been discontinued. All the staff has been transferred to other services (mainly to the Greek Cadastral) and the data and information can only be obtained through personal contacts, as is also the case with most of the above mentioned data.

COUNTRY REPORT FOR ICELAND

Thröstur Eysteinsson
Iceland Forest Service, Iceland

OVERVIEW

Birch *(Betula pubescens)* is the only native forest-forming tree species in Iceland. The birch woodlands and scrublands in Iceland are quite variable; many are dominated by low, under 2 m, scrubby plants, but woodlands with trees up to 10-12 meter also occur. Today they cover about 1200 km^2. Exotic species are used in afforestation, but the area of established plantations is very limited and most are less than 25 years old. In fact, only since 1990 has planting exceeded 2 million trees per year. The cover of forests in Iceland is today about 1.2 % of total land area.

With the exception of a few stand level and individual tree measurements for research purposes, no forest inventories have been carried out in plantation forests in Iceland and no regular inventories are carried out in the native birch woodlands. Some inventorying is expected to start within the next few years in areas with the oldest plantations.

The birch woodlands of Iceland have been surveyed twice: 1972—1975 and again 1987-1991. The 1972-1975 survey was the first real attempt to survey the distribution and status of all Icelandic birch woodlands. The woodlands were mapped by drawing their boundaries on aerial photos and transferring them to maps by hand. The area of the woodlands was then measured on the maps. Height above sea level, slope, geologic substrate and soil depth, were recorded for each woodland. Tree density (stems/ha, in forests) or crown cover (%, in scrubland) and mean height were estimated as well as whether the woodland was advancing, in decline or stable. None of these parameters were measured. Instead, because of the size of the task and the small budget available, they were estimated using standardised groupings even though it was obvious that considerable error would be involved.

FOREST STATISTICS IN ICELAND

The Iceland Forest Service has not published an annual report since 1988. This is about to change, with the annual report for 1995 almost complete. We expect to publish an annual report from now on, which will include statistics on plant production, planting, Christmas trees, wood production, forest planning and inventories, tourism and more. The report will be in Icelandic with an English summary.

DELIVERY OF THE STATISTICS TO UN AND COMMUNITY INSTITUTIONS

a) FRA 1990 - Arnór Snorrason, the Iceland Forest Service
b) Eurostat - the Statistical Bureau of Iceland and Thröstur Eysteinsson, the Iceland Forest Service
c) OECD - Thröstur Eysteinsson, the Iceland Forest Service

BIRCH WOODLANDS SURVEY

In 1987-1991, the Icelandic Forest Service in co-operation with the Agricultural Research Institute, conducted a survey of all birch woodlands in Iceland in order to document their distribution, characteristics and condition.

NOMENCLATURE

Length

Icelandic birch is often crooked, leaning or even prostrate. Therefore, length from root collar to terminal leader was measured rather than height from the ground.

Diameter

Diameter was measured at a height of 50 cm rather than breast height since many Icelandic woodlands do not get much taller than breast height. On trees with more than one stem, the thickest one was measured.

Terminal leader length

It gives an idea of current status.

Distance to the next tree

It was used to get a rough measure of woodland density.

Growth form

Trees were categorised into one of four growth form classes: conical crown (vigorous, usually straight trees), spherical crown (less vigorous, often older trees), flat crown (non-vigorous, often crooked trees) and "krumholtz" (many-stemmed, less than 2 m tall).

Crown %

The crown as a % of the entire length was recorded in one of four categories: 0-25%, 26-50%, 51-75% and 76-100%.

Leaf area

Leaf area was categorised into three categories: good, fair and poor. This is usually an indication of the tree's position with respect to other trees, i.e. dominant or not

Dead shoots

The presence of dead shoots on the tree was recorded as well as whether they were near the top of the tree, at the base of the crown or throughout the entire crown. Dead shoots at the crown base are an indication of self pruning but dead shoots in the crown are an indication of disease or environmental stress.

Stump sprouts

Icelandic birch regenerates either by seed or stump (root collar) sprouts. Birch seedlings are seldom found within woodlands because of poor germination conditions. The main mode of regeneration within woodlands is by stump sprouts. The number of stump sprouts, if any, was recorded for each sample tree and the number of seedlings within a 1 m wide belt between the sample tree and its nearest neighbour was counted.

Crown cover, willow cover and vegetation cover

These were estimated visually in a l m wide and 4 m long belt in the opposite direction from the sample tree of the nearest neighbour.

ASSESSMENT TECHNIQUES

The 1987-1991 survey consisted of two parts:

1) remapping the distribution of birch woodlands along with description of their ecological status and
2) stand level mensuration of representative types.

1) Remapping

The outlines of each of 360 wooded areas were redrawn and in addition, larger woodlands were divided into plots based on differences in attributes. These new maps were plotted and stored in digital form, which allows GIS handling of the data collected. Geographic information recorded includes height above sea level, slope and slope direction.

2) Sample plots

Sampling was carried out within all large woodlands and within other representative woodlands in all parts of the country. Sampling was done along evenly spaced transects in each woodland, the number of transcects depending on the area and landscape diversity of the woodland. Along each transect, 10-25 trees were selected at random and measured. Length, diameter at 50 cm height, distance to the next tree and previous year's terminal leader length were measured. Growth form, leaf area, crown % of total height, amount of dead shoots, stump sprouts, % crown cover, % cover of willows and % cover of other vegetation were assessed.

DATA ANALYSIS

The extensive data collected in this survey are being analyzed at the Iceland Forest Research Station. The results will be published by county or district, but the database of characteristics of each birch stand will later be linked with the maps using a Geographical Information System.

INVENTORY REPORTS

Aradóttir, Á.L., Thorsteinsson, I. & Sigurdsson, S. 1995. Birkiskógar Islands; Könnun 1987-1991 I.(Birch woodlands in Iceland; Survey 1987-1991 I.). Research report 11, Iceland Forest Research Station, Mógilsá. 64 p.

COUNTRY REPORT FOR IRELAND

Peter Dodd
Pat Farrington
Gary Williamson
Coillte Teoranta, Ireland

Summary

2.1 Nomenclature

Coillte's inventory covers state owned forest managed by Coillte Teoranta only. The attributes are mainly categorical and are assessed by field survey at stand (sub-compartment) level. The minimum area for assessment purposes for forest land is 0.5 ha.

2.2 Data sources

The data sources are; field survey forms completed by inventory assessors, OS topographical maps at 1:10560 scale and forest level planting records.

2.3 Assessment techniques

The sample frame covers land managed by Coillte, approximately 427,000 ha. The entire estate is surveyed.

2.4 Data storage and analysis

The data are owned and stored by Coillte in its Arc/Info GIS. Validation programs are run at various stages of the inventory process. Data are disseminated to interested parties on request. There is no charge to government agencies, universities or the general public. However, commercial companies pay for the information.

2.5 Reliability of data

Information from field surveys is checked by an independent auditor (10% check). In addition there are validation programs run at data input stage on PCs locally. Further checks and validations are run by HQ inventory. Special surveys are conducted occasionally to check assessments (5% random sample).

2.6 Models

The models used in Ireland are those developed by the BFC with the exception of the models used for Lodgepole pine which were constructed using a similar methodology but derived from Irish research data. The volume assortment tables for Sitka spruce are also derived from Irish research data. All other species use the BFC volume assortment tables.

2.7 Inventory reports

The inventory reports, analyses and stand maps remain the property of Coillte and are not published externally. Coillte operates a continuous inventory; afforestation, reforestation, fellings, windblows and burned areas are updated annually. Other crops are assessed at intervals; age 5 and age 14 years. Older crops where there is a significant change are resurveyed as required. The main users are Coillte staff, government departments, universities and the wood processing industries.

2.8 Future development and improvement plans

It is planned to introduce aerial photography as an additional GIS layer, supplement existing Arc/Info GIS software with ArcView 2.1 software. The use of GPS in ground survey is also being evaluated. Irish yield models for Sitka spruce are being developed.

3.1 Inventory of private woodland, 1973, nomenclature

The geographic extent of this inventory is the Republic of Ireland. Phase I covered private estates of over 40 ha of forest. Phase II used an aerial photography sample (9%) to locate other private forest areas. BFC top height/age curves were used to determine yield class. BFC mensuration conventions were used in assessing standing volume. Information was recorded at stand level and aggregated to produce national statistics.

3.2 Data sources

OS maps at 1:10560 scale, planting records, aerial photography and field assessment were used.

3.3 Assessment techniques

Phase I covered private estates of over 40 ha of forest. Phase II used an aerial photography sample (9%) to locate other private forest areas.

3.4 Data storage and analysis

Data was stored on IBM mainframe computer. In-house produced COBOL software programs were used to organise and retrieve the data. National level statistics were published. Estate owners were given stand level reports. Subunits were county, forest estate and stand.

3.5 Reliability of data

Field survey accuracy checks were conducted and validation programs run by HQ inventory.

3.7 Inventory reports

Inventory reports published by government and used by forest managers for planning.

3.8 Future development

Similar methodology unlikely to be repeated. A private woodlands survey is planned for 1997 which will utilise remotely sensed data combined with sample survey.

1. OVERVIEW CHAPTER

1.1 GENERAL FORESTRY AND FOREST INVENTORY DATA

Irish forestry, which is 99% man-made, accounts for 7.6% of Ireland's land area and this is projected to rise to 15% by the year 2010. To date the country has a total of 646,000 hectares of land under forest. Annual plantings for the last two years has averaged just short of 20,000 ha, carried out equally by public and private plantings. Under Government policy an annual planting target of 30,000 ha is set. State-owned forests, managed by Coillte, amount to some 80% of Ireland's forests. Of the Coillte estate Sitka spruce accounts for 61% of the total area, with Lodgepole pine accounting for 22%. Broadleaves currently account for 3% of the total area.

The estate is extremely scattered and composed of many small properties with an average stand size of 3 ha. This reflects the pattern of acquisition over the previous fifty years. Acquisitions, with the exception of the blanket peat areas in the western half of the country, were composed of small farms or holdings which in turn were comprised of many small fields. The scattered nature of the estate with relatively few large forest blocks is at variance with plantation forestry in other countries, e.g. New Zealand, where large blocks of more or less uniform composition is the norm. This has implications for the subsequent management of the estate and for economies of scale.

Another feature of the Coillte estate (and indeed of forestry in general in Ireland) which distinguishes it from other countries is its predisposition to windblow. This arises from a combination of factors including soil types which are wet and difficult to drain, high rainfall and the frequency and severity of high wind speeds - Ireland is one of the windiest countries in the world.

The age structure of the estate is far from "normal", with 33% of the total area planted since 1980. The average increment per hectare and the average volume per hectare are 9 m^3 and 105 m^3 respectively.

At present Ireland does not carry out a national forest inventory, however the Forest Service, Department of Agriculture are currently preparing to do a forest inventory.

Within Coillte the forest inventory is carried out with the following objectives:-

1. To collect data on Coillte's forest resource in order to facilitate its proper management.
2. To make up-to-date forecasts of timber production.

The most recently completed forest inventory of private woodlands was completed in 1973.

This inventory had the following objectives:-

1. To estimate the total of area of private woodland in the country,
2. To obtain a breakdown of the forest types in private ownership,
3. To assess present volume and potential future production in the private sector,
4. To provide the larger woodland owners (40 ha) with basic maps and stand descriptions of their forest properties,

5. To facilitate the Forest and Wildlife Service of the Department of Fisheries and Forestry in shaping private woodland policy.

All plantations, aged one year or older at time of Inventory were surveyed. The Inventory was divided into two phases.

Phase I was a complete assessment of woodlands of 40 ha or more in one estate and under a single ownership. This included semi-State bodies and the State itself where the woodlands were not under the direct control of the then Forest and Wildlife Service (now Coillte). The total number of estates was 151 in the following size categories;

Phase II was a 9% strip sample of the country covering all woodlands with the exception of those covered in phase I and those owned by the Forest and Wildlife Service. In this phase the country was divided into four regions and for each region strips 1.8 miles wide and 20 miles apart were marked on a map. Aerial photography was used to locate the woodland in each strip. In some cases it was possible to map individual stands from the photographs. Maps and photographs were then sent to the assessor in the field, who visited the stand, checked the mapping and filled in a stand form.

The Forest Condition Surveys

In compliance with EU Council Regulation number 3528/86 terrestrial forest condition surveys have been conducted in Ireland since 1987. The primary survey consists of 22 plots which were selected using the official EU 16km*16km grid. Three species are represented in the plots: Sitka spruce (*Picea sitchensis*); Lodgepole pine (*Pinus contorta*) and Norway spruce (*Picea abies*).

The objectives of the survey are to describe the health status of Irish forests and to indicate where possible the causes of any forest damage observed. In order to do this individual crowns are assessed annually in each plot for crown vitality which is expressed by the extent of defoliation and discoloration observed.

In addition to the primary forest condition survey, a pilot demonstration study has been conducted in Ireland since 1990 to investigate methods of improving the observation and measurement of damage in forests. As part of this demonstration study the vitality of crowns in 33 plots have been monitored annually using the same methods as those used for the primary forest condition survey. The trees in each of these plots are all located in internal positions of stands.

1.2 OTHER IMPORTANT FOREST STATISTICS

1.2.1 Other forest data and statistics on the national level

Title

CORINE Land Cover (Ireland) Project

Responsible body:

National Resources Development Centre, Trinity College
Dublin 2, Ireland

Contents:

The CORINE Land Cover database covers the Republic of Ireland. The database consists of 44 hierarchical land cover categories. Three of these, Broadleaved, Coniferous and Mixed forest are the only classes involving woodland. The smallest mapping unit is 25 ha.

Land cover maps were produced at a scale 1:100000 and 1:500000 and are publicly available. The database is also available in digital format.

Methodology:

Computer assisted photo-interpretation of Landsat TM imagery from 1989 to 1992 was used to classify the imagery. Ancillary data such as aerial photography were also used to assist interpretation.

The Forest Service have statistics on the number of afforestation proposals in relation to:-

- National Heritage Areas
- Sites and monuments
- Fisheries
- Landscape

These statistics are available on a county basis.

They are classified into four categories

- Refused
- Modified
- Granted
- Not dealt with yet

The Forest Service also have statistics on the number of felling licences granted on a county basis.

1.2.1.1 Non-wood goods and services

Information leaflets produced by a number of other organisations;

The Tree Council of Ireland
The Office of Public Works, Department of Finance
Ecco

1.2.1.2 Removals

Removals are recorded by Coillte's timber sales system and are used as a "prompt" for the Inventory assessor.

1.2.2 Delivery of the statistics to UN and Community institutions

1.2.2.1 Responsibilities for international assessments.

The Forest Service submit forest resource statistics to the FAO/ECE for the 1990 Forest Resource Assessment. Statistics are also given to EUROSTAT.

No forests resource data are supplied to OECD or other inter-governmental bodies.

The person responsible for submission of the above statistics to the various organisations is:

Mr. James Lavelle
Higher Executive Officer
Forest Service
Leeson Lane
Dublin 2
Ireland

1.2.2.2 Data compilation

These statistics are based on data analysis.

2. COILLTE'S FOREST INVENTORY

2.1 NOMENCLATURE

2.1.1 List of attributes directly assessed

A) Geographic regions

Attribute	Data Source	Object	Measurement Unit
Region code	Coillte GIS Map	Stand	Categorical
District Code	Coillte GIS Map	Stand	Categorical
Forest Code	Coillte GIS Map	Stand	Categorical
Property name	OS Maps	Stand	Categorical
Compartment code	Sequential Compartment list	Stand	Alpha-numeric code
Sub-compartment code	Sequence number	Stand	Number
County number	OS Maps	Stand	Number
Map number	OS Maps	Stand	Number

B) Ownership

This inventory covers Coillte land only.

C) Wood production

Attribute	Data Source	Object	Measurement Unit
Stocking level	Field Assessment	Stand	% of stand area
Established- Spacing	Field Assessment	Stand	Metres
Stand area	Field Assessment	Stand	Hectares
Broadleaf volume	Field Assessment	Stand	m^3
Year of Survey	Date of survey	Stand	Year
Quality Class	Field Assessment	Stand	Categorical
Stems	Field Assessment	Stand	Number / ha
Thinning Status	Field Assessment	Stand	Categorical
Year of First Thinning	Field Assessment	Stand	Year
Thinning Explanation	Field Assessment + yield models	Stand	Categorical
Rotation Type	Field Assessment	Stand	Categorical
Felling Year	Field Assessment + yield models	Stand	Year
Rotation Explanation	Field Assessment	Stand	Categorical

D) Site and soil

Attribute	Data Source	Object	Measurement Unit
Site Fertility	OS Map	Stand	Categorical
Soil Type	Field Assessment	Stand	Categorical
Elevation	OS Map	Stand	Metres
Exposure	Field Assessment	Stand	Categorical
Drainage	Field Assessment	Stand	Categorical

E) Forest structure

Attribute	Data Source	Object	Measurement Unit
Land Use Type	Field Assessment	Stand	Categorical
Species code	Field Assessment	Stand	Categorical
Planting Year	Planting file records	Stand	Year
Mixture Type	Field Assessment	Stand	Categorical
Percentage Canopy	Field Assessment	Stand	% of area

F) Regeneration

Attribute	Data Source	Object	Measurement Unit
Planting Year Type	Planting file	Stand	Categorical

G) Forest condition

Attribute	Data Source	Object	Measurement Unit
Stocking level	Field Assessment	Stand	%
Wind Hazard	Field Assessment	Stand	Categorical

H) Accessibility and harvesting

Attribute	Data Source	Object	Measurement Unit
County Road Present	OS Map	Property	Categorical
County Road Improve	Field Assessment	Property	Categorical
ROW Present	Legal title document.	Property	Categorical
ROW Improve	Field Assessment	Property	Categorical
Forest Road Present	Field Assessment	Property	Categorical
Forest Road Improve	Field Assessment	Property	Categorical

I) Attributes describing forest ecosystems

No directly assessed attributes.

J) Non-wood goods and services

Attribute	Data Source	Object	Measurement Unit
Constraint	Field Assessment	Stand	Categorical

K) Forest functions other than production (protection, recreation, etc.)

Attribute	Data Source	Object	Measurement Unit
Constraint	Field Assessment	Stand	Categorical

2.1.2 List of derived attributes

Attribute	Measurement Unit	Input Attributes
Yield Class	$m^3ha^{-1}yr^{-1}$	Top height, Age, Species

Yield Class

Yield class is a measure of the rate of volume growth per hectare based on the maximum mean annual increment using two cubic metres between classes. It is defined as the average annual volume increment per hectare of a stand over its rotation.

2.1.3 Measurement rules for measurable attributes

For each of the measureable attributes the following information is given:

a) measurement rule
b) threshold values
c) measurement scale
d) rounding rules
e) instrument used for measurement
f) data source

Top Height

a) The average total height of the 100 trees of largest diameter at breast height (dbh) per hectare.
b) Minimum height of 3
c) Metres
d) Rounded to the nearest metre
e) Hypsometer
f) Field assessment

2.1.4 Definitions for attributes on nominal or ordinal scale

The following information is provided for each attribute:

a) the definition
b) the categories (classes)
c) the data sources
d) remarks

Region code

a) The Coillte estate is divided into seven regions numbered 1 to 7.
b) 1 to 7
c) Coillte GIS map
d) Administrative units

District Code

a) The district code is a two digit code, made up of the
b) Region code + District number
c) Coillte GIS map
d) Administrative units

Forest Code

a) The forest code is a four digit code that uniquely identifies a forest.
b) It is made up of three distinct parts, as follows
   X Region code
   9 District code
   99 Forest number. (within District)
c) Coillte GIS map
d) Administrative units

Property name

a) This is the local area name for a contiguous block of forest.
b) Town-land name
c) OS map
d) Generally a town-land name.

Compartment code

a) A compartment code is a unique identifier for each compartment.
b) The compartment code consists of a five digit number followed by an alphabetic check-character.
c) Sequential Compartment list
d) List maintained by Inventory manager

Sub-compartment code

a) Code to uniquely to identify a stand within a compartment.
b) 1 to 50
c) Sequential number
d) Attributed to sub-compartments in a clockwise fashion

County Number

a) Counties are numbered sequentially based on alphabetical order.
b) 1- 26
c) OS map
d) Numbered in alphabetical order

Map Number

a) Ordnance Survey Six-inch map series sheet number.
b) 1 - 160
c) OS map

C) Wood production

Stocking level

a) A stocking level adjustment factor is recorded where there are permanent un-stocked areas within a stand which are too small (or scattered) to form a separate stand. This adjustment factor is subsequently used to adjust the area of the stand.
b) 20, 30...100
c) Field assessment
d) To facilitate management prescription and production forecasting

Established-Spacing

a) The average square spacing at the time of planting expressed in linear meters.
b) 1.2 to 4.5
c) Field assessment
d) To facilitate management prescription and production forecasting

Stand area

a) Inventory data from before 1990 approx. contained an estimate of the area of each stand calculated by using a planimeter or an area grid. Since the advent of the Geographic Information system (GIS), a more reliable and accurate estimate of area is available not just for newly acquired areas but also for all existing stands, irrespective of when they were inventoried
b) 0.5 ha minimum
c) Field Assessment and GIS calculation

Broadleaf volume

a) Standing volume of broadleaf species
b) m^3 per ha
c) Ocular assessment

Quality Class

a) Timber Quality Class.
b) Straightness
 1 Very Straight(100% Straight Trees)
 2 Straight (66% Straight Trees)
 3 Crooked (33% Straight Trees)
 4 Very crooked(No Straight Trees)
 Taper
 1 Very slow, almost cylindrical
 2 Medium
 3 Very fast taper
 Branchiness
 1 Pruned to 6 metres
 2 Low Pruned, lightly branched
 3 Unpruned, medium/light branching
 4 Unpruned, heavily branched
c) Field assessment
d) Conifers only. Used to facilitate management prescription and production forecasting

Thinning Status

a) A prescription as to the type and number of thinnings a stand will receive throughout its rotation.
b) 2TH03 2 Thinnings 3 years apart
 2TH03 2 Thinnings 3 years apart
 2TH04 2 Thinnings 4 years apart
 2TH05 2 Thinnings 5 years apart
 2TH06 2 Thinnings 6 years apart
 2TH07 2 Thinnings 7 years apart
 2TH08 2 Thinnings 8 years apart
 2TH09 2 Thinnings 9 years apart
 2TH10 2 Thinnings 10 years apart
 2TH11 2 Thinnings 11 years apart
 2TH12 2 Thinnings 12 years apart
 2TH13 2 Thinnings GE 13 years apart
 3TH03 3 Thinnings, 3 year average cycle.
 3TH04 3 Thinnings, 4 year average cycle.
 3TH05 3 Thinnings, 5 year average cycle.
 3TH06 3 Thinnings, 6 year average cycle.
 3TH07 3 Thinnings, 7 year average cycle.
 LOC Standard Thinning GT 2 years early or late
 NOTH No thin
 SOL One Thinning only
 STD Regular Thinnings to marginal Int.
c) Field assessment
d) Conifers only. Used for production forecasting.

Thinning Explanation

a) An explanation as to why a particular thinning status has been chosen
b) 0 Standard thinning
 ACF Access Forest roads
 ACO Access County roads
 BYD Beyond Thinning
 CGY Contiguity
 CHK Checked crop
 IN1 Very high risk-15m max. top ht
 IN2 High risk-18m max. top ht

IN3	Mod-high risk-21m max. top ht
IN4	Mod risk-24m max. top ht
INS	No longer a valid option
LYC	Low Yield Class
PFM	Poor form
STK	Low stocking level

c) Field assessment
d) Conifers only. Used for production forecasting.

Rotation Type

a) A prescription of the rotation length.
b)

BLOWN	GE50% of sub Blown
BURNT	GE 50% of sub Burned
EXPT	Expt not recorded by Inventory
EXT	Rotation. GT MMAI + 2 years
FELLD	GE 50% of sub Felled
LOC	Crop cannot grow to MMAI
STD	See Instructions!

c) Field assessment
d) Conifers only. Used for production forecasting.

Felling Year

a) The year in which the stand will be clear-felled.
b) Year
c) Field assessment, yield models
d) Conifers only. Used for production forecasting.

Rotation Explanation

a) An explanation as to why a particular rotation length has been chosen.
b)

0	Standard Rotation
ACO	Access County Roads
BLW	Windblow
CGY	Contiguity
DIS	Disease
ENV	Environment
INS	Instability
MKT	Market
PCP	Poor crop
SED	Seed Stands

c) Field assessment
d) Conifers only. Used for silvicultural analysis.

D) Site and soil

Site Fertility

a) An indication of site fertility for forestry based on former land use.
b)

A	Fields and ornamental ground	High fertility areas which have been in intensive agricultural use up to relatively recent times. Vegetation: pasture grasses, herbaceous plants, sometimes rushes.
B	Furze or bracken	Intermediate fertility areas most but not all of which were formerly enclosed for agriculture but which have long been abandoned. Evidence of old walls and ditches. Vegetation: furze, bracken, rough grasses.
C	Rough pasture with/without rock outcrop	Low fertility areas of un-enclosed ground which have never been used intensively for agriculture. Vegetation: heath or peatlandassociations - Calluna, Molinia etc.
X	Old woodland	Old woodland on a range of sites of varying levels of fertility. Includes broadleaved, mixed and coniferous woodland.

c) OS Map
d) Use the OS six inch series map ornament to indicate the former land-use.

Soil Type

a) A broad soil classification.

b) A Brown earth
 B Podsol
 C Gley
 D Alluvium
 E Lithosol
 F Coast Sand
 G High level virgin blanket peat
 H High level cut-over blanket peat
 I Low level virgin blanket peat
 J Low level cut-over blanket peat
 K Virgin raised peat
 L Cut-over raised peat
 M Fen peat
 N Marl

c) Field assessment

d) Site indicators uses are typical associated vegetation for soil type.

Elevation

a) Elevation range in which the stand lies.

b) 1 0 to 100 metres
 2 101 to 200 metres
 3 201 to 300 metres
 4 301 metres

c) OS Map - contour detail

d) Used in species selection

Exposure

a) Level of exposure of the stand.

b) 1 Sheltered
 Very little exposure, e.g. low-lying valley floors, sheltered flat terrain.
 2 Moderate
 Average exposure, e.g. low hills and mid slopes of mountains, flat terrain in East.
 3 High
 Above average exposure, e.g. higher mountain slopes, flat terrain in West.
 4 Severe
 Extremes of Code 3, e.g. high west-facing mountain slopes, sites on western sea-board.

c) Field assessment

d) Gathered to enhance decision making process in relation to thinning.

Drainage

a) Drainage potential of the site

b) 1 Good
 Unrestricted rooting in excess of 45 cm (18"); e.g. brown earth's, podsols.
 2 Moderate
 Restricted rooting in excess of 45 cm (18"); e.g. as above but less satisfactory.
 3 Poor
 Restricted rooting in excess of 25 cm (10"); e.g. deep peats, loamy and O.R.S. gleys.
 4 Very poor
 Very restricted rooting under 25 cm (10"); e.g. surface water and peaty gleys, sites with insurmountable outfall problems.

c) Field assessment.

d) Used to highlight problem areas.

E) Forest structure

Land Use Type

a) The land within each compartment is classified in terms of the use to which it is currently put e.g. a Land Use Type classification of "Conifer High Forest" indicates that the land is growing a crop of conifers.

b) 14 Land use types, eight Forest classes and six non-forest classes:
 1 CHF (Conifer High Forest)
 2 MHF(Mixed High Forest)
 3 BHF(Broadleaf High Forest)
 4 UNDEV(Undeveloped)
 5 SCRUB
 6 FELLED
 7 BLOWN
 8 BURNED
 9 BAREPL(Bare Plantable)
 10 BAREMG(Bare Marginal)
 11 BAREUP(Bare Unplantable)
 12 SWAMP
 13 WATER
 14 MISC(Miscellaneous and Buildings)

c) Field assessment

d) All areas coming under the scope of the

Inventory must be classified into one of the above categories.
For LUT 1-11 minimum size of sub-compartment 0.5ha
For LUT 12-14 minimum size of sub-compartment 0.2ha

Species code

a) Twenty different conifer and ten broadleaf species are distinguished by the forest inventory. The proper identification of species forms part of the training process. The occurrence of up to four species within a stand can be recorded but more detailed information is recorded for the first two.
b) 20 conifer and nine broadleaf tree species, nine vegetation
c) Field assessment
d) Vegetation is only recorded for non forest areas.

Planting Year

a) The year in which a stand was planted
b) Ranging from 1800 to present day.
c) Planting file records
d) Year of origin of the first species for land use types; CHF, MHF, BHF, UNDEV. Important input in determining yield class (potential growth rate).

Mixture Type

a) Indication as to whether the various species composing a mixture are mixed throughout the sub-compartment
b) Two types (INT - intimate, NINT - non intimate).
c) Field assessment
d) Applies to mixtures only.

Percentage Canopy

a) The Canopy percentage is the percentage of the area of each species in a stand.
b) minimum 20% 25% – 95% 100%
c) Field assessment.
d) Differentiation between up to four species within sub-compartment

F) Regeneration

Planting Year Type

a) Four afforestation and four reforestation letter codes which describe the nature of establishment.
b) 1996R = 1996 Reforestation
1996A = 1996 Adjusted afforestation
1996S = 1996 Adjusted reforestation
1996E = 1996 Estimated afforestation
1996T = 1996 Estimated reforestation
1996M = 1996 Mean planting year afforestation
1996U = 1996 Mean planting year reforestation
1996W = 1996 Natural regeneration
For normal situation, 1996 - i.e. none of the above.
c) Field assessment

G) Forest condition

Stocking Level

a) Applies to high forest only, is a measure of productive area expressed as a percentage. A sub-compartment is considered fully stocked if the level of stocking is 96%. < 20% is not considered to be High Forest.
b) 20% – 100%
c) Field assessment
d) The purpose of stocking percentage is to normalise mapped compartment areas for production forecasting.

Wind Hazard

a) Overall windthrow hazard of a stand that has been established using optimum techniques
b) Four categories increasing in severity from one to four including;
Slight
Moderate
High
Severe.
c) Field assessment
d) Used as a guide to potential windblow problems.

H) Accessibility and harvesting

County Road Present

a) The accessibility of the stand as affected by the present condition of county roads leading to it
b) Three categories including;
   Satisfactory
   Barely adequate
   Inadequate.
c) OS maps
d) Information used as a layer on GIS to optimise routing to sawmills.

County Road Improve

a) Indicator of the feasibility/cost of improving a road
b) Four improvement categories
   Not required
   Normal cost
   High cost
   Not viable.
c) Field survey.
d) Driveability considering type of haulage to be used and how much timber is to be transported to the mill via the route.

ROW Present

a) Right of way: a legal entitlement to access to ones property through another landowners property.
b) Three categories including;
   Satisfactory
   Barely adequate
   Inadequate.
c) OS maps and deeds of title.
d) Important to identify to for legal reasons and to ensure extraction damage to road is minimised.

ROW Improve

a) Whether access needs to be improved or not.
b) Four categories including;
   Not required
   Normal cost
   High cost
   Not viable
c) Field assessment
d) Indicates if maintenance or upgrade is required.

Forest Road Present

a) Concerned with internal forest roads constructed and maintained by Coillte.
b) Three categories including;
   Satisfactory
   Barely adequate
   Inadequate.
c) New forest roads mapped by surveyors or field assessors.
d) Forest roads allocated a road number, length estimated using GIS.

Forest Road Improve

a) Need to improve forest road established and cost range identified.
b) Four categories including;
   Not required
   Normal cost
   High cost
   Not viable
c) Field assessment.
d) Indicator of road improvement prior to harvesting operations.

I) Attributes describing forest ecosystems

None

J) Non-wood goods and services

Constraint

a) A "Constraint" is defined as any factor, of human origin, which interferes with the site being managed as a normal forest production area. Three broad classes;
   Environmental
   Experimental
   Legal.
b) There are 16 constraints,* indicates a use other than forestry.
   NONE
   AMNTY (amenity and recreation).
   WLDLFE (wildlife)
   CONSVN (conservation)
   ARBOR (arboriculture)
   EXPT (experiment)
   SEEDST (seed stand)
   NURSERY
   PIT (sand pit)*

ESB (electricity supply board)*
LEGAL*
XMAS (Christmas tree production)
IMAT (Immature forest sales)
NHA (National heritage areas)
BNM (Bord na mona: long term lease arrangement)
OTHER*

c) Field assessment

2.1.5 Forest area definition and definition of "other wooded land"

The Coillte estate can be divided into four broad categories;

High Forest - Commercial forest
Plantable Bare Land - Capable of growing trees
Unplantable Bare Land - Not capable of growing trees.
Other Woodland - Scrub and/or Amenity forests.

Coillte surveys all its land and broadly classifies it by Land Use Type. Unplanted land is easily distinguished from both high forest and scrub. High forest is defined as a forested area with a minimum of 20% stocking and a yield class of 4 m^3 per ha per annum. Scrub is defined as an area of broadleaf crop consisting of stunted trees or shrubs that lack the potential to develop as high forest.

Forest property boundaries are surveyed on the ground at acquisition stage. Woodland boundaries are subsequently surveyed by inventory assessors using OS maps at a scale of 1:10560 by pacing and taking offsets from known positions, with the aid of scale rule. The boundaries and internal polygons are then digitised into the Coillte GIS, which is then used to calculate the area of the property or block.

2.2 DATA SOURCES

A) Field Data

All Coillte land is assessed. All the attributes that are directly assessed (Section 2.1.1)

B) Questionnaire

Not applicable

C) Aerial Photography

Aerial photography has only been used in isolated cases in the past but will be increasingly used in the future. Ireland has been completely flown during 1995 and this photography, which is at a scale of 1:40000, can be purchased from the Ordnance Survey. The aerial interpretation will probably be sub-contracted in the future. A pilot study to evaluate the introduction of aerial photography is underway. Photogrammetry and photo-interpretation will be carried out by contractors. Output will be in digital format compatible with Arc/Info GIS.

Final specifications are not available as a range of scales; 1:10000, 1:20000, 1:40000 and a range of film types; CIR, panchromatic and colour are being evaluated.

D) Spaceborne or airborne digital remote sensing

Has not been used

E) Maps

Paper topographical maps at a scale of 1:10560 are obtained form the Ordnance Survey of Ireland. These maps are circa 90 years old but are quite accurate for upland and forested areas.

Stand mapping is carried out by forest inventory staff using the above topographical maps. This stand mapping is also digitised.

F) Other geo-referenced data

Coillte
Restricted
Stand mapping, Harvest plans
1:10560
hectares
Annually updated
Arc/Info

2.3 ASSESSMENT TECHNIQUES.

Coillte Inventory is a complete assessment inventory, i.e. every stand is visited.

2.3.1 Sampling frame

The sampling frame of the Coillte forest inventory is all land managed by the company within the Republic of Ireland. There are no exclusions.

2.3.2 Sampling units

All Coillte land within the state is surveyed; sample unit forest stands (sub-compartments) 100%.

Field assessments

All stands are visited and information collected as previously outlined. Slope is measured using a hypsometer and calculated with the aid of a statistical calculator. Stands are located on the ground by reference to the OS six inch series 1:10560. Control for the digitising of forest boundaries is supplied by the Ordinance Survey and conforms to Irish National Grid.

2.3.3 Sampling design

Identify all Coillte managed land within Ireland. Sample intensity 100% of stands.

The stands are surveyed at year 1, year 5, year 14. All areas FELLED, BLOWN, and BURNED are updated annually. All afforestation and reforestation are surveyed within a year of planting, all lands acquired, leased, or sold by the Company is surveyed and the GIS updated accordingly. Older crops are resurveyed at the discretion of the Inventory Manager where there is a doubt about the accuracy of the information i.e. crop gone into check, drop in yield class.

Note: from 1997 it is likely that crops aged 25 years will be resurveyed as part of the normal forest inventory, this is in order to improve the input into the production forecasting procedure.

2.3.4 Techniques and methods for combination of data sources

Stand boundaries and roads are digitised and associated attribute data are recorded and analysed using GIS.

2.3.5 Sampling fraction

Sample unit is the forest stand, complete enumeration.

2.3.6 Temporal aspects

The year of survey is recorded for each stand. The year of survey indicates the last year that measurements were taken for a stand. (N.B. Only stands that have changed since the previous year are surveyed).

Inventory Cycle	Time Period of data assessment	Publication of results	Time period between assessments	Reference date
Annual	1st Jan - 31st Dec	Annually	1 year	

2.3.7 Data capturing techniques in the field

Data are recorded on "field forms" and entered into PC computer. Stand boundaries are digitised into Arc/Info GIS.

2.4 DATA STORAGE AND ANALYSIS

2.4.1 Data storage and ownership

The data are stored in a database at Coillte's headquarters in Dublin, Ireland. The data are the responsibility of each region manager within Coillte. The contact person is Peter Dodd, Coillte (Tel: 353-1-2867751, Fax: 353-1-2868126, Email: DODD_P@COILLTE.IE).

2.4.2 Database system used

The data is stored in an Arc/Info GIS and the attributes are also stored in Cognos Powerhouse database.

2.4.3 Data bank design (if applicable)

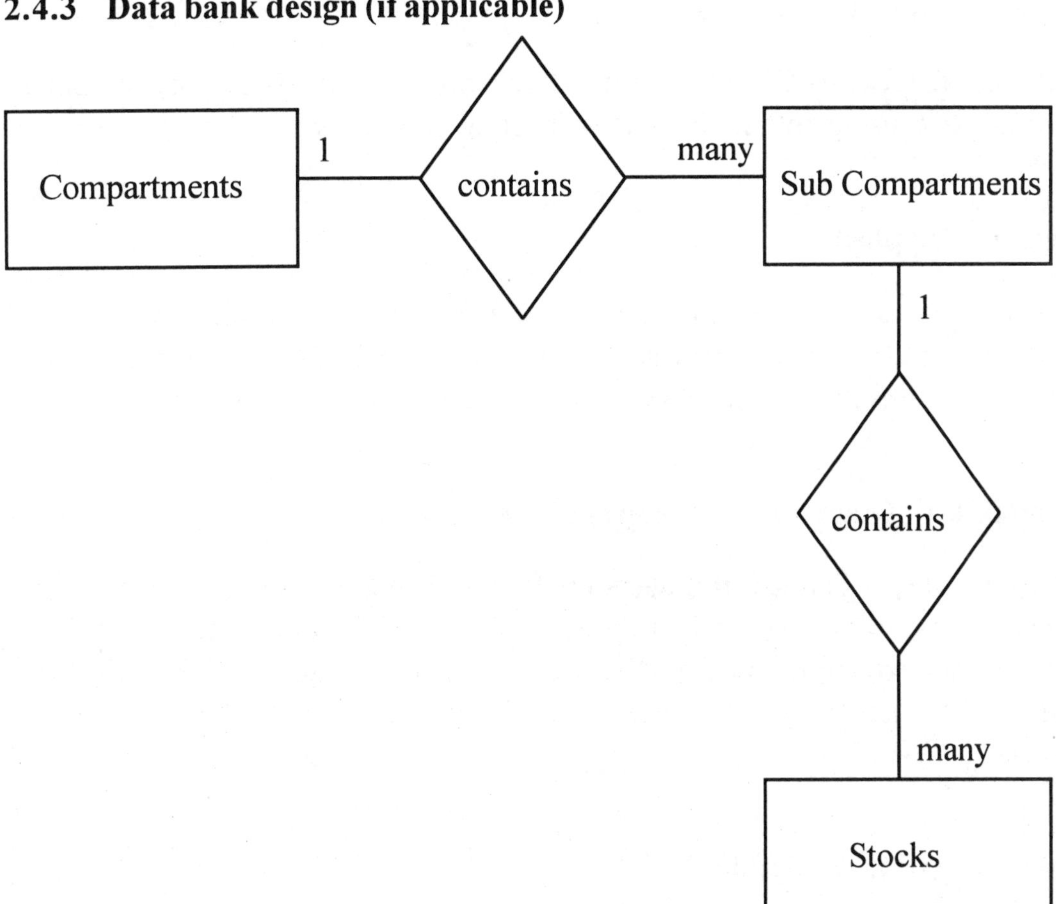

Compartments	(Compartment-code, Forest-number, County-number, Property-name...)
Sub-compartments	(Sub-compartment-code, Compartment-code, Land-use-type, Constraint, Year-survey, stocking, spacing...)
Stocks	(Stock-no, Sub-compartment-code, Species, Planting-year, Yield-class, Canopy-percent, Regeneration-code, Broadleaf-volume...)

Figure 1. Coillte Inventory Entity Relationship Diagram.

2.4.4 Update level

For each record the date, when the data have been assessed, is available. The data are not updated to a common point in time and thus reflect the situation at the time of assessment.

2.4.5 Description of statistical procedures used to analyse data Including Procedures for sampling error estimation

All stands are completely assessed.

2.4.6 Software applied

A combination of Arc/Info GIS and Cognos Powerhouse systems are used. Timber Forecasting is done using internally written Fortran programs.

2.4.7 Hardware applied

Data capture is done on PCs. Analyses are carried out on a Digital Cluster of minicomputers. Arc/Info and Powerhouse run under OpenVMS on Digital Alpha minicomputer. Arc/Info is also installed on Digital Alpha workstation running Unix.

2.4.8 Availability of data (raw and aggregated data)

All recorded raw data and derived attributes stored in the database can be made available to outside persons, provided the use to which the data will be put is specified and acceptable to Coillte. All costs for extracting data from the database and preliminary analysis have to be reimbursed. Analysis according to special requests are possible. It is possible to provide data in digital format.

2.4.9 Subunits (strata) available

Results can be produced by region, district, forest, property and county. Ad hoc queries are also possible.

2.4.10 Links to other information sources

Topic and spatial structure	Age/Period/ Availability	Responsible agency	Kind of data-set
Six-inch map series	Since 1900	Ordnance Survey of Ireland	Paper Maps Scanned Raster and Vector Maps

2.5 RELIABILITY OF DATA

2.5.1 Check assessments

Ten percent of the stands are revisited by an independent Inventory Auditor who also ensures the accuracy of the data through random checking. Each Regional Inventory Manager also carries out random checks and carries out a yield class accuracy survey. The accuracy checks include the accuracy of stand attributes, stand boundaries and forest road centrelines.

All errors are identified to the original assessor who must correct them to the Auditors satisfaction.

The Auditor also suggests modifications to the Inventory code of practice based on the incidence and type of error encountered.

2.5.2 Error budgets

Attribute	Error Source
Top Height	Measurement error
Stocking	Measurement error
Percentage Canopy	Measurement error
Spacing	Measurement error
Area	Measurement error
Broadleaf Volume	Measurement error

2.5.3. Procedures for consistency checks of data

All data that are entered onto PC are validated using a computer validation program. As data are aggregated to the company inventory database, further more extensive validations are carried out. The Inventory Auditor checks data at every stage in the cycle.

2.6 MODELS

2.6.1 Volume (functions, tariffs)

In the Coillte inventory, volume is not currently assessed as part of the field work. However an estimation of total volume production can be made by relating, species, age, and top height. The inputs of species, age and yield class are collected in the inventory field survey and are used in the Coillte forecast system. British Forestry Commission yield models are used (Christie, 1981).

2.6.2 Assortments

a) Sitka spruce volume assortment tables have been derived using height/diameter data collected from various Coillte experiments and permanent sample plots. The models were produced using a methodology based upon Gray's Taper Line (Jordan, 1993). Both single tree and stand assortment tables have been produced for a number of management regimes.

The limited availability of data results in only the single tree volume assortment table, for marginally thinned crops with a DBH of 30 cm or less, having confidence level of 95%. The volume assortment tables used for other species are the British Forestry Commission assortment tables (Hamilton, 1985).

b) ± 2% at 95% confidence level.

c) Coillte experiments and permanent sample plots maintained by our research department.

d) Models Validated by Coillte Research and Dept of Forestry, U.C.D.

2.6.3 Growth components

2.6.4 Potential yield

a) British Forestry Commission yield models are used for potential yield estimation (Christie, 1981). There are over 1,000 yield models representing a range of species, growth rates and silvicultural treatments, the principal input data to the yield models are species, stand age, dominant height, plant spacing and thinning treatment. The yield models are empirical and tabular, and are incorporated into the Coillte forecast programmes, which produce forecasts at national, region, district, forest, compartment and sub-compartment level.

The yield models produce forecasts of potential volume production (7cm top diameter overbark) and basal area production, numbers of trees, mean tree volume, dbh, and future thinning volume.

b) Overview of prediction errors

For individual stands in Irish conditions the models can underestimate volume production by as much as 20%. However at national, regional, and district level prediction errors are negligible. The British Forestry Commission are partners in this project and are providing further information about the yield models.

c) Data material for derivation of model

The yield models are based on periodic measurements from 1,500 permanent mensuration plots .augmented with 1,000 temporary plots established in the UK

d) Validation

Yield models are reviewed and validated as part of an ongoing programme of yield model development and research by the B.F.C. Research branch Coillte also carry out research in this area and have produced models for lodgepole pine grown in Ireland following the same methodology used in the B.F.C.

2.6.5 Forest functions

The Coillte estate is managed with the main purpose the production of wood ,this is given high importance.

Access to the state owned forest is not restrictive, Coillte has an “open forest” policy. So use for recreation is encouraged, both nature conservation and recreation would be given a medium rating.

Hunting is also permitted in certain areas but less people are involved therefore this function where very important to some individuals is of less importance to the vast majority of forest visitors.

2.7 INVENTORY REPORTS

2.7.1 List of published reports and media for dissemination of inventory results

Internal reports are produced for each forest but are not published .

Reports are also available on a Management Information System country wide, about 200 terminals, (reports only).

In addition analysis and query facilities are available in Region and District offices using the Coillte GIS (reports and thematic maps)

2.7.2 List of contents of latest report and update level

Update level of Inventory data used in latest Inventory reports;

Field data 1995
GIS analysis programmes
Area by land use type
Area by age class
Area by age class by yield class by species
Inventory listing
Analysis of forest area
High forest by species
Species group by yield class
Area and mean yield class by species group by age class
Crop survey
Site survey
Harvest programme and forecast
Thinning rotation classification
Other information can be derived through interrogating the data .

There are unlimited possibilities in this area.

2.7.3 Users of the results

All Coillte staff involved in all areas of forest management.
The Forest Service, Department of Agriculture for national forest statistics.
Coillte Customers: sawmills and pulp mill industry.
The Environmental Protection Agency.
The Department of the Environment.
Universities. D.I.T., U.C.D, T.C.D , U.C.C, U.C.G.
Financial institutions. (Irish forest unit trust, IFUT)
Eurostat
UN-FAO
COFORD, Research co-ordinating organisation within Ireland.
I.D.A. Industrial Development Authority.
Society of Irish foresters.
Dail Eireann (Parliamentary questions)
Local government, county councils and regional authorities.
Numerous government agencies
General Public.

Requests by other Institutions (expert estimate):

Coillte	70%
Forest Service	15%
Universities	10%
All others	5%
	100%

Special requests by topic:

Forest structure	70%
Timber volume	10%
Forested area	70%
Forest site	50%
Forest condition	10%
Accessibility	10%
Potential yield	70%

2.8 FUTURE DEVELOPMENT AND IMPROVEMENT PLANS

2.8.1 Next inventory period

1996

2.8.2 Expected or planned changes

2.8.2.1 Nomenclature

The data that are currently being collected on broadleaves and physical site characteristics are being re-evaluated with a view to extension.

2.8.2.2 Data sources

Aerial photography is currently being evaluated for use in stand mapping and will be used in conjunction with ground survey.

2.8.2.3 Assessment techniques

The practicalities and cost/benefit analysis of using aerial photography, for photo interpretation and photogrammetry is being investigated currently. It is probable that we will incorporate black and white photography at 1:20000 scale into our inventory system.

2.8.2.4 Data storage and analysis

ArcView GIS software will be used by forest managers to analyse Inventory and related data.

2.8.2.5 Reliability of data

Additional validation checks will be built into the data entry procedures.

2.8.2.6 Models

Further development of dynamic stand growth models for Sitka spruce will take place.

2.8.2.7 Inventory reports

A new Inventory report structure will be developed utilising digitally stored Ordnance Survey data (maps).

The Coillte GIS will have additional layers of OS maps at 1:10560, scanned aerial photography, and a layer constructed through photo-interpretation/photogrammetry which will be used to validate or update sub-compartments and other features.

3. INVENTORY OF PRIVATE WOODLANDS, 1973

3.1 NOMENCLATURE

3.1.1 List of attributes directly assessed

A) Geographic regions

Attribute	Data Source	Object	Measurement Unit
County Number	OS Map	Stand	Categorical
Map Number	OS Map	Stand	Categorical
Photo-strip	Flight Path and Frame Number	Stand	Categorical
Compartment number	Sequential Compartment list	Stand	Alpha-numeric code
Stand number	Number	Stand	Number

B) Ownership

Attribute	Data Source	Object	Measurement Unit
Estate number	Listing	Estate	Number
Estate name	OS Map	Estate	Categorical

C) Wood production

Attribute	Data Source	Object	Measurement Unit
Stocking level	Field Assessment	Stand	% of stand area
Stand area	Field Assessment	Stand	hectares
Timber per stand	Field Assessment	Stand	m^3
Timber per hectare	Field Assessment	Stand	m^3ha^{-1}
Mean diameter	Field Assessment	Stand	cm
Crop stage and type	Field Assessment	Stand	Categorical
Treatment	Field Assessment	Stand	Categorical

D) Site and soil

None

E) Forest structure

Attribute	Data Source	Object	Measurement Unit
Forest Type	Field Assessment	Stand	Categorical
Species code	Field Assessment	Stand	Categorical
Year of Origin	Planting records	Stand	Year
Mixture Rating	Field Assessment	Stand	Categorical

F) Regeneration

None

G) Forest condition

None

H) Accessibility and harvesting

None

I) Attributes describing forest ecosystems

None

J) Non-wood goods and services

None

K) Forest functions other than production (protection, recreation, etc.)

None

3.1.2 List of derived attributes

Attribute	Measurement Unit	Input Attributes
Yield Class	$m^3ha^{-1}yr^{-1}$	Top height, Age, Species

Yield Class:

Yield class is a measure of the rate of volume growth per hectare based on the maximum mean annual increment using two cubic metres between classes. It is defined as the average annual volume increment per hectare of a stand over its rotation.

3.1.3 Measurement rules for measurable attributes

For each of the measurable attributes the following information is given:

a) measurement rule
b) threshold values
c) measurement scale
d) rounding rules
e) instrument used for measurement
f) data source

Top Height
a) The average total height of the 100 trees of largest diameter at breast height (dbh) per hectare.
b) There is a minimum height of 3 metres.
c) metres
d) rounded to the nearest metre
e) hypsometer
f) field assessment

3.1.4 Definitions for attributes on nominal or ordinal scale

The following information is provided for each attribute:

a) the definition
b) the categories (classes) and
c) the data sources
d) remarks

A) Geographic regions

County Number
a) Counties are numbered sequentially based on alphabetical order.
b) 1 - 26
c) OS map

Map Number
a) Ordnance Survey Six-inch map series sheet number.
b) 1 - 151
c) OS map

Photo-strip
a) Reference number of aerial photography path.
b) Sequential number
c) Flight Path records

Compartment number
a) A sequential number used to uniquely identify each compartment within each forest estate.
b) Number from 1 upwards
c) Listing

Stand number
a) Used to uniquely to identify a stand within a compartment.
b) Number from 1 upwards
c) Listing

B) Ownership

Estate number
a) A three digit alphanumeric code to identify the forest estate.
b) Alphanumeric codes
c) Listing

Estate name

a) The name of the forest estate.
b) Place name
c) OS map

C) Wood production

Stocking level

a) A stocking level adjustment factor is recorded where there are permanent unstocked areas within a stand which are too small (or scattered) to form a separate stand. This adjustment factor is subsequently used to adjust the area of the stand.
b) 20, 30...100
c) Field assessment
d) To facilitate management prescription and production forecasting

Stand area

a) The area of each stand calculated by using a planimeter or an area grid.
b) 0.5 ha minimum
c) Field assessment and planimeter

Timber per stand

a) Measurable volume to 7 cm top diameter for the stand.
b) Number
c) Field assessment

Timber per hectare

a) Measurable volume to 7 cms top diameter per hectare.
b) Number
c) Field assessment

Mean diameter

a) The quadratic mean diameter of the stand.
b) 2 cm class starting from 7 cm
c) Field assessment

Crop stage and type

a) The stage of development of the crop.
b) Five categories;
   Mature
   Thinning stage
   Brashing stage
   Thicket stage
   Pre-thicket stage
c) Field assessment

Treatment

a) Whether a stand required thinning
b) Narrative on treatment required
c) Field assessment
d) To facilitate estate management

D) Site and soil

None

E) Forest structure

Forest Type

a) The broad classification of the forest type
b) Eight categories;
   Conifer High forest
   Broadleaved High forest
   Mixed High forest
   Felled
   Burned
   Blown
   Undeveloped
   Scrub
c) Field assessment

Species code

a) Code indicating the species
b) 17 different conifer and 11 broadleaf species
c) Field assessment

Year of Origin

a) The year in which a stand was planted.
b) Ranging from 1800 to present day.
c) Obtained from planting records

Mixture Rating

a) The percentage area of the first species in a stand.
b) A percentage class; 30, 40...100
c) Field Assessment

F) Regeneration

None

G) Forest condition

None

H) Accessibility and harvesting

None

I) Attributes describing forest ecosystems

None

J) Non-wood goods and services

None

3.1.5 Forest area definition and definition of "other wooded land"

The Private inventory can be divided into four broad categories;

High Forest - Conifer and broadleaf forest
Un-stocked woodland - Capable of growing trees
Undeveloped - Not capable of growing trees.
Scrub - Scrub

3.2 DATA SOURCES

A) Field data

All plantations, aged one year or older at time of Inventory were surveyed.

B) Questionnaire

Not applicable

C) Aerial photography

A number of different contractors were used.

Strips 1.8 miles wide and 20 miles apart covering 9% of the country

D) Spaceborne or airborne digital remote sensing

Not used

E) Maps

Paper topographical maps at a scale of 1:10560 from the Ordnance Survey of Ireland were used. These maps are circa 90 years old but are quite accurate for upland and forested areas.

Stand mapping was carried out by forest inventory staff using the above topographical maps.

F) Other geo-referenced data

None

3.3 ASSESSMENT TECHNIQUES

The Inventory was divided into two phases;

Phase I was a complete assessment of woodlands of 40 ha or more in one estate and under a single ownership.

Phase II was a 9% strip sample of the country covering all woodlands with the exception of those covered in phase I and those owned by the Forest and Wildlife Service.

The only available reference for the Inventory of Private Woodlands 1973 is not sufficiently detailed to provide all of the information requested in the following sections.

3.3.1 Sampling frame

Phase I - Not applicable because every stand was assessed.

Phase II was a 9% strip sample of the entire country covering all woodlands with the exception of those covered in phase I and those owned by the Forest and Wildlife Service. In this phase the country was divided into four regions and for each region strips 1.8 miles wide and 20 miles apart were marked on a map.

3.3.2 Sampling units

Phase I - Stand.

Phase II - 1.8 miles wide strips, 20 miles apart.

3.3.3 Sampling designs

Phase I - Total enumeration.

Phase II - Each strip was interpreted to locate forest areas. The assessor surveyed each forest area identified.

3.3.4 Techniques and methods for combination of data sources

Phase 1 - Stand boundaries were delineated on paper maps and associated attribute data were recorded on field forms and later entered into computer.

3.3.5 Sampling fraction

Phase I

Data source and sampling unit	Proportion of forested area covered by sample	Represented mean area per sampling unit
Forest stand	100%	-

Phase II

Data source and sampling unit	Proportion of forested area covered by sample	Represented mean area per sampling unit
Aerial Photography Strip	9%	-

3.3.6 Temporal aspects

The year of survey is recorded for each stand.

Inventory Cycle	Time Period of data assessment	Publication of results	Time period between assessments	Reference date
1973 Inventory of Private Woodlands	1970- 1973	1979	-	1973

3.3.7 Data capturing techniques in the field

Data were recorded on "field forms" and entered by hand into a mainframe computer. Stand boundaries were delineated on paper maps.

3.4 DATA STORAGE AND ANALYSIS

Data processing and analyses were carried out by the Management Services Unit of the Department of Lands.

3.4.1 Data storage and ownership

Forest Service of the Department of Agriculture.

3.4.2 Data base system used

The data were stored in an IBM mainframe computer in EBCDIC (Extended Binary Coded Decimal Interchange Code) file format.

3.4.3 Data bank design (if applicable)

Flat file structure

3.4.4 Update level

For each stand the date, when the data has been assessed is available. The data are not updated to a current date in time and thus reflect the situation at the time of assessment.

3.4.5 Description of statistical procedures used to analyse data Including Procedures for sampling error estimation

Phase I

a) **area estimation**
The total area is estimated from the aggregation of stand, forest estate and county data.
b) **aggregation of tree and plot data**
Plot data is aggregated to stand level.
c) **estimation of total values**
Stand data is aggregated to forest estate level. Forest estate data is aggregated to County level data. County level data is aggregated to National level.
d) **estimation of ratios**
Timber volume m^3 per ha
e) **sampling at forest edge**
Not applicable
f) **allocation of stand and area related data to single sample plots/points**
g) **hierarchy of analysis: how are sub-units treated**
Data analysis is hierarchical. Units range are in the following hierarchical structure; plot, stand, forest estate, county, national.

Phase II

a) **area estimation**
The total area, for each species in a county was calculated by multiplying the area of woodland in the strip samples in that county by the total area of that county and dividing by the area of the strip in the county. The confidence limit on the total area of forest is ± 29.66% at the 95 % level. The wide confidence limit is due mainly to the small number of samples as lines constituted the basic sampling unit.
b) **aggregation of tree and plot data**
Plot data is aggregated to stand level.
c) **estimation of total values**
Stand data is aggregated to forest estate level. Forest estate data is aggregated to County level data. County level data is aggregated to National level.

d) **estimation of ratios**
Timber volume m^3 per ha

e) **sampling at forest edge**
If 50% or more of the stand falls within the aerial photography strip the entire stand is included, otherwise the stand is ignored.

f) **allocation of stand and area related data to single sample plots/points**

g) **hierarchy of analysis: how are sub-units treated**
Data analysis is hierarchical. Units range are in the following hierarchical structure; plot, stand, forest estate, county, national.

3.4.6 Software applied

Custom written COBOL programs were used to analyse the data.

3.4.7 Hardware applied

IBM mainframe.

3.4.8 Availability of data (raw and aggregated data)

National level statistics are available to the general public. However forest estate information are only available with the written permission of the estate owner. The data is not currently available "on-line" for analyses and report production.

3.4.9 Subunits (strata) available

Results can be produced at a National, county and estate level.

3.4.10 Links to other information sources

Topic and spatial structure	Age/Period/ Availability	Responsible agency	Kind of data-set
Six-inch map series	Since 1900	Ordnance Survey of Ireland	Paper Maps

3.5 RELIABILITY OF DATA

3.5.1 Check assessments

Field data was checked by a group of Forest Inspectors.

3.5.2 Error budgets

Attribute	Error source
Top heigh	Measurement error
Stocking level	Measurement error
Mixture rating	Measurement error
Area	Measurement error
Timber per stand	Measurement error
Timber per ha	Measurement error

3.5.3 Procedures for consistency checks of data

Custom written computer validation programs.

3.7 INVENTORY REPORTS

3.7.1 List of published reports and media for dissemination of inventory results

Inventory	Year of Publication	Citation	Language	Dissemination
Inventory of Private Woodlands	1979	Inventory of Private Woodlands 1973 Department of Fisheries and Forestry Forest And Wildlife Service. 47 pp	English	Printed

3.7.2 List of contents of latest report and update level

Chapter	Title	Number of pages
1	Background, Objectives, Scope and Plan	1
2	Mapping and stand description	1
3	Data Processing and Forecasts of Yield	1
4	Statistics, Acknowledgements and references	1
5	National Analysis Tables	36

Field data: 1970 - 1973

Aerial photography: 1970

3.7.3 Users of the results

The Forest Service, Department of Agriculture for national forest statistics. Estate Owners and forestry consultants.

3.8 FUTURE DEVELOPMENT AND IMPROVEMENT PLANS

3.8.1 Next inventory period

1997

3.8.2 Expected or planned changes

3.8.2.1 Nomenclature

3.8.2.2 Data sources

Remote Sensing (satellite and airborne), Aerial photography

3.8.2.3 Assessment techniques

3.8.2.4 Data storage and analysis

3.8.2.5 Reliability of data

3.8.2.6 Models

3.8.2.7 Inventory Reports

3.8.2.8 Other forestry data

REFERENCES

Christie, J.M., 1981. Yield Models for Forest Management, Forestry Commission Booklet 48. HMSO.

Hamilton, G.J., 1985. Forest Mensuration Handbook. Forestry Commission Booklet 39. HMSO.

Jordan, P., 1993. Masters Thesis

Purcell, T.J., 1979. Inventory of Private Woodlands, 1973

COUNTRY REPORT FOR ITALY

Vittorio Tosi
ISAFA, Trento, Italy

Marco Marchetti
EURIMAGE, Roma, Italy

SUMMARY

1. Overview

1.1 General forestry and forest inventory data

Italian forests cover an area of 8.7 million ha. Of these, 42 % is represented by coppice forest, 25 % by high forest and 25% by bushland and other particular forest area, and the last 8% by specialised wood and non-wood stand and forest areas without vegetation. Stands with deciduous species predominate clearly towards coniferous stands.

The main forest inventory is the national one, carried out in 1985. INDEFO is another inventory conducted on national level according to the International Cooperative Programme on Assessment and Monitoring of Air Pollution Effects on Forests regulations.

At last many Regions, as well as carry out an important forest management action, have developed forest resources inventories on regional level: Lombardia, Veneto, Friuli, Liguria, Emilia Romagna, Toscana, Umbria, Lazio and Valle d'Aosta.

1.2 Other important forest statistics

The main forest statistics other than the NFI and the regional inventories are provided in Italy by ISTAT. The annual reports contain data about forest area, cuts, assortments, fire-wood, non-wood products, forest fires, wood production, prices of wood assortments and penalties.

2. Italian National Forest Inventory

2.1 Nomenclature

The following Attributes are Directly Assessed:

A)Region, Province;
B) Ownership;
C) Diameter at breast height, tree height, average height of stands classified as slow, radial increment, diameter of stumps, type of tree, quality of stems, age of stumps;
D) Elevation, slope, aspect, location, soil depth, soil moisture, texture of soil and presence of stones;
E) Type of stand, cover, crown coverage, age of stands, tree species, main tree species and diffusion, tree species: general classification of stands, planting space, tending of stands;
F) Regeneration: presence, condition, distribution, origin, conifer regeneration;
G) Site degradation: origin and intensity, stand damages: intensity, origin and diffusion;
H) Unevenness of soil morphology, accessibility;
I) Origin of stand;
J) Height, median diameter, thickness, age and type of cork; function of forest area, land use restrictions.

List of Derived Attributes:

C) Single tree volume, high increment, wood increment in high forest, basal area, number of trees, dbh of felled trees, annual felled area in coppice, volume felled in high forest;
E) Frequency of tree main species in particular forest area.

Measurements Rules for following Measurable Attributes are given:

C) Diameter at breast height, tree height, radial increment, diameter of stumps;
D) Elevation, slope, soil depth;
E) Cover, crown coverage, age of stands, planting space;
G) Diffusion of stand damages;
H) Accessibility;
J) Height, median diameter, thickness, age of cork.

Definitions for following attributes on nominal or ordinal scale are given:

B) Ownership;
C) Type of tree, quality of stems;
D) Location, soil moisture, texture;
E) Type of stand, tree species, diffusion level and general classification of stands, tending of stands;
F) Regeneration: presence, condition, distribution and origin;

G) Site degradation: origin and intensity, stand damages: intensity and origin;
H) Unevenness of soil morphology;
I) Origin of stand;
J) Type of cork; function of forest area, land use restrictions.

The forest area is defined by following parameters: minimum size of 2000 m^2, minimum width of 20 m and a minimum canopy coverage of 20%.

2.2 Data sources

Only some information about the topographic maps used for sample points location is given.

2.3 Assessment techniques

A one phase inventory is applied with only ground assessment of systematically distributed plots. The Characters of sampling frame and field sampling unit are given. Data is recorded on different types of paper forms.

2.4 Data storage and analysis

The data of the Italian NFI is stored in a relational database system by the Ministry of Agriculture, Forest and Food in Rome. A 3*3 km systematic point grid is laid over the entire country.

About the statistical procedure used to analyse data, areas are estimated according to the number of points located inside the strata of interest. Total values are estimated using the theory of error propagation. The growth of wood in high forest is estimated by analysis of measured radial increment.

Particular requests about inventory data may be addressed to the SIAN office responsible for data storage.

2.5 Reliability of data

A check assessment was carried out among Province crews during the field inventory. Other checks were then made for consistency of data which was included in forms after their input in PC. A cross check among some parameters was done.

2.6 Models

The tree volume is estimated by the function v = fj(d, h) and using a set (n. 18) of two entry tables assembled specifically for IFNI. The prediction error of the tables is not known. Volume functions reflect the entire stem volume (with dbh over 3 cm) for coniferous species and stem and branches volume for deciduous species and for pines with large crown.

The wood increment estimate procedure is described, starting from dbh and height increment.

Another model is used for estimating volume of felled trees. In this case data measured on reference trees and on stumps are used as input variables.

2.7 Inventory reports

A list of main published reports of inventory results is given: the inventory project, the field instructions, the two entry tables, the methodological summary and results. Of this last one the main contents are described.

2.8 Future development and improvement plans

A decision about the future Inventories has to be taken by the Italian government.

3. Lombardia Region Forest Inventory

The information is related to the Varese and Bergamo Provincial territory.

3.1 Nomenclature

The following attributes are directly assessed

A)Province, Municipality;
B) Ownership;
C) Number of enumerated trees, diameter at breast height, total tree height;
D) Elevation, slope, aspect, location, stoniness, rock outcrops, soil depth, soil development stage;
E) Forest type, silvicultural system and stand development, age, tree species, sstand origin , stand description;
H) Accessibility;
J) Forest function;
K) Survey timing, surveyor and firm name, topographic map, sample plot number, land-use restrictions.

List of derived attributes:

C) Number of stems per ha, basal area, mean diameter, mean height, mean stem volume, mean annual increment;
E) Dominant species, mean age.

The forest area is defined by the following parameters: crown cover >20%; trees growing to more than 5 m. Minimum size of area: 5000 m^2 and minimum width: 20 m.

3.2 Data sources

Only some information about the topographic maps and aerial photographs used for sample plots location and for the preliminary assessment of forest area are given

3.3 Assessment techniques

A one-stage design is applied and attributes are assessed exclusively in the field, on a permanent 1*1 km grid. The sampling frame is given by the forest area definition and from the forested area as mapped on 1:50000 topographic map. Point sampling is used for tree attributes assessment (BAF = 2).

3.4 Data storage and analysis

The data of the Lombardia RFI is stored in a database system at the Lombardia Regional Authority Office - Forest Department. The forest area is assessed on the field. As a single forest point represents 100 ha of forest area, the total forested area is estimated by multiplying the number of plots by 100. In general the total value of a single attribute for a strata is given by multiplying the area related attribute by the estimated area of the strata. Results can be obtained for the two Provinces assessed (Varese and Bergamo) and for territorial units.

3.5 Reliability of data

2-3% of the field plots are visited a second time by check crews. All attributes are assessed a second time independently from the assessment of the regular field crews. The check assessments are used for several purposes.

3.6 Models

Volume functions have been derived using mean d.b.h. and total tree height resulting from stand height curves, constructed for each single species and stem-diameter classes. The form factors are related to total tree height.

3.7 Inventory reports

The list of published reports of the Lombardia RFI is given, with the index of the reports.

4. Veneto Regional Forest Inventory

4.1 Nomenclature

List of Attributes Directly Assessed:

A) Watershed, Province, Mountain Community, Municipality;
B) Ownership;
C) Number of trees selected in the point sample, diameter at breast height, inside bark diameter, stump girth, total height, bole height, crown shape, stem quality, diameters at 4 m, 5 m and 6.5 m height, crown diameter, merchantable height, 10 intermediate diameters, annual radial increment ;
D) Elevation, slope, aspect, location, soil stability;
E) Forest type, stand origin silvicultural system and stand development, stratified stand structure, canopy cover, age, tree species, social standing, trees spatial distribution;
F) Amount, distribution, leading species, health status;
G) Tree health status;
H) Accessibility, forest road network, ground roughness, rock outcrops, breakage and throw trees, undergrowth density;
K) Forest type map number, plan of forest operations, forest type map unit, working plan forest plot number, surveyor and firm name, sample plot number, land-use restrictions.

List of derived attributes:

C) Number of stems per ha, basal area, mean diameter, mean height, stem total volume, tree volume per ha, current diametric increment ;
E) Forest composition.

The forest area is defined by the following parameters: crown cover >10%; trees growing to more than 5 m; extension >5000 m^2; width >20 m.

4.2 Data sources

Field data are assessed on the interpretation area, on the sub-sample of 3P trees, on the small sub-sample of felled trees and on increment cores. A few information from the topographic and cadastral maps used for sample pots location are given.

4.3 Assessment techniques

The sampling frame is given by the forest area definition, but several categories are not field assessed. The entire region is covered by the assessment. Point sampling is used for field data assessment. The sampling unit is a cluster of three field plots. The plots are distributed in a 800*800m grid. The sampling design applied in the Veneto region is different for public and non-public forests.

4.4 Data storage and analysis

The data of the assessment are stored in a database system at the Forest Department of the Veneto Regional Authority Office in Mestre (VE). Area estimation is made by cartographic methods without sampling errors. As the selection of the trees to be recorded is proportional to the individual basal area, the single tree attributes have to be divided by relative basal area. Results can be obtained for the Provinces, Mountain Communities and Watersheds.

4.5 Reliability of data

No data are available for check assessments.

4.6 Models

Volume functions have been derived using d.b.h., diameters at 4 m, 5 m e 6.5 m height and total tree height as input variables. Nine different volume functions for dominant species, silvicultural system and specific species groups have been derived. Trees are distributed in stand height classes, and the growth has been calculated with a growth model based on d.b.h., tree species and age of the single tree.

4.7 Inventory reports

The list of published reports of the Veneto RFI is given, with the index of the reports.

5. Friuli-Venezia Giulia Regional Forest Inventory

5.1 Nomenclature

List of attributes directly assessed:

A) Watershed, Province, Mountain Community, Municipality;
B) Ownership;
C) Number of trees selected in the point sample, Diameter at breast height, bark thickness, stump girth, Total height, Bole height, Crown shape, Stem quality, Diameters at 4 m, 5 m and 6.5 m height, ten intermediate diameters, Annual radial increment;
D) Elevation, slope, aspect, location, soil stability;
E) Forest type, stand origin, silvicultural system and stand development, stratified stand structure, canopy cover, age, tree species, social standing, trees spatial distribution;
F) Amount, distribution, leading species, health status;
G) Tree health status;
H) Accessibility, forest road network, ground roughness, rock outcrops, breakage and throw trees, undergrowth thickness;

K) Surveyor and firm name, sample plot number.

List of derived attributes:

C) Number of stems per ha, basal area per ha, mean diameter, mean height, stem total volume, tree volume per ha, current diametric increment;
E) Forest composition.

The forest area is defined by the following parameters: crown cover >20%; trees growing to more than 5 m; extension >2000 m²; width >20 m.

5.2 Data sources

Field data is assessed on the interpretation area, on the sub-sample of 3P trees, small sub-sample of felled trees and on increment cores. Little information about the topographic map used for sample pots location and about aerial photographs is given.

5.3 Assessment techniques

The sampling frame is given by the forest area definition, but several categories are not field assessed. The entire region is covered by the assessment. Point sampling is used for field data assessment. The sampling unit is a cluster of three field plots. The plots are distributed in a 200*200 m grid. The sampling design applied in the Friuli-Venezia Giulia region is a double-sampling for stratification.

5.4 Data storage and analysis

A lot of data is not available. The forest area is assessed on the aerial photographs. Area estimation is made by cartographic methods without sampling errors. As the selection of the trees to be recorded is proportional to the individual basal area, the single tree attributes have to be divided by relative basal area. Results can be obtained for the Provinces, Mountain Communities and Watersheds.

5.5 Reliability of data

No data is available for check assessments.

5.6 Models

Volume functions have been derived using d.b.h., diameters at 4 m, 5 m e 6.5 m height and total tree height as input variables. Volume functions for dominant species, silvicultural system and specific species groups have been derived. Trees are distributed in stand height classes and the growth has been calculated with a growth model based on d.b.h., tree species and age of the single tree.

5.7 Inventory reports

The list of published reports of the Friuli RFI is given, with the index of the reports.

6. Multipurpose Forest Inventory Of Liguria Region

6.1 Nomenclature

List of attributes directly assessed:

A)Province
B) Ownership;
C) Diameter at breast height, Tree height, Quality of stems, Crown height and wide, Crown class, Exploitation;
D) Elevation, Slope, Aspect, Location, Stoniness, Rockiness, Soil erosion and deposition, Humus, Soil survey;
E) Trace of former wood, Inventory stand types, Cover: height of vegetation levels, Cover: percent of coverage on levels, Stand development, Origin of stand, Sylvicultural system, Conversion to high forest, Tending of stands, Age of stands, Planting space, Cultivation status, Floristic composition, Bush pattern;
F) Regeneration: Condition, Origin, Distribution, Localisation, Transect of regeneration
G) Site degradation: type, origin and intensity; Stand degradation: origin, intensity, distribution and diffusion;
H) Accessibility and harvesting: difference in height-distance between plot and road, difference in height-distance between plot and skidding road/path, road classification distance covering on carriage road: improvements on existing road, Unevenness of soil morphology: types and intensity;
I) Micro-habitat: clearing: extension; margins: type, limits and extension; superficial water: type; Fuel models, Dead fuel survey, Phytosociological survey, Linear transect of mixed areas, Tree limit, Wildlife track: type and diffusion;
J) Infrastructures and facilities, Function of the forest area, Land use restrictions.

List of derived attributes:

C) Single tree volume, Basal area;
I) Actual forest vegetation map, Dynamic vegetation series map.

The forest area is defined by following parameters and divided into six land use classes are:

1) Trees: tree coverage > 20% in polygons with surface $\geq$5000 m^2

2) Bush: bush coverage > 40% with tree coverage < 20% in polygons with surface $\geq$5000 m^2

3) Herbs: permanent meadows, pasture, fallow with herbs coverage $> 20\%$, trees coverage $< 20\%$ and bush $< 40\%$ in polygons with surface $\geq 5000\ m^2$
4) Mixed areas: a mosaic of classes 1/2/3 in small areas (5000 m^2) in polygons with surface not less than 10000 m^2
5) Barren land: areas with herbs coverage $< 20\%$, trees coverage $< 20\%$ and bush $< 40\%$ in polygons with surface $\geq 5000\ m^2$
6) Included areas: surfaces with an homogenous internal land use but not homogenous as regards the surrounding land use with an area $< 5000\ m^2$ and $> 500\ m^2$ for classes 1, 2 and 3, and 40000 m^2 for barren or arable land.

6.2 Data sources

In this survey many sources of information are utilised:

aerial photography, spaceborne or airborne digital remote sensing, maps and other geo-referenced data.

6.3 Assessment techniques

The entire region is covered by the assessment. The sampling unit is a concentric fixed area. The sampling design applied in the Multipurpose Forest Inventory of Liguria Region is double-sampling for stratification. Every data are collected or derive from field assessment. Data recorded on several paper forms filled separately for different survey and types of land-use. The reference year is 1993.

6.4 Data storage and analysis

The data of the assessment are stored in a in a relational ORACLE and DBIII database system on PC. The forest area is assessed on the forested photo plot. Results are obtained for entire Liguria Region and some attributes also for the four Provinces.

6.5 Reliability of data

Check assessment are carried out during the field inventory. Check assessment of Vegetation maps has been carried out after the preliminary mapping. Only sampling errors was calculated for the publication of inventory results. Consistency checks and cross control were done for data collected.

6.6 Models

The tree volume is estimated as a function of dbh and tree height. The IFNI's two entry volume tables (and functions) were used for the IFMR.

6.7 Inventory reports

The list of published reports is given. The content of final report is described.

6.8 Future development and improvement plans

Methodology has been assessed for a continuous updating: every five years for maps and every ten years for the Forest Multipurpose Inventory. Concerning fire assessment it is planned to have a yearly control and mapping of the burned areas using satellite imagery monitoring.

6.9 Miscellaneous

Special investigations have been carried out in the context of the Forest Map 1:50.000, GIS, Multipurpose Forest Inventory of Liguria Region project.

7. Toscana Regional Forest Inventory

7.1 Nomenclature

List of attributes directly assessed:

A)Province, Municipality Association, Municipality;
B) Ownership;
C) Number of trees selected in the point sample, Diameter at breast height, Total height, Crown breadth, Crown shape, Stem quality, Bole height, Intermediate diameters, Logs quality, Annual radial increment;
D) Elevation, Slope, Aspect, Location, Terrain pattern, Rock outcrops, Stoniness, Undergrowth thickness, Soil stability;
E) Canopy cover, Forest type, Stand origin, Silvicultural system, Stand development and structure, Top tree age, Stand age, Tree species, Social standing , Trees spatial distribution;
G) Tree health status;
H) Ground roughness, Lorry-road network, Forest-road network;
K) Sample plot number.

List of derived attributes:

C) Number of stems per ha, Basal area per ha, Mean diameter, Mean height, Stem total volume, Tree volume per ha, Current diametric increment;
E) Forest composition.

The forest area is defined by the following parameters: crown cover >10%; trees usually growing to more than 5 m; extension >5000 m². Included are several peculiar conditions.

7.2 Data sources

The interpretation area for area related data is 0.5 ha. Only a few attributes are assessed on all the enumerated trees; most of them are assessed on a sub-sample of trees. Aerial photography used are 1:33000 and panchromatic. Maps are used for sample plots location and for the assessment of few attributes. They are topographic maps (1:25000).

7.3 Assessment techniques

The sampling frame is given by the forest area definition, but several categories are not field assessed. Only four Provinces are covered by the assessment. Aerial photography sample plots are circular and are systematically distributed in a 400*400 m grid. Point sampling is used for field data assessment. The sampling unit is a cluster of four field plots. The plots are distributed in a 400*400m grid. The sampling design applied in the Toscana region is a double-sampling for stratification. The results of the first IFT have not yet been published.

7.4 Data storage and analysis

The data are stored in a DB III database system at the Toscana Regional Authority Office - Department of Forests. The forest area is assessed on the aerial photographs. As the selection of the trees to be recorded is proportional to the individual basal area, the single tree attributes have to be divided by relative basal area. The total value of attribute for a strata is given by multiplying the area related attribute by the calculated area of the strata. For plots not falling entirely into forests, but extend partially into non forested areas, tree attributes are doubled on an equivalent specula area. Some data is not available as results of the first IFT have not yet been published

7.5 Reliability of data

5% of the field plots are visited a second time by check crews. All attributes are assessed a second time independently from the assessment of the regular field crews. 10% of the photo sample plots have been interpreted. 100% of the photo sample plots have been field-checked and more precise attributes have been added for the definition of strata.

7.6 Models

As results have not yet been published no information are given about models.

7.7 Inventory reports

Information are given on published reports.

8. Emilia-Romagna Regional Forest Inventory

8.1 Nomenclature

List of attributes directly assessed:

A) Watershed, Province, Mountain Community, Municipality;
B) Ownership;
C) Crown closure and average diameter, Number of trees selected in the point sample, Edge case location, Diameter at breast height, bark thickness, stump girth, Total height, Merchantable height, Crown point, Crown shape, Stem quality, Intermediate diameters at 4.5 m or 6.5 m height, 10 intermediate diameters, Annual radial increment;
D) Elevation, slope, aspect, location, ground roughness, soil stability, Rock outcrops and stoniness;
E) Forest type, stand origin, silvicultural system and stand development, vertical structure, Age, tree species, social standing, spatial distribution, undergrowth density;
F) Regeneration features, Leading forest species in regeneration;
G) Forest health, Damage origin;
H) Accessibility, Forest road network and condition;
K) Assessment date, Crew members names and roles, Field plot access traverse.

List of derived attributes:

A) Fitoclimatic zones;
C) Number of stems per ha, Basal area per hectare, Individual basal area, Mean basal area, Mean height, Lorey mean height, Stand top height, Stem total volume, Volume tables, Tree volume per hectare, Mean annual increment;
E) Leading species, Species composition.

The forest area is defined by the following parameters: crown cover >10%; trees growing to more than 5 m; extension >5000 m^2; width >20 m.

8.2 Data sources

Information is given on responsible bodies from whom aerial photographs and maps are obtained. The former are in scale 1:13000, the latter 1:10000 and are used for the location of the field plot. Georeferenced data are used for the sharp identification of rivers and administrative boundaries.

8.3 Assessment techniques

The Emilia-Romagna Forest Inventory concerns only the mountainous and hilly areas, as defined by the administrative boundaries of the Mountain Communities. On aerial photographs 0.5 ha circular plots systematically distributed in a 200*200 m grid have been interpreted. Sample pointing has been carried out in the field. The sampling unit is a cluster of 4 recording units, one in the central point, systematically distributed on a 1km*1km grid. The sampling design applied in the Emilia-Romagna region is a double-phase sampling for stratification. Attributes for stratification are assessed in photo plots and are used for estimating strata sizes. Attributes assessed in the field plots are weighted with the strata sizes derived from the photo-interpretation.

8.4 Data storage and analysis

The data of the assessment are stored in a database system at the Emilia-Romagna Regional Authority Office - Forest resource Department - Forest Inventory Unit. The data is stored in a DBIII - DBIV database. The data bank design is presented. The forest area is assessed on aerial photographs. Two different sampling errors have to be considered: the sampling error due to the estimation of the strata sizes and the sampling error of the attribute of interest. Results can be obtained for watershed, Province, Mountain Community and Municipality.

8.5 Reliability of data

10% of the field plots are visited a second time by check crews. All attributes are assessed a second time independently from the assessment of the regular field crews. 3% of the photo sample plots are regularly checked by regional forest officers; when the interpretation does not agree the field check is carried out.

8.6 Models

Volume is expressed as a function of dbh and total height. 1,230 individual trees distributed among the main species and groups of minor species are used. A form factor is calculated for deriving the merchantable stem volume.

8.7 Inventory reports

The list of published reports of the Emilia-Romagna RFI is given, with the index of the reports.

8.8 Future development and improvement plans

The next inventory period should possibly start in the year 2000 and the acquisition of forest and land use cover types from aerial photographs and maps should be accomplished by the year 2005.

9. Umbria Regional Forest Inventory

9.1 Nomenclature

List of attributes directly assessed:

A) Mountain Community, Municipality;
B) Ownership;
C) Diameter at breast height, total tree height;
D) Elevation, slope, aspect, location, stoniness, rock outcrops, soil depth, soil development stage, texture;
E) Forest type, silvicultural system and stand development, canopy cover, age, tree species, number of stools, stand description;
H) Accessibility;
J) Forest function;
K) Survey timing, surveyor and firm name, topographic map, sample-plot number, land-use restrictions.

List of derived attributes:

C) Top height, number of stems per ha, basal area, mean diameter, mean height, form factor, mean stem volume, mean annual increment;
E) Dominant species.

The forest area is defined by the following parameters: crown cover >20%; trees growing to more than 5m; extension >2000m^2; width >20m.

9.2 Data sources

Only some information about the topographic maps and the cadastral maps used for sample plots location are given.

9.3 Assessment techniques

A one-stage inventory design is applied and attributes are assessed exclusively in the field, on a permanent 1*1 km grid. The sampling frame is given by the forest area definition. The sampling units have a fixed area and circular shape.

9.4 Data storage and analysis

The data of the Umbria RFI are stored in a database system at the Umbria regional Authority Office. A 1*1 km systematic point grid is laid over the entire region. A single forest point represents 100 ha of forest area. Total values are estimated using the theory of error propagation. The estimation of growth is calculated with the mean annual increment.

9.5 Reliability of data

Check assessments are made on a restricted number of randomly selected plots and are aimed at several purposes.

9.6 Models

Tree volume is estimated by the function v=fj(d,h) accordingly with a two entry tables set assembled for the IFNI. The prediction error of the tables is not known. Volume functions reflect the entire stem volume (with dbh over 3 cm) for coniferous species and stem and branch volume for deciduous species and for pine with large crown. The mean annual increment has been calculated in each single plot and for every single tree species and type.

9.7 Inventory reports

The list of published reports of the Umbria RFI is given, with the index of the reports.

10. Lazio Regional Forest Inventory

All the information are related to the Frosinone Provincial Forest Inventory.

10.1 Nomenclature

List of attributes directly assessed:

A)Mountain Community, Municipality;
B) Ownership;
C) Diameter at breast height, total tree height;
D) Elevation, slope, aspect, location, stoniness, rock outcrops, soil depth, soil development stage, texture;
E) Forest type, silvicultural system and stand development, canopy cover, age, tree species, number of stools, stand description;
H) Accessibility;
J) Forest function, land-use restrictions;
K) Survey timing, surveyor and firm name, topographic map, sample-plot number.

List of derived attributes:

C) Top height, number of stems per ha, basal area, mean diameter, mean height, form factor, mean stem volume, mean annual increment;

E) Dominant species.

The forest area is defined by the following parameters: crown cover >20%; trees growing to more than 5m; extension >2000m²; width >20m.

10.2 Data sources

Only some information about the topographic maps and the cadastral maps used for sample plots location are given.

10.3 Assessment techniques

A one-stage design is applied and attributes are assessed exclusively in the field, on a permanent 1*1 km grid. The sampling frame is given by the forest area definition. The sampling units are concentric circular plots.

10.4 Data storage and analysis

The data of the Umbria RFI are stored in a database system at the Umbria regional Authority Office. A 1*1 km systematic point grid is laid over the entire region. A single forest point represents 100 ha of forest area. Total values are estimated using the theory of error propagation. The estimation of growth is calculated as mean annual increment.

10.5 Reliability of data

Check assessments are made on a restricted number of randomly selected field plots. Their purposes are given. A cross check among some parameters was done.

10.6 Models

The tree volume is estimated by the function $v = f_j(d, h)$ accordingly with a set of two entry tables assembled for the IFNI. The prediction error of the tables is not known. Volume functions reflect the entire stem volume (with dbh over 3 cm) for coniferous species and stem and branches volume for deciduous species and for pines with large crown. The mean annual increment has been calculated in each single plot and for every single tree species and type.

10.7 Inventory reports

The list of published reports of the Lazio RFI is given, with the index of the reports.

11. Valle D'Aosta Regional Forest Inventory

11.1 Nomenclature

List of attributes directly assessed:

A) Mountain Community, Municipality;
B) Ownership;
C) Diameter at breast height, total height, tree type, radial increment;
D) Elevation, slope, aspect;
E) Forest type, silvicultural system, stand development stage, crown canopy, age, tree species, tending of stands: type and priority;
F) Number of young trees;
G) Number of dead trees, predominant damage, damage intensity, defoliation and discolouration, grazing damage;
H) Accessibility, hauling system;
J) Forest function;
K) Forest compartment number, plot number.

List of derived attributes:

C) Basal area per ha, top tree height, number of stems per ha, growing stock per ha.

The forest area definition is not available.

11.2 Data sources

Aerial photography and maps are the data sources. Information on the responsible bodies from whom they can be obtained, the scale and the date is given.

11.3 Assessment techniques

The entire Region is covered by the assessment. Data are assessed on permanent and temporary fixed area circular plot. The sampling design applied in the Valle d'Aosta RFI is characterised by a preliminary phase in which a forest/non-forest decision is made on the aerial photographs/map and land-use strata assigned to the photo plots. In the second phase forested plots are field assessed. Strata areas are calculated on the maps as plots represents respectively 25 ha, 50 ha (temporary plots) e 450 ha (permanent plots). Data recorded in the field by handheld computer (FW60) with immediate plausibility check.

11.4 Data storage and analysis

The data of the assessment are stored in a database system at the Forest Department of the Valle d'Aosta Regional Authority in Aosta. A lot of data is not available.

11.5 Reliability of data

No data is available for check assessments.

11.6 Models

Volume functions have been derived using d.b.h., and total height as input variables. Models on the assortments, potential yield, forest functions are not assessed.

11.7 Inventory reports

The list of published reports of the Valle d'Aosta RFI is given, with the index of the reports.

12. Italian National Inventory Of Forest Damage

12.1 Nomenclature

The Italian National Inventory of Forest Damage is carried out on the plots of the INFI. For each visited plot the following attributes are directly assessed:

A) Geographic Regions: Region, Province, Municipality;
B) Ownership;
D) Site and Soil: elevation, aspect, slope, site class fertility;
E) Forest Structure: wood type, stand type, stand age, tree species, d.b.h.;
F) Forest Condition: tree damage: defoliation, de-coloration, origin;
K) Miscellaneous: INFI plot number;

Such attributes are used for deriving the following ones:

G) Forest Condition: number of damaged trees, percentage of damaged trees, frequency of damage in main groups of species.

12.2 Data sources

Field assessment is the only data source.

12.3 Assessment techniques

The plots are systematically distributed on the 9*9 km grid and surveys are limited to such plots enumerating less than 30 trees. All plots are permanently monumented and documented. The sampling design applied in the INDEFO is a one stage design restricted to field assessment. Since 1984 the surveys are conducted every year and the results are published the successive year.

12.4 Data storage and analysis

The data of assessment is stored in an ORACLE database system in SIAN (*Sistema Informativo Agricolo Nazionale*) by the Direzione Generale delle Risorse Forestali, Montane ed Idriche of Ministero delle Risorse Agricole, Forestali ed Alimentari. Requests of raw data and derived data by outside persons have to be addressed to the secretary's office of SIAN. The reference date is the summer of the assessment year.

12.5 Reliability of data

Information about reliability of data is not available. Models have not been applied.

12.6 Models

Any model is applied.

12.7 Inventory reports

Every year the results are published and edited by the Ministry of Agriculture.

12.8 Future development and improvement plans

The methodology is going to change in 1996 and is foreseeable the reduction of the sampling units.

1. OVERVIEW

1.1 GENERAL FORESTRY AND FOREST INVENTORY DATA

The total area of Italy is 30,127,761 ha. Of this 28.8% or 8.7 million ha is forest land. 80% of forested area is covered by deciduous trees, 16% by coniferous and 4% by mixed stands. The forested area is distributed according the following types:

- high forest	2,178,900 ha
- coppice	3,673,800 ha
- plantation for wood (poplar-stands) or non-wood production (sweet chestnut-, cork oak-stands)	288,900 ha
- particular forest area (bushland/*maquis*, riparian forest and rock-wood area)	2,160,900 ha
- other forest areas without wood-vegetation (areas included, felling or destroyed areas)	372,600 ha

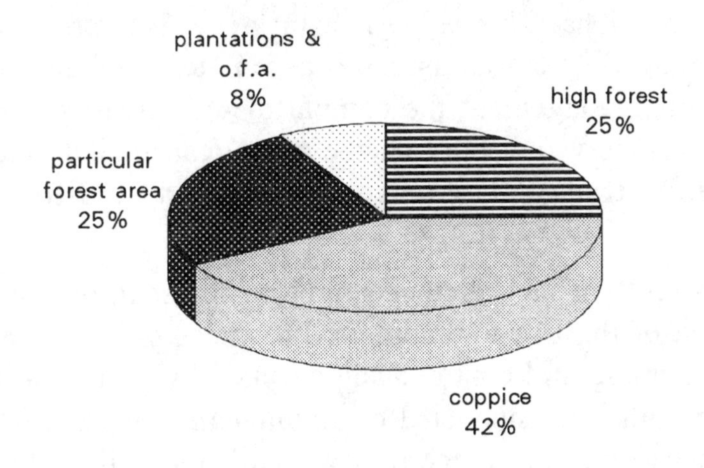

Figure 1. Main types of forest area in Italy.

The main tree species of high forest are represented by spruce (*Picea abies*), silver fir (*Abies alba*), larch (*Larix europaea*), mountain pines (*Pinus sylvestris, P. nigra, P. cembra, P. montana*), Mediterranean pines (*Pinus halepensis, P. pinea* and *P. pinaster*) and beech (*Fagus sylvatica*). The coppice forests are represented especially by sweet chestnut (*Castanea sativa*), beech , oaks (*Quercus cerris, Q. pubescens, Q. ilex, Q. robur*) and hornbeam (*Ostrya carpinifolia*).

The average volume per hectare is 211 m^3 in high forest (with a height > 5 m) and 178 m^3 if we consider also the areas covered by stands of high forest with height less than 5 m (young stands) and that without stocking. The average current increment in high forest is 7.9 m^3. 36 % of the current increment is felled.

The average volume per hectare of coppice (> 5 m) is 115 m^3. The average felled area per year is about 72,000 ha (2 % of total coppice area).

66 % (or 5,730,300 ha) of total forest area is private and 34 % (2,944,800 ha) is public (almost of municipalities).

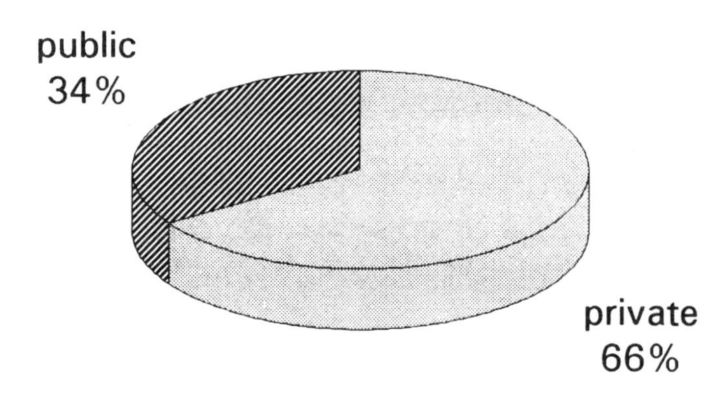

Figure 2. Ownership proportions of forest area in Italy.

The number of inhabitants (57,746,000 inhabitants were counted in year 1991) per ha of forested land is 6.7 (each citizen has therefore 1,500 m² of forested land).

Referring only the managed forest area, the smallest unit or compartment (*particella* or *sezione*) has an average size of 17 ha. This value is a bit aged: it refers to the situation of 1981, when the managed area was assessed as one-tenth of the total forested area. In the Italian forest management the compartment, which is really delimited by boundaries on the territory (not only on the map), can include more stands (groups of homogeneous trees), that may be subjected to different management systems.

The forest management represents in fact another source of forest information. For instance the total forest area of the two autonomous Provinces of Trento and Bolzano is managed by forest planning and continuously checked by an up-to-date information system. Since these plans are conducted on municipality or farm level through integral census of forest resources, afterwards we will not analyse these surveys.

The following surveys were conducted in Italy:

Italian National Forest Inventory (Inventario Forestale Nazionale Italiano - IFNI)

Time of first and last assessment: 1985 -
Number of inventory cycles: 1
Area covered: nation
Institutions and organisations involved in the assessment: National and Regional (or local) Forestry Corps, Forest and Range Management Research Institute (ISAFA)
Tasks: 550 foresters
Resources: 3.5 billion Italian. lire (with an average cost per hectare of forested land of 400 lire)
Legal status: Law 984/1977 about the food and agriculture plan.

Other surveys were carried out at regional level; some of them have given results for the entire regional territory, other only for single provinces or mountain communities. A brief description follows.

Lombardia

The following surveys were conducted in Lombardia:

Varese Provincial Forest Inventory (*Inventario Forestale della Provincia di Varese)*

Time of first and last assessment: 1985
Number of inventory cycles: 1
Area covered: Varese Provincial territory
Institutions and organisations involved in the assessment: Lombardia Regional Authority, S.A.F. *(Società Agricola Forestale)*, independent forest consulting firms, professionals
Budget: data not available

Bergamo Provincial Forest Inventory (*Inventario Forestale della Provincia di Bergamo)*

Time of first and last assessment: 1987
Number of inventory cycles: 1
Area covered: Bergamo Provincial territory
Institutions and organisations involved in the assessment: Lombardia Regional Authority, S.A.F. *(Società Agricola Forestale)*, independent forest consulting firms, professionals
Budget: 261 millions Italian lire

Plantations rows inventory of the Lombardia lowland *(Inventario dei filari di piante da legno della pianura Lombarda)*

Time of first assessment: 1982
Number of cycles: 1
Area covered: the lowland of the region
Institutions and organisations involved in the assessment: data not available
Budget: data not available

Veneto

The following surveys were conducted in Veneto:

Veneto Regional Forest Inventory - Public Forests *(Inventario Forestale dei boschi pubblici della Regione Veneto)*

Time of first and last assessment: 1986
Number of inventory cycles: 1
Area covered: regional territory
Institutions and organisations involved in the assessment: Veneto Regional Authority, independent forest consulting firms, professionals
Budget: data not available

Veneto Regional Forest Inventory - Non-Public Forests *(Inventario Forestale dei boschi non pubblici della Regione Veneto)*

Time of first and last assessment: 1985
Number of inventory cycles: 1

Area covered: regional territory
Institutions and organisations involved in the assessment: Veneto Regional Authority, independent forest consulting firms, professionals
Budget: 2,450 million Italian lire

Friuli

The following survey was conducted in this region:

Friuli-Venezia Giulia Forest Inventory *(Inventario Forestale della Regione Autonoma Friuli-Venezia Giulia)*

Time of first and last assessment: 1987
Number of inventory cycles: 1
Area covered: regional territory
Institutions and organisations involved in the assessment: Friuli-Venezia Giulia Regional Authority, independent forest consulting firms, professionals
Budget: 3 billion Italian lire
Legal status: Regional Regulation 22/1982

Liguria

The following surveys were conducted in this region:

Forest Map 1:50000, GIS, Multipurpose Forest Inventory of Liguria Region *(Carta Forestale d'Italia 1:50000 e Sistema Informativo Geografico Forestale della Regione Liguria).*

Time of first and last assessment: 1993-1995
Number of inventory cycles: 1
Area covered: Liguria Region
Institutions and organisations involved in the assessment: Ministry of Agricultural - National Forestry Corp public purchaser; the operative working group is made by a Consortium of engineering Companies (Agrisiel, Aquater, Fisia, Italeco and Nomisma
Resources: 3.5 billion Italian lire (with an average cost per hectare of forested land of 12,700 lire) for
Legal status: Pilot project for the new national forest map (Law no. 47/1975)

Toscana

The following surveys were conducted in Toscana Region:

- Toscana Forest Inventory *(IFT - Inventario Forestale della Toscana)*

Number of inventory cycles: 1
Area covered: 4 Provinces (Firenze, Pistoia, Arezzo, Siena)
Institutions and organisations involved in the assessment: Toscana Regional Authority, independent forest firms, professionals.
Budget: data not available
Legal status: Regional Regulation 56/1980

Toscana West-Side Inventory: a new project, with a new methodology, was conducted

1993-95 in the territory not previously covered. The results are not yet available.

The description refers only to the first inventory.

Emilia Romagna

The following survey was conducted in Emilia-Romagna:

Emilia-Romagna Forest Inventory - *(Inventario Forestale della Regione Emilia Romagna)*

Time of first and last assessment: 1990
Number of inventory cycles: 1
Area covered: the mountain and hilly areas of the region, as defined by the administrative boundaries of the Mountain Communities regional territory
Institutions and organisations involved in the assessment: Emilia-Romagna Regional Authority, independent forest consulting firms, professionals
Budget: 1,630 million Italian lire
Legal status: Regional Regulation 30/1981

Umbria

The following survey was conducted in this region:

Umbria Regional Forest Inventory *(Inventario Forestale della Regione Umbria)*

Time of first and last assessment: 1993
Number of inventory cycles: 1
Area covered: regional territory
Institutions and organisations involved in the assessment: Umbria Regional Authority, S.A.F. *(Società Agricola Forestale)*, independent forest consulting firms, professionals
Budget: 1,850 millions Italian lire

Lazio

The following surveys were conducted in Lazio:

Frosinone Provincial Forest Inventory (*Inventario Forestale della Provincia di Frosinone)*

Time of first and last assessment: 1994
Number of inventory cycles: 1
Area covered: Frosinone Provincial territory
Institutions and organisations involved in the assessment: Lazio regional Authority, S.A.F. *(Società Agricola Forestale)*, independent forest consulting firms, professionals
Budget: 832 million Italian lire

Simbruini Mountain Natural Park Forest Resources Inventory (*Inventario delle risorse forestali del Parco Naturale Regionale dei Monti Simbruini)*

Time of first and last assessment: 1987
Number of inventory cycles: 1

Area covered: Natural Park area
Institutions and organisations involved in the assessment: Lazio regional Authority, S.A.F. *(Società Agricola Forestale)*, independent forest consulting firms, professionals
Budget: data not available

Alta Tuscia Mountain Community Forest Inventory (*Inventario Forestale della Comunità Montana dell'Alta Tuscia)*

Time of first and last assessment: 1987
Number of inventory cycles: 1
Area covered: Alta Tuscia Mountain Community territory
Institutions and organisations involved in the assessment: Lazio regional Authority, S.A.F. *(Società Agricola Forestale)*, independent forest consulting firms, professionals
Budget: data not available

Valle d'Aosta

The following survey was conducted in this region:

Valle d'Aosta Regional Forest Inventory - *(Inventario Forestale delle risorse forestali e del territorio regionale della Regione autonoma Valle d'Aosta)*

Time of first and last assessment: 1994
Number of inventory cycles: 1
Area covered: regional territory
Institutions and organisations involved in the assessment: Valle d'Aosta Regional Authority, independent forest consulting firms, professionals
Budget: 800 million Italian lire

INDEFO

The survey refers to the Italian National Inventory of Forest Conditions (*Indagine sul deperimento forestale - INDEFO*)

Time of first and last assessment: 1984-1995
Number of inventory cycles: 12
Area covered: nation
Institutions and organisations involved in the assessment: Ministry of Agriculture - National Forestry Corp
Task: internal staff - forest guards
Resources: internal budget - unassessed

Another region, Sardegna, has carried out a regional inventory, but publications are not available and therefore no information should be given for the moment. Calabria and Marche too have started their regional inventories.

The regional survey of Sicilia was broken off after starting. At least a survey of poplar plantations in the Po valley was carried out and repeated every year since 1987.

1.2 OTHER IMPORTANT FOREST STATISTICS

1.2.1 Other forest data and statistics on the national level

As explained in Chapter 1.1 the forest management represents an important source of forest information at local level. The trend of managed forest area in Italy has be studied in the 1970s and 1980s by ISAFA (the last update refers to the situation of 1981), but non later, because of the transferring of many technical concerns from national to regional Forestry Services.

In Italy, every year a report about forest statistics (*statistiche forestali*) will be made. This report contains data about forest area, cuts (area felled and volume of wood), assortments, firewood, non-wood products, forest fires, regeneration in tree nurseries, penalties for forestry laws or game, wood production, prices of wood assortments. This data is collected through formulas distributed to the Provincial Forest Service and to the local Forestry Corps. Some of this information is collected by visual assessment.

The responsible body of forest statistics is the 'Istituto Nazionale di Statistica' (ISTAT). The last update was published in 1995 and gives information about the 1993.

Other particular surveys carried out by the National Forestry Corp concern the trash unloading in forests, the mines, the afforestation area, the census of monumental and historical trees.

1.2.2 Delivery of statistics to UN and Community institutions

1.2.2.1. Responsibilities for international assessment

ISTAT provides forest resource data. Dr. Pedicini is now the head of forestry statistics office.

By Ministry of Agriculture, Food and Forests there is an office for statistics which gives the necessary information to ISTAT. Because the recent reform of the Ministry, it's not possible to give a responsible for each international institution (FAO/ECE, EUROSTAT, OECD, etc.).

Actually for any information dr. Cavalensi should be considered as reference person by Ministry of Agriculture (tel.+39 6 4665 7047, fax +39 6 6754060, e-mail cfs@flashnet.it).

1.2.2.2 Data compilation

About forest fires, ISTAT compiles periodical reports for EUROSTAT and FAO/ECE. The data is collected by the local Forestry Corps. Another source of fire data is the Division XII of Ministry of Agriculture, which provides information almost in real time on fire communication.

As explained in point 1.2.1. the data about non-wood goods and services so as the cuts in forest are collected every three months by the local Forestry Corps.

About the INDEFO survey data see chapter 12.

2 ITALIAN NATIONAL FOREST INVENTORY
(INVENTARIO FORESTALE NAZIONALE ITALIANO - IFNI)

2.1. NOMENCLATURE

2.1.1. List of attributes directly assessed

A) Geographic regions

Attribute	Data source	Object	Measurement unit
Region (Regione)	map	plot centre	categorical
Province (Provincia)	map	plot centre	categorical

B) Ownership

Attribute	Data source	Object	Measurement unit
Ownership (Proprietà)	local information cadastral	stand	categorical

C) Wood production

Attribute	Data source	Object	Measurement unit
Diameter at breast height *(Diametro a petto d'uomo)*	field assessment	tree	cm
Tree height (*Altezza arborea*)	field assessment	tree	m
Average height of stands classified as low (*Altezza media del soprassuolo nei popolamenti classificati bassi*)	field assessment	stand	m
Radial increment *(Incremento radiale)*	field assessment	last 5 year of tree radial growth	mm
Diameter of stumps *(Diametro delle ceppaie)*	field assessment	stumps of trees felled during the last year	cm
Type of tree (*Dendrotipo*)	field assessment	tree	categorical
Quality of stems *(Qualità dei fusti)*	field assessment	tree	categorical
Age of stumps (*Età delle ceppaie)*	field assessment	stump	year

D) Site and soil

Attribute	Data source	Object	Measurement unit
Elevation (*Altitudine*)	field (altimeter)	plot centre	m
Slope (*Pendenza*)	field (clisimeter)	plot	1 %
Aspect (*Esposizione*)	field (compass)	plot	degree
Location (*Giacitura*)	field (visual observation)	plot	categorical
Soil depth (*Profondità del suolo)*	field	plot	cm
Soil moisture (*Umidità del suolo)*	field	plot	categorical
Texture of soil and stoniness*(Tessitura/ Pietrosità del suolo)*	field	plot	categorical

E) Forest structure

Attribute	Data source	Object	Measurement unit
Type of stand *(Tipo di popolamento = tipo inventariale di riferimento)*	field	area of 0.5 ha	categorical
Cover (*Copertura*)	field	plot	percent
Canopy cover *(Grado di copertura)*	field	plot (stands with height <5m)	percent
Age of stand *(Età del soprassuolo)*	field	plot	year
Tree species (*Composizione*)	field	tree	categorical
Main tree species and diffusion *(Composizione specifica e diffusione)*	field	plot (particular forest area only)	categorical
Tree species: general classification of stand *(Composizione: inquadramento tipologico generale)*	field	stand	categorical
Planting space (*Sesto d'impianto)*	field	plot (specialised forest area only)	m
Tending of stand (*Cure colturali)*	field	plot (specialised forest area only)	categorical

F) Regeneration

Attribute	Data source	Object	Measurement unit
Presence (*Presenza*)	field	plot (high forest > 5 m only)	categorical
Condition (*Stato vegetativo*)	field	plot (")	categorical
Distribution (*Distribuzione*)	field	plot (")	categorical
Origin (*Origine*)	field	plot (")	categorical
Conifer regeneration (*Rinnovazione di conifere)*	field	plot (coppice only)	categorical

G) Forest condition

Attribute	Data source	Object	Measurement unit
Site degradation: origin *(Degrado della stazione: causa)*	field	plot	categorical
Site degradation: intensity *(Degrado della stazione: intensità)*	field	plot	categorical
Stand damages: effect *(Danni al soprassuolo: effetto)*	field	plot	categorical
Stand damages: origin *(Danni al soprassuolo: origine)*	field	plot	categorical
Stand damages: diffusion *(Danni al soprassuolo: diffusione)*	field	plot	percent of trees

H) Accessibility and harvesting

Attribute	Data source	Object	Measurement unit
Unevenness of soil morphology *(Accidentalità)*	field	area around the plot centre	categorical
Accessibility (*Densità stradale)*	field or topographic maps	distance between plot centre and the nearest road	100 m

I) Attributes describing forest ecosystems

Attribute	Data source	Object	Measurement unit
Origin of stand *(Origine del soprassuolo)*	field	stand (high forest only)	categorical

J) Non-wood goods and services

Attribute	Data source	Object	Measurement unit
Height of cork *(Altezza di decortica)*	field	tree (cork oak only)	dm
Median diameter of cork *(Diametro mediano all'altezza di decortica)*	field	tree (")	cm
Thickness of cork *(Spessore del sughero)*	field	tree (")	mm
Age of cork (*Età del sughero*)	field	tree (")	year
Type of cork (*Tipo di sughero)*	field	tree (")	categorical
Function of the forest area *(Funzione dell'area forestale)*	field	forest area	categorical
Land use restrictions (*Vincoli*)	field, forest service	forest area	categorical

K) Miscellaneous

Attribute	Data source	Object	Measurement unit
Survey timing *(Tempi di lavoro)*	clock	//	minutes
Sample plot number *(Punto di campionamento)*	map	sampling unit	integer
Sample plot coordinates *(Coordinate UTM del punto di campionamento)*	map	sampling unit	m
Number of tally sheets filled in *(Numero di modelli compilati)*	office	sampling unit	integer

2.1.2. List of derived attributes

C) Wood production

Attribute	Measurement unit	Input attributes
Single tree volume o. b. *(Volume legnoso individuale)*	m^3	d.b.h., total tree height, tree species
Height increment *(Incremento ipsometrico)*	m	d.b.h., total tree height
Wood increment in high forest *(Incremento legnoso nelle fustaie)*	m^3	d.b.h., average annual radial increment, total tree height, annual height increment, tree species
Basal area (*Area basimetrica*)	m^2	d.b.h.
Number of trees per hectare *(Numero di alberi per ettaro)*	integer	number of trees per plot, tree species
D.b.h. of felled trees (*Diametro a petto d'uomo degli albcri prelevati)*	cm	diameter of stumps, d.b.h. and basal diameter of reference trees
Annual felled area in coppice *(Superficie interessata da utilizzazione nei cedui)*	ha	age of coppice shoot, tree species
Volume felled in high forest *(Massa prelevata nelle fustaie)*	m^3	diameter of stumps, d.b.h., basal area and volume of reference tree, d.b.h. and basal area of felled tree

E) Forest structure

Attribute	Measurement unit	Input attributes
Dominant species in particular forest area *(Principali specie censite nelle formazioni particolari)*	%	presence of tree or shrub species

2.1.3 Measurements rules for measurable attributes

For each of the measureable attributes the following information is given:

a) measurement rule
b) threshold values
c) measurement scale
d) rounding rules
e) instrument
f) data source

C) Wood production

Diameter at breast height (Diametro a petto d'uomo)

a) Measured in 1.3 m height above ground, on slopes measured from uphill side. One reading in (all) coppice forest and in high forest with d.b.h. less than 17.5 cm. Two readings in high forests with d.b.h. of more than 17.5 cm. First reading with calliper pointing the plot centre, the second perpendicular. Big or irregular stems: the girth is measured and divided by p.
b) Minimum dbh: 3 cm
c) cm, recorded in 1 cm classes
d) rounded to the nearest (lower or upper) class limit
e) calliper
f) field assessment

Tree height (Altezza totale)

a) Length of tree from ground level to top of tree
b) minimum tree height: 5 m
c) m
d) rounded to the closest dm class limit
e) Blume-Leiss or Suunto height meter, Suunto clinometer
f) field assessment

Radial increment (Incremento radiale)

a) Measured the depth of last 5 years in radial cores extracted on b.h.
b) No threshold values
c) mm
d) rounded to closest 0.5 mm
e) Pressler borer
f) field assessment

Diameter of stumps (Diametro delle ceppaie)

a) Measured each stump of trees felled inside the plot during the last solar year.
b) No threshold values
c) cm
d) rounded to the closest cm class limit
e) calliper or tape
f) field assessment

D) Site and soil

Elevation (Altitudine s. l. m.)

a) measured in plot centre
b) no threshold values
c) m
d) rounded to 10 m classes
e) altimeter or reading of map
f) field assessment

Slope (Pendenza media del versante)

a) The average slope of soil profile is measured on the maximum slope line through the plot centre, ranging two opposite stakes located uphill and downhill at 25 m distance from centre.
b) threshold values: 0% -
c) %,
d) rounded to closest 1% class limit
e) Suunto clinometer
f) field assessment

Soil depth (Profondità del suolo)

a) The average depth of soil is measured or assessed by ocular estimates in natural soil profiles or knowledge of forest service.
b) no threshold
c) cm, recorded in three classes: less than 25 cm, 25 - 60 cm and more than 60 cm
d) no rounding
e) ocular estimate or tape
f) field assessment

E) Forest structure

Cover (Grado di copertura)

a) The ratio of tree and shrub vegetation covered area and total area of sample plot is assessed by ocular estimates. In case of doubt a check is carried out through a 8*8 m grid of 24 points split in four equivalent sectors. The central point coincides with the sample plot centre.
b) Minimum cover: 20 %. The minimum of 5 covered points has to be found in at least two different sectors.
c) %, 30% classes (20-50%, 51- 80%, > 80%)
d) no rounded
e) ocular estimate
f) field assessment

Canopy cover (Grado di copertura)

a) This attribute is assessed only in stands with height lower than 5 m
b) Minimum coverage value: 20%
c) %, 10% classes (20-29%, 30-39%, ..., 90-100%)
d) rounded to closest class limit
e) ocular estimate
f) field assessment

Age of stand (Età del popolamento)

a) Five possible types of assessment could be selected by surveyors:
   · historical estimate of stand planting year
   · counting of branches in some trees (conifer only)
   · counting of annual ring number on stumps in case of recent felled trees
   · counting of annual ring number on radial cores extracted by some trees
   · ocular estimate of age based on comparing the situation with other known
b) no threshold
c) year
d) no rounded
e) Pressler borer or ocular estimate
f) field assessment

Planting space (Sesto d'impianto)

a) Average distance between trees measured in specialised forest stands only.
b) No threshold
c) m
d) rounded to closest 0.5 m
e) tape
f) field assessment

G) Forest condition

Stand damages: diffusion (Danni al soprassuolo: effetto)

a) Per cent diffusion of trees interested by damages within the sample plot.
b) No threshold
c) %, three classes: - 20%, 21- 50%, 51% -
d) No rounded
e) Ocular estimate
f) Field assessment

H) Accessibility and harvesting

Accessibility (Livello di accessibilità o esboscabilità)

a) Evaluation of slope, distance and difference in height from plot centre and the nearest road.
b) No threshold
c) slope: %, distance: km, difference in height: m, three classes of accessibility:
   1) slope ≤ 10% and distance ≤ 1 km or slope > 10% and distance ≤ 0.5 km and difference in height ≤ 100 m
   2) slope ≤ 10% and distance > 1 km

and ≤ 4 km or slope > 10% and distance > 0.5 and ≤ 2 km and difference in height > 100 and ≤ 400 m

3) slope ≤ 10% and distance > 4 km or slope > 10% and distance > 2 km and difference in height > 400 m

d) no rounded
e) slope: clinometer, distance: telemeter or tape or reading on map, difference in height: reading on map or calculated by distance and slope
f) field assessment

J) Non-wood goods and services

Height of cork (Altezza di decortica)

a) Measured the height of stems above ground corresponding to the cork level.
b) no threshold
c) m
d) rounded to the closest dm class
e) tape
f) field assessment

Median diameter of cork (Diametro a metà dell'altezza di decortica)

a) Measure of median diameter of the half cork height
b) no threshold
c) cm
d) rounded to the closest cm class
e) calliper
f) field assessment

Thickness of cork (Spessore del sughero)

a) A cork core is extracted and the thickness measured
b) Minimum threshold: 0
c) mm
d) rounded to the closest mm class
e) rule
f) field assessment

Age of cork (Età dello strato suberoso)

a) Estimated the age of last cork extraction
b) Minimum threshold: 0
c) year
d) no rounded
e) ocular estimate or information by forest service
f) field assessment

K) Miscellaneous

Sample plot number and U.T.M. coordinates (Codice del punto di campionamento e coordinate U.T.M.)

a) time zone, topographic map number and plot number, U.T.M. coordinates
b) no threshold
c) integer, m
d) no rounded
e) map reading
f) topographic map

2.1.4 Definitions for attributes on nominal or ordinal scale

The following information is provided for each attribute:

a) the definition
b) the categories (classes)
c) the data sources
d) remarks

A) Geographic regions

Region (Regione)

a) The administrative Region of the plot-centre area.
b) One of the following 20 normal or autonomous Regions:
 1) Piemonte
 2) Valle D'Aosta (autonomous Region)
 3) Lombardia
 4) Trentino - Alto Adige (two autonomous Provinces)
 5) Veneto
 6) Friuli - Venezia Giulia (auton. R.)
 7) Liguria
 8) Emilia- Romagna
 9) Toscana
 10) Umbria
 11) Marche
 12) Lazio
 13) Abruzzi
 14) Molise
 15) Campania
 16) Puglia
 17) Basilicata
 18) Calabria
 19) Sicilia (auton. R.)
 20) Sardegna (auton. R.)
c) local forest service and cadastral register
d) not relevant

Province (Provincia)

a) The Province of the plot-centre area.
b) 95 Provinces
c) local forest service and cadastral register
d) not relevant

B) Ownership

Ownership (Tipo di proprietà)

a) The owner of the forest area under concern
b) 1) state, region and autonomous provinces *(Stato, Regioni e Province autonome)*
 2) province and municipalities *(Province e Comuni)*
 3) other public communities *(Altri Enti pubblici)*
 4) private ownership (society) *(Proprietà privata: persona giuridica)*
 5) private ownership (person) *(Proprietà privata: persona fisica)*
c) local forest service and cadastral register
d) if the precise category is not known, the classification of public or private ownership is almost required.

C) Wood production

Type of tree (Dendrotipo)

a) Classification of trees regarding the gamic or agamic reproduction
b) two classes: gamic reproduced tree *(albero proveniente da seme)* or shoot *(pollone)*
c) field assessment
d) only for high forest

Quality of stems (Caratteristiche qualitative dei fusti)

a) Each tree is classified by two characters: the general condition of stem and the main external characters of stem.
b) Four classes of stem general condition:
 1) whole intact stem
 2) broken off stem
 3) dry stem

4) uprooted stem
Three classes of the main stem characters:
1) · conifer with cone-shaped crown: upright and regular stem
· broadleaf and conifer with spherical crown: upright and regular stem without main branches for almost 2/3 of height
2) · conifer with cone-shaped crown: stem moderately curved in the lower part or tapered stem, or stem with main branches or truncated over 3/4 of height or up to 1.3 m
· broadleaf and conifer with spherical crown: stem moderately curved in the lower part or with main branches over 2/3 and up to 1/2 height, or with branches up to 1.3 m
3) stems with more than one defect described in point 2
c) field assessment
d) assessment made only in stands classified as high forest and specialised forest area for wood production

D) Site and soil

Location (Giacitura)
a) The sample plot location in relation to the surrounding terrain morphology.
b) 1) high slope, top *(alto versante, dorsale)*
2) intermediate slope *(medio versante)*
3) low slope *(basso versante)*
4) sinking of the ground, depression *(avvallamento, depressione aperta)*
5) bottom of the valley *(fondovalle)*
6) plain *(pianura)*
7) basin, closed depression *(conca, depressione chiusa)*
8) alluvial cone *(cono di deiezione)*
c) field assessment
d) not relevant

Soil moisture (Umidità del suolo)
a) Assessment of the annual average water condition of the soil.
b) 1) dry soil *(terreno arido o asciutto)*
2) fresh soil *(terreno fresco)*
3) moist soil *(terreno umido o paludoso)*
c) field assessment
d) not relevant

Texture of soil and stoniness (Tessitura-pietrosità)
a) Assessment of the average soil grain dimension
b) 1) clayey soil *(terreno argilloso)*
2) soil of middle-mixture *(terreno franco o di medio impasto)*
3) sandy or slimy soil *(terreno sabbioso o limoso)*
4) stone-soil *(terreno ghiaioso o sassoso)*
5) rocky soil *(terreno roccioso)*
c) field assessment
d) not relevant

E) Forest structure

Type of stand (Tipo di popolamento o tipo inventariale)
a) Assessment made considering cover, system and crop type
b) 18 types of stand:
· even-aged high forest (six crop types) *(fustaia coetanea)*
· uneven aged " " *(fustaia disetanea)*
· combined " " *(fustaia articolata)*
· irregular " " *(fustaia irregolare)*
· transitory high forest *(fustaia transitoria)*
· simple coppice *(ceduo semplice o senza matricine)*
· coppice of two rotations or with limited number of standards *(ceduo matricinato)*
· coppice with standards *(ceduo composto)*

· plantation for wood production *(popolamento specializzato a produzione legnosa)*
· plantation for non-wood production *(popolamento specializzato a produzione non legnosa)*
· bushland/maquis *(arbusteto)*
· riparian wood *(formazione riparia)*
· rock-wood area *(formazione rupestre)*

c) field assessment
d) not relevant

Tree species (Composizione specifica)

a) Scientific name (code) of selected list of 31 coniferous and 59 broadleaf tree species and two additional classes (other coniferous and other broadleaf species). Another list of 70 shrub species is used for particular forest area only.
b) 90 tree species and two superior classes; 70 shrub species.
c) field assessment
d) not relevant

Main tree species and diffusion level (Composizione specifica: valutazione sintetica e grado di diffusione)

a) Assessment of the three main species in the sample plot and the diffusion level of each one.
b) The same tree and shrub species previously examined. Four diffusion (presence) levels:
   1) rare *(sporadica)*
   2) secondary *(subordinata)*
   3) prevalent *(prevalente)*
   4) sole *(esclusiva)*
c) field assessment
d) assessed only for stands with height < 5 m.

Tree species: general classification of stands (Inquadramento tipologico generale delle formazioni boschive)

a) Synthetic assessment made considering the main species of vegetation and according with the climatic condition, the geographic location and the elevation of the site.
b) A selected list of 13 vegetation types is used (for instance 'Mediterranean deciduous evergreen stands with prevalence of *Quercus ilex*', ...)
c) field assessment
d) not relevant

Tending of stands (Cure colturali)

a) Each type of tending is registered if made or not.
b) The following types of tending are evaluated:
   1) cleaning and thinning *(sfollamenti e diradamenti)*
   2) pruning *(potature)*
   3) sanitary treatment *(controllo fitosanitario)*
   4) soil fertilisation *(concimazione del terreno)*
   5) soil cultivation *(lavorazione del terreno)*
   6) weeding, cutting of weed *(diserbo, falciatura e controllo della vegetazione infestante)*
   7) irrigation *(irrigazione)*
c) field assessment
d) assessed only specialised forest area

F) Regeneration

Presence (Presenza di rinnovazione)

a) The natural or artificial tree vegetation with height over 30 cm and with dbh not more than 2.5 cm is assessed.
b) three classes:
   1) absent *(assente)*
   2) scarce *(scarsa)*
   3) abundant *(abbondante)*
c) field assessment
d) assessed only for high forests with height > 5 m and for all the coppice forests with regard to the coniferous regeneration.

Condition (Stato vegetativo della rinnovazione)

a) The natural or artificial tree vegetation with height over 30 cm and with dbh not more than 2.5 cm is assessed.

b) three classes:

1) poor *(scadente)*

2) good *(buono)*

3) excellent *(ottimo)*

c) field assessment

d) assessed only for high forests with height > 5 m and for all the coppice forests with regard to the coniferous regeneration.

Distribution (Distribuzione della rinnovazione)

a) The natural or artificial tree vegetation with height over 30 cm and with dbh not more than 2.5 cm is assessed.

b) two classes:

1) continuous *(andante)*

2) by groups *(a gruppi)*

c) field assessment

d) assessed only for high forests with height > 5 m and for all the coppice forests with regard to the coniferous regeneration.

Origin (Origine della rinnovazione)

a) The natural or artificial tree vegetation with height over 30 cm and with dbh not more than 2.5 cm is assessed.

b) three classes:

1) natural *(naturale)*

2) artificial *(artificiale)*

3) mixed *(mista)*

c) field assessment

d) assessed only for high forests with height > 5 m and for all the coppice forests with regard to the coniferous regeneration.

G) Forest condition

Site degradation: origin (Degrado della stazione: causa)

a) Synthetic assessment of origin of site degradation by human activity.

b) seven classes:

1) fire *(incendio)*

2) grazing *(pascolamento)*

3) irrational cutting of stand *(utilizzazione irrazionale del soprassuolo)*

4) excessive use by tourism *(eccessiva utenza turistico-ricreativa)*

5) alteration of filtering and downflow of water by infrastructures building *(alterazione della percolazione idrica e del deflusso in seguito alla costruzione di infrastrutture)*

6) composite origins *(cause combinate)*

7) other origins *(altre cause)*

c) field assessment

d) not relevant

Site degradation: intensity (Degrado della stazione: intensità)

a) Synthetic assessment of intensity of site degradation by human activity.

b) three classes:

1) slight degradation *(degrado debole)*

2) middle-degradation *(degrado medio)*

3) heavy degradation *(degrado intenso)*

c) field assessment

d) not relevant

Stand damages: intensity (Danni al soprassuolo: effetto)

a) Assessed the average intensity level of damaged trees inside the sample plot.

b) three classes:

1) temporary damage *(danno temporaneo)*

2) permanent damage *(danno permanente)*

3) death-damage *(danno letale)*

c) field assessment

d) particular forest area and other forest areas not included.

Stand damages: origin (Danni al soprassuolo: origine)

a) Assessed the main origin of damage for trees inside the sample plot.

b) five classes:

1) grazing or game *(da pascolo o selvaggina)*

2) pests *(da parassiti)*

3) fire *(da incendio)*

4) meteoric phenomenon *(da agenti meteorici)*

5) direct human action (cutting, timber extraction, road opening) *(per azione diretta dell'uomo)*

c) field assessment

d) particular forest area and other forest areas not included.

H) Accessibility and harvesting

Unevenness of soil morphology (Grado di accidentalita)

a) Assessed the presence of obstacles that could obstruct the timber extraction.

b) three classes:

1) obstacles absent *(non accidentato)*

2) difficult obstacles *(scarsamente accidentato)*

3) insuperable obstacles *(molto accidentato)*

c) field assessment

d) assessed after evaluating of slope, distance of roads and difference in height for estimating hauling level.

I) Attributes describing forest ecosystems

Origin of stand (Origine del soprassuolo)

a) Assessed the natural or artificial origin of stand.

b) three classes:

1) natural stand *(di origine spontanea)*

2) < 50 % of trees of artificial origin *(meno del 50% degli individui introdotto artificialmente)*

3) ≥50 % of trees of artificial *origin (più del 50% degli individui introdotto artificialmente)*

c) field assessment

d) assessed only high forest

J) Non-wood goods and services

Type of cork (Tipo di sughero)

a) Assessed if cork is of first or another generation.

b) two classes:

1) first generation cork *(sugherone)*

2) later generations *(sughero gentile)*

c) field assessment

d) assessed only non-wood specialised cork oak stands.

Function of the forest area (Funzione dell'area forestale)

a) The main functions of the forest area are given, in connection to goods or services derived.

b) A maximum of three functions, in order of decreasing importance, may be selected by the following list:

1) wood production *(produttiva legnosa)*

2) non-wood goods production (cork, pine-seeds, sweet chestnut*) (produttiva non legnosa)*

3) direct protection (against avalanches, stone-falls towards roads or human settlements) *(protettiva diretta)*

4) indirect protection (against soil erosion, floods, etc.) *(protettiva indiretta)*

5) nature protection (parks, etc.) *(naturalistica)*

6) recreation *(turistico-ricreativa)*

c) field assessment

d) not relevant

Land use restrictions (Vincoli)

a) Assessed if the area has particular land use restrictions, given by specific laws.

b) A maximum of six constraints should be registered by a list of 12 classes (military, protective, scenic, management, naturalist constraints).

c) forest service information and field assessment

d) not relevant

2.1.5. Forest area definition and definition of "other wooded land"

Forest area: a territory with one or more of following characters:

- purpose to wood or non-wood goods productions currently regarded as forestal;
- contain tree or bush stands with direct or indirect function of protection;
- contain spontaneous tree or bush stands with naturalist, scenic or recreation function.

Included also areas temporarily without a stand because cutting or exceptional occurrence.

Not included: city parks, gardens, botanical gardens and other areas with only aesthetic function. Likewise not considered: forest nurseries, fruit cultivation of walnut and filbert, manna ash stands, carob tree stands, and every fruit tree stand. Excluded also the tree rows and scattered trees in agricultural territory and along the roads.

Summarising the forest area comprises:

forest area:	normal:	with a stand	
		temporarily without stand:	because a clear cutting
			because exceptional occurrence
	specialised:	in wood production	
		in non-wood goods production	
	particular:	riparian forest, rock-wood areas	
		bushland/maquis	

The **minimum size** *(superficie minima)* of a forest area is 2,000 m^2. Trees in smaller groups than this are not assessed by the INFI. In case of narrow forest areas, a **minimum width** *(larghezza minima)* of 20 m is prescribed. Regarding the **canopy coverage** *(grado di copertura)*, a minimum of 20 % is requested. In case of forest areas temporarily without stand, the standard of real canopy coverage is replaced by the conventional canopy coverage. This is assessed through a check procedure, that regards as conventionally covered the circular areas, given by a fixed radius around young trees or stumps.

Not forest areas with size not more than 2,000 m^2 and included in forest areas are classified as included areas.

In plots containing both, forest and non forest areas, the forest boundary is to fix exactly on the ground to permit the plot division. The forest boundary is fixed by measuring the following distances from stems located in forest edge:

	distance (m) from	
dbh (cm)	trees with cone-shaped crown	trees with spherical crown
10	0.8	1.0
15	1.0	1.3
20	1.1	1.5
25	1.3	1.7
30	1.4	1.8
35	1.5	2.0
40	1.6	2.1
45	1.7	2.6
50	1.8	2.8
55	1.9	2.9
60	1.9	3.0
65	2.0	3.6
70	2.1	3.8
75	2.2	3.9
80	2.2	4.0
85	2.3	4.1
90	2.4	4.3
95	2.4	4.4

The boundary has to be drawn as straight as possible.

For consecutive computing, in the margin plots the size of non-forest area (or the area interested by another stratum) has to be estimated. The ratio a/r helps to attribute the size (S) to the fraction (expressed in one-tenth of total plot area), as given by following values, which aid the field assessment:

a/r	S
0.000 - 0.078	5/10
0.079 - 0.237	4
0.238 - 0.404	3
0.405 - 0.585	2
0.586 - 0.805	1
0.806 - 1.000	0

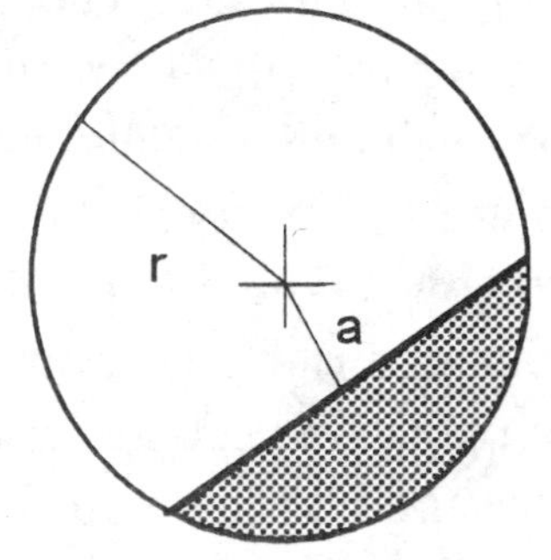

Figure 3. Example of subdivided plot for estimating the fraction size.

2.2. DATA SOURCES

A) Field data

The characteristics of sample plots will be described in chapter 2.3.

Type of stand: assessed on a visual area of approximate 0.5 ha. This because a normal plot area of 600 m^2 is not big enough, in many cases, for assessing stand structures as even-aged, or uneven aged by groups, or as irregular high forest.

E) Maps

Plot-centre are located on maps of the Italian Military Geographic Institute (I.G.M.). The dates are very different.
Scale: 1:25000
Printed topographic maps. Some are made through direct measurement and some through photo-grammetric procedure.

2.3. ASSESSMENT TECHNIQUES

2.3.1. Sampling frame

The sampling frame of the Italian National Forest Inventory is given by the forest area definition. The entire country is covered by the assessment. Some sample points (roughly 0.2 %) were not classified because not accessible due to the terrain condition as well not evaluable at distance. They were excluded from the assessment.

2.3.2. Sampling units

Field assessment: Fixed area, circular plot.

The normal size is 600 m^2, but in particular cases of tree cover more than 50 % it could be reduced to:

- 400 m^2 in coppice stands with an average height of dominant trees between 10 and 15 m and in high forest with an average height of dominant trees between 10 and 20 m;
- 200 m in coppice and high forest with an average height of dominant trees between 5 and 10 m.

Slope correction is done by increasing the plot radii according to slope. All plots are permanently monumented and documented (but marked hidden without giving an evident impact to the plot location). Azimuth and distance of reference marks from plot centre are recorded. The plots are systematically distributed in a 3*3 km grid. The plot location coordinates are recorded. The plot centre is located with a polygon quickly measured by tape (or telemeter) and a compass and with the aid of a programmed pocket computer. The operative distance and the azimuth from a start-point to the plot (grid) centre are measured on the map. The plot centre is marked by a metal stake hidden in the ground.

2.3.3. Sampling designs

The sampling design applied in the INFI is one phase inventory with only a ground assessment of systematically distributed field plots on a grid of 3*3 km. The sample plots are laid on the maps (1:25000) over the 1*1 km grid.

2.3.4. Techniques and methods for combination of data sources

Since a one phase inventory is applied, all data is collected or derive from field assessment. Therefore no combination of data sources is applied.

2.3.5. Sampling fraction

Data source and sampling unit	proportion of forested area covered by sample	represented mean area per sampling unit
field assessment, fixed area sample plots	0.00007	900 ha

2.3.6. Temporal aspects

Inventory cycle	Time period of data assessment	Publication of results	Time period between assessment	Reference date
1st NFI	1984-1985	1988	-	1 January 1985

2.3.7. Data capturing techniques in the field

Data is recorded on paper forms filled out separately for different types of forest: one for topographic location, a general one for every grid-point, one for general site parameters and three for forest parameters (coppice, high forest and specialised stands).

2.4. DATA STORAGE AND ANALYSIS

2.4.1. Data storage and ownership

The data of the assessment is stored in a database system on PC by the Direzione Generale dei Servizi e del Personale of Ministero delle Risorse Agricole, Forestali e Alimentari, Via XX Settembre, 20, Roma. A new office for Information System is going to be created by that Direction, the SIAN (*Sistema Informativo Agricolo Nazionale*). The coordinator of the Technical Secretary's Office of SIAN, Dr. Giuseppe Serino, Tel. +39-6-4665-5060, should be considered as contact person for information.

2.4.2. Database system used

The data are stored in a relational ORACLE database.

2.4.3. Data-bank design

//

2.4.4. Update level

The reference date is January, 1, 1985. The data assessed is used without any specific updating.

2.4.5. Description of statistical procedures used to analyse data including procedures for sampling error estimation

a) area estimation:

The forest area is assessed on the field. A 3*3 km systematic point grid is laid over the entire country and for each point a forest/non-forest decision is made.

The estimation of the proportion of forested area is done according to

$$p = \frac{a}{n}$$

$$v(p) = sp^2 = \frac{pq}{n}$$

and the standard error is

$$s_p = \sqrt[2]{v(p)}$$

where

p = proportion of forest land
q = 1- p = proportion of non-forested land
v(p) = variance of p
s_p = standard error of p
n = total number of points on the point grid
a = number of forested points on the point grid

As a single forest point represents 900 ha of forest, the area estimation is made by multiplying the number of points of the desired unit by 900:

$$A_w = n_w 900 = n_w \frac{A}{n}$$

with variance $v(A_w)$ and standard error $s(A_w)$

$$v(A_w) = A^2 s_p^{\ 2}$$

$$s(A_w) = \sqrt[2]{v(A_w)}$$

The per cent standard error is consequently

$$s(A_w)\% = 100s \frac{(A_w)}{p}$$

b) aggregation of tree and plot data

The aggregation of single tree data is done by weighting each single tree attribute Y_{ij} by

$$w_{ij} = \frac{A}{a_{ij}} = \frac{10{,}000m^2}{600m^2} \left(\frac{10{,}000}{400} \text{or} \frac{10{,}000}{200} \right)$$

where ij stands for tree i on plot j.

The plot expansion factor has to be adjusted for a plot not lying entirely in the forested area or for a plot interesting more than one stratum.

Once the single tree attributes have been related to unit area by multiplication with the plot expansion factor, the total values for plot i, Yi, can be calculated by summing the individual single tree attributes:

$Y_i = S(Y_{ij} w_{ij})$

c) estimation of total values

In general the total value of attribute Y for a stratum h (of area $A_{h)}$ is given by summing the attribute value for each plot:

$$Y_h = 900 \sum \frac{Y_i}{a_i} = A_h \ddot{y}$$

where $\ddot{y}$ = mean value of total value Y_h

Since both, A_h and $\ddot{y}$, are affected respectively by the standard errors s_{Ah} and $s_{\ddot{y}}$, we get an error propagation evaluable with the expression

$$s(Y_h) = (A_h^2 s_{Ah}^2 + \ddot{y}^2 s_{\ddot{y}}^2)^{1/2}$$

d) estimation of ratios

//

e) sampling at forest edge

For area estimation each plot centre located into the forest represents 900 ha of forest, and also if the plot is not entirely covered by forest.

For attributes related to the area (basal area, volume, growth) only the part of plot interested by the stratum is computed for calculating mean and total values.

For example if a sample plot of 600 m² is covered by 8/10 of high forest and 2/10 of coppice, the expansion factors for basal area (or another attribute) estimation are

respectively given by: $\frac{10,000}{(600 * 0.8)}$ and $\frac{10,000}{(600 * 0.2)}$

f) estimation of growth and growth components (mortality, cut, ingrowth)

In the high forest and in the specialised forest area for wood production, the growth of wood of a forest stand is estimated by a procedure based on the direct radial increment extracted by trees and on the high increment. For each tree the per cent increment was calculated (Hellrigl, 1986).

The average per cent increment for each diameter class was than calculated and applied to the tree volumes. The current increment per plot area was derived by summing the tree values.

No growth was estimated in coppice forest.

g) allocation of stand and area related data to single sample plots/sample points.

The class of categorical, area related attributes, in which the plot centre is located, is assigned to the entire plot.

h) hierarchy of analysis: how are sub-units treated?

Estimates for the entire area and regional sub-units (Regions) are obtained independently.

2.4.6. Software applied

A particular software was developed for the analysis and calculation procedure. The language used was COBOL and FORTRAN. For the maps a graphic interface program for AUTOCAD was developed.

2.4.7. Hardware applied

A computer TANDEM was used for data storage and data analysis. The operative system was GUARDIAN.

2.4.8. Availability of data (raw and aggregated data)

The INFI data are stored in a relational database on PC. Requests of raw data and derived data by outside persons have to be addressed to the secretary's office of SIAN. Data requested by public institutions are normally available without particular costs if the purpose for which the data will be used is specified. In any case a regulation of data-concession is going to be made.

2.4.9. Sub-units (strata) available

Results are obtained for entire Italy and some attributes also for the 19 Regions and the two Autonomous Provinces. In addition other results for individual Regions and A.P. can be provided.

2.4.10. Links to other information sources

//

2.5. RELIABILITY OF DATA

2.5.1. Check assessment

a) Check assessments are carried out during the field inventory. A cross control was organised among two or more Province crews: one crew checked some field plots made by another crew and vice versa. About 5 % of the randomly selected field plots was visited a second time. The measuring procedures for locating the plot centre was repeated. All attributes was assessed a second time and the filled forms was checked.

b) The check assessments are used for the following purposes:

- feed-back to field crews to improve the data quality;
- if important differences were found among first and second assessment a new inventory should be made;
- results will not be used to set up error budgets and to quantify the total error.

2.5.2. Error budgets

Only sampling errors was calculated for the publication of inventory results.

2.5.3. Procedures for consistency checks of data

A field assessment was carried out and forms was filled with the data.

The data was than put into a computer and after editing the data, a consistency check was done. First was checked if the correct type of forms were filled and then if any data was missing in the different forms filled in connection to forest type. In some cases a forcing of data was provided. For some parameters a minimum and maximum value was given and checked. Later a cross check among different forms and some parameters was carried out. For plots with wrong or lacking data, that could not be corrected or forced, a new field assessment was provided.

2.6. MODELS

2.6.1. Volume functions

a) outline of the model

In each single plot and for every single tree the tree volume is estimated using dbh and tree height as input variables. Volume functions of following type was drawn up

$v = f_j(d, h)$

where
v = tree volume of a species j
d = dbh
h = total tree height

The dbh is measured for each tree up to 3 cm, while the height value may be derived by individual mensuration for each tree (e.g. standards in coppice) or by height curve. In general one height curve was made for each sample plot. In some cases of mixed stands of coniferous and deciduous two height functions per plot was derived. A protocol gave instructions on the number of sample trees to select in the field for height measurement in order to the different stand composition. So, for example, in pure stands or stands with only coniferous or only deciduous species, almost 15 trees (if present of course) over 17.5 cm of dbh and 5 trees with dbh in the range 7.5 - 17.5 cm had to be collected. In mixed stands of coniferous and deciduous species the selection had to concern almost 10 trees over 17.5 cm per each group and 5 in the range 7.5 - 17.5 of dbh. In any case the trees had to be fairly distributed in the lower, middle and upper dbh range.

Two entry volume tables (and functions) were derived for the INFI: 12 for high forest and six for coppice species or species-groups. These tables were carried out by a procedure of homogenisation, assembling and equalisation of pre-existing two entry tables. Most of these basic tables are national or local and in some cases also of neighbouring countries. For example the volume table of spruce (*Picea abies*) was derived by eight basic tables; this table will be applied also to estimate the volume of other similar species (*Cupressus sp., Chamaecyparis sp.*).

For coppice forest, the functions cover the entire diametric range, while for high forest the volume values are given only over the threshold of 17.5 cm. The volume of trees with dbh < 17.5 cm is estimated by one independent variable volume functions, as

$$v = f_j(d)$$

which links two volume values corresponding to the diametric limits of 3 - 18 cm. The lower value is fixed to 0.002 m^3 (average volume of tree with dbh = 3 cm). The upper value of anchorage is given by the

$$v = f_j(d, h)$$

where d = 18 cm and h is the height given by the height curve in correspondence to that diameter. If no height curve is available the height value is derived by a linear estimated proportion based on the average dbh and height values of trees with dbh less than 18 cm.

b) overview of prediction errors

The volume functions reflect the entire stem volume of single tree for coniferous species (except the pines with large crown), and stem and branches volume (diameter > 3 cm) for deciduous species and pines with large crown. The high forest upper dbh threshold is 80 cm but functions permit also extrapolations above that limit. Since their origin, the prediction error of the tables is not known.

c) data material for derivation of model

Many Italian two entry base-tables are used to draw up the INFI volume functions. The number of origin trees is very big but not exactly known.

d) methods applied to validate the model

The interpreting attitude of models could be verified only on single trees of some species, the volume of them was known. The best validation of functions has be obtained later: in many cases the INFI volume functions was applied with favourable outcome to regional or local management uses.

2.6.2. Assortments

Not assessed.

2.6.3. Growth components

a) outline of the models

For each sample tree in high forest and specialised stands for wood production the annual volume increment is estimated as difference between the volume at the time of assessment and the volume one year earlier. The wood increment is estimated using the method of Hellrigl (1969):

$$p = 100\left[\left(\frac{2\partial d}{d}\right) + \left(\frac{\partial h}{h}\right)\right]$$

where:

p = per cent volume increment
∂d = dbh annual increment
∂h = height annual increment
d = average dbh of the period (year)
h = average height of the period

The annual increment was directly measured by radial boring in sample trees, while the height increment was derived because the measuring difficulties in the top of trees. Therefore a regression study of total height over dbh for each species or species group was carried out.

The model used in interpreting this regression is

$$h = 1.3 + \frac{1}{\left(\frac{a+b}{d}\right)^3}$$

where:

h = total height in meters
d = dbh in centimetres
a, b = regression coefficients

Consequently the height increment could be derived as height difference related to dbh and (dbh-∂d).

b) overview over prediction errors

//

c) data material for the derivation of the model

29,000 observations of height over dbh were used and 29 general regression functions were studied.

d) methods applied to validate the model

//

2.6.4. Potential yield

Not applied.

2.6.5. Forest functions

Not applied.

2.6.6. Other models applied

a) outline of the models

The volume of felled trees is assessed by following relation

$$v_u = \left(\frac{d_u^2}{d_r^2}\right) v_r = \left(\frac{g_u}{g_r}\right) v_r$$

where
v_u = volume of felled tree
v_r = volume of reference tree
d_u = dbh of felled tree
d_r = dbh of reference tree
g_u = basal area of felled tree
g_r = basal area of reference tree

d_u was estimated by two models, one for coniferous and one for deciduous species, of following form

$d_u = a + bd_s$

with d_s = diameter of felled tree stump

b) overview of prediction errors

//

c) data material for the derivation of the model

A sample of 650 couple of data, foot diameter and dbh, collected on reference trees during the INFI field assessment, was used for the study of this model.

d) methods applied to validate the model

//

2.7. INVENTORY REPORTS

2.7.1. List of published reports and media for dissemination of inventory results

Inventory	Year of publication	Citation	Language	Dissemi-nation
Italian NFI	1983	ISAFA-MAF, 1983. Inventario forestale nazionale italiano - Progetto operativo. Trento. 272 p.	Italian	printed
Italian NFI	1983	MAF-ISAFA, 1983. Inventario forestale nazionale italiano - Istruzioni per le squadre di rilevamento.Trento. 172 p.	Italian	printed
Italian NFI	1984	MAF-ISAFA, 1984. Inventario forestale nazionale italiano - Tavole di cubatura a doppia entrata.Trento. 116 p.	Italian	printed
Italian NFI	1988	MAF-ISAFA, 1988. Inventario forestale nazionale 1985 - Sintesi metodologica e risultati - Rappresentazioni cartografiche. Trento. 464 + 14 p.	Italian	printed

2.7.2. List of contents of report and update level

Reference of latest report:

MAF-ISAFA, 1988. Inventario forestale nazionale 1985 - Sintesi metodologica e risultati - Rappresentazioni cartografiche. Trento. 464 + 14 p.

Update level of inventory data used in this report: Field data: 1984-1985
Content:

Chapter	Title	Number of pages
	Presentation	5
	Introduction	6
	The INFI project and the main steps for carrying out	4
	Part One	
1	Methodology of survey	19
2	Estimate of inventory quantities	17
3	Methodology of carrying out of inventory results maps	4
	Part Two	
4	Forest area data	83
5	Standing volume and increment of high forests	181
	Standing volume of coppice forests	100
	Standing volume and increment of specialised stands for wood production	7
6	Cuts and volume felled	4
7	Main regional data	25
	Thematic maps of inventory results	14

2.7.3. Users of the results

Foresters, public or private agencies or bodies, professionals and other interested people.

2.8. Future development and improvement plans

A decision about the future Inventories has to be taken by the It. government.

A proposal for creating an inventories structure have been made by ISAFA. This structure should involve multiple function, of a permanent nature, capable of checking the current conditions and evolving dynamics. The acquisition and integration of the information from different sources at various levels (national, regional, local), according to a multi-resource philosophy, should be possible.

2.9. MISCELLANEOUS

//

REFERENCES

Hellrigl, B., *et al.* 1986. Nuove metodologie nella elaborazione dei piani di assestamento dei boschi. I.S.E.A., Bologna. 1133 p.

MAF-CFS, 1987. 1° Inventario Forestale Nazionale (IFN). Sintesi provvisoria curata dalla Direzione Generale per l'Economia Montana e per le Foreste. Roma. 44 p.

MAF-ISAFA, 1983. Inventario Forestale Nazionale Italiano (IFNI). Progetto operativo. Trento. 272 p.

MAF-ISAFA, 1983. Inventario Forestale Nazionale Italiano (IFNI). Istruzioni per le squadre di rilevamento. Trento. 172 p.

MAF-ISAFA, 1984. Inventario Forestale Nazionale Italiano (IFNI). Tavole di cubatura a doppia entrata. Trento. 116 p.

MAF-ISAFA, 1988. Inventario Forestale Nazionale (IFN) 1985. Sintesi metodologica e risultati. Rappresentazioni cartografiche. Trento. 464 + 14 p.

Pardé, J., 1989. L'inventaire forestier national Italian. Revue Forestier Francaise, XLI, 3: pp. 245-248.

Scrinzi, G., 1995. Inventario Forestale Nazionale Italiano: cenni sui principali aspetti metodologici. RAISA, Copertura Forestale e Territorio a cura di Bagnaresi U. e Vianello G. Franco Angeli: pp. 127-154.

Tosi, V. & Caruso, C., 1985. Italianische Forstinventur. Forstliche Nationalinventuren in Europa by Pelz, D. R. & Cunia, T. Mitteilungen der Abteilung für forstliche Biometrie, Freiburg: pp. 179-187.

Tosi, V., 1995. The Italian Forest Inventory: Methods, Results and Perspectives. The Monte Verita' Conference on Forest Survey Designs. Köhl M. et al. Eds. WSL, ETH, Zurich: pp. 171-176.

ANNEX

Forest tree species

010 *Abies sp.*
012 *Abies cephalonica*
030 *Chamaecyparis sp.*
040 *Cupressus sp.*
042 *Cupressus sempervirens*
051 *Larix decidua*
060 *Picea sp.*
070 *Pinus sp.*
072 *Pinus cembra*
074 *Pinus leucodermis*
076 *Pinus nigra austriaca*
078 *Pinus nigra italica*
080 *Pinus pinea*
082 *Pinus strobus*
090 *Pseudotsuga sp.*
100 *Taxus sp.*
110 *Juniperus sp.*
200 *Acacia sp.*
211 *Acer campestre*
213 *Acer monspessulanum*
215 *Acer platanoides*
220 *Alnus sp.*
222 *Alnus glutinosa*
230 *Betula sp.*
241 *Carpinus betulus*
250 *Castanea sp.*
260 *Ceratonia sp.*
270 *Eucalyptus sp.*
272 *Eucayptus globulus*
280 *Fagus sp.*
290 *Fraxinus sp.*
292 *Fraxinus ornus*
301 *Juglans nigra*
310 *Ostrya sp.*
320 *Populus sp.*
322 *Populus* - - (hybrid poplar)
324 *Populus tremula*
331 *Prunus avium*
341 *Quercus borealis*
343 *Quercus frainetto*
345 *Quercus macrolepis*
347 *Quercus robur (peduncolata)*
349 *Quercus suber*

360 *Robinia sp.*
370 *Salix sp.*
011 *Abies alba*
020 *Cedrus sp.*
031 *Chamaecyparis lawsoniana*
041 *Cupressus arizonica*
050 *Larix sp.*
052 *Larix leptolepis*
061 *Picea abies*
071 *Pinus brutia*
073 *Pinus halepensis*
075 *Pinus montana uncinata*
077 *Pinus nigra calabrica*
079 *Pinus pinaster*
081 *Pinus radiata*
083 *Pinus sylvestris*
091 *Pseudotsuga menziesii*
101 *Taxus baccata*
199 coniferous
210 *Acer sp.*
212 *Acer opulifolium*
214 *Acer obtusatum*
216 *Acer pseudoplatanus*
221 *Alnus cordata*
223 *Alnus incana*
240 *Carpinus sp.*
242 *Carpinus orientalis*
251 *Castanea sativa*
261 *Ceratonia siliqua*
271 *Eucalyptus camaldulensis*
273 *Eucalyptus trabutii*
281 *Fagus sylvatica*
291 *Fraxinus excelsior*
300 *Juglans sp.*
302 *Juglans regia*
311 *Ostrya carpinifolia*
321 *Populus alba*
323 *Populus nigra*
330 *Prunus sp.*
340 *Quercus sp.*
342 *Quercus cerris*
344 *Quercus ilex*
346 *Quercus petraea*
348 *Quercus pubescens*

350	*Quercus trojana*
361	*Robinia pseudoacacia*
371	*Salix alba*
372	*Salix caprea*
380	*Sorbus sp.*
390	*Tilia sp.*
400	*Ulmus sp.*
410	*Corylus sp.*
420	*Celtis sp.*
421	*Celtis australis*
430	*Platanus sp.*
440	*Ailanthus sp.*
499	broadleaves

Bushland species

Alnus minor (viridis)
Amelanchier ovalis
Arbutus unedo
Berberis vulgaris
Buxus sempervirens
Calicotome spinosa et villosa
Cercis siliquastrum
Chamaerops humilis
Cistus sp.
Clematis sp.
Cornus mas
Corylus avellana
Crataegus oxyacantha
Crataegus azarolus
Cytisus alpinus
Cytisus laburnum
Cytisus scoparius
Eleagnus angustifolia
Erica arborea
Erica carnea
Erica scoparia
Evonymus europaeus
Euphorbia sp.
Genista aetnensis
Genista cinerea
Genista germanica
Genista radiata
Gleditsia triacanthos
Hippophae rhamnoides
Ilex aquifolium
Juniperus communis
Juniperus macrocarpa
Juniperus phoenicea
Juniperus sabina
Laurus nobilis
Ligustrum vulgare
Lonicera sp.
Mespilus germanica
Myrtus communis
Nerium oleander
Olea europea oleaster
Paliurus spina-christi
Phillirea angustifolia
Phillirea latifolia
Pinus mugo
Pistacia lentiscus
Pistacia terebinthus
Prunus sp.
Pirus communis
Pirus malus
Pirus torminalis
Quercus coccifera
Rhamnus alaternus
Rhamnus frangula
Rhododendron sp.
Rhus coriaria
Rosa sp.
Rosmarinus officinalis
Rubus fruticosus
Salix sp.
Sambucus nigra
Sambucus racemosa
Spartium junceum
Tamarix gallica
Ulex europaeus
Vaccinium myrtillus
Vaccinium vitis-idaea
Viburnum lantana
Viburnum tinus

3 LOMBARDIA REGION FOREST INVENTORY

(INVENTARIO FORESTALE DELLA REGIONE LOMBARDIA)

3.1. NOMENCLATURE

3.1.1. List of attributes directly assessed

A) Geographic regions

Attribute	Data source	Object	Measurement unit
Province *(Provincia)*	map	sampling unit	categorical
Municipality *(Comune)*	map	sampling unit	categorical

B) Ownership

Attribute	Data source	Object	Measurement unit
Ownership *(Proprietà)*	local information, land register	sampling unit	categorical

C) Wood production

Attribute	Data source	Object	Measurement unit
Number of enumerated trees *(Numero di piante del campione relascopico)*	field assessment	sampling unit	integer
Diameter at breast height *(Diametro a petto d'uomo)*	field assessment	tree	cm
Total height *(Altezza dendrometrica)*	field assessment	tree	dm

D) Site and soil

Attribute	Data source	Object	Measurement unit
Elevation *(Quota)*	field assessment	sampling unit	dam
Aspect *(Esposizione)*	field assessment	sampling unit	categorical
Slope *(Pendenza)*	field assessment	sampling unit	degree
Location *(Giacitura)*	field assessment	sampling unit	categorical
Stoniness *(Pietrosità)*	field assessment	sampling unit	categorical
Rock outcrops *(Affioramenti rocciosi)*	field assessment	sampling unit	percent

Attribute	Data source	Object	Measurement unit
Soil depth *(Profondità del suolo)*	field assessment	sampling unit	categorical
Soil development stage *(Grado di evoluzione del suolo)*	field assessment	sampling unit	categorical

E) Forest structure

Attribute	Data source	Object	Measurement unit
Forest type *(Tipo fisionomico)*	field assessment	stand (interpretation area)	categorical
Silvicultural system and stand development *(Stadio evolutivo)*	field assessment	stand (interpretation area)	categorical
Age *(Età)*	field assessment, local information	tree	years
Tree species *(Specie forestali)*	field assessment	sampling unit	categorical
Stand origin *(Origine fustaia)*	field assessment	stand (interpretation area)	categorical
Stand description *(Osservazioni selvicolturali)*	field assessment	stand (interpretation area)	categorical

H) Accessibility and harvesting

Attribute	Data source	Object	Measurement unit
Accessibility *(Accessibilità)*	field assessment	sampling unit	categorical

J) Non-wood goods and services

Attribute	Data source	Object	Measurement unit
Forest function *(Funzione prevalente)*	field assessment	stand	categorical
Land-use restrictions *(Vincoli)*	local information, map	stand	categorical

K) Miscellaneous

Attribute	Data source	Object	Measurement unit
Survey timing *(Tempi di lavoro)*	clock	///	minutes
Surveyor and firm name *(Rilevatore e Ditta)*	///	sampling unit	///
Topographic map *(Tavoletta IGM)*	map list	///	///
Sample plot number *(Punto UTM)*	map	sampling unit	geographic coordinates

3.1.2. List of derived attributes

C) Wood production

Attribute	Measurement unit	Input attributes
Basal area *(Area basimetrica)*	m^2	BAF, n of enumerated trees, tree species
Number of stems per ha *(N°di piante per ettaro)*	integer	dbh, tree species
Mean diameter *(Diametro medio)*	cm	d.b.h., tree species
Mean height *(Altezza media)*	m	h, d.b.h., tree species
Mean stem volume *(Massa dendrometrica)*	m^3	h, d.b.h., tree species
Mean annual increment *(Incremento medio)*	m^3	volume, age, tree species

E) Forest structure

Attribute	Measurement unit	Input attributes
Dominant species *(Specie prevalente)*	categorical	basal area
Mean age *(Età media)*	years	volume, mean annual increment

3.1.3. Measurement rules for measurement attributes

For each of the measureable attributes the following information is given:

a) measurement rule
b) threshold values
c) measurement scale
d) rounding rules
e) instrument used for measurement
f) data source

C) Wood production

Diameter at breast height

a) Measured in 1.3 m height above ground, on slopes measured from uphill side. One reading, calliper pointing to the plot centre.
b) Minimum dbh: 3 cm
c) cm, recorded every single cm
d) alternatively rounded off and up
e) calliper
f) field assessment

Total height

a) Length of tree from ground level to top of tree.
b) no threshold
c) m
d) rounded to 0.5 m
e) Suunto, Haga, Bitterlich, Blume-Leiss hypsometer
f) field assessment

D) Site and soil

Elevation

a) measured in the plot centre as altitude
b) no threshold
c) m
d) rounded to 10 m
e) altimeter
f) field assessment

Slope

a) measured from the plot centre along the ruling gradient line, average of uphill and downhill values
b) 5°
c) degree
d) rounded to closest degree
e) clisimeter
f) field assessment

Rock outcrops

a) measured as a percent out of the sample plot area
b) no threshold
c) in 5%-classes
d) rounded to nearest class limit
e) ocular estimate
f) field assessment

E) Forest structure

Age

a) measured for every single tree species and in all forest types, including those with dominant trees mean height below 5 m.
b) no threshold
c) For uneven-aged stands three different ages for low, medium and high stem diameter classes
d) no rounding rules
e) local information about the time of last cut and of plantation; count out of wood rings; count out of branches for coniferous species
f) field assessment

K) Miscellaneous

Survey timing

a) time of each single field survey stage
b) no threshold
c) minutes
d) no rounding rules
e) watch
f) field assessment

Sample plot number
a) UTM coordinates of the sample plot
b) no threshold
c) km
d) no rounding rules
e) map reading
f) map

3.1.4. Definitions for attributes on nominal or ordinal scale

The following information is provided for each attribute:

a) the definition
b) the categories (classes)
c) the data sources
d) remarks

A) Geographic regions

Province
a) administrative unit joining several close municipalities
b) two provinces
c) map

Municipality
a) administrative unit
b) data not available
c) map

B) Ownership

Ownership
a) owner of the forest area under concern
b) 1) private ownership
   2) municipality
   3) state
   4) regional
   5) other public
c) local information, land register
d) not relevant

C) Wood production

Number of enumerated trees
a) number of trees selected in the point sample with the critical angle instrument (Bitterlich relascope) which permits the observer to decide to include or exclude the tree. BAF = 2.
b) no categories
c) field assessment
d) minimum d.b.h.: 2.5 cm. Starting from North in clockwise direction; trees with a breast-height diameter larger than the fixed angle chord corresponding to Basal Area Factor = 2 are tallied “in”; borderline trees are tallied in if D/d (D = d.b.h.; d = distance to the plot centre) is > 0.028284.

D) Site and soil

Aspect
a) direction of exposure of the slope measured in the sample plot centre
b) nine classes
c) field assessment
d) “all” is used when the slope is less than 5°.

Location
a) morphologic site
b) 1) slope
   2) top/ridge
   3) hollow
   4) plain
c) field assessment
d) not relevant

Stoniness
a) assessed as presence of superficial stones
b) 1) absent
   2) scanty
   3) abundant
c) field assessment
d) not relevant

Soil depth
a) average depth of the soil
b) 1) superficial
   2) deep
c) field assessment
d) not relevant

Soil development stage
a) soil features
b) 1) poor
   2) medium
   3) high
c) field assessment
d) not relevant

E) Forest structure

Forest type
a) based on dominant species
b) eight categories
c) field assessment
d) not relevant

Silvicultural system and stand development
a) based on silvicultural system and on age related to rotation period
b) 15 categories
c) field assessment
d) not relevant

Tree species
a) Scientific, Italian name and acronym of selected list of ten coniferous and 44 broadleaf tree species
b) 54 tree species
c) field assessment
d) shrub species not included

Stand origin
a) according to the type of regeneration generating the stand
b) three categories:
   1) artificial
   2) natural dissemination
   3) mixed
c) field assessment
d) no remarks

Stand description
a) short summary reporting on main stand attributes and on other attributes not directly assessed (forest condition, stem quality, regeneration, undergrowth,....)
b) no categories
c) ocular estimate
d) sole survey for stands with height of the dominant trees below 5 m

H) Accessibility and harvesting

Accessibility
a) assessed as the opportunity to approach the sample plot with a cross-country motor vehicle
b) 1) not necessary
   2) necessary
   3) insufficient
c) field assessment
d) not relevant

J) Non-wood goods and services

Forest function
a) main forest function
b) 1) wood production
   2) non-wood production
   3) direct protection
   4) indirect protection
   5) nature conservation
   6) recreation
c) field assessment
d) not specified

Land use restrictions
a) if the sampling unit is located in an area with use restrictions
b) 12 different classes
c) local information, map
d) not specified

3.1.5. Forest area definition and definition of "other wooded land"

Forest area has a tree crown cover of more than about 20% of the total area and has trees usually growing to more than 5m in height. Minimum size of area: 5,000 m^2 and minimum width: 20 m.

3.2. DATA SOURCES

C) aerial photography

- Mapping Unit of the Regional Authority Office
- 1:20000 average scale colour reversal film, 23*23 cm photo format, no ortophotographs
- 1980-1981
- the whole inventory interested area
- stereoscopic pairs
- data not available
- the cost is approximately 25,000 Italian Lire for each photograph

E) maps

- Italian Military Geographic Institute *(Istituto Geografico Militare Italiano)*; Lombardia Regional Authority
- data not available;
- 1:25000; 1:50000
- analogue printed topographic map; analogue printed land-use map

3.3. ASSESSMENT TECHNIQUES

3.3.1. Sampling frame

The sampling frame of the Lombardia Regional Forest Inventory is given by the forest area definition and from the forested area as mapped on 1:50000 topographic map. It means that sampling units in forested areas < 6 ha are not assessed. Only the Varese and Bergamo provincial territories have been assessed. However, areas higher than 1,500 m and with slope over 100% are not visited. Some plots (roughly 1%) are not accessible due to the terrain conditions. They are excluded from the assessment. Plots, that in the field assessment, turn out to be a different land use are excluded from the assessment, too. This is due to the minimum area extension of 6 ha in the map used. If the plot lies in non-forested linear inclusions (roads, etc.) broader than 3 m or in not-forested inclusions wider than 100 m^2 is not assessed.

3.3.2. Sampling units

Point sample with Bitterlich relascope, standing in the plot centre. Interpretation area for area related data (forest type, stand description, ...) of 0.5 ha. All plots permanently monumented and azimuth and distance of tallied trees from plot centre recorded. The plot centre is marked by a round metal plate inserted in a metal pipe hidden in the ground. The plots are permanent and distributed in a 1*1 km grid.

3.3.3. Sampling designs

The sampling design applied in the Lombardia Regional Forest Inventory is a one-stage design. All the plots are preliminary interpreted on the aerial photographs for the mapping of forest-types map on which a forest/non-forest decision is made. All the forest plots, on a 1*1 km grid, are field assessed.

3.3.4. Techniques and methods for combination of data sources

Aerial photographs are used for the drawing of the forest map on which a forest/ non-forest decision is made. Attributes are directly and exclusively assessed in the field.

3.3.5. Sampling fraction

Data source and sampling unit	Proportion of forested area covered by sample	Represented mean area per sampling unit
field assessment, mean area sample plots	///	100 ha

3.3.6. Temporal aspects

Inventory cycle	Time period of data assessment	Publication of results	Time period between assessments	Reference date
1st LRFI (Varese)	1985	1986		not defined
1st LRFI (Bergamo)	1987	1988		not defined

3.3.7. Data capturing techniques in the field

Data recorded on tally sheets and edited by hand into the computer.

3.4. DATA STORAGE AND ANALYSIS

3.4.1. Data storage and ownership

The data of the assessment are stored in a database system at the Lombardia Regional Authority Office - Forest Department

3.4.2. Database system used

Data not available

3.4.3. Data bank design

Data not available

3.4.4. Update level

Data not available

3.4.5. Description of statistical procedures used to analyse data including procedures for sampling error estimation

a) area estimation

The forest area is assessed on the field. A 1*1 km systematic point grid is laid over the map and for each point a forest/non-forest decision is made. The estimation of forested area is done according to the following formulas:

$$p = \frac{a}{n}$$

$$v(p) = s_p^{\ 2} \approx \frac{pq}{n}$$

and the standard error is

$$s_p = \sqrt[2]{v(p)}$$

where

p = proportion of forest land
q = 1-p = proportion of non-forested land
$v(p)$ = variance of p
s_p = standard error of p
n = total number of points on the point grid
a = number of forested point on the point grid

As a single forest point represents 100 ha of forest area, the total forested area A_w is estimated by multiplying the number, n_w of points by 100.

$$A_w = n_w 100 = n_w \frac{A}{n}$$

with variance $v(A_w)$ and standard errors $s(A_w)$

$$v(A_w) = A^2 s_p^2$$

$$s(A_w) = \sqrt[2]{v(A_w)}$$

b) aggregation of tree and plot data

As the selection of the trees to be recorded is proportional to the individual basal area, the single tree attributes have to be divided by relative basal area.

The total values for plot i; Y_i, can be calculated in the following way:

$$Y_i = k \sum_{1}^{n} Y_i / g_i$$

where:
k = BAF
g = basal area
n = number of enumerated trees

c) estimation of total values

In general the total value of attribute Y for a strata h is given by multiplying the area related attribute, Y_h, with the estimated area, A_h, of the strata:

$$Y = A_h Y_h$$

Since both, A_h and Y_h, are affected respectively by the standard errors S_{ah} and S_{yh} we get an error propagation valuable with the expression

$$s(Y) = \sqrt{A_h{}^2 S_{ah}{}^2 + Y_h{}^2 S_{yh}^2}$$

d) estimation of ratios

///

e) sampling at the forest edge

////

f) estimation of growth and growth components (mortality, cut, ingrowth)*

The mean annual increment has been calculated for every single tree species. This is the only attribute provided for the estimation of growth.

g) allocation of stand and area related data to single sample plots/sample points.

The class of categorical, area related attributes, in which the plot centre is located, is assigned to the entire plot.

h) Hierarchy of analysis: how are sub-units treated?*

In the Lombardia RFI results have to be presented for the Provinces assessed and for territorial units. This is done for metric attributes as well as for categorical (area related) attributes.

3.4.6. Software applied

Data not available

3.4.7. Hardware applied

Data not available

3.4.8. Availability of data (raw and aggregated data)

Data not available

3.4.9. Sub-units (strata) available

Results can be obtained for the two Provinces assessed (Varese and Bergamo) and for territorial units.

3.4.10. Links to other information sources

////

3.5. RELIABILITY OF DATA

3.5.1. Check assessment

a) 2-3% of the field plots are visited a second time by check crews. The check crew is a well trained and experienced crew, mainly dealing with check assessments, but not regular field work. The field plots to be checked are randomly selected form the list of already assessed plots. All attributes are assessed a second time independently from the assessment of the regular field crews, i.e., no data recorded by the field crews are available for the check crews. The measurement procedures for locating the plot centre are repeated.

b) The check assessments are used for the following purposes:

- feed-back to field crews to improve the data quality
- results are used for the interpretation of inventory results
- if the field crews and the check crews differ, the results of the check crews instead of the field crews are used for further analysis

3.5.2. Error budgets

///

3.5.3. Procedures for consistency checks of data

See 2.5.3.

3.6. MODELS

3.6.1. Volume functions

a) outline of the model

Volume functions have been derived using mean d.b.h. and total tree height resulting from stand height curves, constructed for each single species and stem-diameter classes. The form factors are related to total tree height, on the basis of the following outline:

- d.b.h. below 12.5 = 2/3
- d.b.h. over 12.5 (coniferous species) = 1/2
- d.b.h. over 12.5 (broad-leaved species) - <200 stems per ha = 3/4
- d.b.h. over 12.5 (broad-leaved species) - 201-400 stems per ha = 2/3
- d.b.h. over 12.5 (broad-leaved species) - >400 stems per ha = 1/2

b) overview of prediction errors

See 2.6.1 b)

c) data material for derivation of model

See 2.6.1 c)

d) methods applied to validate the model

See 2.6.1 d)

3.6.2. Assortments

Not assessed.

3.6.3. Growth component

a) outline of the model

The mean annual increment has been calculated in each single plot and for every single tree species and type

b) overview over prediction errors

///

c) data material for the derivation

///

d) methods applied to validate the model

///

3.6.4. Potential yield

Not assessed.

3.6.5. Forest functions

///

3.6.6. Other models applied

Not assessed.

3.7. INVENTORY REPORTS

3.7.1. List of published reports and media for dissemination of inventory results

Inventory	Year of publication	Citation	Language	Dissemina tion
LRFI (Varese)	1986	Indagine conoscitiva sui boschi della provincia di Varese. 1. Relazione e Allegati. 1986. SAF. 122 pp. 2. Elaborati al calcolatore I 3. Elaborati al calcolatore II 4. Cartografia	Italian	printed

3.7.2. List of contents of latest report and update level

Reference of latest report:

SAF (Società Agricola e Forestale), 1986. Indagine conoscitiva sui boschi della provincia di Varese. Relazione e allegati.

Update level of inventory data used in this report:
Field data and questionnaire: 1985
Aerial photography: 1980-81

Content:

1. Relazione e allegati

Chapter	Title	Number of pages
1	Introduction	3
2	Purpose of the forest inventory	13
3	Ecological features	5
4	Data processing on map	22
5	Field assessment results	32
6	Accessibility, improvement plans of the road network	12
7	Forestry guidelines	11
8	Major findings	2
	Annexes	20

3.7.3. Users of the results

Not directly assessed. Local administrations, agencies and bodies. Professionals.

3.8. FUTURE DEVELOPMENT AND IMPROVEMENT PLANS

///

REFERENCES

Regione Lombardia, SAF, 1986. Indagine conoscitiva sui boschi della provincia di Varese. A.R.F.

4. VENETO REGIONAL FOREST INVENTORY - PUBLIC AND PRIVATE WOODLANDS

(*INVENTARIO FORESTALE DELLA REGIONE VENETO - BOSCHI PUBBLICI E NON PUBBLICI*)

4.1. NOMENCLATURE

4.1.1. List of attributes directly assessed

A) Geographic regions

Attribute	Data source	Object	Measurement unit
Watershed (*Bacino idrografico*)	maps	sampling unit	categorical
Province *(Provincia)*	maps	sampling unit	categorical
Mountain Community (*Comunità Montana)*	maps	sampling unit	categorical
Municipality *(Comune)*	maps	sampling unit	categorical

B) Ownership

Attribute	Data source	Object	Measurement unit
Ownership *(Proprietà)*	local information, land register	sampling unit	categorical

C) Wood production

Attribute	Data source	Object	Measurement unit
Number of trees selected in the point sample *(Numero di piante del campione relascopico)*	field assessment	sampling unit	integer
Diameter at breast height *(Diametro a petto d'uomo)*	field assessment	tree	cm
Inside bark diameter *(Diametro sotto corteccia)*	field assessment	tree (increment core)	mm
Stump girth *(Circonferenza alla ceppaia)*	field assessment	3P tree, felled tree	cm
Total height *(Altezza dendrometrica)*	field assessment	3P tree, felled tree	m
Bole height *(Altezza di inserzione della chioma)*	field assessment	3P tree	m

Crown shape *(Morfologia della chioma)*	field assessment	tree	categorical
Stem quality *(Morfologia del fusto)*	field assessment	tree	categorical
Diameters at 4m, 5m and 6.5m height *(Diametri intermedi a 4, 5 e 6.5 m)*	field assessment	3P tree	cm
Crown diameter *(Diametro della chioma)*	field assessment	felled tree	cm
Merchantable height *(Altezza cormometrica)*	field assessment	felled tree	cm
10 Intermediate diameters *(10 diametri intermedi)*	field assessment	felled tree	cm
Annual radial increment *(Incremento diametrico annuo)*	field assessment	felled tree, tree (increment core)	mm

D) Site and soil

Attribute	Data source	Object	Measurement unit
Elevation *(Quota)*	map	sampling unit	m
Aspect *(Esposizione)*	field assessment	sampling unit	categorical
Slope *(Pendenza)*	field assessment	sampling unit	categorical
Location *(Posizione topografica)*	field assessment	sampling unit	categorical
Soil stability *(Assetto superficiale del suolo)*	field assessment	sampling unit	categorical

E) Forest structure

Attribute	Data source	Object	Measurement unit
Forest type *(Tipologia forestale)*	map	sampling unit	categorical
Stand origin *(Origine del soprassuolo)*	field assessment	sampling unit	categorical
Silvicultural system and stand development stage *(Forma di governo e di trattamento)*	field assessment	sampling unit	categorical
Stratified stand structure *(Tipo di struttura verticale)*	field assessment	sampling unit	categorical
Canopy cover *(Grado di copertura)*	field assessment	sampling unit	categorical

Age *(Età)*	field assessment,	tree (increment core)	integer
Tree species *(Classificazione botanica delle specie)*	field assessment	sampling unit	categorical
Social standing *(Posizione sociale)*	field assessment	tree	categorical
Trees spatial distribution *(Distribuzione spaziale)*	field assessment	tree	degree and cm

F) Regeneration

Attribute	Data source	Object	Measurement unit
Amount *(Consistenza della rinnovazione)*	field assessment	sampling unit	categorical
Distribution *(Distribuzione)*	field assessment	sampling unit	categorical
Leading species *(Specie prevalente)*	field assessment	sampling unit	categorical
Health status *(Stato vegetativo)*	field assessment	sampling unit	categorical

G) Forest condition

Attribute	Data Source	Object	Measurement unit
Tree health status *(Stato vegetativo)*	field assessment	tree	categorical

H) Accessibility and harvesting

Attribute	Data source	Object	Measurement unit
Accessibility *(Accessibilità)*	field assessment, map	sampling unit	categorical
Forest road network *(Viabilità)*	field assessment, map	sampling unit	categorical
Ground roughness *(Asperità del terreno)*	field assessment	sampling unit	categorical
Rock outcrops *(Rocciosità e/o pietrosità)*	field assessment	sampling unit	categorical
Breakage and throw trees *(Presenza di schianti o sradicamenti)*	field assessment	sampling unit	categorical
Undergrowth density *(Densità del sottobosco)*	field assessment	sampling unit	categorical

K) Miscellaneous

Attribute	Data source	Object	Measurement unit
Forest type map number *(Tavoletta Carta Forestale Regionale)*	map list	sampling unit	ordinal
Plan of forest operations *(Piano di riassetto forestale)*	Regional Forest Department	sampling unit	ordinal
Forest type map unit *(Particella cartografica)*	Forest types map	sampling unit	ordinal
Working plan forest plot number *(Particella assestamentale)*	Regional Forest Department	sampling unit	ordinal
Sample plot number *(Coordinate UTM dell'area di saggio principale)*	map	sampling unit	ordinal
Surveyor name *(Rilevatore)*	////	sampling unit	////

4.1.2. List of derived attributes

C) Wood production

Attribute	Measurement unit	Input attributes
Basal area per hectare *(Area basimetrica per ettaro)*	m²	BAF, no. of trees selected in the point sample, tree species
Number of stems per hectare *(Numero di piante per ettaro)*	integer	basal area/ha, dbh, tree species
Mean diameter *(Diametro medio)*	cm	basal area, n of stems, tree species
Mean height *(Altezza media)*	m	mean diameter, tree heights, tree species
Stem total volume *(Volume del fusto intero)*	m³	intermediate diameters, log lengths
Tree volume per hectare *(Volume per ettaro)*	m³	no. of stems/ha, d.b.h., h, tree species
Current diametric increment *(Incremento corrente diametrico)*	cm	mean increment on cores for 5-years classes, tree species, d.b.h., top height

E) Forest structure

Attribute	Measurement unit	Input attributes
Forest composition *(Composizione specifica)*	categorical	species, basal area, no. of trees

4.1.3. Measurement rules for measurable attributes

C) Wood production

Diameter at breast height

a) measured in 1.3m height above ground, on slopes measured from uphill side. Two readings, for each selected tree, the first with calliper pointing to plot centre, the second perpendicular to the first
b) minimum dbh: 2.5 cm
c) cm, recorded every single cm
d) rounded off and up to closest cm
e) calliper
f) field assessment

Inside bark diameter

a) measured on increment cores extracted at breast height, detracting the bark thickness from the outside bark d.b.h.
b) no threshold
c) cm
d) rounded off and up to closest cm
e) tape-line
f) field assessment on the closest tree to the plot centre

Stump girth

a) measured at 30 cm height
b) no threshold
c) cm
d) rounded off and up to closest cm
e) tape-line
f) field assessment

Total height

a) length of tree from ground level to top of tree.
b) no threshold
c) m
d) rounded to closest 0.5 m
e) Suunto, Haga, Blume-Leiss hypsometer, Bitterlich relascope
f) field assessment: ocular estimate on enumerated trees, hypsometer for 3P trees, tape for felled trees

Bole height

a) distance between ground level and the origin of the lowest crown-forming branches (crown point)
b) no threshold
c) m
d) rounded to closest 0.5 m
e) Suunto, Haga, Blume-Leiss hypsometer, Bitterlich relascope
f) field assessment

Diameters at 4 m, 5 m and 6.5 m height

a) on slopes measured from uphill side with calliper pointing to plot centre and perpendicular to the stem
b) minimum tree height: 5 m
c) cm
d) rounded to closest cm
e) Finnish calliper
f) field assessment

Crown diameter

a) measured only on felled trees, minimum and maximum diameter
b) no threshold
c) cm
d) rounded to closest cm
e) tape
f) field assessment

Merchantable height

a) measured on felled trees as the distance between ground level and the stem point to which the diameter is 2.5 cm, for broad-leaved species, and 7.5 cm for coniferous species
b) no threshold
c) dm
d) rounded to closest dm
e) tape
f) field assessment

10 Intermediate diameters

a) measured only on felled trees: at 1/20, 1/10, 2/10, ...9/10, two readings
b) no threshold
c) cm
d) rounded to closest cm
e) calliper
f) field assessment

Annual radial increment

a) measured both on logs of the felled trees and on increment cores extracted on standing trees; in the former assessed at 0.30, 1.30, 4.50, 6.50m, merchandable height (Hc) and half the difference between merchandable height and 6.50 m [(Hc - 6.50)/2] ; in the latter at breast height. Length of every five increment rings.
b) minimum height: 5 m; minimum d.b.h.: 7.5 cm
c) mm
d) rounded to closest mm
e) magnifying glass, calliper, Pressler increment borer
f) field assessment on the closest tree to the plot centre, laboratory

D) Site and soil

Elevation

a) assessed on the map on the grid point as altitude
b) no threshold
c) m
d) rounded to closest m
e) ocular estimate
f) 1:25000 topographic map

E) Forest structure

Age

a) measured on increment cores counting increment rings
b) no threshold
c) years
d) no rounding rules
e) ocular estimate (magnifying glass, calliper)
f) field assessment on the closest tree to the plot centre; laboratory

Trees spatial distribution

a) horizontal distance and azimuth of selected trees measured from plot centre
b) no threshold
c) distance in cm, azimuth in degrees
d) rounded to closet cm and degree
e) tape, compass, tripod
f) field assessment

4.1.4. Definitions for attributes on nominal or ordinal scale

A) Geographic regions

Watershed

a) geomorphologic unit
b) 47 watershed units
c) map

Province

a) administrative unit joining several closed Municipalities
b) seven Provinces
c) map

Mountain Community

a) administrative unit joining several closed municipalities in the mountain area aimed at planning and managing mountain regions
b) 18 Communities
c) map

Municipality

a) administrative unit
b) data not available
c) map

B) Ownership

Ownership

a) owner of the forest area under concern
b) 1) private ownership
2) public ownership
c) map

C) Wood production

Number of trees selected in the point sample

a) number of trees enumerated with the critical angle instrument (Bitterlich relascope) which permits the observer to decide to include or exclude the tree. BAF = 2.
b) no categories
c) field assessment
d) minimum d.b.h.: 2.5 cm. Starting from North in clockwise direction; trees with a breast-height diameter larger than the fixed angle chord corresponding to Basal Area Factor = 2 are tallied "in"; mean diameter and distance of borderline and hidden trees are checked with tape and calliper.

Crown shape

a) development in height and width of the crown related to the total tree height
b) nine categories according to crown point and size
c) field assessment
d) no remarks

Stem quality

a) commercial value of the stem
b) eight categories
c) field assessment
d) assessed on the stem portion included between 0.3 and 4/8 m from the ground

D) Site and soil

Aspect

a) direction of exposure of the slope
b) nine categories
c) field assessment
d) "all" is used when the slope is less than 5%

Slope

a) along the ruling gradient line
b) six categories
c) field assessment
d) no remarks

Location

a) morphological site
b) 24 classes
c) field assessment
d) no remarks

Soil stability

a) susceptibility to erosion
b) seven main classes and 22 sub-classes
c) field assessment
d) no remarks

E) Forest structure

Forest type

a) the forest types are defined on the basis of the Fitoclimatic zones
b) seven main categories, sub-divided with hierarchical classification in 27 types
c) field assessment, Forest types map
d) no remarks

Stand origin

a) according to the type of regeneration generating the stand
b) three categories:
1) artificial
2) natural dissemination
3) agamic regeneration
c) field assessment
d) no remarks

Silvicultural system and stand development

a) the silvicultural system applied and the stand development stage
b) five categories:
1) simple coppice (three sub-classes)
2) coppice selection system
3) coppice with standards (two sub-classes)
4) even-aged high forest (six sub-classes)
5) storied high forest (two sub-classes)
c) field assessment
d) no remarks

Stratified stand structure

a) crown storeys distribution
b) three categories:
 1) one-storey
 2) two-storeys
 3) more than two storeys or irregular
c) field assessment
d) no remarks

Canopy cover

a) crown cover on the ground
b) six categories
c) field assessment
d) no remarks

Tree species

a) Scientific and Italian name of selected list of tree and major shrub species
b) 83 categories, some of which grouping more than one species
c) field assessment
d) no remarks

Social standing

a) the way the individual tree growth is performing in comparison with the closest trees and the stand development stage
b) six categories for even-aged stands; three categories for uneven-aged stands
c) field assessment
d) no remarks

E) Regeneration

Amount

a) expressed as seedlings cover on the sample plot
b) four categories
c) field assessment
d) seedlings height < 1.30 m

Distribution

a) spatial distribution
b) six categories
c) field assessment
d) no remarks

Leading species

a) most represented tree or shrub species
b) 83 categories, some of which grouping more than one species
c) field assessment
d) no remarks

Health status

a) assessment of regeneration quality, external defects and damage type
b) eight classes
c) nine field assessment
d) no remarks

G) Forest condition

Tree health status

a) external defects, damage, and/or decay assessment of enumerated trees
b) nine classes
c) field assessment
d) no remarks

H) Accessibility and harvesting

Accessibility

a) accessibility of the sampling unit from closest all-weather motor road
b) seven classes according to distance and location
c) field assessment, map
d) no remarks

Forest road network

a) type and density of roads and paths within the forest
b) seven classes
c) field assessment
d) paths wide less than 1.5 m are not considered

Ground roughness

a) frequency of obstacles according to their relief amplitude
b) six classes
c) field assessment
d) no remarks

Rock outcrops

a) frequency of rock outcrops with a relief amplitude exceeding 50 cm
b) six classes
c) field assessment
d) no remarks

Breakage and throw trees
a) frequency of felled trees for natural causes
b) three classes
c) field assessment
d) no remarks

Undergrowth thickness
a) regeneration and shrub density
b) six classes
c) field assessment
d) height between 0.5 m and 3-4 m

4.1.5. Forest area definition and definition of "other wooded land"

Forest area is defined as a land under natural or planted stands of trees, not tilled, naturally developing and liable to forestry activities related to the production of wood and other goods and services of the forest. The extension is at least 5,000 m², the width 20 m, crown cover of more than about 10 % and stand height over 5 m.

Other wooded land has the same characteristics of forest land but has a minimum extension of 1,250 m² and crown cover > 40%. It has to be at least 20 m far from forest land.

Excluded are areas not meeting the conditions of forests as described above and shrublands.

The forest edge is defined by the line of demarcation between different cover types or land-use. A special procedure has been used for attributes assessed in borderline trees taking into account the missing area (not forested, etc.) of the field plot. Borderline trees are tallied on tally sheets.

4.2. DATA SOURCES

A) field data

- the interpretation area for area related data (Regeneration, Canopy cover, ground roughness....) is included in the girth determined by the most external enumerated tree.
- sub-sample of 3P trees (Probability Proportional to Prediction) among enumerated trees selected with probability proportion to total stand height: total tree height, bole height, stump girth, diameters at 4, 5 and 6.5 m height
- small sub-sample of felled trees: total tree height, merchantable height, stump girth, 10 intermediate diameters, crown diameter
- increment cores extracted from the closest tree to the plot centre: age, annual radial increment, inside bark diameter

E) maps

- Italian Military Geographic Institute; Veneto Regional Authority Office - Forest Department; Provincial Land Office:
- 1982 and earlier
- 1:25000; 1:25000; 1:10000
- analogue printed topographic map; analogue printed "Forest Type Map"; analogue printed cadastral maps

4.3. ASSESSMENT TECHNIQUES

4.3.1. Sampling frame

The sampling frame is given by the forest area definition. The entire Region is covered by the assessment. Excluded are: young natural stands and all plantations which have not yet reached a stand height of 5 m and a canopy cover >10%; sweet chestnut woodlands; plantations.

4.3.2. Sampling units

Field plot: point sampling. Basal Area Factor = 2. The sampling unit is a cluster of three field plots: one in the central point, the other two, 100m far from the central point, are randomly selected among the six vertexes of the equilateral triangles composing an hexagon. These latter may be moved in the case that fall in a not-forested area. All plots permanently monumented and azimuth and distance of tallied trees from plot centre recorded. The plots are systematically distributed in a 800*800 m grid. The field plots have been randomly selected within units of the forest types map chosen with Probability Proportional to Size from a sorted list. The two field plots are 50 m far from the central point.

4.3.3. Sampling designs

The sampling design applied in the Veneto RFI is different for public and non-public forests. In the former field plots are systematically distributed on a 800*800 m grid after that a forest/non-forest decision is made on the "Forest Type Map" implemented by forest working plans.

In non-public forests strata are assigned to "Forest Type Map" units which have been chosen with Probability Proportional to Size from a sorted list. Field plots have been randomly selected in such units on a 800*800m grid.

4.3.4. Techniques and methods for combination of data sources

Attributes are directly and exclusively assessed on the field plots. Therefore there is no combination of data sources.

4.3.5. Sampling fraction

Data source and sampling unit	proportion of forested area covered by sample	represented mean area per sampling unit
field assessment, conventional fixed area associated to sampling point	////	////

4.3.6. Temporal aspects

Inventory cycle	time period of data assessment	publication of results	time period between assessments	reference date
1st VRFI (non-public forests)	1984	1985		not quoted
1st VRFI (public forests)	1985	1986		not quoted

4.3.7. Data capturing techniques in the field

Data recorded on tally sheets and edited by hand into the computer.

4.4. DATA STORAGE AND ANALYSIS

4.4.1. Data storage and ownership

The data of the assessment are stored in a database system at the Forest Department of the Veneto Regional Authority Office in Mestre (VE).

4.4.2. Database system used

Data not available.

4.4.3. Data bank design

Data not available

4.4.4. Update level

Not quoted

4.4.5. Description of statistical procedures used to analyse data including procedures for sampling error estimation

a) area estimation

Area estimation is made by cartographic methods without sampling errors.

b) aggregation of tree and plot data

As the selection of the trees to be recorded is proportional to the individual basal area, the single trre attributes have to be divided by relative basal area.

The total values for plot i; Y_i, can be calculated in the following way:

$$Y_i = k\sum_{1}^{z} Y_i / g_i$$

where:
k = BAF
g = basal area
z = number of enumerated trees

c) estimation of total values

The estimation of total value of single tree attributes, Y, can be calculated as follows:

$$Y_h = {}^{k}\!/_{n} \sum_{j=1}^{n} \sum_{i=1}^{z} {}^{Y_i}\!/_{g_i}$$

where
n = number of plots

In general the total value of attribute Y for a strata h is given by multiplying the area related attribute, Y_h, with the measured area, A_h,of the strata:

$$Y = A_h Y_h$$

Since the strata area is planimetred on the map, the only error taken into account is the standard error of the total value of basal area attribute Y_g

$$s(Y_g) = \sqrt{\frac{(\sum_{1}^{n} g^2) - \frac{(\sum_{1}^{n} g)^2}{n}}{n(n-1)}}$$

As regard to volume estimation, based on 3P trees, it has been considered the standard error of the 3P trees as well:

$$s(Y_{3P}) = \sqrt{\frac{(\sum_{1}^{n} R^2) - \frac{(\sum_{1}^{n} R)^2}{n}}{n(n-1)}}$$

where:
R = ratio between calculated and field assessed attribute

The total error is given by:

$$s = \sqrt{s(Y_i)^2 + s(Y_{3P})^2}$$

d) estimation of ratios

///

e) sampling at the forest edge

At the forest edge attributes are assessed in borderline trees taking into account the missing area (not forested, etc.) of the field plot. Borderline trees are tallied on tally sheets.

f) estimation of growth and growth components (mortality, cut, ingrowth)*

Trees are distributed in stand height classes and the growth has been calculated with a growth model based on d.b.h., tree species and age of the single tree.

g) allocation of stand and area related data to single sample plots/sample points.

The class of categorical, area related attributes, in which the plot centre is located, is assigned to the entire plot.

h) Hierarchy of analysis: how are sub-units treated?*

In the Veneto RFI results have to be presented for Provinces, Mountain Communities and Watershed. This is done for metric attributes as well as for categorical (area related) attributes.

4.4.6. Software applied

Data not available

4.4.7. Hardware applied

Data not available

4.4.8. Availability of data (raw and aggregated data)

Data not available

4.4.9. Sub-units (strata) available

Results can be obtained for the Provinces, Mountain Communities and Watersheds

4.4.10. Links to other information sources

Not applied.

4.5. RELIABILITY OF DATA

4.5.1. Check assessment

Data not available

4.5.2. Error budgets

Data not available

4.5.3. Procedures for consistency checks of data

Data not available

4.6. MODELS

4.6.1. Volume functions

a) outline of the model

Volume functions have been derived using d.b.h., diameters at 4 m, 5 m e 6.5 m height and total tree height as input variables. nine different volume functions for dominant species, silvicultural system and specific species groups have been derived. The felled trees are used in order to check and to adjust volume functions. Based on the obtained volumes double-entry tables are constructed.

b) overview of prediction errors

Not stated

c) data material for derivation of model

Input data, for non-public forest, were 5161 3P trees and 402 felled trees

d) methods applied to validate the model

Not stated

4.6.2. Assortments

Not assessed.

4.6.3. Growth components

a) outline of the model

Trees are distributed in stand height classes and the growth has been calculated with a growth model based on d.b.h., tree species and age of the single tree.

b) overview over prediction errors

Not stated

c) data material for the derivation

Input data, for non-public forest, were 2,387 increment cores extracted from the closest tree to the plot centre

d) methods applied to validate the model

Not stated

4.6.4. Potential yield

Not assessed.

4.6.5. Forest functions

Not assessed

4.6.6. Other models applied

////

4.7. INVENTORY REPORTS

4.7.1. List of published reports and media for dissemination of inventory results

Inventory	Year of publication	Citation	Language	Dissemination
VRFI	1985	Inventario dei boschi non pubblici - Risultati del primo inventario. 1984 Regione del Veneto - Dipartimento Foreste	Italian	printed
VRFI	1984	Inventario dei boschi non pubblici. Regione del Veneto - Dipartimento Foreste 1. Linee progettuali 2. Vademecum per i rilievi di campagna	Italian	printed
VRFI	1985	Inventario dei boschi pubblici. Regione del Veneto - Dipartimento Foreste	Italian	printed

4.7.2. List of contents of latest report and update level

Reference of latest report:
Regione del Veneto - Dipartimento Foreste, 1985. Inventario dei boschi non pubblici. Risultati del primo Inventario.

Update level of inventory data used in this report:
Field data: 1984

Content:

Chapter	Title	Number of pages
1	Historical background	3
2	Why a forest survey of non-public woodlands?	16
3	Data processing	4
4	Forested area	2
5	Forest types and strata	10
6	Location attributes	2
7	Forest improvement guidelines	2
8	Socio-economical features	4
	Annexes	20
	Tables	110
	Reference list	2

4.7.3. Users of the results

Not directly assessed. Local administrations, agencies and bodies. Professionals.

4.8. Future development and improvement plans

///

REFERENCES

Bianchi, M., 1982. Il campionamento dendrometrico per aree di saggio riunite a gruppi. Monti e Boschi 1/2: pp. 63-66

Bianchi, M. & Preto, G., 1982. La localizzazione delle aree di saggio sul terreno in un inventario forestale continuo. Monti e Boschi, 6: pp. 53-57

Cochran, W.G., 1963. Sampling techniques. J.Wiley & Son. Inc. New York. 413 p.

Costantini, B., 1983. In corso di attuazione la Legge Forestale Regionale del Veneto- Economia Montana 5: pp. 18-21.

Costantini, B., 1983. Organizzazione del Sistema Informativo Forestale del Veneto.Atti del Convegno Nazionale Informatica. 27-29 Aprile 1983. Padova: pp. 23-32.

Costantini, B., Crespi, M. & Santocono, A. 1985. Problematiche di Integrazione fra Banche Dati Tradizionali e Banche Dati Derivanti da Piattaforme Satelittarie, ai Fini Della Gestione di un Servizio Regionale Territoriale ed Ambientale. atti del V Convegno di Informatica Enti Locali e Territorio. Padova: pp. 89-95.

Loetsch, F. & Haller, K.E., 1964. Forest Inventory-vol. I. BLV - München. Basel. Wien. 435 p.

Loetsch, F., Haller, K.E. & Zohrer, F., 1973. Forest Inventory-vol. II. BLV -München. Basel. Wien. 469 p.

Preto, G., 1983. Un inventario prmanente delle risorse forestali europee. Economia Montana 3: pp. 11-21.

Santocono, A., 1984. Decision-Making and Management in Forest Land. Atti del Convegno Data Urban Management. Simposio Europeo dei Sistemi Informativi Urbani. Padova: pp. 97-106.

Stage, A.R., 1971. Sampling with Probability Proportional to Size from a Sorted List. -Intermountain For. Range Exp. Stn. Res. Paper INT. 88.

Susmel, L., 1983. La Carta Regionale Forestale del Veneto - Regoine del Veneto.

5. FRIULI-VENEZIA GIULIA REGIONAL FOREST INVENTORY

(INVENTARIO FORESTALE DELLA REGIONE AUTONOMA FRIULI-VENEZIA GIULIA - IFFVG)

5.1. NOMENCLATURE

5.1.1. List of attributes directly assessed

A) Geographic regions

Attribute	Data source	Object	Measurement unit
Watershed *(Bacino idrografico)*	maps	sampling unit	categorical
Province *(Provincia)*	maps	sampling unit	categorical
Mountain Community *(Comunità Montana)*	maps	sampling unit	categorical
Municipality *(Comune)*	maps	sampling unit	categorical

B) Ownership

Attribute	Data source	Object	Measurement unit
Ownership *(Proprietà)*	land register	sampling unit	categorical

C) Wood production

Attribute	Data source	Object	Measurement unit
Number of trees selected in the point sample *(Numero di piante del campione relascopico)*	field assessment	sampling unit	integer
Diameter at breast height *(Diametro a petto d'uomo)*	field assessment	tree	cm
Bark thickness *(Spessore della corteccia)*	field assessment	3P tree, felled tree	mm
Stump girth *(Circonferenza alla ceppaia)*	field assessment	3P tree, felled tree	cm
Total height *(Altezza dendrometrica)*	field assessment	3P tree, felled tree	m
Bole height *(Altezza di inserzione della chioma)*	field assessment	3P tree	m

Crown shape *(Morfologia della chioma)*	field assessment	tree	categorical
Stem quality *(Morfologia del fusto)*	field assessment	tree	categorical
Diameters at 4m, 5m and 6.5m height *(Diametri intermedi a 4, 5 e 6.5 m)*	field assessment	3P tree	cm
10 Intermediate diameters *(10 diametri intermedi)*	field assessment	felled tree	cm
Annual radial increment *(Incremento diametrico annuo)*	field assessment	felled tree, tree (increment core)	mm

D) Site and soil

Attribute	Data source	Object	Measurement unit
Elevation *(Quota)*	field assessment, map	sampling unit	m
Aspect *(Esposizione)*	field assessment, map	sampling unit	categorical
Slope *(Pendenza)*	field assessment, map	sampling unit	categorical
Location *(Posizione topografica)*	field assessment	sampling unit	categorical
Soil stability *(Assetto superficiale del suolo)*	field assessment	sampling unit	categorical

E) Forest structure

Attribute	Data source	Object	Measurement unit
Forest type *(Tipologia forestale)*	map	sampling unit	categorical
Stand origin *(Origine del soprassuolo)*	field assessment	sampling unit	categorical
Silvicultural system and stand development stage *(Forma di governo e di trattamento)*	field assessment	sampling unit	categorical
Stratified stand structure *(Tipo di struttura verticale)*	field assessment	sampling unit	categorical
Canopy cover *(Grado di copertura)*	field assessment	sampling unit	categorical

Age *(Età)*	field assessment	tree (increment core)	integer
Tree species *(Classificazione botanica delle specie)*	field assessment	sampling unit	categorical
Social standing *(Posizione sociale)*	field assessment	tree	categorical
Trees spatial distribution *(Distribuzione spaziale)*	field assessment	tree	degree and cm

F) Regeneration

Attribute	Data source	Object	Measurement unit
Amount *(Consistenza della rinnovazione)*	field assessment	sampling unit	categorical
Distribution *(Distribuzione)*	field assessment	sampling unit	categorical
Leading species *(Specie prevalente)*	field assessment	sampling unit	categorical
Health status *(Stato vegetativo)*	field assessment	sampling unit	categorical

G) Forest condition

Attribute	Data Source	Object	Measurement unit
Tree health status *(Stato vegetativo)*	field assessment	tree	categorical

H) Accessibility and harvesting

Attribute	Data source	Object	Measurement unit
Accessibility *(Accessibilità)*	field assessment, map	sampling unit	categorical
Forest road network *(Viabilità)*	field assessment, map	sampling unit	categorical
Ground roughness *(Asperità del terreno)*	field assessment	sampling unit	categorical
Rock outcrops *(Rocciosità e/o pietrosità)*	field assessment	sampling unit	categorical
Breakage and throw trees *(Presenza di schianti o sradicamenti)*	field assessment	sampling unit	categorical
Undergrowth thickness *(Densità del sottobosco)*	field assessment	sampling unit	categorical

K) Miscellaneous

Attribute	Data source	Object	Measurement unit
Sample plot number *(Coordinate UTM dell'area di saggio principale)*	map	sampling unit	ordinal
Surveyor name *(Rilevatore)*	////	sampling unit	////

5.1.2. List of derived attributes

C) Wood production

Attribute	Measurement unit	Input attributes
Basal area per hectare *(Area basimetrica per ettaro)*	m^2	BAF, no of trees selected in the point sample, tree species
Number of stems per hectare *(Numero di piante per ettaro)*	integer	basal area/ha, dbh, tree species
Mean diameter *(Diametro medio)*	cm	basal area, no. of stems, tree species
Mean height *(Altezza media)*	m	mean diameter, tree heights, tree species
Stem total volume *(Volume del fusto intero)*	m^3	intermediate diameters, log lengths
Tree volume per hectare *(Volume per ettaro)*	m^3	no. of stems per ha, d.b.h., height, tree species
Current diametric increment *(Incremento corrente diametrico)*	cm	mean increment on cores for 5-years classes, tree species, d.b.h., top height

E) Forest structure

Attribute	Measurement unit	Input attributes
Forest composition *(Composizione specifica)*	categorical	species, basal area, no. of trees

5.1.3. Measurement rules for measurable attributes

C) Wood production

Diameter at breast height

a) measured in 1.3 m height above ground, on slopes measured from uphill side. Two readings, for each selected tree, the first with calliper pointing to plot centre, the second perpendicular to the first
b) minimum dbh: 2.5 cm
c) cm, recorded every single cm
d) rounded off and up to closest cm
e) calliper
f) field assessment

Bark thickness

a) measured at d.b.h.
b) no threshold
c) mm
d) rounded to closest mm
e) bark thickness instrument
f) field assessment

Stump girth

a) measured at 30 cm height
b) no threshold
c) cm
d) rounded off and up to closest cm
e) tape-line
f) field assessment

Total height

a) length of tree from ground level to top of tree.
b) no threshold
c) m
d) rounded to closest 0.5 m
e) Suunto, Haga, Blume-Leiss hypsometer, Bitterlich relascope
f) field assessment: ocular estimate on enumerated trees, hypsometer for 3P trees, tape for felled trees

Bole height

a) distance between ground level and the origin of the lowest crown-forming branches (crown point)
b) no threshold
c) m
d) rounded to closest 0.5 m
e) Suunto, Haga, Blume-Leiss hypsometer, Bitterlich relascope
f) field assessment

Diameters at 4 m, 5 m and 6.5 m height

a) on slopes measured from uphill side with calliper pointing to plot centre and perpendicular to the stem
b) minimum tree height: 5 m
c) cm
d) rounded to closest cm
e) Finnish calliper
f) field assessment

Ten Intermediate diameters

a) measured only on felled trees: at 1/20, 1/10, 2/10,... 9/10 two readings
b) no threshold
c) cm
d) rounded to closest cm
e) calliper
f) field assessment

Annual radial increment

a) measured both on logs of the felled trees and on increment cores extracted on standing trees; in the former assessed at 0.30, 1.30, 4.50, 6.50 m, merchandable height (Hc) and half the difference between merchantable height and 6.50 m [(Hc - 6.50)/2]; in the latter at breast height. Length of every five increment rings.
b) minimum height: 5 m; minimum d.b.h.: 7.5 cm
c) mm
d) rounded to closest mm
e) magnifying glass, calliper, Pressler increment borer
f) field assessment, laboratory

D) Site and soil

Elevation

a) assessed on the map on the grid point as altitude
b) no threshold
c) m

d) rounded to closest m
e) ocular estimate
f) 1:25000 topographic map

E) Forest structure

Age
a) measured on increment cores counting increment rings
b) no threshold
c) years
d) no rounding rules
e) ocular estimate (magnifying glass, calliper)
f) field assessment; laboratory

Trees spatial distribution
a) horizontal distance and azimuth of selected trees measured from plot centre
b) no threshold
c) distance in cm, azimuth in degrees
d) rounded to closet cm and degree
e) tape, compass, tripod
f) field assessment

K) Miscellaneous

Sample plot number
a) UTM coordinates of the sample plot
b) no threshold
c) m
d) no rounding rules
e) map
f) map

5.1.4. Definitions for attributes on nominal or ordinal scale

A) Geographic regions

Watershed
a) geomorphologic unit
b) data not available
c) map

Province
a) administrative unit joining several closed Municipalities
b) Four Provinces
c) map

Mountain Community
a) administrative unit joining several closed municipalities in the mountain area aimed at planning and managing mountain regions
b) data not available
c) map

Municipality
a) administrative unit
b) data not available
c) map

B) Ownership

Ownership
a) owner of the forest area under concern
b) Seven classes
c) land register, map

C) Wood production

Number of trees selected in the point sample
a) number of trees enumerated with the critical angle instrument (Bitterlich relascope) which permits the observer to decide to include or exclude the tree. BAF = 2.
b) no categories
c) field assessment
d) minimum d.b.h.: 2.5 cm. Starting from North in clockwise direction; trees with a breast-height diameter larger than the fixed angle chord corresponding to Basal Area Factor = 2 are tallied "in"; mean diameter and distance of borderline and hidden trees are checked with tape and calliper.

Crown shape
a) development in height and width of the crown related to the total tree height
b) nine categories according to crown point and size
c) field assessment
d) no remarks

Stem quality
a) commercial value of the stem
b) eight categories
c) field assessment
d) assessed on the stem portion included between 0.3 and 4/8m from the ground

D) Site and soil

Aspect
a) direction of exposure of the slope
b) nine categories
c) map, field assessment
d) "all" is used when the slope is less than 5%

Slope
a) along the ruling gradient line
b) six categories
c) map, field assessment
d) no remarks

Location
a) morphological site
b) 24 classes
c) map, field assessment
d) no remarks

Soil stability
a) susceptibility to erosion
b) seven main classes and 22 sub-classes
c) field assessment
d) no remarks

E) Forest structure

Forest type
a) the forest types are defined on the basis of latitudinal belt, coniferous or broad-leaved, leading species
b) five main categories, sub-divided with hierarchical classification in nine groups and 19 types
c) field assessment, aerial photographs
d) no remarks

Stand origin
a) according to the type of regeneration generating the stand
b) three categories:
 1) artificial
 2) natural dissemination
 3) agamic regeneration
c) field assessment
d) no remarks

Silvicultural system and stand development
a) the silvicultural system applied and the stand development stage
b) five categories:
 1) simple coppice (three sub-classes)
 2) coppice selection system
 3) coppice with standards (two sub-classes)
 4) even-aged high forest (six sub-classes)
 5) storied high forest (two sub-classes)
c) field assessment
d) no remarks

Stratified stand structure
a) crown storeys distribution
b) three categories:
 1) one-storey
 2) two-storeys
 3) more than two storeys or irregular
c) field assessment
d) no remarks

Canopy cover
a) crown cover on the ground
b) six categories
c) field assessment
d) no remarks

Tree species
a) Scientific and Italian name of selected list of tree and major shrub species
b) 83 categories, some of which grouping more than one species
c) field assessment
d) no remarks

Social standing
a) the way the individual tree growth is performing in comparison with the closest trees and the stand development stage
b) six categories for even-aged stands; three categories for uneven-aged stands
c) field assessment
d) no remarks

F) Regeneration

Amount
a) expressed as seedlings cover on the sample plot
b) four categories
c) field assessment
d) seedlings height < 1.30 m

Distribution
a) spatial distribution
b) six categories
c) field assessment
d) no remarks

Leading species
a) most represented tree or shrub species
b) 83 categories, some of which grouping more than one species
c) field assessment
d) no remarks

Health status
a) assessment of regeneration quality, external defects and damage type
b) eight classes
c) field assessment
d) no remarks

G) Forest condition

Tree health status
a) external defects, damage, and/or decay assessment of enumerated trees
b) nine classes
c) field assessment
d) no remarks

H) Accessibility and harvesting

Accessibility
a) accessibility of the sampling unit from closest all-weather motor road
b) seven classes according to distance and location
c) field assessment, map
d) no remarks

Forest road network
a) type and density of roads and paths within the forest
b) seven classes
c) field assessment
d) paths wide less than 1.5 m are not considered

Ground roughness
a) frequency of obstacles according to their relief amplitude
b) six classes
c) field assessment
d) no remarks

Rock outcrops
a) frequency of rock outcrops with a relief amplitude exceeding 50 cm
b) six classes
c) field assessment
d) no remarks

Breakage and throw trees
a) frequency of felled trees for natural causes
b) three classes
c) field assessment
d) no remarks

Undergrowth thickness
a) regeneration and shrub density
b) six classes
c) field assessment
d) height between 0.5 m and 3-4 m

5.1.5. Forest area definition and definition of "other wooded land"

Forest area is defined as a land under natural or planted stands of trees, whether productive or not. The extension is at least 5,000 m², the width 20 m, crown cover of more than about 10% and stand height over 5 m. Other wooded land has the same characteristics of forest land but has a minimum extension of 1,250 m² and crown cover > 40%. It has to be at least 20 m far from forest land.

The forest edge is defined by the line of demarcation between different cover types or land-use. A special procedure has been used for attributes assessed in borderline trees taking into account the missing area (not forested, etc.) of the field plot. Borderline trees are tallied on tally sheets.

5.2. DATA SOURCES

A) field data

- the interpretation area for area related data (regeneration, canopy cover, ground roughness...) is included in the girth determined by the most external enumerated tree.
- sub-sample of 3P trees (Probability Proportional to Prediction) among enumerated trees selected with probability proportion to total stand height: total tree height, bole height, stump girth, diameters at 4, 5 and 6.5 height, bark thickness
- increment cores extracted from the closest tree to the plot centre and from the mean diameter tree: age, annual radial increment

C) aerial photography

- 1:18000/1:30000, no ortophotographs
- 1988/1980-84
- the entire Region
- stereoscopic pairs
- not available
- not available

E) maps

- Italian Military Geographic Institute
- 1986
- 1:25000
- analogue printed topographic

5.3. ASSESSMENT TECHNIQUES

5.3.1. Sampling frame

The sampling frame is given by the forest area definition. The entire Region is covered by the assessment. Excluded are: plain areas. Field assessment are not carried out in plantations which have not yet reached a stand height of 5 m and a canopy cover >10%; arboriculture plantations; shrublands.

5.3.2. Sampling units

Aerial photography sample plots: fixed area with centre in sampling point. Interpretation square area of 1 ha. Plots are systematically distributed in a 200*200 m grid.

Field plot: point sampling. Basal Area Factor = 2. The sampling unit is a cluster of three field plots: one in the central point, the other two, 50 m far from the central point, are randomly selected among the fourth vertexes of a square. These latter may be moved in the case that fall in a not-forested area. All plots permanently monumented and azimuth and distance of tallied trees from plot centre recorded. The plots are distributed in a 200*200 m grid.

5.3.3. Sampling designs

The sampling design applied in the Friuli-Venezia Giulia region is a double-sampling for stratification.

In the first phase (photo-interpretation) systematically distributed photo-plots are interpreted, land use assessment and strata assigned to the forested photo-plots. In the second phase a smaller, randomly selected among a sub-sample of photo-plots chosen on the basis of the strata extension and their forest management relevance, sub-sample of field plots is assessed.

5.3.4. Techniques and methods for combination of data sources

The attributes assessed in field and photo sample plots are combined by a double sampling for stratification approach. Attributes for stratification are assessed in photo-plots and used for estimating strata sizes.

Topographic maps are used for field plot localisation and for the assessment of the administrative unit, watershed, elevation, aspect, slope and accessibility

5.3.5. Sampling fraction

Data source and sampling unit	proportion of forested area covered by sample	represented mean area per sampling unit
field assessment, conventional fixed area associated to sampling point	////	////

5.3.6. Temporal aspects

Inventory cycle	time period of data assessment	publication of results	time period between assessments	reference date
1st IFFVG	1985-87	not available		////

5.3.7. Data capturing techniques in the field

Data recorded on tally sheets and edited by hand into the computer.

5.4. DATA STORAGE AND ANALYSIS

5.4.1. Data storage and ownership

Not available

5.4.2. Database system used

Data not available.

5.4.3. Data bank design

Data not available

5.4.4. Update level

Not quoted

5.4.5. Description of statistical procedures used to analyse data including procedures for sampling error estimation

The forest area is assessed on the aerial photographs. A 200*200 m systematic point grid is laid over the entire regional area and each point is assigned a forest or other land cover/use. The estimation of forest and other land cover types area is done counting the proportional number of point then multiplying by the area:

$$a = \left(\frac{n}{N}\right) * A$$

where
a = forest or other land cover type area
n = number of points assigned to forest type
N = total number of points
A = area

The standard error, s(a), is calculated according to one of the formulas provided in publications.

b) aggregation of tree and plot data

As the selection of the trees to be recorded is proportional to the individual basal area, the single tree attributes have to be divided by relative basal area.

The total values for plot i, Y_i, can be calculated in the following way:

$$Y_i = k\sum_{1}^{z} Y_i / g_i$$

where:
k = BAF
g = basal area
z = number of enumerated trees

c) estimation of total values

The estimation of total value of single tree attributes, Y, can be calculated as follows:

$$Y_h = \frac{k}{n}\sum_{j=1}^{n}\sum_{i=1}^{z}\frac{Y_i}{g_i}$$

where
n = number of plots

In general the total value of attribute Y for a strata h is given by multiplying the area related attribute, Y_h, with the calculated area, A_h, of the strata:

$$Y = A_h Y_h$$

The standard error of the total value of basal area attribute Yg is:

$$s(Y_g) = \sqrt{\frac{(\sum_{1}^{n} g^2) - \frac{(\sum_{1}^{n} g)^2}{n}}{n(n-1)}}$$

As regard to volume estimation, or other attributes based on 3P trees, it has been considered the standard error of the 3P trees as well:

$$s(Y_{3P}) = \sqrt{\frac{(\sum_{1}^{n} R^2) - \frac{(\sum_{1}^{n} R)^2}{n}}{n(n-1)}}$$

where:
R = ratio between calculated and field assessed attribute

The total error is given by:

$$s = \sqrt{s(Y_i)^2 + s(Y_{3P})^2 + s(a)^2}$$

d) estimation of ratios

///

e) sampling at the forest edge

At the forest edge attributes are assessed in borderline trees taking into account the missing area (not forested, etc.) of the field plot. Borderline trees are tallied on tally sheets.

f) estimation of growth and growth components (mortality, cut, ingrowth)*

///

g) allocation of stand and area related data to single sample plots/ sample points.

The class of categorical, area related attributes, in which the plot centre is located, is assigned to the entire plot.

h) Hierarchy of analysis: how are sub-units treated?*

///

5.4.6. Software applied

Data not available

5.4.7. Hardware applied

Data not available

5.4.8. Availability of data (raw and aggregated data)

Data not available

5.4.9. Sub-units (strata) available

Results can be obtained for the Provinces, Mountain Communities and Watersheds

5.4.10. Links to other information sources

////

5.5. RELIABILITY OF DATA

5.5.1. Check assessment

Data not available

5.5.2. Error budgets

Data not available

5.5.3. Procedures for consistency checks of data

Data not available

5.6. MODELS

5.6.1. Volume functions

a) outline of the model

Volume functions have been derived using d.b.h., diameters at 0.3 m, 4 m, 5 m e 6.5 m height and total tree height as input variables. Volume functions for dominant species, silvicultural system and specific species groups have been derived. The felled trees are used in order to check and to adjust volume functions. Based on the obtained volumes double-entry tables are constructed.

b) overview of prediction errors

Not stated

c) data material for derivation of model

Input data were 9665 3P trees and 1,000 felled trees.

d) methods applied to validate the model

Not stated

5.6.2. Assortments

Not assessed.

5.6.3. Growth components

a) outline of the model

Data not available

b) overview over prediction errors

Not stated

c) data material for the derivation

5,967 increment cores were extracted from two trees in each plot.

d) methods applied to validate the model

Not stated

5.6.4. Potential yield

Not assessed.

5.6.5. Forest functions

Not assessed

5.6.6. Other models applied

////

5.7. INVENTORY REPORTS

5.7.1. List of published reports and media for dissemination of inventory results

Inventory	Year of publication	Citation	Language	Dissemination
IFFVG	1986	Inventario Forestale Regionale. Regione Autonoma Friuli-Venezia Giulia	Italian	printed

5.7.2. List of contents of latest report and update level

Reference of latest report:
Regione Autonoma Friuli-Venezia Giulia, 1986. Inventario Forestale Regionale.

Content:

Chapter	Title	Number of pages
1	Survey guidelines	21
2	Photo-interpretation rules	71
3	Field assessment rules	136

5.7.3. Users of the results

Not directly assessed. Local administrations, agencies and bodies. Professionals.

5.8. FUTURE DEVELOPMENT AND IMPROVEMENT PLANS

///

REFERENCES

Preto, G., 1986. Inventario Forestale Regionale. Regione Autonoma Friuli Venezia Giulia. Direzione Regionale delle Foreste. Omnia Press, Trieste, Udine.

Spazio Verde, IF Friuli Venezia Giulia. Relazione Tecnica. Unpublished.

6. MULTIPURPOSE FOREST INVENTORY OF LIGURIA REGION

(INVENTARIO FORESTALE MULTIRISORSE DELLA REGIONE LIGURIA - IFMR)

6.1. NOMENCLATURE

The following description refers to Forest Map 1:50000, GIS, Multipurpose Forest Inventory of Liguria Region (Carta Forestale d'Italia 1:50.000 e Sistema Informativo Geografico Forestale della Regione Liguria).

6.1.1. List of attributes directly assessed

A) Geographic regions

Attribute	Data source	Object	Measurement unit
Province *(Provincia)*	Maps	Administrative unit	ha

B) Ownership

Attribute	Data source	Object	Measurement unit
Ownership (*Proprietà)*	local information	plot	categorical

C) Wood production

Attribute	Data source	Object	Measurement unit
Diameter at breast height (*Diametro a petto d'uomo*)	field assessment	tree	cm
Tree height (*Rilievo della altezze*)	field assessment	tree	m
Quality of stems (*Qualità dei fusti*)	field assessment	tree	percent of trees
Crown height and wide *(Inserzione e larghezza della chioma)*	field assessment	tree	m, cm
Crown class *(Posizione sociale)*	field assessment	tree	categorical
Exploitation *(Utilizzazioni)*	field + forest service	plot	year, m^3, £

D) Site and soil

Attribute	Data source	Object	Measurement unit
Elevation (*Altitudine)*	field (altimeter + map)	plot centre	10 m
Slope (*Pendenza*)	field (clisimeter)	plot	1 %
Aspect (*Esposizione*)	field (compass)	plot	categorical
Location *(Elemento geomorfologico)*	field (visual observation)	plot	categorical
Stoniness *(Pietrosità)*	field	plot	categorical
Rockiness *(Rocciosità)*	field	plot	categorical
Soil erosion and deposition *(Erosione e deposizione)*	field	plot	categorical
Humus *(Humus)*	field	plot	categorical
Soil survey *(Rilievo pedologico)*	field	plot	categorical

E) Forest structure

Attribute	Data source	Object	Measurement unit
Trace of former wood *(Tracce di vecchi boschi)*	field	plot	categorical
Inventory stand types (*Tipo di popolamento = tipo inventariale di riferimento*	field	area around the plot (0.5 ha)	categorical
Cover: height of vegetation levels *(Altezza degli strati di vegetazione)*	field	plot	cm, m
Cover: percent of coverage on levels *(Copertura degli strati di vegetazione)*	field	plot	%
Stand development *(Stadio evolutivo)*	field	stand	categorical
Origin of stand (*Origine del soprassuolo*)	field	stand	categorical
Sylvicultural system *(Trattamenti)*	field	area around the plot (0.5 ha)	categorical
Conversion to high forest *(Conversione all'altofusto)*	field	stand	categorical

Tending of stands (*Cure colturali)*	field	area around the plot (0.5 ha)	categorical
Age of stands (*Età del soprassuolo*)	field	plot	categorical
Planting space (*Sesto d'impianto*)	field	area around the plot (0.5 ha)	m
Cultivation status *(Stato di coltivazione)*	field	area around the plot (0.5 ha)	categorical
Floristic composition *(Censimento della flora)*	field	plot	species and %l
Bush pattern *(Rilievo del pattern degli arbusti)*	field	plot	m

F) Regeneration

Attribute	Data source	Object	Measurement unit
Condition (*Stato vegetativo*)	field	plot	categorical
Origin (*Origine*)	field	plot	categorical
Distribution (*Distribuzione*)	field	plot	categorical
Localisation (*Localizzazione)*	field	plot	categorical
Transect of regeneration *(Rilievo della rinnovazione)*	field	plot	categorical

G) Forest condition

Attribute	Data source	Object	Measurement unit
Site degradation: type (*Degrado della stazione: tipo*)	field	plot	categorical
Site degradation: origin (*Degrado della stazione: causa*)	field	plot	categorical
Site degradation: intensity (*Degrado della stazione: intensità)*	field	plot	categorical
Stand degradation: origin (*Degrado del popolamento: origine*)	field	plot	categorical
Stand degradation: intensity (*Degrado del popolamento: effetto*)	field	plot	categorical
Stand degradation: distribution (*Degrado del popolamento:distribuzione*)	field	plot	categorical

Stand degradation: diffusion (*Degrado del popolamento:diffusione*)	field	plot	percent of trees

H) Accessibility and harvesting

Attribute	Data source	Object	Measurement unit
Accessibility and harvesting: difference in height-distance between plot and road *(Accessibilità ed esboscabilità: dislivello-distanza tra area di saggio e viabilità)*	field	area around the plot	categorical
Accessibility and harvesting: difference in height-distance between plot and skidding road/ path *(Accessibilità ed esboscabilità: dislivello-distanza tra area di saggio e viabilità)*	field	area around the plot	categorical
Accessibility and harvesting: road classification *(Accessibilità ed esboscabilità: classificazione della viabilità)*	field	area around the plot	categorical
Accessibility and harvesting: distance covering on carriage road *(Accessibilità ed esboscabilità: percorrenza su rotabile secondaria)*	field	area around the plot	categorical
Accessibility and harvesting: improvements on existing roads *(Accessibilità ed esboscabilità: interventi di miglioramento della viabilità esitente)*	field	area around the plot	categorical
Unevenness of soil morphology: types (*Accidentalità: tipo di ostacolo)*	field	area around the plot (0.5 ha)	categorical
Unevenness of soil morphology: intensity (*Accidentalità: classi di frequenza)*	field	area around the plot (0.5 ha)	categorical

I) Attributes describing forest ecosystems

Attribute	Data source	Object	Measurement unit
Micro-habitat - clearing: extension *(Micro-habitat - Radure: estensione)*	field	area around the plot (0.5 ha)	m^2
Micro-habitat - margins: type *(Micro-habitat - Margini: tipo)*	field	area around the plot (0.5 ha)	categorical
Micro-habitat - margins: limits *(Micro-habitat - Margini: limiti)*	field	area around the plot (0.5 ha)	categorical
Micro-habitat - margins: extension *(Micro-habitat - Margini: estensione)*	field	area around the plot (0.5 ha)	m
Micro-habitat - superficial water: type *(Micro-habitat - Acque superficiali: tipo)*	field	area around the plot (0.5 ha)	categorical
Fuel models *(Modelli di combustibile)*	field	plot	categorical
Dead fuel survey *(Rilievo della necromassa)*	field	plot	categorical
Phytosociological survey *(Rilevamento fitosociologico)*	field	area around the plot (0.5 ha)	categorical
Linear transect of mixed areas *(Rilievo delle aree miste)*	field	area around the plot (0.5 ha)	categorical
Tree limit *(Limite superiore del bosco)*	field	area around the plot (0.5 ha)	categorical
Wildlife track: type *(Tracce di fauna: tipo)*	field	plot	categorical
Wildlife track: diffusion *(Tracce di fauna: diffusione)*	field	plot	categorical

J) Non-wood goods and services

Attribute	Data source	Object	Measurement unit
Infrastructures and facilities *(Manufatti ed infrastrutture)*	field	area around the plot (0.5 ha)	categorical
Function of the forest area (*Funzione dell'area forestale*)	field	forest area	categorical
Land use restrictions (*Vincoli*)	field + forest service	forest area	categorical

6.1.2. List of derived attributes

C) Wood production

Attribute	measurement unit	input attributes
Single tree volume (V*olume legnoso individuale*	m^3	d.b.h., total tree height, tree species
Basal area (*Area basimetrica*)	m^2	d.b.h.

I) Attributes discribing forest ecosystems

Attribute	measurement unit	input attributes
Actual forest vegetation map *(Carta della copertura forestale)*	ha	Floristic composition, aerial infrared photo-interpretation, DEM, TM imagery classification
Dynamic vegetation series map *(Carta delle serie di vegetazione)*	ha	Actual vegetation map, physiological survey, land unit map, bio-climatic map

6.1.3 Measurements rules for measurable attributes

A) Geographic regions

No measurable attribute

C) Wood production

Diameter at breast height

a) Measured in 1.3 m height above ground, on slopes measured from uphill side. One reading, with calliper pointing the plot centre,
b) Minimum dbh: 4.5 cm
c) cm, recorded in 1-cm classes
d) rounded to the nearest (lower or upper) class limit
e) calliper
f) field assessment

Tree height

a) Length of tree from ground level to top of tree
b) No more than 5 trees for a single specie, not less than three trees for all the present species into the sample plot
c) 10 cm
d) rounded to the closest 10 cm class limit
e) Blume-Leiss or Suunto altimeter, Suunto clinometer
f) field assessment

Quality of stems

a) The stand within the sample plot is classified by two characters: the percent of trees interested by defects of the stems and number of defects
b) minimum dbh: 12.5 cm
c) - percent of trees: four classes : < 20%; 20 - 50% ; 50 - 80% ; > 80%
   - defects: three classes:
     - no defect, one defect, two defects:
     - no defect: whole intact stem or pright and regular stem without main branches for almost 2/3 of height
     - defects: stem moderately curved in the lower part or tapered stem, or stem with main branches or truncated over ¾ of height or up to 1.3 m; or uprooted, broken off stem with many dry branch
d) no rounded
e) ocular estimate
f) field assessment

Crown height and wide

a) Crown height is the length of tree from ground level to the crown point; crown wide is assessed measuring the maximum diameter of crown projection and its plumb-line
b) Is assessed the crown height and wide only for trees subject of the tree height survey
c) crown height: m; crown wide: cm
d) rounded to closest 1 cm or m class limit
e) ocular estimate
f) field assessment

Exploitation

a) Assessment of harvested plot in the last two years
b) no threshold values
c) year of exploitation; m3 for removals; stumpage value in Italian lire x 1000
d) rounded to closest 1% class limit
e) no instruments used
f) forest service information and field assessment

D) Site and soil

Elevation

a) measured in plot centre
b) no threshold values
c) 10 m
d) rounded to closest class limit
e) altimeter or reading of map
f) field assessment

Slope

a) The average slope of soil profile is measured on the maximum slope line through the plot centre, ranging two opposite stakes located uphill and downhill at 20 m distance from centre.
b) threshold values: 0% -
c) %,
d) rounded to closest 1% class limit
e) Suunto clinometer
f) field assessment

E) Forest structure

Cover: height of vegetation levels

a) The height of tree, bush and herbs vegetation levels covering respective levels (for trees are surveyed two strata: higher arboreal strata and the lowest one) on total area of sample plot
b) no threshold values
c) cm for shrub and herbs, m for trees
d) rounded to closest 1 cm or m class limit
e) ocular estimate
f) field assessment

Cover: percent of coverage on levels

a) The percentage of tree, shrub and herbs vegetation covering respective strata (for trees are surveyed two strata: higher arboreal stratum and the lowest one) on total area of sample plot is assessed
b) no threshold values
c) %
d) rounded to closest 1% class limit
e) ocular estimate
f) field assessment

Planting space

a) Average distance between trees measured in specialised forest stands only (Macro-category: Plantation forest)
b) No threshold
c) m
d) rounded to closest 1 m
e) tape
f) field assessment

Floristic composition

a) The species of tree, shrub and herbs vegetation present in respective strata (for trees are surveyed two strata: higher arboreal stratum and the lower one) and their coverage are assessed on total area of sample plot
b) No threshold
c) %, four classes: - < 20%, >21- <50%; > 51% - < 80%; > 81%
d) No rounded
e) Ocular estimate
f) Field assessment

Bush pattern

a) Trees and shrubs with dbh < 4.5 cm are assessed on a 10 or 20 m*1m transect, according to cover < 50% (10 m) or > 50% (20 m); total height and crown cover is assessed for each single tree or shrub
b) dbh < 4.5
c) dm
d) rounded to closest 1 dm
e) tape and calliper
f) field assessment

G) Forest Condition

Stand damages: diffusion

a) Per cent diffusion of trees interested by damages within the sample plot.
b) No threshold
c) %, three classes: - 20%, 21- 50%, 51% -
d) No rounded
e) Ocular estimate
f) Field assessment

I) Attributes describing forest ecosystems

Micro-habitat - clearing: extension

a) Evaluation of the clearing extension
b) maximum extension: 20,000 m2
c) nine classes:
$S < 1,000\ m^2$; $1,000\ m^2 < S < 2,500\ m^2$; $2,500\ m^2 < S < 5,000\ m^2$; $5,000\ m^2 < S < 7,500\ m^2$; $7,500\ m^2 < S < 10,000\ m^2$; $10,000\ m^2 < S < 12,500\ m^2$; $12,500\ m^2 < S < 15,000\ m^2$; $15,000\ m^2 < S < 17,500\ m^2$; $17,500\ m^2 < S < 20,000\ m^2$
d) No rounded
e) Ocular estimate
f) Field assessment

Micro-habitat - margins: extension

a) Evaluation of the clearing extension
b) maximum extension: 99 m
c) m, recorded in 1-m classes
d) If the extension is more than 99 m, then you must write 99
e) Ocular estimate or tape
f) Field assessment

6.1.4 Definitions for attributes on nominal or ordinal scale

B) Ownership

Ownership

a) The owner of the forest area under concern
b) 1) private ownership (person and society)
2) public ownership
3) municipality ownership
4) mountain communities ownership
5) region and autonomous provinces ownership
6) state ownership
7) uncertain ownership
c) information from local people and local forest service

C) Wood production

Crown class

a) Is assessed the crown class only for trees subject of the tree height survey
b) six classes:
1) predominant
2) dominant
3) subdominant
4) dominated
5) over-topped
6) single tree
c) field assessment
d) not relevant

D) Site and soil

Aspect

a) Is assessed the sample plot location in relation to the exposure

b)

code	Aspect	Degree	
		centesimal	sexagesimal
0	no exposure	-------------	-------------
1	N	375° - 24°	337° 30' - 22° 29'
2	N - E	24° - 74°	22°30' - 67° 29'
3	E	75° - 124°	67° 30' - 112°29'
4	S - E	125° - 174°	112°30' - 157°29'
5	S	175° - 224°	157°30' - 202° 29'
6	S - W	225° - 274°	202° 30' - 247° 29'
7	W	275° - 324°	247° 30' - 292° 30'
8	N - W	325° - 375°	292° 30' - 337° 30'

c) field assessment

d) not relevant

Location

a) The sample plot location in relation to the surrounding terrain morphology.

b) eight principal classes:

1) top-hills or ridge (with three sub-classes)
2) slopes (with three sub-classes)
3) lower part of slope (with four sub-classes)
4) bottom of the valley (with three sub-classes)
5) plains (minimum area: 80 ha) (with two sub-classes)
6) coasts (with five sub-classes)
7) water bodies (with five sub-classes)
8) other areas (with two sub-classes)

c) field assessment

d) not relevant

Stoniness

a) Is assessed considering the coverage and the distance among the stones

b) six classes

c) field assessment

d) not relevant

Rockiness

a) Is assessed considering the coverage and the distance among the rocks

b) six classes

c) field assessment

d) not relevant

Soil erosion and deposition

a) Is assessed the presence of soil erosion and deposition

b) For soil erosion are considered four classes:

- water erosion (with three sub-classes);
- wind erosion
- landslides (with two sub-classes
- anthropological erosion

For deposition are considered three classes

c) field assessment

d) not relevant

Humus

a) Classification of humus's type placed on sample plot

b) 1) Mor
2) Moder
3) Mull

c) field assessment

d) not relevant

Soil survey

a) Assessment of soil features
b) Many attributes are surveyed for each horizon assessed (i.e. depth, texture of soil and presence of stones, concretion, cutans, speckling, presence of roots, etc.)
c) field assessment
d) only the plot on the IFNI's grid (3*3 km)

E) Forest structure

Trace of former wood

a) Assessment of existing trace of former wood, differing form actual one by species composition and stand type
b) two parameters: species and stand type (two classes: coppice or high forest)
c) field assessment
d) not relevant

Inventory stand types

a) Assessment made considering cover, system and crop type and according with forest and other wooded land definitions (see chapter 2.1.5.)
b) More than 50 types of stand (see chapter 2.1.5.)
c) field assessment
d) not relevant

Stand development

a) Are assessed the different stages in the life of a stand
b) Several classes in connection of the type of stand
c) field assessment
d) not relevant

Origin of stand

a) Is assessed the origin of stand (natural succession, human activities or artificial regeneration)
b) Several classes in connection of the type of stand
c) field assessment
d) not relevant

Sylvicultural system

a) Is assessed the process whereby stand is tended, harvested and replaced
b) 1) Clear cutting
   2) Shelterwoods-Strip cutting system
   3) Group cutting system
   4) Selection cutting system
   5) Seed tree cutting method
   6) Selection cutting's felling
   7) Coppice selection system
   8) Sanitation cutting
c) field assessment
d) only for Managed forest and Plantation forest types

Conversion to high forest

a) Assessment of conversion system to high forest
b) two classes:
   - natural conversion
   - intensive standard system
c) field assessment
d) only for coppice stand

Tending of stands

a) Each type of tending is registered if made or not and the years in which was carried out.
b) The following types of tending are evaluated:
   - cleaning
   - thinning
   - pruning
   - sanitary treatment
   - soil fertilisation
   - soil cultivation
   - weeding, cutting of weed
   - irrigation
   - stone removal
c) field assessment
d) only forestry area

Age of stands

a) Four possible types of assessment could be chosen by surveyors in order to know the age class of the stand:
   - historical estimate of stand planting year
   - counting of branches in some trees (conifer only)

- counting of annual ring number on stumps in case of recent felled trees
- ocular estimate of age based on comparing the situation with other known

b) Classification is based on age classes divided into two categories regarding their gamic or agamic origin.
Coppice: six classes
a) 0 - 3 years
b) 3 - 10 years
c) 10 - 15 years
d) 15 - 20 years
e) 20 - 25 years
f) more than 25 years
High forest: seven classes:
a) 0 - 10 years
b) 10 - 20 years
c) 20 - 30 years
d) 30 - 50 years
e) 50 - 80 years
f) 80 - 120 years
g) more than 120 years

c) field assessment
d) not relevant

Cultivation status

a) Is assessed the presence of tending in specialised forest stands in the area around the plot (0.5 ha)
b) 13 types of tending
c) field assessment
d) only in specialised forest stands (Macro-category: Plantation forest

F) Regeneration

Condition

a) The natural or artificial tree vegetation with height over 30 cm (seedling) or less than 30 cm (effective seedling) and with dbh not more than 4.5 cm of the three main tree species is assessed.
b) three classes:
1) poor
2) good
3) excellent
c) field assessment
d) not relevant

Origin

a) The natural or artificial tree vegetation with height over 30 cm (seedling) or less than 30 cm (effective seedling) and with dbh not more than 4.5 cm of the three main tree species is assessed.
b) three classes:
1) natural
2) artificial
3) mixed
c) field assessment
d) not relevant

Distribution

a) The natural or artificial tree vegetation with height over 30 cm (seedling) or less than 30 cm (effective seedling) and with dbh not more than 4.5 cm of the three main tree species is assessed.
b) two classes: 1) continuous
2) by groups
c) field assessment
d) not relevant

Localisation

a) The natural or artificial tree vegetation with height over 30 cm (seedling) or less than 30 cm (effective seedling) and with dbh not more than 4.5 cm of the three main tree species is assessed.
b) five classes:
1) under the canopy
2) inner edge
3) outer edge
4) big hole
5) little hole
c) field assessment
d) not relevant

Transect of regeneration

a) The regeneration with height less than 30 cm of the tree species is assessed on a transect length 10 or 20 m (if the coverage of regeneration is < or > than 50%) and wide 1 m
b) tree species
c) field assessment
d) not relevant

G) Forest condition

Site degradation: type

a) Synthetic assessment of type of site degradation.
b) five classes:
   1) intensive use of path
   2) soil compassion
   3) animal ploughing (i.e. wild-boar)
   4) pulling litter
   5) alteration of filtering and downflow of water
c) field assessment
d) not relevant

Site degradation: origin

a) Synthetic assessment of origin of site degradation.
b) five classes:
   1) grazing
   2) excessive use by tourism
   3) infrastructures building
   4) irrational cutting of stand
   5) other origins
c) field assessment
d) not relevant

Site degradation: intensity

a) Synthetic assessment of intensity of site degradation.
b) three classes: 1) slight degradation
   2) middle-degradation
   3) heavy degradation
c) field assessment
d) not relevant

Stand damages: origin

a) Assessed the main origin of damage for trees inside the sample plot.
b) 14 types of origin (grazing, pests, fungi, fire, meteoric phenomenon, irrational cutting, etc.)
c) field assessment
d) only forest stands

Stand damages: intensity

a) Assessed the average intensity level of damaged trees inside the sample plot.
b) three classes:
   1) temporary damage
   2) permanent damage
   3) death-damage
c) field assessment
d) only forest stands

Stand degradation: distribution

a) Assessed the distribution into diameter classes of damaged trees inside the sample plot.
b) four classes:
   1) dbh < 7.5 cm
   2) 7.5 cm < dbh < 17.5 cm
   3) dbh > 17.5
   4) all dbh
c) field assessment
d) only forest stands

H) Accessibility and harvesting

Accessibility and harvesting: difference in height-distance between plot and road

a) Evaluation of difference in height-distance between the sample plot and the nearest road
b) six classes:
   a) distance < 1,000 m, difference in height < 100 m; road below sample plot;
   b) distance < 1,000 m, difference in height < 100 m; road above sample plot;
   c) 1,000 m < distance < 3000 m, 100 m < difference in height < 300 m; road below sample plot;
   d) 1,000 m < distance < 3,000 m, 100 m < difference in height < 300 m; road above sample plot;
   e) distance > 3,000 m, difference in height < 300 m; road above sample plot;
   f) distance > 3,000 m, difference in height < 300 m; road below sample plot;
c) field assessment
d) not relevant

Accessibility and harvesting: difference in height-distance between plot and skidding road/ path

a) Evaluation of difference in height-distance between the sample plot and the nearest and skidding road/path wide less than 2 m
b) three classes:
  a) distance < 100 m;
  b) 100 m < distance < 300 m
  c) distance > 300 m
c) field assessment
d) not relevant

Accessibility and harvesting: road classification

a) Description of the nearest road considered in the evaluation of difference in height-distance
b) a) principal road that can be used by lorries, not only for forestry use
  b) secondary road that can be used by lorries, not only for forestry use
  c) cart-way
  d) principal drag road
  e) secondary drag road
c) field assessment
d) not relevant

Accessibility and harvesting: distance covering on carriage road

a) Evaluation of eventual distance covering on carriage road to reach the principal road
b) three classes:
  a) distance < 1,000 m
  b) 1,000 m < distance < 3,000 m
  c) distance > 3,000 m
c) field assessment
d) not relevant

Accessibility and harvesting: improvements on existing road

a) Description of eventual needing improvements needing on the last 1,000 m of existing road
b) Ten types of protecting works (i.e. breast and retaining wall, blind drain, fascine, etc.)
c) field assessment
d) not relevant

Unevenness of soil morphology: types

a) Assessed the main obstacles that could obstruct the timber hauling
b) Four types:
  1) subsidence, ditch, pit,
  2) breast wall
  3) down-tree dead or alive with dbh > 17.5
  4) undergrowth high > 0.5 m but not practicable
c) field assessment
d) not relevant

Unevenness of soil morphology: intensity

a) Assessed the intensity of obstacles that could obstruct the timber hauling.
b) Two classes:: 1) moderate
  2) frequent
c) field assessment
d) not relevant

I) Attributes describing forest ecosystems

Micro-habitat - margins: type

a) Synthetic assessment of the closest different land use
b) Seven classes:
  a) forest
  b) shrub land
  c) grass land
  d) arable land
  e) water bodies - lake, river
  f) rocks
  g) clearing
c) field assessment
d) not relevant

Micro-habitat - margins: limits

a) Description of the shape of margin
b) Three classes:
  a) clear-cut
  b) gradual
  c) not evident
c) field assessment
d) not relevant

Micro-habitat - superficial water: type

a) Assessed the presence of superficial water around the plot

b) Seven classes:

a) temporary superficial stagnation
b) standing superficial stagnation
c) source
d) rill
e) torrent
f) pond
g) peatland

c) field assessment

d) not relevant

Fuel models

a) Identification of fuel models in order to assess the fire behavioural model

b) Four classes:

- pastureland (three sub-classes)
- bushland (four sub-classes)
- wood (three sub-classes)
- slash, lop and top on ground (three sub-classes)

c) field assessment

d) not relevant

Dead fuel survey

a) The dead fuel on ground is assessed on a linear transect length 20 m (Line Intersect Method)

b) four classes:

- $\varnothing < 0.5$ cm
- $0.5 < \varnothing < 2.5$ cm
- $2.5 < \varnothing < 7.5$ cm
- $\varnothing > 7.5$ cm

c) field assessment

d) not relevant

Phytosociological survey

a) Assessment of the species of tree, shrub and herbs vegetation present and the diffusion level of each one with Braun-Blanquet method (presence/abundance) in the area around the sample plot

b) six class of coverage: 5; 4; 3; 2; 1; +; r

c) field assessment

d) not relevant.

Linear transect of mixed areas

a) Assessment is made describing the elements of mixed areas and their width crossed by two ortogonal linear transect length 200 m with departure from centre plot

b) Woodland, bushland and herbs classes are the same as inventory stand type
Urban land (six classes)
Inland waters (six classes)
Farm Land (16 classes)
Rows and hedge with trees or bush (11 classes)

c) field assessment

d) not relevant.

Tree limit

a) Assessment of presence in the area around the plot (0.5 ha) of tree limit, and if is present discribing the type

b) three classes:

a) wood
b) lonely trees
c) lonely twisted trees by meteoric phenomenon

c) field assessment

d) not relevant

Wildlife track: type

a) Assessed the presence of wildlife track: within the plot

b) Five classes:

a) little cavity in trees
b) big cavity in trees
c) underground lair
d) perch of nocturnal bird
e) nest

c) field assessment

d) not relevant

Wildlife track: diffusion

a) Number of the presence of wildlife track: within the plot

b) Four classes: a) $0 < n° < 5$
b) $5 < n°$ 10
c) $10 < n° < 20$
d) $n° > 20$

c) field assessment

d) not relevant

J) Non-wood goods and services

Infrastructures and facilities

a) Assessment of the infrastructures and facilities around the plot
b) 23 types of infrastructures and facilities (i.e. recreation facility, fire line, dam, spring and drinking-trough, electric power line, parking, etc.)
c) field assessment
d) not relevant

Function of the forest area

a) The main functions of the forest area are given, in connection to goods or services derived.
b) A maximum of three functions, in order of decreasing importance, may be selected from the following list:
1) wood production
2) non-wood goods production (cork, pine-seeds, sweet chestnut)
3) direct protection (against avalanches, stone-falls towards roads or human settlements)
4) indirect protection (against soil erosion, floods, etc.
5) naturalist (parks, etc.)
6) excursionist
7) historical and cultural prominence
8) landscape
9) recreation
c) field assessment
d) not relevant

Land use restrictions

a) Assessed if the area has particular land use restrictions, given by specific laws.
b) More constraints should be registered by a list of 13 classes (military, protective, scenic, management, naturalist constraints, hunting reserve, etc.).
c) forest service information and field assessment
d) not relevant

6.1.5. Forest area definition and definition of "other wooded land"

In the forest context nomenclature were of a traditional kind and so they were not operative for the necessity of the definition of non-canonical coenosis from silvicultural aspect.

In the botanical field, the trend was that of producing nomenclature systems linked to physiognomic type only on Floristic base and without structural data, or particularly sintaxonomic on physiological base.

Because of the new emergent situation of multi-functional and intersectoriality of the natural and semi-natural vegetation, nomenclature systems are now organised in an integrated form among different subjects and answers to the requirements of the various instruments and methodology, and among these the necessity to use the typology on a large scale.

In fact in current experiences, in order to achieve requirements as we have already said, there is a particular attention for the hierarchic structure and for the collaboration among Botanist, Forest Engineers and Remote Sensing and Mapping specialists.

The hierarchic structure of the Nomenclature allows flexibility of movement towards superior hierarchic levels for the classes that are difficult to define with the present technologies.

According to multipurpose aims land use classes studied areas are forests and woodlands, bushlands, range lands and mixed areas.

This last class is kept for the Italian territory characteristics that has an alternation of natural and semi-natural environments with different ecosystems (grasslands, woods) and with different utilisation of the agriculture land-use.

Definitions of assessed land use classes are (The minimum size of a forest area is 5,000 m^2)

1) Trees: tree coverage > 20% in polygons with surface ≥5,000 m^2

2) Bush: bush coverage > 40% with tree coverage < 20% in polygons with surface ≥5,000 m^2

3) Herbs: permanent meadows, pasture, fallow with herbs coverage > 20%, trees coverage < 20% and bush < 40% in polygons with surface ≥ 5,000 m^2

4) Mixed areas: a mosaic of classes 1/2/3 in small areas (< 5,000m^2) in polygons with surface not less than 10,000 m^2

5) Barren land: areas with herbs coverage < 20%, trees coverage < 20% and bush < 40% in polygons with surface ≥5,000 m^2

6) Included areas: surfaces with an homogenous internal land use but not homogenous as regards the surrounding land use with an area < 5,000 m^2 and > 500 m^2 for classes 1/2/3 and 40,000 m^2 for barren or arable land

For this reason it is often necessary to sample even crop strata that, because of the desertion of the marginal land; there is also a large diffusion of natural systems of new formation and successions (for example most of all in olive trees, grasslands and arable hill-lands), in addition to the wide presence of woodland environments out of the forest.

All the unites we have described can be included into the Dynamic Vegetation Series identified through physiological surveys; thanks to these unites we can define affinities and the sintaxonomic position for the different typologies, in a different level of classification (association, alliance, order, class).

The proposed nomenclature, surely perfectible, has the characteristics of completeness and simplicity and more over it has the advantage of being 'open' and this means the possibility of incidental addition of new typology unites, if they appear as a consequence of further researches and/or technological improvement.

Inventory stand types

Woodland and bushland

Macro-category	Category	Type	Subtype	Code
Unmanaged forest	Woodland	Consolidated	Undisturbed	NAC1
			Other	NAC2
		New stands		NAN0
	Bushland	Consolidated	Thicket communities	NBC1
			Edge mantle	NBC2
			Successional bush	NBC3
		New stands		NBN0
Managed forest	High forest	Even aged		GFM0
		Double-storey		GFB0
		Storied/uneven aged	All aged	GFP1
			Groups	GFP2
		Articulated (different groups uneven aged with a wide between 1,500 and 5,000 m^2)		GFA0
		Confused (more than 2 types in 0.5 ha)		GFC0
		Conversion derived		GFT0
	Coppice	On conversion		GCI0
		Without standards	Single coppice system	GCS1
			Coppice selection system	GCS2
		Coppice of two rotations	By broad-leaved	GCM1
			By conifers	GCM2
			Mixed	GCM3
		With standards		GCC0
		Under conifers high forest		GCF0

Macro-category	Category	Type	Subtype	
Plantation forest	High forest	Wood production	Poplar	CFL1
			Timber production	CFL2
		No wood production	Cork-tree wood	CFN1
			Chestnut orchard	CFN2
			Truffle wood	CFN3
			Other production	CFN4
	Coppice	Wood production	Timber production	CCL1
			Other production	CCL2
		No wood production	Hazel-nut orchard	CCN1
			Other production	CCN2
	Parks and gardens			CPX0
	Nurseries			CVX0

Herbaceous associations

Macro-category	Category	Type	Subtype	
Grassland	Pastureland	Simple	Natural	PPS1
			Improved	PPS2
		with shrubs	Natural	PPA1
			Improved	PPA2
		with trees	Natural	PPB1
			Improved	PPB2
	Meadow	Simple	Natural	PRS1
			Improved	PRS2
		with shrubs	Natural	PRA1
			Improved	PRA2
		with trees	Natural	PRB1
			Improved	PRB2
Uncultivated		Simple		IXS0
		with shrubs		IXA0
		with trees		IXB0

Mixed areas

Macro-category	Category	Type	Subtype	
Natural and semi-natural				MNX0
Agro-forestry				MAX0
With urban elements				MUX0

Particular land use

Macro-category	Category	Type	Subtype	
Nude and barren land		Littoral		NXL0
		Dune		NXP0
		Riparian		NXR0
		Rocks		NXF0
		Scree, landslide, gravel		NXM0
		Wetland, petland		NXT0
		Alpine, glacial		NXA0
		Other sites		NXS0

Included areas

Macro-category	Category	Type	Subtype
Included areas	Natural and semi-natural	6 classes	
	Cultivated	13 classes	
	Permanent barren land	8 classes	

6.2. DATA SOURCES

A) Field data

Type of stand: assessed on a visual area of approximately 0.5 ha.

C) aerial photography

- "Volo Italia" 1990, b/w at scale 1:75000, for land use pre-stratification (from CGR-*Compagnia Generale Riprese Aeree*, Parma, Italy); total coverage; orthophotomap at 1:50000; Wild Aviopret stereoscope
- CIR 1:33000 1989, for forest mapping, (from IGMI- *Istituto Geografico Militare Italiano*, Florence, Italy); total coverage; Wild Aviopret stereoscope

D) spaceborne or airborne digital remote sensing

- Landsat 5TM, system corrected (*) and raw data, seven bands, 30 m of ground resolution.

  date of images: multi-temporal data

Frame	Date
195-29*	25/6/91
194-29*	5/8/91
193-29*	31/7/92
193/29	4/3/90
193/29	3/8/93
193/29	4/7/91
193/29	16/2/92
193/29	26/10/92
193/30	6/11/90
194/3029	17/4/92
193/30	10/8/93

- applied techniques: radiometric correction with LOWTRAN6 software and Forester algorithm
  geometric correction with ERDAS software for system corrected images and Precision Correction System software for raw data
- classification: a KAS (Knowledge Aquisition System) has been prepared using the Forest Information System; Bayesian thresholds and non parametric supervised classification algorithm modifying path of Kidmore and Turner has been applied

E) maps

- Plot-centre are located on topographic detailed maps of Liguria Region Techinical Cartography (C.T.R., 1991) at scale 1:10000.
- Topographic maps ,at scale 1:25000 and 1:50000, variable dates (from IGMI- *Istituto Geografico Militare Italiano*, Florence, Italy) has been used.
- Geologic map of Italy at scale 1:100000, variable dates (from SGI- *Servizio Geologico Italiano, Rome, Italy*),
- Various land use, vegetation and forest maps, at different scale for local collection of the ground truth

F) other geo-referenced data

- Available digital terrain model from Liguria Region (from C.T.R 1:10000), TIFF format
- Available historical climatic data from Ministry of Public Works (1959-1985), ASCII format

6.3. ASSESSMENT TECHNIQUES

6.3.1. Sampling frame

The sampling frame of the Multipurpose Forest Inventory of Liguria Region is given by the forest area definition.

The entire region is covered by the assessment. Some sample points (roughly 1.5 %) were not classified because not accessible due to the terrain condition as well not evaluative at distance. They were excluded from the assessment.

6.3.2. Sampling units

Field assessment: concentric fixed area, circular plot sizing 600 m^2: smaller circle of 200 m^2 for all the trees (dbh > 4.5 cm), medium circle (400 m^2) for trees with dbh > 12.5 cm, larger circle of 600 m^2 for trees with dbh > 17.5 cm.

Slope correction is done by increasing plot radii according to slope. All plots permanently monumented and documented (but marked hidden without giving an evident impact to the plot location). Azimuth and distance of reference marks from plot centre are recorded. The plot are systematically distributed in a 1*1 km grid. The plot centre is located with GPS and the plot location coordinates are recorded by it.

6.3.3. Sampling designs

The sampling design applied in the Multipurpose Forest Inventory of Liguria Region is double-sampling for stratification. In the first phase (assessment level) systematically distributed photo-plots are interpreted, a forest/non-forest decision is made and strata assigned to the forested photo-plots. In the second phase a systematically distributed sub-sample of field plots is assessed.

6.3.4. Techniques and methods for combination of data sources

Every data are collected or derive from field assessment. Therefore no combination of data sources is applied.

6.3.5. Sampling fraction

Data source and sampling unit	proportion of forested area covered by sample	represented mean area per sampling unit
field assessment, fixed area sample plots	0.0004	150 ha

6.3.6. Temporal aspects

Inventory cycle	time period of data assessment	publication of results	time period between assessment	reference date
1st IFMR	1993	1995	-	1 August 1993

6.3.7. Data capturing techniques in the field

Data is recorded on several paper forms filled in separately for different survey and types of land-use: two for topographic location (A.1-A.2), one general for every grid-point describing the site of the sample plot (B), two for stand parameters, one for Floristic survey (D), one for dead fuel survey, one other forest parameters (data about recent logging, tending of stands, parameter of specialised stands, etc.) (F), three for the physiological survey (G.1-G.2-G.3), one for tree height survey (H), two for shrub and renovation pattern survey (I.1-I.2), two for mixed area survey (L.1-L.2) and two for soil survey (M.1-M.2).

6.4. DATA STORAGE AND ANALYSIS

6.4.1. Data storage and ownership

The data of the assessment are stored in a database system in PC by the Direzione Generale delle Risorse Forestali, Montane e Idriche of Ministero delle Risorse Agricole, Forestali e Alimentari, Via Carducci 5, Roma. The contact person for information is Ing. A. Lalle - Tel.+39-6-46657082

6.4.2. Database system used

The data are stored in a relational ORACLE and DBIII database.

6.4.3. Data-bank design

Data-bank is based on ARCVIEW2 and ACCESS

6.4.4. Update level

The reference date is January, 1, 1994. The data assessed are used as they are without any specific updating.

6.4.5. Description of statistical procedures used to analyse data including procedures for sampling error estimation

a) area estimation:

The forest area is assessed on the forested photo plot. A 1*1 km systematic point grid is laid over the entire region and for each point a forest/ non forest decision is made.

The estimation of the proportion of forested area is done according to

$$p = \frac{a}{n}$$

$$v(p) = s_p^{\,2} = \frac{pq}{n}$$

and the standard error is

$$s_p = \sqrt{v(p)}$$

where
p = proportion of forest land
q = 1- p = proportion of non-forested land
$v(p)$ = variance of p
s_p = standard error of p
n = total number of points on the point grid
a = number of forested points on the point grid

As a single forest point represents 150 ha of forest, the area estimation is made by multiplying the number of points of the desired unit by 150:

$$A_w = n_w 150 = n_w \frac{A}{n}$$

with variance $v(A_w)$ and standard error $s(A_w)$

$$v(A_w) = A^2 s_p^{\,2}$$

$$s(A_w) = \sqrt{v(A_w)}$$

The per cent standard error is consequently

$$s(A_w)\% = \frac{100 s(A_w)}{p}$$

b) aggregation of tree and plot data

The aggregation of single tree data is done by weighting each single tree attribute Y_{ij} by

$$w_{ij} = \frac{A}{a_{ij}} = \frac{10{,}000\text{m}^2}{600\text{m}^2} \left(\text{or} \frac{10{,}000}{400} \text{or} \frac{10{,}000}{200}\right)$$

where ij stands for tree i on plot j.

The plot expansion factor has to be adjusted for plot not lying entirely in the forested area or for plot interesting more than one stratum.

$$Y_i = \sum Y_{ij} w_{ij}$$

Once the single tree attributes have been related to unit area by multiplication with the plot expansion factor, the total values for plot i, Y_i, can be calculated by summing the individual single tree attributes.

c) estimation of total values

In general the total value of attribute Y for a stratum h (of area $A_{h)}$ is given by summing the attribute value for each plot:

$$Y_h = 150 \sum \frac{Y_i}{a_i} = A_h \ddot{y}$$

where $\ddot{y}$ = mean value of total value Y_h

Since both, A_h and $\ddot{y}$, are affected respectively by the standard errors s_{Ah} and $s_{\ddot{y}}$, we get an error propagation evaluable with the expression

$$s(Y_h) = {A_h}^2 {s_{Ah}}^2 + \ddot{y}^2 {s_{\ddot{y}}}^2$$

$$s(Y_h) = \sqrt{A_h^2 s_{Ah}^2 + \ddot{y}^2 s_{\ddot{y}}^2}$$

d) estimation of ratios

//

e) sampling at forest edge

For area estimation each plot centre located into the forest represents 1,486 ha of forest, also if the plot is not entirely covered by forest.

For attributes related to the area (basal area, volume) only the part of plot interested by the stratum is computed for calculating mean and total values.

f) estimation of growth and growth components (mortality, cut, ingrowth)

//

g) allocation of stand and area related data to single sample plots/sample points.

//

h) hierarchy of analysis: how are sub-units treated?

//

6.4.6. Software applied

The software applied for the analysis and calculation procedure was ORACLE and ACCESS.

6.4.7. Hardware applied

The ORACLE database is installed under UNIX on a SUN-Server. ACCESS is installed on PC. The analysis is done on SUN SPARC 20 workstations under UNIX.

6.4.8. Availability of data (raw and aggregated data)

The IFMR data are stored in a relational database on a PC based on ACCESS. Requests of raw data and derived data by outside persons have to be addressed to the Direzione Generale delle Risorse Forestali, Montane e Idriche of Ministero delle Risorse Agricole, Forestali e Alimentari, Via Carducci 5, Roma.

6.4.9. Sub-units (strata) available

Results are obtained for entire Liguria Region and some attributes also for the 4 Provinces and also for all the III order watersheds. In addition other results for individual Provinces can be provided.

6.4.10. Links to other information sources

//

6.5. RELIABILITY OF DATA

6.5.1. Check assessment

a) Field plots:
Check assessment are carried out during the field inventory.

About 10 % of the field plots are visited by check crews either together with the field crews or after the field crews had assessed them. The organisation and the analysis of the check assessments are done at the main office. The field plots to be checked are randomly selected form the list of already assessed plots. Only qualitative attributes are assessed a second time independently from the assessment of the regular field crews, i.e. no data recorded by the field crews are available for the check crews. The measurement procedures for locating the plot centre are repeated. For each field crew an identical proportion of plots is checked. Maps:

Vegetation map

Check assessment of Vegetation maps has been carried out after the preliminary mapping. The cross control has been organised as in the CORINE LAND COVER PROJECT

- random extraction of polygons of a number of polygons proportional to total area of each class;

- superimposition for each polygon of a 500m*500 m grid, dividing internal and border points;
- verification in the field;
- calculation of accuracy index.

b) The check assessments are used for the following purposes:

- feed-back to field crews to improve the data quality
- the results will not be used to set up error budgets and to quantify the total error.

6.5.2. Error budgets

Only sampling errors was calculated for the publication of inventory results.

6.5.3. Procedures for consistency checks of data

Data assessed in the field:

A field assessment was carried out and forms was filled with the data.

The data were than put into a computer and after editing the data a consistency check was done. First was checked if the correct type of forms were fill ed and than if data were missing in the different forms filled in connection to forest type. In some cases a forcing of data was provided. For some parameters a minimum and maximum value was given and checked and for many attributes was done the plausibility tests. Later a cross check among different forms and some parameters was carried out. For plots with wrong or lacking data, that could not be corrected or forced, a new field assessment was provided.

Vegetation Map:

After verification of accuracy the map has been modified for that classes and areas which accuracy was not good.

6.6. MODELS

6.6.1. Volume functions

a) outline of the model

The tree volume is estimated using dbh and tree height as input variables. Volume functions of following type was drawn up

$v = f_j(d, h)$

where
v = tree volume of a species j
d = dbh
h = total tree height

The dbh is measured for each tree up to 5 cm, while the height value may be derived by height curve. In general height curve were made for some groups of main species and main groups of inventory stand type. In some cases of mixed stands of coniferous and deciduous two height functions were derived.

The IFNI's two entry volume tables (and functions) were used for the IFMR: nine for high forest and five for coppice species or species-groups.

For coppice forest and high forest the functions cover the entire diametric range. The volume of trees with dbh < 17.5 cm is estimated by one independent variable volume functions, as

$v = f_j(d)$

which links two volume values corresponding to the diametric limits of 3 - 18 cm. The lower value is fixed to 0.002 m^3 (average volume of tree with dbh = 3 cm). The upper value of anchorage is given by the

$v = f_j(d, h)$

where d = 18 cm and h is the height given by the height curve in correspondence to that diameter. If no height curve is available the height value is derived by a linear estimated proportion based on the average dbh and height values of trees with dbh less than 18 cm.

b) overview of prediction errors

//

c) data material for derivation of model

For IFMR were used the INFI volume functions.

d) methods applied to validate the model

//

6.6.2. Assortments

//

6.6.3. Growth components

//

6.6.4. Potential yield

//

6.6.5. Forest functions

//

6.6.6. Other models applied

//

6.7. INVENTORY REPORTS

6.7.1. List of published reports and media for dissemination of inventory results

Year of publication	Citation	Language	Dissemination
1993	Ministry of Agriculture - National Forestry Corp - Consortium of Agrisiel, Aquater, Fisia, Italeco and Nomisma: *Carta Forestale d'Italia 1:50.000 e Sistema Informativo Geografico Forestale della Regione Liguria: manuale per le squadre di rilevamento*	Italian	printed
1995	Ministry of Agriculture - National Forestry Corp - Consortium of Agrisiel, Aquater, Fisia, Italeco and Nomisma: *Carta Forestale d'Italia 1:50.000 e Sistema Informativo Geografico Forestale della Regione Liguria: relazione finale*	Italian	printed

6.7.2. List of contents of report and update level

Reference of latest report:

Ministry of Agriculture - National Forestry Corp - Consortium of Agrisiel, Aquater, Fisia, Italeco and Nomisma, 1995. *Carta Forestale d'Italia 1:50.000 e Sistema Informativo Geografico Forestale della Regione Liguria: relazione finale*

Update level of inventory data used in this report:
Field data: 1993-1995

Content:

Chapter	Title	Number of pages
	Introduction	2
	Part One	
	Work organisation	1
	Methodological introduction	7

Part Two	
A short account of historical and cultural aspects of ligurian woods	25
Part three	
Methodology of survey of Multipurpose Forest Inventory	51
Part four	
Actual forest vegetation map	90
Part five	
Analysis of vegetation and land use alteration	59
Forest Health Monitoring	17
Stumpage evaluation	27
Fire risk	15
Quality of habitat evaluation	15
Soil erosion protection	15
Carbon balance of ligurian forest ecosystems	7
Economical evaluation of wood production and non-wood goodsand services of ligurian woods	149
Disposition for recreational use	5
Silvicultural operations and objective	15
Part six	
HW/SW prototype and adjournment system	9

Thematic maps of inventory and other analysis and studies results

Actual forest vegetation map at scale 1:50000
Dynamic vegetation series map at scale 1:50000
Stumpage value map at scale 1:250000
Fire risk map at scale 1:250000
Disposition for recreational use map at scale 1:250000
Fuel model map at scale 1:250000
Quality of habitat map at scale 1:250000
Silvicultural operations and objective map at scale 1:250000

6.7.3. Users of the results

//

6.8. FUTURE DEVELOPMENT AND IMPROVEMENT PLANS

Methodology has been assessed for a continuous updating: each five years for maps and ten years for the Forest Multipurpose Inventory. For what concern fire assessment it is planned to have a yearly control and mapping of the burned areas with satellite imagery monitoring.

6.9. MISCELLANEOUS

Special investigations have been carried out in the context of the Forest Map 1:50000, GIS, Multipurpose Forest Inventory of Liguria Region project.

- Forest Health Monitoring has been conducted across a portion territory of Liguria Region during 1993. Geocoded permanent plot on PPS photographic transects on C.I.R. Aerial photographs (1:8000) have been assessed.
- Questionnaires on recreational value of Liguria forests and data processing
- Biomass assessment: Biomass functions derived from crown diameter and total tree/shrub height for CO_2 content

REFERENCES

Dacquino,C., Lozupone, G. & Marchetti, M., 1995. Nomenclatura e mappatura dei sistemi forestali da dati telerilevati: un confronto quantitativo nel bacino dell'Aveto (Appennino Ligure). Atti del Congresso A.I.T. 1995. Chieri (TO)

Lalle, A. & Marchetti, M., 1994. Integrazione di tecniche di telerilevamento nella realizzazione della Carta Forestale d'Italia. Atti del Congresso dell'A.I.T. "Il telerilevamento per lo studio e la pianificazione forestale". Bressanone (BZ), 1-2 Dicembre 1994.

Marchetti, M., Raggi, L. & Garavoglia, S., 1993. Metodi di rilievo per un inventario multiobiettivo dei boschi e di altre aree naturali e seminaturali. Atti del seminario di studi curato dall'UNIF: "Ricerca ed esperienze nella pianificazione multifunzionale del bosco", 23-24 Novembre 1993, Centro Ricerche ENEA Brasimone (BO): pp. 136-155.

Marchetti, M., 1994. Georeferenced Forest Informations for Italy. Proceedings of the Intenational workshop "Designing a system of nomenclature for European Forest Mapping", European Forest Institute, Joensuu (Finland), 13-15 June 1994: pp. 241-255.

7. TOSCANA REGIONAL FOREST INVENTORY
(IFT - INVENTARIO FORESTALE DELLA TOSCANA)

7.1. NOMENCLATURE

7.1.1. List of attributes directly assessed

A) Geographic regions

Attribute	Data source	Object	Measurement unit
Province *(Provincia)*	maps	sampling unit	categorical
Municipality Association *(Associazione Intercomunale)*	maps	sampling unit	categorical
Municipality *(Comune)*	maps	sampling unit	categorical

B) Ownership

Attribute	Data source	Object	Measurement unit
Ownership *(Proprietà)*	land register	sampling unit	categorical

C) Wood production

Attribute	Data source	Object	Measurement unit
Number of trees selected in the point sample *(Numero di piante del campione relascopico)*	field assessment	sampling unit	integer
Diameter at breast height *(Diametro a petto d'uomo)*	field assessment	tree	cm
Total height *(Altezza dendrometrica)*	field assessment	tree (sub-sample)	m
Crown breadth *(Larghezza della chioma)*	field assessment	tree (sub-sample)	categorical
Crown shape *(Conformazione-Morfologia della chioma)*	field assessment	tree (sub-sample)	categorical
Stem quality *(Conformazione-Morfologia del fusto)*	field assessment	tree (sub-sample)	categorical
Bole height *(Altezza del livello inferiore della chioma)*	field assessment	tree (sub-sample)	m

Intermediate diameters *(Stima degli assortimenti legnosi)*	field assessment	tree (sub-sample)	cm
Logs quality *(Classificazione dei toppi)*	field assessment	tree (sub-sample)	categorical
Annual radial increment *(Incremento diametrico annuo)*	field assessment	tree (sub-sample)	mm

D) Site and soil

Attribute	Data source	Object	Measurement unit
Elevation *(Quota)*	map	sampling unit	m
Location *(Forma topografica)*	field assessment	sampling unit	categorical
Aspect *(Esposizione)*	field assessment, map	sampling unit	categorical
Slope *(Pendenza)*	field assessment	sampling unit	percent
Terrain pattern *(Caratteristiche del terreno)*	field assessment	sampling unit	categorical
Rock outcrops *(Rocciosità)*	field assessment	sampling unit	categorical
Stoniness *(Pietrosità)*	field assessment	sampling unit	categorical
Undergrowth thickness *(Sottobosco arbustivo)*	field assessment	sampling unit	categorical
Soil stability *(Assetto del terreno)*	field assessment	sampling unit	categorical

E) Forest structure

Attribute	Data source	Object	Measurement unit
Canopy cover *(Grado di copertura)*	aerial photographs	sampling unit	categorical
Forest type *(Tipologia forestale)*	aerial photographs, field assessment	sampling unit	categorical
Stand origin *(Origine del soprassuolo)*	field assessment	sampling unit	categorical
Silvicultural system *(Forma di governo e di trattamento)*	field assessment	sampling unit	categorical
Stand development and structure *(Descrizione strutturale)*	field assessment	sampling unit	categorical

Top tree age *(Eta dell'albero dominante)*	field assessment	tree (increment core)	integer
Stand age *(Eta del bosco)*	field assessment	sampling unit	integer
Tree species *(Specie)*	field assessment	sampling unit	categorical
Social standing *(Posizione sociale)*	field assessment	tree (sub-sample)	categorical
Trees spatial distribution *(Distribuzione spaziale)*	field assessment	tree	degree and cm

G) Forest condition

Attribute	Data Source	Object	Measurement unit
Tree health status *(Stato vegetativo)*	field assessment	tree (sub-sample)	categorical

H) Accessibility and harvesting

Attribute	Data source	Object	Measurement unit
Ground roughness *(Asperità del terreno)*	field assessment	sampling unit	categorical
Lorry-road network *(Strade camionabili)*	field assessment	sampling unit	categorical
Forest-road network *(Strade forestali)*	field assessment	sampling unit	categorical

K) Miscellaneous

Attribute	Data source	Object	Measurement unit
Sample plot number *(Coordinate UTM dell'area di saggio principale)*	map	sampling unit	ordinal

7.1.2. List of derived attributes

C) Wood production

Attribute	Measurement unit	Input attributes
Basal area per hectare *(Area basimetrica per ettaro)*	m^2	BAF, no. of trees selected in the point sample, tree species
Number of stems per hectare *(Numero di piante per ettaro)*	integer	basal area/ha, dbh, tree species
Mean diameter *(Diametro medio)*	cm	basal area, no. of stems, tree species
Mean height *(Altezza media)*	m	mean diameter, tree heights, tree species
Stem total volume *(Volume del fusto intero)*	m^3	intermediate diameters, log lengths
Tree volume per hectare *(Volume per ettaro)*	m^3	no. of stems/ha, d.b.h., h, tree species
Current diametric increment *(Incremento corrente diametrico)*	cm	mean increment on cores for 5-years classes, tree species, d.b.h., top height

E) Forest structure

Attribute	Measurement unit	Input attributes
Forest composition *(Composizione specifica)*	categorical	species, basal area, no. of trees

7.1.3. Measurement rules for measurable attributes

C) Wood production

Diameter at breast height

a) measured in 1.3 m height above ground, on slopes measured from uphill side. One reading with calliper pointing to plot centre. Two readings for irregular stems.
b) minimum dbh: 2.5 cm
c) cm, recorded every single cm
d) rounded off and up to closest cm
e) calliper
f) field assessment

Total height

a) length of tree from ground level to top of tree.
b) no threshold
c) m
d) rounded to closest 0.5 m
e) Suunto, Haga, Blume-Leiss hypsometer, Bitterlich relascope
f) field assessment: ocular estimate on enumerated trees, hypsometer for 3P trees, tape for felled trees

Bole height

a) distance between ground level and the origin of the lowest crown-forming branches (crown point)
b) no threshold
c) m
d) rounded to closest 0.5 m
e) Suunto, Haga, Blume-Leiss hypsometer, Bitterlich relascope
f) field assessment

Intermediate diameters

a) on slopes measured from uphill side with calliper pointing to plot centre and perpendicular to the stem. Measured in 2, 6, 10, 14, 18m.....height above ground, e.g. in middle height of log sections of equal length (4 m).
b) minimum diameter: 20 cm
c) cm
d) rounded to closest cm
e) Finnish calliper
f) field assessment

Annual radial increment

a) measured on three increment cores extracted at d.b.h. on standing trees at azimuth 120° each other; number of rings in the last cm; the last increment ring is not considered
b) minimum d.b.h.: 12.5 cm
c) integer
d) no rounding rules
e) magnifying glass, Pressler increment borer
f) field assessment,

D) Site and soil

Elevation

a) assessed on the map on the grid point as altitude
b) no threshold
c) m
d) rounded to closest m
e) ocular estimate
f) 1:25000 topographic map

Slope

a) measured from the plot centre along the ruling gradient line, average of uphill and downhill values
b) 5°
c) degree
d) rounded to closest degree
e) clisimeter
f) field assessment

E) Forest structure

Top tree age

a) measured on increment cores extracted on top tree in d.b.h.; assessed counting increment rings; last increment ring has not been considered
b) no threshold
c) years
d) no rounding rules
e) ocular estimate (magnifying glass)
f) field assessment

Stand age

a) assessed on evenaged stands:
- counting increment rings on increment cores extracted on 3-4 standing trees in d.b.h.;
- counting increment rings on stumps of recent felled trees;
- counting increment rings on felled trees (d.b.h.<15cm);
- counting branches in young coniferous stands;
- local information about last cut or plantation;
b) no threshold
c) years
d) no rounding rules
e) ocular estimate (magnifying glass, calliper)
f) field assessment

Trees spatial distribution

a) horizontal distance and azimuth of enumerated trees; measured from plot centre
b) no threshold
c) distance in m, azimuth in degrees
d) rounded to closet cm and degree
e) tape, compass, tripod
f) field assessment

K) Miscellaneous

Sample plot number
a) UTM coordinates of the sample plot
b) no threshold
c) m
d) no rounding rules
e) map
f) map

7.1.4. Definitions for attributes on nominal or ordinal scale

A) Geographic regions

Province
a) administrative unit joining several closed Municipalities
b) 4 Provinces
c) map
Municipality Association
a) administrative unit joining several closed municipalities;
b) 32 Municipality Associations
c) map

Municipality
a) administrative unit
b) data not available
c) map

B) Ownership

Ownership
a) owner of the forest area under concern
b) seven classes
c) land register, map

C) Wood production

Number of trees selected in the point sample
a) number of trees enumerated with the critical angle instrument (Bitterlich relascope) which permits the observer to decide to include or exclude the tree. BAF = 2.
b) no categories
c) field assessment
d) minimum d.b.h.: 2.5 cm. Starting from North in clockwise direction; trees with a breast-height diameter larger than the fixed angle chord corresponding to Basal Area Factor = 2 are tallied "in"; mean diameter and distance of borderline and hidden trees are checked with tape and calliper.

Crown breadth
a) crown breadth related to crown height
b) six categories
c) field assessment, ocular estimate
d) no remarks

Crown shape
a) development in height and width of the crown related to the total tree height
b) six categories according to crown point and size
c) field assessment, ocular estimate
d) no remarks

Stem quality
a) commercial value of the stem
b) five categories
c) field assessment, ocular estimate
d) no remarks

Logs quality
a) health status, shape and amount of branching of each single 4 m length log
b) three categories and ten sub-categories
c) field assessment, ocular estimate
d) no remarks

D) Site and soil

Location
a) morphological site
b) six categories and 29 sub-categories
c) field assessment
d) no remarks

Aspect
a) direction of exposure of the slope
b) nine categories
c) map, field assessment
d) "all" is used when the slope is less than 5%

Terrain pattern
a) water content and dead litter
b) eight categories
c) field assessment
d) no remarks

Rock outcrops
a) measured as a percent out of the interpretation area
b) four categories
c) field assessment
d) > 50 cm

Stoniness
a) measured as a percent out of the interpretation area
b) four categories
c) field assessment
d) < 50 cm

Undergrowth thickness
a) presence of shrub species with height > 1m and unable to grow over 5 m, measured as a percent out of the interpretation area
b) four categories
c) field assessment
d) seedlings and coppice shoots are not included

Soil stability
a) susceptibility to erosion, landslides, etc..
b) eight main classes and 22 sub-classes
c) field assessment
d) no remarks

E) Forest structure

Canopy cover
a) tree species cover on the ground
b) three categories
c) aerial photographs
d) no remarks

Forest type
a) the forest types are defined on the basis of latitudinal belt, coniferous or broad-leaved species, and stand height
b) seven main categories, sub-divided with hierarchical classification in eight groups and 36 types
c) field assessment, aerial photographs
d) no remarks

Stand origin
a) according to the type of regeneration generating the stand
b) six categories
c) field assessment
d) no remarks

Silvicultural system
a) the silvicultural system applied and the stand development stage
b) three categories:
1) coppice (six sub-classes)
2) high forest (six sub-classes)
5) irregular stands (12 sub-classes)
c) field assessment
d) no remarks

Stand development and structure
a) stand development stage and crown storeys distribution
b) four categories and 20 sub-classes
c) field assessment
d) no remarks

Tree species
a) Scientific and Italian name of selected list of tree and major shrub species
b) 67 categories, some of which grouping more than one species
c) field assessment
d) no remarks

Social standing
a) the way the individual tree growth is performing in comparison with the closest trees and the stand development stage
b) six categories for even-aged stands; six categories for uneven-aged stands
c) field assessment
d) no remarks

G) Forest condition

Tree health status
a) tree defects or damages
b) six classes
c) field assessment
d) no remarks

H) Accessibility and harvesting

Ground roughness
a) frequency of obstacles according to their relief amplitude and distance
b) four classes
c) field assessment
d) no remarks

Lorry-road network
a) vertical descent and horizontal distance between sample plot and closest lorry-road
b) three classes
c) field assessment
d) roads with slope <15%, width > 4 m

Forest-road network
a) horizontal distance between sample plot and closest forest road
b) six classes
c) field assessment
d) roads with slope <20%, width >3 m

7.1.5. Forest area definition and definition of "other wooded land"

Forest area is defined as a land under natural or planted stands of trees, whether productive or not, with tree crown cover of more than about 10%, extension > 5,000 m^2 and with trees usually growing more than about 5 m in height.

Included are:
a) areas normally forming part of the forest area which are unstocked as a result of human intervention or natural causes but which are expected to revert to forest;
b) young natural stands and all plantations established for forestry purposes which have not yet reached a crown density of more than 10%;
c) areas of windbreak and shelterbelt trees larger than 20m;
d) other wooded land with a minimum extension of 1,250 m^2 and crown cover > 40%;
e) sweet chestnut woods
f) shrublands

The forest edge is defined by the line of demarcation between different cover types or land-use.

7.2. DATA SOURCES

A) field data

- the interpretation area for area related data (Rock outcrops, Stand origin, Ground roughness, etc.) is 0.5 ha
- sub-sample of trees: selected from the North in clockwise direction; at least 25% of enumerated trees (four trees as a minimum).

C) aerial photography

- Italian Military Geographic Institute
- 1:33000, panchromatic photographs (Kodak XX); no ortophotographs
- 1978-79

- the entire Region
- stereoscopic pairs
- not available

E) maps:

- Toscana Regional Authority Office
- 1978
- 1:25000
- analogue printed topographic map

7.3. ASSESSMENT TECHNIQUES

7.3.1. Sampling frame

The sampling frame is given by the forest area definition. Excluded from field assessment are stands with height < 5m, areas which are unstocked as a result of human intervention or natural causes, young natural stands and all plantations established for forestry purposes which have not yet reached a crown density of more than 10%, sweet chestnut woods and shrublands. Four Provinces (Firenze, Pistoia, Arezzo, Siena) are covered by the assessment.

7.3.2. Sampling units

Aerial photography sample plots: fixed area with centre in sampling point. Circular plot of 0.5 ha. Plots are systematically distributed in a 400*400 m grid.

Field plot: point sampling. Basal Area Factor = 2. The sampling unit is a cluster of four field plots: one in the central point, the other three, 50 m far from the central point, are selected among the six vertexes of an hexagon with azimuth 0, 120 e 240 degrees. These latter may be moved in the case that fall in a not-forested area. All plots permanently monumented and azimuth and distance of tallied trees from plot centre recorded. The plots are distributed in a 400*400 m grid.

7.3.3. Sampling designs

The sampling design applied in the Toscana region is a double-sampling for stratification. In the first phase (photo-interpretation) systematically distributed photo-plots are interpreted, land use assessment and strata assigned to the forested photo-plots. In the second phase a smaller, randomly selected among a sub-sample of photo-plots chosen on the basis of the strata extension and their forest management relevance, sub-sample of field plots is assessed.

7.3.4. Techniques and methods for combination of data sources

The attributes assessed in field and photo sample plots are combined by a double sampling for stratification approach. Attributes for stratification are assessed in photo-plots and used for estimating strata sizes.

Topographic maps are used for field plot localisation and for the assessment of the administrative unit, and elevation.

7.3.5. Sampling fraction

Data source and sampling unit	Proportion of forested area covered by sample	Represented mean area per sampling unit
aerial photography, fixed area sample plot	0.3125	16 ha
field assessment, conventional fixed area associated to sampling point	/////	16 ha

7.3.6. Temporal aspects

Inventory cycle	time period of data assessment	publication of results	time period between assessments	reference date
1st IFT	1981-1987	not published	////	////

7.3.7. Data capturing techniques in the field

Data recorded on tally sheets and edited by hand into the computer.

7.4. DATA STORAGE AND ANALYSIS

7.4.1. Data storage and ownership

The data of the assessment are stored in a database system at the Toscana Regional Authority Office - Department of Forests.

7.4.2. Database system used

The data are stored in a DB III database.

7.4.3. Data bank design

Data not available

7.4.4. Update level

Not quoted

7.4.5. Description of statistical procedures used to analyse data including procedures for sampling error estimation

a) area estimation

The forest area is assessed on the aerial photographs. A 400*400 m systematic point grid is laid over the entire regional area and each point is assigned a forest or other land cover/use. The estimation of forest and other land cover types area is done according one of the following formulas:

- counting the proportional number of point multiplied by the area:

$$a = \left(\frac{n}{N}\right) * A$$

where
a = forest or other land cover type area
n = number of points assigned to forest type
N = total number of points
A = area

- counting the number of points multiplied by the represented area (16 ha):

$$a = N * 16$$

The standard error, s(a), is calculated according to one of the formulas provided in publications.

b) aggregation of tree and plot data

As the selection of the trees to be recorded is proportional to the individual basal area, the single tree attributes have to be divided by relative basal area.

The total values for plot i; Y_i, can be calculated in the following way:

$$Y_i = k\sum_{1}^{z} Y_i / g_i$$

where:
k = BAF
g = basal area
z = number of enumerated trees

c) estimation of total values

The estimation of total value of single tree attributes, Y, might be calculated as follows:

$$Y_h = \frac{k}{n}\sum_{j=1}^{n}\sum_{i=1}^{z}\frac{Y_i}{g_i}$$

where
n = number of plots

In general the total value of attribute Y for a strata h is given by multiplying the area related attribute, Y_h, with the calculated area, A_h,of the strata:

$$Y = A_h Y_h$$

The standard error of the total value of basal area attribute Yg is:

$$s(Y_g) = \sqrt{\frac{(\sum_1^n g^2) - \frac{(\sum_1^n g)^2}{n}}{n(n-1)}}$$

The total error is given by:

$$s = \sqrt{s(Y_i)^2 + s(a)^2}$$

d) estimation of ratios

///

e) sampling at the forest edge

For plots not falling entirely into forests, but extend partially into non forested areas, tree attributes are doubled on an equivalent specula area.

f) estimation of growth and growth components (mortality, cut, ingrowth)*

///

g) allocation of stand and area related data to single sample plots/sample points.

The class of categorical, area related attributes, in which the plot centre is located, is assigned to the entire plot.

h) Hierarchy of analysis: how are sub-units treated?*

///

7.4.6. Software applied

Data not available

7.4.7. Hardware applied

Data not available

7.4.8. Availability of data (raw and aggregated data)

At present, all data are not available.

7.4.9. Submits (strata) available

Results would be obtained for Provinces, Municipality Associations

7.4.10. Links to other information sources

////

7.5. RELIABILITY OF DATA

7.5.1. Check assessment

Field plots:

5% of the field plots are visited a second time by check crews. The check crew is a well trained and experienced crew. The field plots to be checked are randomly selected from the list of already assessed plots. All attributes are assessed a second time independently from the assessment of the regular field crews. i.e., no data recorded by the field crews are available for the check crews. The measurement procedures for locating the plot are not repeated. The field crews do not know in advance, which plots will be checked.

Photo interpretation:

10% of the photo sample plots have been interpreted. Due to the high workload by orienting the stereo models, not single photo-plots but groups of plots (plots on the same sheet of the topographic map) are selected.

100% of the photo sample plots have been field-checked and more precise attributes have been added for the definition of strata (silvicultural system and stand development stage)

b)

The check assessments are used for the following purposes:
- the results will not be used to set up error budgets and to quantify the total error
- if the field crews and the check crews differ the results of the check crews instead of the field crews are used for further analysis
- the results will not be used to correct any raw data or results

7.5.2. Error budgets

Data not available

7.5.3. Procedures for consistency checks of data

Data not available

7.6. MODELS

7.6.1. Volume functions

a) outline of the model

As the inventory results haven't be derived until now, the ISAFA is going to cooperate with Toscana Region . The IFNI volume models will be applied for wood volumes estimating.

b) overview of prediction errors

Not available

c) data material for derivation of model

See IFNI

d) methods applied to validate the model

///

7.6.2. Assortments

Not assessed.

7.6.3. Growth components

a) outline of the model

Not available

b) overview over prediction errors

Not available

c) data material for the derivation

Not available

d) methods applied to validate the model

Not available

7.6.4. Potential yield

Not assessed.

7.6.5. Forest functions

Not assessed

7.6.6. Other models applied

////

7.7. INVENTORY REPORTS

7.7.1. List of published reports and media for dissemination of inventory results

Inventory	Year of publication	Citation	Language	Dissemination
1st IFT	1984	Inventario Forestale della Toscana. Regione Toscana Giunta Regionale. 1. Progetto generale 2. Manuale di fotointerpretazione 3. Istruzioni per il rilevamento a terra	Italian	printed

7.7.2. List of contents of latest report and update level

Reference of latest report:
Regione Toscana - Giunta Regionale, 1984. Inventario Forestale della Toscana.
1. Progetto generale
2. Manuale di fotointerpretazopne
3. Istruzioni per il rilevamento a terra

Content:
1. Progetto generale

Chapter	Title	Number of pages
1	Sampling design of the 1st IFT	19
2	Administrative, site and legal data processing	10
3	Land and cover use categories definition	20
4	Grid points location	14
5	Sample plots location on aerial photographs	10
6	Accessibility and road network	11
7	Field assessment general guidelines	23
8	Tree and plot data aggregation	9
9	Sampling units	30

2. Manuale di fotointerpretazione

Chapter	Title	Number of pages
1	Introduction to photo-interpretation	12
2	Opportunities and constraints to photo-interpretation of Regional aerial photographs	3
3	Vegetation features	6
4	Land use and cover categories	45
5	Canopy cover categories	15
6	Photo-interpretation form	15

3. Istruzioni per il rilevamento a terra

Chapter	Title	Number of pages
1	Introduction	6
2	Field plots location	10
3	Field plot cluster	15
4	Site and stand description	20
5	Measurement rules for measurable attributes	40
6	Field instruments	4
7	Field assessment general guidelines	23
	Annexes	10

7.7.3. Users of the results

Not directly assessed. Local administrations, agencies and bodies. Professionals.

7.8. FUTURE DEVELOPMENT AND IMPROVEMENT PLANS

///

REFERENCES

Bianchi, M. & Preto, G., 1984. Inventario forestale della Toscana. Progetto generale. Regione Toscana, Giunta Regionale.

Preto, G., 1984. Inventario forestale della Toscana. Manuale di fotointerpretazione. Regione Toscana, Giunta Regionale.

Bianchi, M., 1984. Inventario forestale della Toscana. Istruzioni per le squadre di rilevamento. Regione Toscana, Giunta Regionale.

Dream Italia, 1993. Inventario Forestale della Toscana. Progetto esecutivo.Relazione tecnica (not published).

8. EMILIA-ROMAGNA REGIONAL FOREST INVENTORY
(*INVENTARIO FORESTALE EMILIA-ROMAGNA - IFER*)

8.1. NOMENCLATURE

8.1.1. List of attributes directly assessed

A) Geographic regions

Attribute	Data source	Object	Measurement unit
Watershed (*Bacino idrografico*)	maps	sampling unit	categorical
Province *(Provincia)*	maps	sampling unit	categorical
Mountain Community (*Comunità Montana)*	maps	sampling unit	categorical
Municipality *(Comune)*	maps	sampling unit	categorical

B) Ownership

Attribute	Data source	Object	Measurement unit
Ownership *(Proprietà)*	local information, cadastral map	sampling unit	categorical

C) Wood production

Attribute	Data source	Object	Measurement unit
Crown closure and average diameter *(Copertura e diametro delle chiome)*	aerial photographs	sampling unit	%
Number of trees around point sample centre *(Numero di piante del campione relascopico)*	field assessment	sampling unit	integer
Edge case location *(Situazione di margine)*	field assessment	sampling unit tree	categorical
Diameter at breast height *(Diametro a petto d'uomo)*	field assessment	tree	cm
Bark thickness *(Spessore della corteccia)*	field assessment	tree	mm
Stump girth (*Circonferenza alla ceppaia)*	field assessment	tree	cm
Total height *(Altezza dendrometrica)*	field assessment	tree	m

Merchantable height *(Altezza cormometrica)*	field assessment	felled tree	m
Crown point *(Altezza di inserzione della chioma)*	field assessment	tree	m
Crown shape *(Morfologia della chioma)*	field assessment	tree	categorical
Stem quality *(Morfologia del fusto)*	field assessment	tree	categorical
Intermediate diameters at 4.5m or 6.5m *(Diametri intermedi a 4.5m o 6.5 m)*	field assessment	tree	cm
10 Intermediate diameters *(10 diametri intermedi)*	field assessment	felled tree	cm
Annual radial increment *(Incremento diametrico annuo)*	field assessment	tree felled tree	mm

D) Site and soil

Attribute	Data source	Object	Measurement unit
Elevation *(Quota)*	map	sampling unit	m
Aspect *(Esposizione)*	field assessment	sampling unit	categorical
Slope *(Pendenza)*	field assessment	sampling unit	categorical
Location *(Posizione topografica)*	field assessment	sampling unit	categorical
Ground roughness (*Morfologia superficiale*)	field assessment	sampling unit	categorical
Rock outcrops and stoniness *(Rocciosità e pietrosità)*	field assessment	sampling unit	categorical
Soil stability *(Assetto superficiale del suolo)*	field assessment	sampling unit	categorical

E) Forest structure

Attribute	Data source	Object	Measurement unit
Forest type *(Tipologia forestale)*	aerial photographs, field assessment	sampling unit	categorical
Stand origin *(Origine del soprassuolo)*	field assessment	sampling unit	categorical
Silvicultural system and stand development *(Stadio evolutivo)*	field assessment	sampling unit	categorical

Vertical structure *(Tipo di struttura verticale)*	field assessment	sampling unit	categorical
Age *(Età)*	field assessment,	tree	integer
Tree species *(Specie forestali)*	field assessment	sampling unit	categorical
Social standing *(Posizione sociale)*	field assessment	tree	categorical
Spatial distribution *(Distribuzione spaziale)*	field assessment	tree	degree and cm
Undergrowth density *(Densità del sottobosco)*	field assessment	sampling unit	categorical

F) Regeneration

Attribute	Data source	Object	Measurement unit
Regeneration features *(Caratteri della rinnovazione)*	field assessment	sampling unit	categorical
Leading forest species in regeneration *(Specie forestale prevalente nella rinnovazione)*	field assessment	sampling unit	categorical

G) Forest condition

Attribute	Data Source	Object	Measurement unit
Forest health *(Stato fitosanitario)*	field assessment	sampling unit	categorical
Damage origin *(Origine dei danni)*	field assessment	sampling unit	categorical

H) Accessibility and harvesting

Attribute	Data source	Object	Measurement unit
Accessibility *(Accessibilità)*	field assessment, map	sampling unit	categorical
Forest road network and condition *(Viabilità)*	field assessment, map	sampling unit	categorical

K) Miscellaneous

Attribute	Data source	Object	Measurement unit
Assessment date *(Data del rilievo)*	field assessment,	sampling unit	categorical
Crew members names and roles *(Nomi dei rilevatori e ruoli)*	field assessment,	sampling unit	categorical
Field point access traverse *(Poligonale di accesso al punto)*	field assessment	sampling unit	degree, m, cm

8.1.2. List of derived attributes

A) Geographic regions

Attribute	Measurement unit	Input attributes
Phitoclimatic zones *(Zone fitoclimatiche)*	categorical	forest types, latitude, longitude, elevation, aspect, climatic data

C) Wood production

Attribute	Measurement unit	Input attributes
Basal area per hectare *(Area basimetrica a ettaro)*	m^2	BAF, no. of enumerated trees
Individual basal area *(Area basimetrica individuale)*	m^2	d.b.h of enumerated trees
Number of trees per hectare *(Numero di piante a ettaro)*	integer	BAF, stems dbh
Mean basal area (*Area basimetrica media)*	m^2	basal area/ha, no. of trees/ha
Mean height *(Altezza media)*	m	heights, no. of trees/ha
Lorey's mean height *(Altezza di Lorey)*	m	heights, basal areas/ha
Stand top height *(Altezza dominante)*	m	heights
Stem total volume *(Volume del fusto intero)*	m^3	intermediate diameters, stem lengths
Volume tables *(Tavole dendrometriche)*	m^3	dbh, heights, tree volumes
Tree volume per hectare *(Volume a ettaro)*	m^3	basal areas, heights, volume tables
Mean annual increment *(Incremento medio annuo)*	m^3	volume, age, tree species and type

E) Forest structure

Attribute	Measurement unit	Input attributes
Leading species *(Specie prevalenti)*	categorical	species, basal area, no. of trees
Species composition *(Composizione specifica)*	categorical	species, basal area, no. of trees

8.1.3. Measurement rules for measurable attributes

C) Wood production

Crown closure and average diameter

a) measured on a circular area corresponding to 5,000 m² on the ground with centre in the sampling point
b) minimum closure: 5%, minimum diameter: 2.5 m, maximum diameter: 10m
c) 20 classes according to combination of crown closure % (5, 10-40, 40-70, 70-100) and crown average diameter in m (2.5, 3.5, 5.0, 7.0, 10.0)
d) rounding to the closest 5% for closure and to the closest half m for diameter
e) Topcon MSS stereoscope, crown closure template, crown diameter gauge template, circular area template
f) aerial photographs

Number of trees around point sample centre

a) standing on the sampling point a 360° sweep is made looking through Bitterlich relascope, starting from North in clockwise direction; trees with a breast-height diameter larger than the fixed angle chord corresponding to Basal Area Factor = 2 are tallied "in"; mean diameter and distance of borderline and hidden trees are checked with tape and calliper
b) minimum diameter at breast height: 2.5cm
c) integer
d) rounded off to the nearest full number
e) Bitterlich Spiegel Relascope
f) field assessment

Diameter at breast height

a) measured in 1.30 m height above ground, on slopes measured from uphill side. Two readings, for each enumerated tree, the first with calliper pointing to plot centre, the second perpendicular to the first
b) minimum dbh: 2.5 cm
c) cm, recorded every single cm
d) rounded up or down to the nearest cm
e) calliper
f) field assessment

Bark thickness

a) measured only on 3P (Probability Proportional to Prediction) sub sample of enumerated trees, at 1.30 m height, at the point tangent to the right arm of the calliper, while pointing it toward the sampling point
b) no thresholds
c) mm
d) rounded up or down to the nearest mm
e) bark thickness measurement instrument
f) field assessment

Stump girth

a) measured only on 3P sub sample of enumerated trees, at 30 cm height
b) no thresholds
c) cm
d) rounded up or down to the nearest cm
e) tape
f) field assessment

Total height

a) measured only on 3P sub sample of enumerated trees; length of tree from ground level to top of tree, taking at least two readings each time, from uphill in sloping terrain, at a distance approximately pair to the tree height; visual estimate for all enumerated trees
b) no threshold
c) m
d) rounded up or down to the nearest 0.1 m (to the nearest m when assessed only by visual estimate)
e) Suunto, Haga, Blume-Leiss hypsometer, Bitterlich relascope
f) field assessment

Merchantable height

a) measured only on felled trees (which are a sub sample of 3P trees) length of stem from the base to the point where diameter = 2.5 cm for broad-leaved species and diameter = 7.5 cm for coniferous species
b) no thresholds
c) m
d) rounded up or down to the nearest cm
e) tape and calliper
f) field assessment

Crown point

a) measured only on 3P sub sample of enumerated trees, as length of the stem from ground level to lower point where living branches fit in
b) no thresholds
c) m
d) rounded up or down to the nearest 0.1 m
e) Suunto, Haga, Blume-Leiss hypsometer, Bitterlich relascope
f) field assessment

Intermediate diameters (at 4.5 m and 6.5 m)

a) measured only on 3P trees, the diameter at 4.5 m height from ground level is recorded if the total height is less than 15 m, at 6.5 m height from ground level if total height is equal or exceeding 15 m
b) minimum tree height: 5 m
c) cm
d) rounded up or down to the nearest cm
e) Finnish calliper, Bitterlich Spiegel relascope
f) field assessment

Ten Intermediate diameters

a) measured only on felled trees: starting from the tree base the diameters at 1/20, 1/10, 2/10,... 9/10 of the total height and those at 4.5 m or 6.5 m from the base are recorded, with two perpendicular readings made respectively in North-South and East-West direction referring to the original orientation of the stem (the trees are marked on purpose before being felled)
b) no thresholds
c) cm
d) rounded up and down to the nearest half cm
e) tape and calliper
f) field assessment

Annual radial increment

a) measured on pieces of the stem cut perpendicularly to the length of the felled tree at 0.30 m, 1.30m, 4.50m, 6.50m, merchandable height (Hc) and half the difference from merchandable height and 6.50m [(Hc - 6.50)/2] from the stem base, taking as reference along the stem the 1.30 height, which is marked on the tree before felling it; also measured on sample standing tree with increment borer
b) the most recent increment ring is not considered
c) mm
d) rounded up or down to the nearest 1 mm
e) lens, calibre, Pressler increment borer
f) field and laboratory assessment

D) Site and soil

Elevation
a) measured in the sampling point as altitude
b) no threshold
c) m
d) rounded up or down to the nearest m
e) rule, calculator
f) 1:10000 map

Slope
a) measured for a circular area of approximately 500 m (10 -15 m radius) with sampling point as centre along the ruling gradient line, average of uphill and downhill values
b) six classes: ≤5, 6-30, 31-45, 46-60, 61-80, > 81
c) %
d) rounded up or down to closest 1%
e) clisimeter
f) field assessment

Aspect
a) measured for a circular area of approximately 500 m (10 - 15 m radius) with sampling point as centre
b) nine classes: indefinite, N, NE, E, SE, S, SW, W, NW
c) degree
d) rounded up or down to closest degree
e) compass
f) field assessment

E) Forest structure

Age
a) measured counting increment rings on cross sectional areas of felled tree and on increment core extracted from sample standing tree
b) the current increment ring is not counted
c) integer number of years
d) rounded up or down to closest full number
e) lens, calibre
f) field and laboratory assessment

Spatial distribution
a) horizontal distance and azimuth of enumerated trees measured from sampling point, while effecting the sweep
b) no thresholds
c) distance in cm, azimuth in degree
d) rounded up or down to closest cm and degree
e) tape, compass, tripod

Point access traverse
a) survey consisting of a set of connecting lines of known length, meeting each other at measured angles
b) no thresholds
c) azimuth and slope in degree, distance in cm
d) rounded up or down to closest cm and degree
e) tape, compass, tripod, programmable pocket calculator, stakes

8.1.4. Definitions for categorical attributes

A) Geographic regions

Watershed
a) geomorphologic unit; the watersheds of the main regional rivers, which are those rising in the Apennine ridge, single or grouped to have a minimum area of 20,000 hectares each
b) 12 watershed units
c) map

Province
a) administrative unit joining several closed Municipalities and one or few Mountain Communities
b) seven assessed
c) map

Mountain Community

a) administrative unit joining several closed municipalities in the mountain area aimed at planning and managing mountain regions

b) 17 assessed

c) map

Municipality

a) administrative unit

b) 114 assessed

c) map

B) Ownership

Ownership

a) owner of the forest area under concern

b) 1) private ownership
2) public institutions ownership
3) Municipality
4) Province
5) Region
6) State
7) unknown to crew members

c) local information, land register

C) Wood production

Edge case location

a) specification about the location of the sampling point and the enumerated trees with reference to the forest edge

b) one category: edge

c) field assessment

Crown shape

a) development in height and width of the crown

b) nine categories according to crown height (> ½ the total tree height, ½-¼, < ¼) and dimension (small, normal, big)

c) field assessment

Stem quality

a) commercial value of the stem

b) categories:
1) straight and regular
2) straight but with defects
3) twisted or forked
4) truncated or bent
5) small or drawn up

D) Site and soil

Location

a) terrain mapping unit following a geo-morphologic classification

b) 24 classes

c) aerial photo-interpretation and field assessment

Ground roughness

a) differences in elevation within the sampling area

b) six classes according to the proportion of area interested (<1/3, 1/3-2/3, >2/3) and the prevailing elevation differences (< 0.5 or > 0.5 m)

c) field assessment

Rock outcrops and stoniness

a) Presence and distribution of rocks and stones

b) six classes according to the proportion of area interested (< 5%, 5%-20%, 20%-40%, > 40%) and the distribution (spread or concentrate)

Soil stability

a) Susceptibility to surface morphologic changes

b) seven main classes:
1) stable
2) water erosion (five sub-classes)
3) wind erosion (two sub-classes)
4) landslides (five sub-classes)
5) unstable rocks (four sub-classes)
6) sedimentation
7) water bodies at surface (four sub-classes)

c) field assessment

E) Forest structure

Forest type

a) the forest types are defined on the base of the altitude belt, the phytoclimatic classification, the species composition, the height, the anthropic influence

b) ten main categories according to height (< 15 m, > 15 m, composite) and species composition (broad-leaved, coniferous, mixed), sub-divided with hierarchical classification in 27 types

c) aerial photographs, field assessment

Stand origin
a) the way the stand has been born
b) categories:
1) artificial (planted)
2) natural dissemination
3) agamic regeneration
4) mixed: agamic and from seeds
5) uncertain
c) field assessment

Silvicultural system and stand development
a) the silvicultural system applied and the stand development stage
b) categories:
1) coppice (six sub-classes)
2) high forest (eight sub-classes)
c) field assessment

Vertical structure
a) crowns storeys distribution
b) categories:
1) single storey
2) two storeys
3) more than two storeys or irregular
c) field assessment

Tree species
a) Scientific, Italian name and acronym of selected list of tree and major shrub species
b) 98 categories, some of which grouping more than one species
c) field assessment

Social standing
a) the way the individual tree growth is performing in comparison with the closest trees and the stand development stage
b) categories:
1) predominant
2) dominant
3) intermediate
4) dominated
5) subordinate
6) regeneration
c) field assessment

Undergrowth density
a) distribution of shrubs in the sampling area
b) six categories according to proportion of area occupied (<1/3, 1/3-2/3, >2/3) and homogeneity (spread or concentrated)
c) field assessment

F) Regeneration

Regeneration features
a) abundance and location of seedlings and/ or sprouts
b) 19 categories according to abundance (absent, scarce, frequent, abundant), distribution (in groups, homogenous) and light condition (under cover, uncovered, at stand edge)
c) field assessment

Leading forest species in regeneration
a) most frequent tree or major shrub species present in the regeneration storey
b) categories: same 98 codes used for forest species
c) field assessment

G) Forest condition

Forest health
a) vegetative vigour and presence of damages on trees
b) five classes according to proportion of area interested (< 1%, 1%-10%, 11%-30%, 31%-50%, >50%) five classes according to proportion of crowns interested (0%-10%, 10%-25%, 25%-60%, > 60%, dry)
c) field assessment

Damage origin
a) main causes of observed damages
b) categories:
1) unknown (five sub-classes according to proportion of crowns interested (0%-10%, 10%-25%,25% 60%, > 60%, dry)
2) direct antropic (11 sub-classes)
3) indirect antropic (nine sub-classes)

4) climatic (13 sub-classes)
5) biologic (54 sub-classes)
c) field assessment

H) Accessibility and harvesting

Accessibility
a) accessibility of the sampling area from nearest all-weather motor road
b) seven classes according to distance (< 500 m, 500 m- 2000 m, > 2000 m) and location (downward, upward)
c) field assessment, map

Forest road network and condition
a) type and density of ways within the forest
b) seven classes according to type (motor road, large track, small track), and distance (< 100 m, 100 -300 m, > 300 m)
c) field assessment, map

K) Miscellaneous

Assessment date
a) day, month, year of assessment
b) integers
c) field assessment

Crew member names and roles
a) family name and own name of crew members
b) two categories: crew leader, crew member
c) field assessment

8.1.5. Forest area definition and definition of "other wooded land"

Forest is defined here as an area not used for agricultural purposes, with a woody vegetation capable of influencing both abiotic and biotic factors of the environment, primarily used for forestry purposes, larger than 0.5 ha and 20 m width for linear formations, having tree species crown closure of more than 10% and average height of more than 5 m.

The inventory in object extends the ground sample to "small woods" (*"boschetti"*), defined as areas smaller than 0.5 ha, of more than 20 m width, having tree species crown closure of more than 40% and height more than 5 m, chestnut woods, defined as areas larger than 2,500 m^2 with prevalent chestnut trees cover (more than 1/3 of the area or more than 50% of individual trees), and fast growing plantation for wood production outside the forest larger than 1,250 m^2.

Excluded from the ground sample, but classified as other wooded lands on the photo-plots are shrublands, defined as areas with shrubs cover of at least 40%, having height of less than 5 m and tree species cover of less than 10%, reforestation, defined as areas planted with trees having special interest not reaching height and density of forest and larger than 1,250 m^2 and areas that are temporarily uncovered (cuttings, regeneration, fired, slides etc.) but that are expected to become forest areas again in a reasonable time, larger than 1,250 m^2.

The forest edge is defined by the vertical projection of the canopy on the ground, or, if not applicable, by the limit with other land cover types.

8.2. DATA SOURCES

A) field data: treated under 8.3

C) aerial photography

- responsible bodies from whom aerial photographs have been obtained:

1) Servizio Regionale Cartografico e Geologico, Regione Emilia-Romagna
   Viale Silvani 4, 40121 Bologna, Italy
2) Compagnia Generale di Riprese Aeree di Parma
   Via Cremonese 35/a, 43100 Parma, Italy

- 1:13000 approximate scale colour reversal film, 23*23 cm photo format
- dates: early summer 1976; early summer 1978
- coverage: whole inventory area interested
- instruments for preparing aerial photographs: Galileo stereo-micrometer, OMI stereo-restitutor, punchers;
- instrument for photo-interpretation: Topcon mirror stereoscope, crown closure and gauge templates, circular area templates
- number of stereo pairs used in the assessment: approximately 1,500
- cost of aerial photographs: 25,000 Italian lire each photograph

E) maps

- responsible bodies from whom map have been obtained:
  Servizio Regionale Cartografico e Geologico, Regione Emilia-Romagna
  Viale Silvani 4, 40121 Bologna, Italy
- date span: 1976- 1980
- scale: 1:10000
- type and format: topographic map, printed

F) other georeferenced data

- responsible bodies from whom geo-referenced data have been obtained:
  Servizio Regionale Cartografico e Geologico, Regione Emilia-Romagna
  Viale Silvani 4, 40121 Bologna, Italy
- availability: restricted
- themes: rivers, municipality boundaries, soil types
- spatial resolution: various: topography and administrative boundaries from 1:10000 map, roads and rivers from 1:100000 map, soil types from 1:250000
- area covered: Regione Emilia-Romagna
- date span 1985-1995
- format: Arc/Info

8.3. ASSESSMENT TECHNIQUES

8.3.1. Sampling frame

The Emilia-Romagna Forest Inventory concerns the mountain and hilly area of the region, as defined by the administrative boundaries of the Mountain Communities.

The ground sampling phase concerns the forests and the other wooded lands defined in 8.1.5, within this territory.

A small number of ground sampling unit (roughly 1%) has not been visited because terrain conditions make them inaccessible without special mountain climbing equipment; less than 0.5% of trees that have been selected for being felled was left untouched because belonging to natural reserves of protected areas.

8.3.2. Sampling units

Aerial photography

Permanent circular plots with centre in sampling point corresponding to a ground area of 5,000 m^2 (corrected for slope), systematically distributed in a 200*200 m grid over the corresponding nodes of the UTM coordinate system

Field assessment

Permanent points with variable area plots for relascope sampling with basal area factor = 2. The sampling unit is a cluster of four recording units: one in the central point, the other three at 40 m distance in the 0°, 120° and 240° directions.

Area related assessment other than tree measurements (site and soil, ground roughness, forest health, etc.) are made for every recording unit assuming as reference a circular area of approximately 500 m^2 (10 - 15 m radius) around the sampling point.

The points are distributed systematically on a 1*1 km grid over the forest area, in the corresponding centres of the photo-plots.

8.3.3. Sampling designs

The sampling design applied in the Emilia-Romagna region is a double-phase sampling for stratification.

In the first phase (photo-interpretation) the land use and cover is recorded for the systematically distributed photo-plots, along with the geographic and site attributes assessment; crown-cover, forest type and stand height class is further assessed for forest and other wooded land plots.

In the second phase a smaller, systematically distributed sub-sample of field plots is assessed.

8.3.4. Techniques and methods for combination of data sources

The attributes assessed in field and photo sample plots are combined by a double sampling for stratification approach.

Attributes for stratification are assessed in photo-plots and used for estimating strata sizes. Attributes assessed in the field plots are weighted with the strata sizes derived from the photo-interpretation.

Photo and field attributes are digitised and stored in a georeferenced Database and combined with other georeferenced data to analyse both forest and other land attributes with the help of a GIS.

Topographic maps are used to transfer by stereo-restitution the sampling points to photograms, to measure photo-plots slope, elevation, accessibility, to delineate administrative and geographic boundaries and to reach the field plot in the ground phase.

8.3.5. Sampling fraction

Data source and sampling unit	proportion of forested area covered by sample	represented mean area per sampling unit
aerial photography, fixed area sample plot	0.125	4 ha
field assessment, conventional fixed area associated to sampling point	0.005	100 ha

8.3.6. Temporal aspects

Inventory cycle	time period of data assessment	publication of results	time period between assessments	reference date
1st I.F.E.R.	1984 - 1995	1997 (expected)	20 years (expected)	1 June 1990

8.3.7. Data capturing techniques in the field

Data are recorded on tally sheets and edited by hand into the computer.

8.4. DATA STORAGE AND ANALYSIS

8.4.1. Data storage and ownership

The data of the assessment are stored in a database system at the Emilia-Romagna Regional Authority Office - Forest resource Department - Forest Inventory Unit. The person responsible for the data is located in the Regional Forest Inventory Unit (Dr. L. Gherardi, IFER, I-40121 Bologna, Tel: +39-51-6396851, Fax: +39-51-6396829, email: sig@iperbole.it).

8.4.2. Database system used

The data are stored in a DBIII - DBIV database and in a self-developed software operating in the mentioned environment.

8.4.3. Data bank design

a) Aerial photograph and map sample data bank structure

Input	Relation	System	Relation	Data Banks
Photo-interpretation data files				Consistency principles and main structures
	⇒		⇔	
		Aerial photographs sample data bank (DIFI)	⇔	Georeference d Database of validated photo-interpretation data
	⇒		⇔	
Map data-files: geography topography UTM coordinates				Area estimates and errors processing models

b) Field sample data-bank structure with reference to carried out calculations

Input	Relation	System	Relation	Data Banks
Field data files: Geographic data (RCRT) Cluster configuration (RCCP) Point access traverse (RCPS) Point site characteristics (RSAS) Relascopic sweep (RDRE) 3P trees measured attributes (RD3P) Volume increment measurments (RAIN) Notes on volume increments (RANO)	⇒	Field data bank	⇔	Consistency principles Height and volume processing data
			⇔	Georeferenced database of validated field data
			⇔	Georeferenced database of forest measurment data Diameter distribution Plot volume by species

c) Entity-relationship diagram

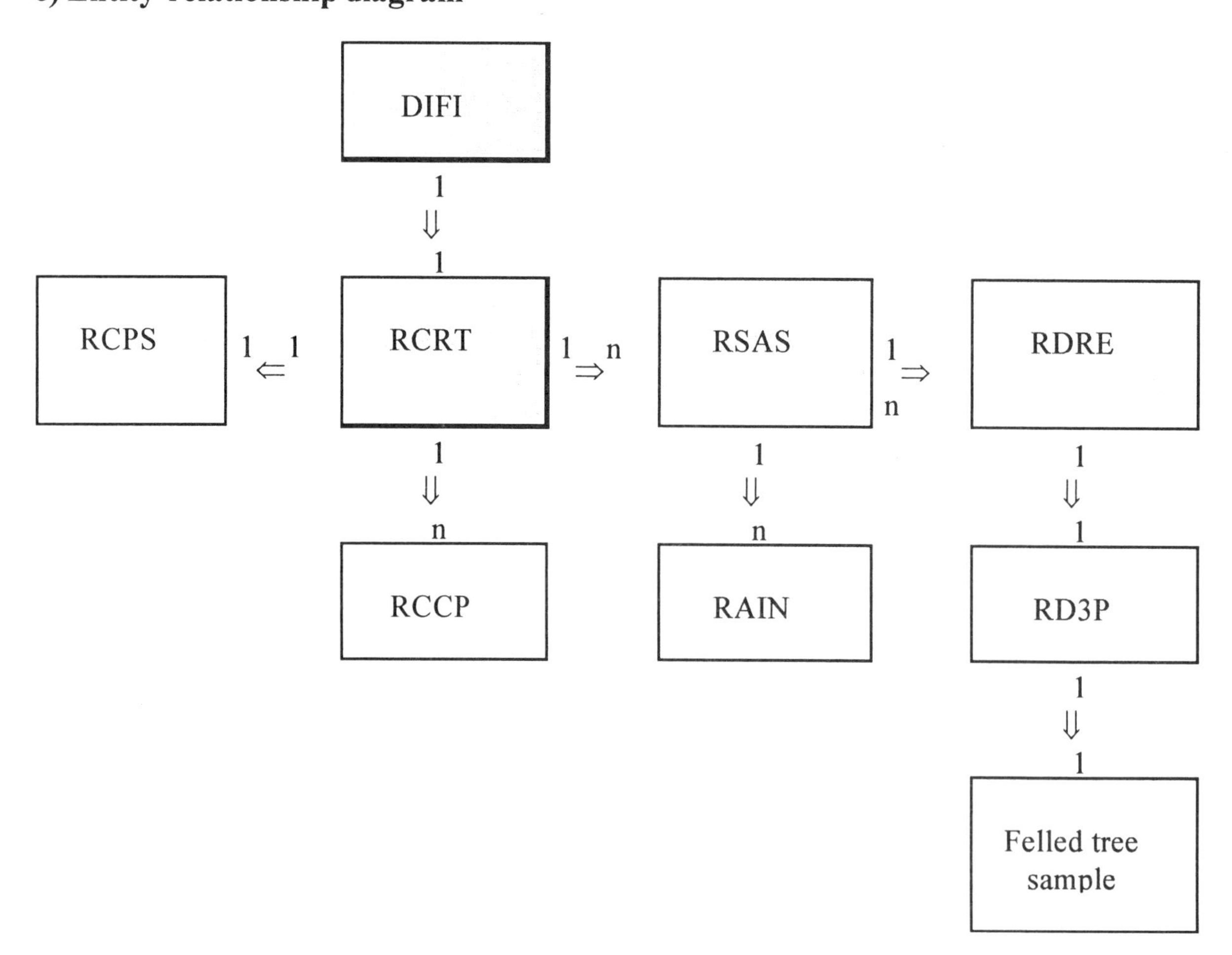

8.4.4. Update level

For each watershed the year, when the data have been assessed, is consistent. The data are expected to be updated to a common point in time as regards each single watershed; for what concerns data from aerial photographs and maps the reference date has better to be regarded as 1976-1978, or 1980, for simplicity; attributes values which are estimated for the regional area as a whole from field data are considered to represent the average situation in the year 1990.

8.4.5. Description of statistical procedures used to analyse data including procedures for sampling error estimation

a) area estimation

The forest area is assessed on the aerial photo-pairs. A 200*200 m systematic point grid is laid over the entire regional area and each point is assigned a forest or other land cover/use category. The estimation of forest and other land cover types area is done counting the proportional number of point then multiplying by the total watershed area, which is assumed to be free from error:

$$a = \left(\frac{n}{N}\right) * A$$

where
a = forest or other land cover type area
n = number of points assigned to forest type
N = total number of points
A = total area

and the estimated sampling error (following Matern, 1961), is derived from the summation of all the cross differences between each point and the four closest ones in the inventory grid:

$$s\% = \pm t * 100 * \left[\left(\frac{\Sigma D_{ij}^{2}}{4n^{2}}\right)^{1/2}\right]$$

where
S% = standard error %
t = Student t value (for a = 95% probability level)
and

$$D_{ij} = w(_{ij}) + w(i+1, j+1) - w(i+1, j) - w(i, j+1)$$

is the summation of the cross differences between the point having coordinates: i,j pertaining to the considered forest or other land cover type: w = 1; for points with other type than the one considered: w = 0

and also, following Bonnor (1975):

$$\log\ S\% = \pm(1.759 - 0.75 * \log n) \qquad \text{(for } t = 1)$$

After the field phase, the area adjustment for misinterpretation is carried out by reallocation of plots to strata, following Loetsch-Haller (1964), and the standard error of the adjusted stratum area proportion is calculated as a combination of photo-interpretation standard error and field check standard error.

b) aggregation of tree and plot data

As the relascopic sweep gives directly the basal area per hectare, the diameters of the enumerated trees are recorded and the individual basal area used to calculate the virtual number of trees represented by each single tree "in"; the resulting weight factor has to be applied for the aggregation of the other attributes of interest.

c) estimation of total values

The statistical design of the IFER can be characterised as double sampling for stratification. The size of the strata is not known, but is estimated by means of aerial photo interpretation. Thus two kinds of sampling error have to be considered: the sampling error due to the estimation of the strata sizes and the sampling error of the attribute of interest.

Total values and errors are expected to be calculated for each watershed unit following Loetsch-Haller (1973) for stratified population.

Appropriate tests (t test, F test, analysis of variance) have to be carried out before attribute values from different watersheds are combined for statistical treatment.

Overall means, variances and standard errors have to be calculated.

d) estimation of ratios

The area of the unit of reference (entire region or watershed (= sub-unit)) is assumed to be free from error.

As strata area are affected by errors, this has to be taken into consideration and attributes mean and total estimates standard error have to be calculated of ratios, following Loetsch-Haller (1973)

e) sampling at the forest edge

The correction applied to sampling points located at the forest edge is given by Bitterlich (1980).

A weighting factor is calculated for each tree as the quotient of the total area of its imaginary circle and the difference from total area and the one of the marginal circle.

f) estimation of growth and growth components (mortality, cut, ingrowth)*

Strata mean annual increments are calculated using mean volume and mean age of all trees, current annual volume and % increment are calculated from felled trees and increment cores. The regression errors of the derived functions along with the correlation coefficient and the significance of the correlation have to be calculated.

g) allocation of stand and area related data to single sample plots/ sample points

Each of the four points of the cluster are assigned the class of categorical, area related attributes, related to a conventional circular area of approximately 500 m^2 around their centres (10-15 m radius).

h) Hierarchy of analysis: how are sub-units treated?*

The between/variance of the four measuring units forming the cluster in the field sampling point has to be analysed. Errors are calculated consequently.

8.4.6. Software applied

The analysis software consists of the following programmes: Lotus 1-2-3, Microsoft Excel and a set of self-developed programmes based on DBIII - DBIV. The Arc/View software is used to extract and combine information on different attributes.

8.4.7. Hardware applied

The hardware applied consists of PC using DOS system, Windows environment and a graphic tablet operating under AUTOCAD and Arc/Cad software.

8.4.8. Availability of data (raw and aggregated data)

All raw data and derived attributes stored in the database can be made available for outside persons as long as the purpose for which the data will be used is laid down. However, data on ownership are not provided. The exchange of data is laid down in a special regulation obtainable from the Forest Inventory Unit of the Emilia-Romagna Region.

8.4.9. Sub-units (strata) available

Results can be obtained for watershed, Province, Mountain Community and Municipality.

8.4.10. Links to other information sources

Topic and spatial structure	age/ period	responsible agency	kind of data-set	availability
administrative boundaries, principal roads, rivers	since 1992	Servizio Regionale Cartografico e Geologico Emilia-Romagna tel. + 39 51 284483	digital	limited
volume tables Italy	since 1984	I.S.A.F.A.- Trento tel. + 39 461 924248	pre-processed	available data
phytoclimatic types Emilia-Romagna geological and hydrological maps	since 1993	Istituto di Biologia evolutiva. Università di Bologna tel. + 39 51 351280	map	limited

8.5. RELIABILITY OF DATA

8.5.1. Check assessment

a) the way how check assessments are organised and carried out

Field plots:

10% of the field plots are visited a second time by check crews. The check crew is a well trained and experienced crew, mainly dealing with check assessments, but not regular field work. The field plots to be checked are randomly selected from the list of already assessed plots. All attributes are assessed a second time independently from the assessment of the regular field crews. The measurement procedures for locating the plot centre are also repeated. For each field crew an identical proportion of plots is checked. The field crews do not know in advance, which plots will be checked. The check crews visit the plots after the field crews had assessed them. The organisation and the analysis of the check assessments is done at the main office.

Photo-interpretation:

The point restitution and the photo-interpretation work is done by two outside companies. After completion of successive lots of photo-pairs 3% of the photo sample plots are regularly checked by regional forest officers; when the interpretation does not agree the field check is carried out. Due to the high workload by orienting the stereo models, not single photo-plots but groups of plots (plots on the same sheet of the topographic map) are selected.

Digitisation:

10% of successive groups of the records in the digitised files are systematically checked by regional officers during the digitising work; errors have not to be present in the 98% of the records; automated consistency check routine programs are always employed when digitising.

Analysis of check assessments:

All data that are assessed by both companies crews and regional crews are compared, and accuracy tolerances checked for. When errors exceed the values and quantities always stated in the contracts (as a general rule the relative accuracy is fixed at the 2%), the companies crew have to do the all the work again. This work will be also checked the same way.

b) how the results of check assessments are used in the inventory system.

The check assessments are used for the following purposes:
- feed-back to field crews to improve the data quality
- results are used to revise the assessment instructions
- results are used for the interpretation of inventory results
- the results will be used to correct raw data or results.

8.5.2. Error budgets

Major non-sampling errors affecting I.F.E.R. data are concerned with the time gap between aerial photographs and field data; methodologies to provide corrections have not been examined thoroughly yet, but this will be a topic of central interest as soon as the elaboration results will be available.

8.5.3. Procedures for consistency checks of data

Data assessed in the field:

The data are recorded on card files on the field plot and checked a first time by field crews before returning them to the regional office. A second consistency check is done by computer programs while digitising them, using data from the tables of the National Forest inventory and cross-checks of different attributes for plausibility tests.

Data assessed on photo-plots and maps:

The data are recorded on card files on the field plot and checked a first time by field crews before returning them to the regional office. A second consistency check is done by computer programs while digitising them, using cross-checks of different attributes for plausibility tests.

8.6. MODELS

8.6.1. Volume (functions, tariffs)

a) brief outline of the models

Volume is expressed as a function of dbh and total height.

After the volume is measured on felled trees, resulting values are used to assess 3P trees sample volume, and the values used to derive the coefficient of the following functions:

1) $V = a + bD^2H$
2) $V = a + bD^3H$
3) $V = a + bD^2H^2$
4) $V = a + bD^2H + cD$
5) $V = a + bD^2H + cD^2$
5) $V = a + bD^2H + cD^3$
6) $V = a + bD^2H + cD^3 + dH$
7) $V = a + bD^2H + cD + dH$
8) $V = a + bD^2H + cD^2 + dH$
9) $V = a + bD^2H + cD^2 + dD^2$
10) $V = a + bD^2H + cD^2 + dD^2 + eH$
11) $V = a + bD^2H + cD^2 + dD^2 + eH + fH^2$

where
V = stem total volume
D = diameter at breast height
H = total height
a, b, c, d, e, f = coefficients

b) overview over prediction error

The regression error and correlation coefficient are used to select the preferred model for each species or group of species

c) the data-material for the derivation of the model

1,230 individual trees distributed among the main species and groups of minor species

d) methods applied to validate the model

The significance of the regression has to be usually tested by F (following Snedecor and Cochran, 1971) compared with regression mean square and error mean square ratio, or t compared with regression coefficient values and standard errors ratios.

8.6.2. Assortments

A form coefficient is calculated for deriving the merchantable stem volume.

From 3P sample trees, stem volume can be obtained for three different assortments: basal billet (from the base to 1.30 m), central billet (from 1.30 m to 4.50 m or 6.50 m) and top billet (from 4.50 m or 6.50 m to top).

8.6.3. Growth components

Growth components are expected to be investigated with the next inventory period.

8.6.4. Potential yield

///

8.6.5. Forest functions

///

8.6.6. Other models applied

///

8.7. INVENTORY REPORTS

8.7.1. List of published reports and media for dissemination of inventory results

Inventory	Year of publication	Citation	Language	Dissemination
I.F.E.R.	1989 1995 (reprint)	I popolamenti forestali del bacino del Panaro Risultati della fase sperimentale dell'inventario forestale regionale Regione Emilia-Romagna, Bologna. 191 p.	Italian	printed
I.F.E.R.	1992	Inventario Forestale Unità inventariale n.11 Fiumi Uniti vol. I:Sintesi metodologica. 43p. vol. II: Relazione Generale. 67p. vol. III: Tabelle e grafici. 92p. Azienda Regionale Foreste Emilia-Romagna, Bologna	Italian	printed
I.F.E.R.	1992	Inventario Forestale Emilia-Romagna Primi risultati da telerilevamento Bozza Azienda Regionale Foreste Emilia-Romagna, Bologna. 114p.	Italian	printed
I.F.E.R.	1996	Un sistema informativo per la pianificazione a scala regionale: teoria ed applicazione alla gestione dell'Inventario Forestale della Regione Emilia-Romagna, F. Ferretti, L. Gherardi, R. Scotti, Monti e Boschi n. 1, Bologna. 5p.	Italian	printed

8.7.2. List of contents of latest report and update level

Reference of latest report:
Azienda Regionale Foreste Emilia-Romagna, 1992. Inventario Forestale Emilia-Romagna -
1) Sintesi metodologica (vol. I)
2) Primi risultati da telerilevamento - Bozza

Update level of inventory data used in this report:
Field check from 1986 to 1992
Aerial photography: 1976-78

Content:

1)

Chapter	Title	Number of pages
1	Introduction	half
2	Origin of the Regional Forest Inventory	2
3	Delimitations of inventory units	2
4	Sampling frame and grid	1
5	Data digitisation	1
6	Photo-restitution and photo-interpretation	2
7	Checking procedures for aerial photographs data	1
8	Field data	3
9	Checking procedures for field data	2
10	Elaboration	2
11	The areas	half
11.1	Land uses categories and area estimated	6
12	Analysis of forest mensuration variables	
12.1	Volume tables	5
12.2	Correction for visual estimated height	1
12.3	Correction for forest edge points	2
13	Final estimated values	
13.1	Forest mensuration data	half
13.2	Site typology data	half
Appendix	Volume tables results for watershed n. 11	9

2)

Tables	Title	Number of pages
0	List of tables	1
1	Land use and cover distribution with error estimates	1
2	Forest and other wooded land distribution with error estimate	1
3A	Distribution of forest area by height categories	half
3B	Distribution of main forest types by species composition	half
4	Distribution of forest area in crown cover classes	1
5A	Land use and cover distribution in altitude belts	2
5B	Forest and other wooded land distribution in altitude belts	1

6A	Land use and cover distribution in aspect classes	1
6B	Forest and other wooded land distribution in aspect classes	1
7A	Land use and cover distribution in slope classes	1
7B	Forest and other wooded land distribution in slope classes	1
U.I. no.1	Tables from 1 to 7B for watershed unit no. 01	9
U.I. no.2	Tables from 1 to 7B for watershed unit no. 02	9
U.I. no. 3	Tables from 1 to 7B for watershed unit no. 03	9
U.I. no.4	Tables from 1 to 7B for watershed unit no. 04	9
U.I. no.5	Tables from 1 to 7B for watershed unit no. 05	9
U.I. no.6	Tables from 1 to 7B for watershed unit no. 06	9
U.I. no.7	Tables from 1 to 7B for watershed unit no. 07	9
U.I. no.8	Tables from 1 to 7B for watershed unit no. 08	9
U.I. no.9	Tables from 1 to 7B for watershed unit no. 09	9
U.I. no.10	Tables from 1 to 7B for watershed unit no. 10	9
U.I. no.11	Tables from 1 to 7B for watershed unit no. 11	9
U.I. no.12	Tables from 1 to 7B for watershed unit no. 12	9

8.7.3. Users of the results

The Regional Departments, the Watershed Government Authority of the Po river, the Provinces, the Mountain Communities, the Municipalities will use the results for land evaluation and planning purposes, public and private environmental agencies for environment related matters, universities for research topics and education, professionals for their reports.

8.8. FUTURE DEVELOPMENT AND IMPROVEMENT PLANS

8.8.1. Next inventory period

The next inventory period should possibly start in the year 2000 and the acquisition of forest and land use cover types from aerial photographs and maps should be accomplished by the year 2005.

8.8.2. Expected or planned changes

8.8.2.1. Nomenclature

No major changes are expected in nomenclature, except from the addition of new categories, part of which should derive from the results of the first period and the combination with other data sources.

8.8.2.2. Data sources

At present, a regional forest map at 1:10000 scale is carried on with the I.F.E.R. classification system, which will represent the major input for area changes information and spatial resolution; other data are acquired in vector format relating to natural protected areas, road network, forest management plans and "monument trees" specimens; a special study project regarding forest health and air pollution within the regional area will also be combined with the inventory results.

The integration with satellite and other remote sensing data is highly desirable, and should be deeply investigated with the help of the research community.

8.8.2.3. Assessment techniques

The assessment techniques will change according to the sake for the reduction of costs and time involved as compared to the first inventory period; the introduction of differential real-time Global Positioning System is also under examination at this purpose.

8.8.2.4. Data storage and analysis

The development of a specialised Forestry Information System is strongly supported which operates in GIS environment, capable of combining forest related data from different sources.

8.8.2.5. Reliability of data

Check assessment and procedures for consistency check of data will be added and changed according with the assessment techniques employed.

8.8.2.6. Models

Growth and growth component model are expected to be applied in the next inventory period.

Satellite data processing models are wished to become integral part of the Forestry Information System since the beginning of the next inventory period

8.8.2.7. Inventory reports

Reports should be better made on a regular time schedule base during the course of the next inventory period, with a final report at the end.

Error calculation procedures, equations used for calculations and figures should always be included.

Time and costs analysis should be considered, along with a brief overview on advantages and disadvantages of the techniques applied.

8.8.2.8. Other forestry data

Non-wood forest product and services data will become part of the Forestry Information System

8.9. MISCELLANEOUS

///

REFERENCES

Baratozzi, L., 1984. Classificazione dell'utilizzazione reale del suolo. Servizio Inform. e Statistica, Uff. Analisi e Ricerche Territoriali, Regione Emilia-Romagna, fasc. I e II, Bologna. mimeo 66 p. 36 p.

Bitterlich, W., 1984. The Relaskope Idea. Relative Measurements in Forestry. Commonwealth Agricultural Bureaux: 242 p.

Credraro,V., Pirola, A., Speranza, M. & Ubaldi, D., 1980. Carta della vegetazione del crinale appenninico del M.te Giovo al Corno delle Scale (Appennino Tosco-Emiliano). Progetto finalizzato "Promozione qualità dell'ambiente" AQ/1/81 CNR, Roma.

Credraro,V. & Pirola, A., 1974-75. Note sulla vegetazione ipsofila dell'Appennino Tosco-Emiliano. Atti Ist.Bot. Lab. Crittog. Univ. Pavia, 10 (6): pp. 35-58.

Ferrari, C. Pirola, A., Puppi, G., Speranza, M. & Ubaldi, D., 1980. Flora e vegetazione dell'Emilia-Romagna. Regione Emilia-Romagna, Bologna. 356 p.

Kolbl, O. & Tracshler, H., 1980. Regional Land Use Survey Based on Point Sampling on Aerial Photographs. XVI Congress of the International Society of Photogrammetry, Hamburg 1980: 14p.

Loetsch, F. & Haller. K.E., 1964. Forest Inventory-vol. I. BLV -München. Basel. Wien. 435 p.

Loetsch, F., Haller, K. E. & Zohrer, F., 1973. Forest Inventory-vol. II. BLV München. Basel. Wien. 469 p.

Matern, B., 1982. The precision of Basal Area Estimates. -For. Sci. 18(2): pp. 123 - 125.

Ministero Agricoltura E Foreste, ISAFA, 1988. Inventario Forestale nazionale 1985, Sintesi metodologica e visultati. Trento. 464+14 p.

Paiero, P., 1984. La vegetazione forestale dell' Emilia Romagna . Introduzione allo studio tipologico dei boschi dell'Appennino emiliano-romagnolo e rispettivo avanterra come base per l' inventario forestale e la gestione del patrimonio boschivo regionale. Mines, ARF Emilia Romagna, Bologna: 60 p.

Preto, G., 1982. La determinazione delle aree col metodo della conta dei punti. Monti e Boschi 3/5: pp. 65-72.

Preto, G., 1983. La fotoiterpretazione per punti. Monti e Boschi 6: pp. 55-60.

Preto, G., 1983. Gli inventari forestali della Toscana e dell'Emilia-Romagna. Economia Montana 1: pp. 2-13.

Preto, G., 1987. Combined Use of Aerial Photographs, Maps and Field Sampling for Management Planning. Poceedings 18th IUFRO World Congress (Div. 4), Ljubjana, Sept. 7-21, 1986.

Regione Emilia-Romagna, Servizio Cartografico E Dei Suoli, 1976. Carta della utilizzazione reale del suolo: Carta dei suoli. Centro Stampa Regione Emilia-Romagna, Bologna.

Wolf, U., 1989. Norme per la classificazione delle forme fisiografiche con illustrazione esplicativa. Az. Reg. Foreste Emilia-Romagna, Bologna, dattiloscritto.

9. UMBRIA REGIONAL FOREST INVENTORY
(INVENTARIO FORESTALE DELLA REGIONE UMBRIA)

9.1. NOMENCLATURE

9.1.1. List of attributes directly assessed

A) Geographic regions

Attribute	Data source	Object	Measurement unit
Mountain Community (*Comunita' Montana)*	map	sampling unit	categorical
Municipality *(Comune)*	map	sampling unit	categorical

B) Ownership

Attribute	Data source	Object	Measurement unit
Ownership *(Proprietà)*	local information, land register	sampling unit	categorical

C) Wood production

Attribute	Data source	Object	Measurement unit
Diameter at breast height *(Diametro a petto d'uomo)*	field assessment	tree	cm
Total height *(Altezza dendrometrica)*	field assessment	tree	m

D) Site and soil

Attribute	Data source	Object	Measurement unit
Elevation *(Quota)*	field assessment	sampling unit	dam
Aspect *(Esposizione)*	field assessment	sampling unit	categorical
Slope *(Pendenza)*	field assessment	sampling unit	degree
Location *(Giacitura)*	field assessment	sampling unit	categorical
Stoniness *(Pietrosità)*	field assessment	sampling unit	categorical
Rock outcrops *(Affioramenti rocciosi)*	field assessment	sampling unit	percent
Soil depth *(Profondità del suolo)*	field assessment	sampling unit	categorical
Soil development stage *(Gradi di evoluzione del suolo)*	field assessment	sampling unit	categorical
Texture *(Tessitura del suolo)*	field assessment	sampling unit	categorical

E) Forest structure

Attribute	Data source	Object	Measurement unit
Forest type *(Tipo fisionomico)*	field assessment	sampling unit	categorical
Silvicultural system and stand development *(Stadio evolutivo)*	field assessment	sampling unit	categorical
Canopy cover *(Grado di copertura)*	field assessment	sampling unit	percent
Age *(Età)*	field assessment, local information	tree	years
Tree species *(Specie forestali)*	field assessment	sampling unit	categorical
Number of stools *(Numero delle ceppaie)*	field assessment	sampling unit	integer
Stand description *(Osservazioni selvicolturali)*	field assessment	stand (interpretation area)	categorical

H) Accessibility and harvesting

Attribute	Data source	Object	Measurement unit
Accessibility *(Accessibilità)*	field assessment, map	sampling unit	categorical

J) Non-wood goods and services

Attribute	Data source	Object	Measurement unit
Forest function *(Funzione prevalente)*	field assessment	stand	categorical
Use restrictions *(Vincoli)*	local information, map	stand	categorical

K) Miscellaneous

Attribute	Data source	Object	Measurement unit
Survey timing *(Tempi di lavoro)*	clock	///	minutes
Surveyor and firm name *(Rilevatore e Ditta)*	///	sampling unit	///
Topographic map *(Tavoletta IGM)*	map list	///	///
Sample plot number *(Punto UTM)*	map	sampling unit	geographic coordinates

9.1.2. List of derived attributes

C) Wood production

Attribute	Measurement unit	Input attributes
Top height *(Altezza dominante)*	m	h, d.b.h
Number of stems per ha *(N° di piante per ettaro)*	integer	no. of stems, tree species and type
Basal area *(Area basimetrica)*	m²	d.b.h., number of stems, tree species and type
Mean diameter *(Diametro medio)*	cm	d.b.h., tree species and type
Mean height *(Altezza media)*	m	h, d.b.h., tree species and type
Form factor *(Coefficente di riduzione)*	integer	h, d.b.h., tree species and type, volume
Mean stem volume *(Volume dell'albero medio)*	m³	h, d.b.h., tree species and type
Mean annual increment *(Incremento medio)*	m³	volume, age, tree species and type

E) Forest structure

Attribute	Measurement unit	Input attributes
Dominant species *(Specie prevalente)*	categorical	basal area

9.1.3. Measurement rules for measurement attributes

C) Wood production

Diameter at breast height

a) Measured in 1.3m height above ground, on slopes measured from uphill side. One reading, for each tree, specifying the tree type (seedtree, shoot, coppice even-aged seedling, standard). Two readings for d.b.h. over 17.5 cm and for irregular stems.
b) Minimum dbh: 2.5 cm
c) cm, recorded every single cm
d) alternatively rounded off and up
e) calliper
f) field assessment

Total height

a) Length of tree from ground level to top of tree. Total height for every tree species of 10% of the 3-12 cm diameter class trees; 50% of the 13-22 cm class; 100% of the >23.
b) no threshold
c) m
d) rounded to 0.5 m
e) Suunto, Haga, Bitterlich, Blume-Leiss hypsometer
f) field assessment

D) Site and soil

Elevation
a) measured in the plot centre as altitude
b) no threshold
c) m
d) rounded to 10 m
e) altimeter
f) field assessment

Slope
a) measured from the plot centre along the ruling gradient line, average of uphill and downhill values
b) 5°
c) degree
d) rounded to closest degree
e) clisimeter
f) field assessment

Rock outcrops
a) measured as a percent out of the sample plot area
b) no threshold
c) in 5%-classes
d) rounded to nearest class limit
e) ocular estimate
f) field assessment

E) Forest structure

Canopy cover
a) number of covered points out of a total of 24, plotted by on a 8*8 m grid and split in four equivalent sectors. The central point (not assessed) coincides with the sample plot centre. No slope correction
b) 20%. A minimum of 5 covered points has to be found in at least two different sectors
c) three classes (20-50%; 51-80%; >80%) based on the number of points covered (5-12; 13-20; >20)
d) no rounding rules
e) ocular estimate
f) field assessment

Age
a) measured in all forest types, including those with dominant trees mean height below 5 m.
b) no threshold
c) for every single tree species and type (shoot, standard, seedtree), one age for each diameter class (3-12; 13-22; 23-27; 28-32;....)
d) no rounding rules
e) local information about the time of last cut and of plantation; count out of wood rings; count out of branches for coniferous species
f) field assessment

Number of stools
a) Stools number in the sample plot.
b) no threshold
c) integer
d) stools on the border are counted as half
e) ocular estimate
f) field assessment

K) Miscellaneous

Survey timing
a) time of each single field survey stage
b) no threshold
c) minutes
d) no rounding rules
e) watch
f) field assessment

Sample plot number
a) UTM coordinates of the sample plot
b) no threshold
c) km
d) no rounding rules
e) map reading
f) map

9.1.4. Definitions for attributes on nominal or ordinal scale

A) Geographic regions

Mountain Community

a) administrative unit joining several close municipalities aimed at planning and management of mountain regions
b) nine units
c) map

Municipality

a) administrative unit
b) 94 municipalities
c) map

B) Ownership

Ownership

a) owner of the forest area under concern
b) 1) private ownership
 2) municipality or province
 3) state or regional
 4) other public agencies
c) local information, land register
d) not relevant

D) Site and soil

Aspect

a) direction of exposure of the slope measured in the sample plot centre
b) nine classes
c) field assessment
d) "all" is used when the slope is less than 5°.

Location

a) morphologic site
b) 1) slope
 2) top/ridge
 3) hollow
 4) plain
c) field assessment
d) not relevant

Stoniness

a) assessed as presence of superficial stones
b) 1) absent
 2) scanty
 3) abundant
c) field assessment
d) not relevant

Soil depth

a) average depth of the soil
b) 1) superficial
 2) little deep
 3) deep
c) field assessment
d) not relevant

Soil development stage

a) soil features
b) 1) poor
 2) medium
 3) high
c) field assessment
d) not relevant

Texture

a) assessment of the average soil grain dimension
b) 1) clayey
 2) medium
 3) sandy or silty
c) field assessment
d) not relevant

E) Forest structure

Forest type

a) based on dominant species
b) 15 categories
c) field assessment
d) not relevant

Silvicultural system and stand development

a) based on silvicultural system and on age related to rotation period
b) 17 categories
c) field assessment
d) not relevant

Tree species

a) Scientific, Italian name and acronym of selected list of 33 coniferous, 67 broadleaf tree species and 46 shrub species
b) 110 tree species and 46 shrub species
c) field assessment
d) not relevant

Stand description

a) short summary reporting on main stand attributes and on other attributes not directly assessed (forest condition, stem quality, regeneration, undergrowth,....)
b) no categories
c) ocular estimate
d) sole survey for stands with height of the dominant trees below 5 m

H) Accessibility and harvesting

Accessibility

a) assessed as the distance between the plot and the nearest road and taking into account the ground roughness of the soil
b) 1) good
   2) scanty
   3) poor
c) field assessment, map
d) not relevant

J) Non-wood goods and services

Forest function

a) main forest function
b) 1) wood production
   2) non-wood production
   3) direct protection
   4) indirect protection
   5) nature conservation
   6) recreation
c) field assessment
d) up to three different functions, quoted for decreasing relevance

Land use restrictions

a) if the sampling unit is located in an area with use restrictions
b) 14 different classes
c) local information, map
d) up to six different restrictions

9.1.5. Forest area definition and definition of "other wooded land"

Forest area has a tree crown cover of more than about 20% of the area and has trees usually growing to more than 5 m in height. The extension is at least 2,000 m² and the width is over 20 m.

If the plot is close to the forest edge and more than 50% of the 600 m² sample plot has a different land use (pasture or agricultural land, urban area, etc.) the plot is not assessed. On the contrary, the missing forest area has to be taken into account by doubling the tree attributes of an equivalent sector. The same procedures are used if the plot has two different forest types or stand development stages and in the case of non-forest inclusions wider than 100 m² or, for linear inclusions (roads, rivers, etc.), broader than 3 m.

9.2. DATA SOURCES

A) field data

Stand description: made on an interpretation area of 0.5 ha.

E) maps

- Italian Military Geographic Institute *(Istituto Geografico Militare Italiano)*; Provincial Land Office *(Ufficio Tecnico Erariale)*; Umbria Regional Authority
- 1977; 1977; 1977
- 1:25000; 1:10000; 1:10000
- analogue printed topographic map; analogue printed cadastral map; analogue printed ortophotomap

9.3. ASSESSMENT TECHNIQUES

9.3.1. Sampling frame

The sampling frame of the Umbria Regional Forest Inventory is given by the forest area definition. Field assessment is carried out for sampling units with shrub cover > 20%, too. Sampling units that in the field check show a cover below 20% are not assessed but the plot land use has to be stated (pasture or agricultural land, urban area, etc.). Such sampling units have to be listed. The entire Region is covered by the assessments. 49 plots (roughly 1.6 %) are not accessible due to the terrain conditions. The plot land-use should be given (forest, shrub, pasture, agricultural land....) as a result of ocular estimate. Such plots have to be listed.

9.3.2. Sampling units

Field plot: fixed area, concentric plot. Smaller circle of 200 m2 for dominant trees height between 5 and 10 m, medium circle of 400 m2 for dominant trees height between 10 and 15 m, larger circle of 600 m2 for dominant trees height >15 m and when the plot is split. Interpretation area for area related data (stand description) of 0.5 ha. Slope correction is done by increasing plot radii according to slope. All plots are permanently monumented and azimuth and distance of tallied trees from plot centre are recorded. A colour slide is taken of the plot centre. The plot centre is marked by a round metal plate inserted in a metal pipe hidden in the ground. The plots are permanent and distributed in a 1*1 km grid.

9.3.3. Sampling designs

The sampling design applied in the Umbria Regional Forest Inventory is a one-stage design. All the plots are preliminary interpreted on the aerial photos and a forest/ non-forest decision is made. All the forest plots, on a 1*1 km grid, are field assessed.

9.3.4. Techniques and methods for combination of data sources

Aerial photographs are used in order to take a forest/non-forest decision and for the drawing of the forest map. Attributes are directly and exclusively assessed on the field plots. Therefore there is no combination of data sources.

9.3.5. Sampling fraction

Data source and sampling unit	proportion of forested area covered by sample	represented mean area per sampling unit
field assessment, mean area sample plots	0.06	100 ha

9.3.6. Temporal aspects

Inventory cycle	time period of data assessment	publication of results	time period between assessments	reference date
1st URFI	1992 - 1993	1995		1993

9.3.7. Data capturing techniques in the field

Data recorded on tally sheets and edited by hand into the computer.

9.4. DATA STORAGE AND ANALYSIS

9.4.1. Data storage and ownership

The data of the assessment are stored in a data-base system at the Umbria Regional Authority Office - Forest Department in Perugia, Italy (Regione Umbria - Ufficio Foreste ed Economia Montana)

9.4.2. Database system used

The data are stored in a self-created ASCII file-system (IFRUMB93).

9.4.3. Data bank design

///

9.4.4. Update level

The data are updated to a common point in time (1993).

9.4.5. Description of statistical procedures used to analyse data including procedures for sampling error estimation

a) area estimation

The forest area is assessed on the field. A 1*1 km systematic point grid is laid over the assessed area and for each point a forest/non-forest decision is made. The estimation of forested area is done according to the following formulas:

$$p = \frac{a}{n}$$

$$v(p) = s_p^{\ 2} \approx \frac{pq}{n}$$

and the standard error is

$$S_p = \sqrt[2]{v(p)}$$

where

p = proportion of forest land
q = 1-p = proportion of non-forested land
v(p) = variance of p
s_p = standard error of p
n = total number of points on the point grid
a = number of forested point on the point grid

As a single forest point represents 100 ha of forest area, the total forested area A_w is estimated by multiplying the number, n_w of points by 100.

$$A_w = n_w 100 = n_w \frac{A}{n}$$

with variance $v(A_w)$ and standard errors $s(A_w)$

$$v(A_w) = A^2 s_p^2$$

$$s(A_w) = \sqrt[2]{v(A_w)}$$

b) aggregation of tree and plot data

The aggregation of single tree data is done by weighting each single tree attribute Y_{ij} by

$$w_{ij} = \frac{A}{a_{ij}} = \frac{10{,}000\text{m}^2}{600\text{m}^2} \text{ or } \frac{10{,}000}{400} \text{ or } \frac{10{,}000}{200}$$

where ij stands for tree i on plot j.

The plot expansion factor has not to be adjusted for plots not lying entirely in the forested area (plots at the forest edge) as already treated below under e. 'Sampling at the forest edge'.

Once the single tree attributes have been related to unit area by multiplication with the plot expansion factor, the total values for plot i, Y_i, can be calculated by summing the individual single tree attributes.

$$Y_i = \sum Y_{ij} w_{ij}$$

c) estimation of total values

In general the total value of attribute Y for a strata h (of area A_h) is given by summing the attribute value for each plot:

$$Y_h = 100 \sum Y_i / a_i = A_h \ddot{y}$$

where $\ddot{y}$ = mean value of total value

Since both, A_h and $\ddot{y}$, are affected respectively by the standard errors S_{ah} and $S_{\ddot{y}}$ we get an error propagation valuable with the expression

$$s(Y_h) = \sqrt{A_h^2 S_{ah}^2 + \ddot{y}^2 S\ddot{y}^2}$$

d) estimation of ratios

///

e) sampling at the forest edge

As each single forest plot is representative for 100 ha, the plot expansion factor has not to be adjusted for plots that do not fall entirely into forests. Adjustment is not needed as a consequence of the procedures reported in 9.1.5. about plots at the forest edge.

f) estimation of growth and growth components (mortality, cut, ingrowth)*

The mean annual increment has been calculated in each single plot and for every single tree species and type. This is the only attribute provided for the estimation of growth.

g) allocation of stand and area related data to single sample plots/ sample points.

The class of categorical, area related attributes, in which the plot centre is located, is assigned to the entire plot.

h) Hierarchy of analysis: how are sub-units treated?*

In the Umbria RFI results have to be presented for the entire Region, the two Provinces (Perugia and Terni) and for the Mountain Communities (9). This is done for metric attributes as well as for categorical (area related) attributes.

9.4.6. Software applied

A particular software was developed for the analysis and calculation procedure. The language used was BASIC.

9.4.7. Hardware applied

Personal computer.

9.4.8. Availability of data (raw and aggregated data)

Data not available

9.4.9. Sub-units (Strata) available

Results can be obtained for the entire Region, the two Provinces (Perugia and Terni) and for the Mountain Communities (9).

9.4.10. Links to other information sources

///

9.5. RELIABILITY OF DATA

9.5.1. Check assessment

a) 2-3% of the field plots are visited a second time by check crews. The check crew is a well trained and experienced crew, mainly dealing with check assessments, but not regular field work. The field plots to be checked are randomly selected form the list of already assessed plots. All attributes are assessed a second time independently from the assessment of the regular field crews, i.e. no data recorded by the field crews are available for the check crews. The measurement procedures for locating the plot centre are repeated.

b) The check assessments are used for the following purposes:

- feed-back to field crews to improve the data quality
- results are used for the interpretation of inventory results
- if the field crews and the check crews differ, the results of the check crews instead of the field crews are used for further analysis

9.5.2. Error budgets

///

9.5.3. Procedures for consistency checks of data

See 2.5.3.

9.6. MODELS

9.6.1. Volume functions

a) outline of the model

The mean stem volume has been calculated in each single plot and for every single tree species and type. Volume functions have been derived using mean diameter and mean height (from stand height curve). The mean stem volume is derived from the INFI two-entry volume tables.

b) overview of prediction errors

See 2.6.1 b)

c) data material for derivation of model

see 2.6.1 c)

d) methods applied to validate the model

See 2.6.1 d)

9.6.2. Assortments

Not assessed.

9.6.3. Growth components

a) outline of the model

The mean annual increment has been calculated in each single plot and for every single tree species and type.

b) overview over prediction errors

///

c) data material for the derivation

///

d) methods applied to validate the model

///

9.6.4. Potential yield

Not assessed.

9.6.5. Forest functions

///

9.6.6. Other models applied

Not assessed.

9.7. INVENTORY REPORTS

9.7.1. List of published reports and media for dissemination of inventory results

Inventory	Year of publication	Citation	Language	Dissemination
URFI	1995	Inventario Forestale Regionale. 1. Relazione. 1995: SAF. 74 p. 2. Allegati. 1995. 3. Tabelle di stima. 1995. 4. Manuale d'uso IFRUMB93.	Italian	printed

9.7.2. List of contents of latest report and update level

Reference of latest report:
SAF (Società Agricola e Forestale), 1995. Inventario Forestale Regionale. Relazione.

Update level of inventory data used in this report:
Field data and questionnaire: 1992-93
Aerial photography: 1977

Content:

1. Relazione

Chapter	Title	Number of pages
1	Introduction	1
2	Purpose of the forest inventory	1
3	Ecological frame	9
4	Sampling frame and methods	9
5	Definition of the sampling area	5
6	Data processing methods	6
7	Results	21
8	Major findings	2
	Annexes	20

9.7.3. Users of the results

Not directly assessed. Local administrations, agencies and bodies. Professionals.

9.8. FUTURE DEVELOPMENT AND IMPROVEMENT PLANS

///

9.9. MISCELLANEOUS

///

REFERENCES

S.A.F., Regione Umbria, 1993. Inventario forestale regionale. Relazione. Not published.

10. LAZIO REGIONAL FOREST INVENTORY
(INVENTARIO FORESTALE REGIONALE DEL LAZIO)

Foreword

So far, the only territory assessed is the Frosinone provincial territory. In 1987 there was the Inventory of the forest resources of the Simbruini Mountain Natural Park and of the Alta Tuscia Mountain Community. The methodology applied was slightly different from the one applied in the Frosinone Inventory which has to be considered as the improvement of the first one. Thus, all the coming information are related to the Frosinone Provincial Forest Inventory.

10.1. NOMENCLATURE

10.1.1. List of attributes directly assessed

A) Geographic regions

Attribute	Data source	Object	Measurement unit
Mountain Community *(Comunità Montana)*	map	sampling unit	categorical
Municipality *(Comune)*	map	sampling unit	categorical

B) Ownership

Attribute	Data source	Object	Measurement unit
Ownership *(Proprietà)*	local information, land register	sampling unit	categorical

C) Wood production

Attribute	Data source	Object	Measurement unit
Diameter at breast height *(Diametro a petto d'uomo)*	field assessment	tree	cm
Total height *(Altezza dendrometrica)*	field assessment	tree	dm

D) Site and soil

Attribute	Data source	Object	Measurement unit
Elevation *(Quota)*	field assessment	sampling unit	dam
Aspect *(Esposizione)*	field assessment	sampling unit	categorical

Slope *(Pendenza)*	field assessment	sampling unit	degree
Location *(Giacitura)*	field assessment	sampling unit	categorical
Stoniness *(Pietrosità)*	field assessment	sampling unit	categorical
Rock outcrops *(Affioramenti rocciosi)*	field assessment	sampling unit	percent
Soil depth *(Profondità del suolo)*	field assessment	sampling unit	categorical
Soil development stage *(Gradi di evoluzione del suolo)*	field assessment	sampling unit	categorical
Texture *(Tessitura del suolo)*	field assessment	sampling unit	categorical

E) Forest structure

Attribute	Data source	Object	Measurement unit
Forest type *(Tipo fisionomico)*	field assessment	sampling unit	categorical
Silvicultural system and stand development *(Stadio evolutivo)*	field assessment	sampling unit	categorical
Canopy cover *(Grado di copertura)*	field assessment	sampling unit	percent
Age *(Età)*	field assessment, local information	tree	year
Tree species *(Specie forestali)*	field assessment	sampling unit	categorical
Number of stools *(Numero delle ceppaie)*	field assessment	sampling unit	integer
Stand description *(Osservazioni selvicolturali)*	field assessment	stand (interpretation area)	categorical

H) Accessibility and harvesting

Attribute	Data source	Object	Measurement unit
Accessibility *(Accessibilità)*	field assessment, map	sampling unit	categorical

J) Non-wood goods and services

Attribute	Data source	Object	Measurement unit
Forest function *(Funzione prevalente)*	field assessment	stand	categorical
Land-use restrictions *(Vincoli)*	local information, map	stand	categorical

K) Miscellaneous

Attribute	Data source	Object	Measurement unit
Survey timing *(Tempi di lavoro)*	clock	///	minutes
Surveyor and firm name *(Rilevatore e Ditta)*	///	sampling unit	///
Topographic map *(Tavoletta IGM)*	map list	///	///
Sample plot number *(Punto UTM)*	map	sampling unit	geographic coordinates

10.1.2. List of derived attributes

C) Wood production

Attribute	Measurement unit	Input attributes
Top height *(Altezza dominante)*	m	h, d.b.h.
Number of stems per ha *(N°di piante per ettaro)*	integer	no. of stems, tree species and type
Basal area *(Area basimetrica)*	m^2	d.b.h., no. of stems, tree species and type
Mean diameter *(Diametro medio)*	cm	d.b.h., tree species and type
Mean height *(Altezza media)*	m	h, d.b.h., tree species and type
Form factor *(Coefficente di riduzione)*	integer	h, d.b.h., tree species and type, volume
Mean stem volume *(Volume dell'albero medio)*	m^3	h, d.b.h., tree species and type
Mean annual increment *(Incremento medio)*	m^3	volume, age, tree species and type

E) Forest structure

Attribute	Measurement unit	Input attributes
Dominant species *(Specie prevalente)*	categorical	basal area

10.1.3. Measurement rules for measurement attributes

C) Wood production

Diameter at breast height

a) Measured in 1.3m height above ground, on slopes measured from uphill side. One reading, for each tree, specifying the tree type (seedtree, shoot, coppice even-aged seedling, standard). Two readings for trees with d.b.h. over 17.5 cm and for irregular stems.
b) Minimum dbh: 2.5 cm
c) cm, recorded every single cm
d) alternatively rounded off and up
e) calliper
f) field assessment

Total height

a) Length of tree from ground level to top of tree. Total height for every tree species of 10% of the 3-12 cm diameter class trees; 50% of the 13-22 cm class; 100% of the >23.
b) no threshold
c) m
d) rounded to 0.5 m
e) Suunto, Haga, Bitterlich, Blume-Leiss hypsometer
f) field assessment

D) Site and soil

Elevation

a) measured in the plot centre as altitude
b) no threshold
c) m
d) rounded to 10 m
e) altimeter
f) field assessment

Slope

a) measured from the plot centre along the ruling gradient line, average of uphill and downhill values
b) 5°
c) degree
d) rounded to closest degree
e) clisimeter
f) field assessment

Rock outcrops

a) measured as a percent out of the sample plot area
b) no threshold
c) in 5%-classes
d) rounded to nearest class limit
e) ocular estimate
f) field assessment

E) Forest structure

Canopy cover

a) number of covered points out of a total of 24, plotted by on a 8*8 m grid and split in four equivalent sectors. The central point (not assessed) coincides with the sample plot centre. No slope correction
b) 20%. A minimum of five covered points has to be found in at least two different sectors
c) three classes (20-50%; 51-80%; >80%) based on the number of points covered (5-12; 13-20; >20)
d) no rounding rules
e) ocular estimate
f) field assessment

Age

a) measured in all forest types, including those with dominant trees mean height below 5 m.
b) no threshold
c) for every single tree species and type (shoot, standard, seedtree), one age for each diameter class (3-12; 13-22; 23-27; 28-32;....)
d) no rounding rules
e) local information about the time of last cut and of plantation; count out of wood rings; count out of branches for coniferous species
f) field assessment

Number of stools
a) Stools number in the sample plot.
b) no threshold
c) integer
d) stools on the border are counted as half
e) ocular estimate
f) field assessment

K) Miscellaneous

Survey timing
a) time of each single field survey stage
b) no threshold
c) minutes
d) no rounding rules
e) watch
f) field assessment

Sample plot number
a) UTM coordinates of the sample plot
b) no threshold
c) km
d) no rounding rules
e) map reading
f) map

10.1.4. Definitions for attributes on nominal or ordinal scale

A) Geographic regions

Mountain Community
a) administrative unit joining several close municipalities aimed at planning and management of mountain regions
b) data not available
c) map

Municipality
a) administrative unit
b) data not available
c) map

B) Ownership

Ownership
a) owner of the forest area under concern
b) 1) private ownership
2) municipality or province
3) state or regional
4) other public agencies
c) local information, land register
d) not relevant

D) Site and soil

Aspect
a) direction of exposure of the slope measured in the sample plot centre
b) nine classes
c) field assessment
d) "all" is used when the slope is less than 5°.

Location
a) morphologic site
b) 1) slope
2) top/ridge
3) hollow
4) plain
c) field assessment
d) not relevant

Stoniness
a) assessed as presence of superficial stones
b) 1) absent
2) scanty
3) abundant
c) field assessment
d) not relevant

Soil depth
a) average depth of the soil
b) 1) superficial
2) little deep
3) deep
c) field assessment
d) not relevant

Soil development stage
a) soil features
b) 1) poor
2) medium
3) high
c) field assessment
d) not relevant

Texture
a) assessment of the average soil grain dimension
b) 1) clayey
2) medium
3) sandy or silty
c) field assessment
d) not relevant

E) Forest structure

Forest type
a) based on dominant species
b) 15 categories
c) field assessment
d) not relevant

Silvicultural system and stand development
a) based on silvicultural system and on age related to rotation period
b) 17 categories
c) field assessment
d) not relevant

Tree species
a) Scientific, Italian name and acronym of selected list of 33 coniferous, 67 broadleaf tree species and 46 shrub species
b) 110 tree species and 46 shrub species
c) field assessment
d) not relevant

Stand description
a) short summary reporting on main stand attributes and on other attributes not directly assessed (forest condition, stem quality, regeneration, undergrowth, ...)
b) no categories
c) ocular estimate
d) sole survey for stands with height of the dominant trees below 5 m

H) Accessibility and harvesting

Accessibility
a) assessed as the distance between the plot and the nearest road and taking into account the ground roughness of the soil
b) 1) good
2) scanty
3) poor
c) field assessment, map
d) not relevant

J) Non-wood goods and services

Forest function
a) main forest function
b) 1) wood production
2) non-wood production
3) direct protection
4) indirect protection
5) nature conservation
6) recreation
c) field assessment
d) up to three different functions, quoted for decreasing relevance

Land use restrictions
a) if the sampling unit is located in an area with use restrictions
b) 14 different classes
c) local information, map
d) up to six different restrictions

10.1.5. Forest area definition and definition of "other wooded land

Forest area has a tree crown cover of more than about 20% of the area and has trees usually growing to more than 5 m in height. The extension is at least 2,000 m² and the width is over 20 m.

If the plot is close to the forest edge and more than 50% of the 600 m² sample plot has a different land use (pasture or agricultural land, urban area, etc.) the plot is not assessed. On the contrary, the missing forest area has to be taken into account by doubling the tree attributes of an equivalent sector. The same procedures are used if the plot has two different forest types or stand development stages and in the case of non-forest inclusions wider than 100 m² or for roads, rivers broader than 3 m.

10.2. DATA SOURCES

A) field data

Stand description: made on an interpretation area of 0.5 ha.

E) maps

- Italian Military Geographic Institute *(Istituto Geografico Militare Italiano)*; Provincial Land Office *(Ufficio Tecnico Erariale)*
- 1942/1957; 1960
- 1:25000; 1:2000
- analogue printed topographic map, analogue printed cadastral map

10.3. ASSESSMENT TECHNIQUES

10.3.1. Sampling frame

The sampling frame of the Lazio Regional Forest Inventory is given by the forest area definition. The assessment is made also for sampling units with a shrub cover > 20%. Sampling units that in the field check show a cover below 20% are not assessed but the plot land use has to be stated (pasture or agricultural land, urban area, etc.). Such sampling units have to be listed. Some plots (roughly 2%) are not accessible due to the terrain conditions. However the plot land use should be given (forest, shrub, pasture, agricultural land....) as a result of ocular estimate. Such plots have to be listed.

10.3.2. Sampling units

Field plot: concentric plot. Smaller circle of 200 m² for dominant trees height between 5 and 10 m, medium circle of 400 m² for dominant trees height between 10 and 15 m, larger circle of 600 m² for dominant trees height >15 m and when the plot is split. Interpretation area for area related data (stand description) of 0.5 ha. Slope correction is done by increasing plot radii according to slope. All plots are permanently monumented and azimuth and distance of tallied trees from plot centre are recorded. A colour slide is taken of the plot centre. The plot centre is marked by a round metal plate inserted in a metal pipe hidden in the ground. The plots are permanent and distributed in a 1*1 km grid.

10.3.3. Sampling designs

The sampling design applied in the Lazio Regional Forest Inventory is a one-stage design. All the plots are preliminary interpreted on the aerial photos and a forest/ non-forest decision is made. All the forest plots, on a 1*1 km grid, are field assessed.

10.3.4. Techniques and methods for combination of data sources

Aerial photographs are used in order to take a forest/non-forest decision and for a the drawing of the forest map. Attributes are directly and exclusively assessed on the plots. Therefore there is no combination of data sources.

10.3.5. Sampling fraction

Data source and sampling unit	proportion of forested area covered by sample	represented mean area per sampling unit
field assessment, mean area sample plots	0.06	100 ha

10.3.6. Temporal aspects

Inventory cycle	time period of data assessment	publication of results	time period between assessments	reference date
1st LRFI	1994	not published		not available

10.3.7. Data capturing techniques in the field

Data is recorded on tally sheets and edited by hand into the computer.

10.4. DATA STORAGE AND ANALYSIS

10.4.1. Data storage and ownership

The data of the assessment is not yet available.

10.4.2. Database system used

The data are stored in a self-created ASCII file-system (IFFRO94).

10.4.3. Data bank design

///

10.4.4. Update level

The data are not yet available

10.4.5. Description of statistical procedures used to analyse data including procedures for sampling error estimation

a) area estimation

The forest area is assessed on the field. A 1*1 km systematic point grid is laid over the assessed area and for each point a forest/non-forest decision is made. The estimation of forested area is done according to the following formulas:

$$p = \frac{a}{n}$$

$$v(p) = s_p^{\ 2} \approx \frac{pq}{n}$$

and the standard error is

$$S_p = \sqrt[2]{v(p)}$$

where
p = proportion of forest land
q = 1-p = proportion of non-forested land
$v(p)$ = variance of p
s_p = standard error of p
n = total number of points on the point grid
a = number of forested point on the point grid
As a single forest point represents 100 ha of forest area, the total forested area A_w is estimated by multiplying the number, n_w of points by 100.

$$A_w = n_w 100 = n_w \frac{A}{n}$$

with variance $v(A_w)$ and standard errors $s(A_w)$

$$v(A_w) = A^2 s_p^2$$

$$s(A_w) = \sqrt[2]{v(A_w)}$$

b) aggregation of tree and plot data

The aggregation of single tree data is done by weighting each single tree attribute Y_{ij} by

$$w_{ij} = \frac{A}{a_{ij}} = \frac{10{,}000m^2}{600m^2} \text{ or } \frac{10{,}000}{400} \text{ or } \frac{10{,}000}{200}$$

where ij stands for tree i on plot j.

The plot expansion factor has not to be adjusted for plots not lying entirely in the forested area (plots at the forest edge) as already treated below under e. 'Sampling at the forest edge'.

Once the single tree attributes have been related to unit area by multiplication with the plot expansion factor, the total values for plot i, Y_i, can be calculated by summing the individual single tree attributes.

$$Y_i = \sum Y_{ij} w_{ij}$$

c) estimation of total values

In general the total value of attribute Y for a strata h (of area A_h) is given by summing the attribute value for each plot:

$$Y_h = 100 \sum Y_i / a_i = A_h \ddot{y}$$

where $\ddot{y}$ = mean value of total value

Since both, A_h and $\ddot{y}$, are affected respectively by the standard errors S_{ah} and $S_{\ddot{y}}$ we get an error propagation valuable with the expression

$$s(Y_h) = \sqrt{A_h^2 S_{ah}^2 + \ddot{y}^2 S\ddot{y}^2}$$

d) estimation of ratios

///

e) sampling at the forest edge

As each single forest plot is representative for 100 ha, the plot expansion factor has not to be adjusted for plots that do not fall entirely into forests. Adjustment is not needed as a consequence of the procedures reported in 10.1.5. about plots at the forest edge.

f) estimation of growth and growth components (mortality, cut, ingrowth)*

The mean annual increment has been calculated in each single plot and for every single tree species and type. This is the only attribute provided for the estimation of growth.

g) allocation of stand and area related data to single sample plots/sample points.

The class of categorical, area related attributes, in which the plot centre is located, is assigned to the entire plot.

h) Hierarchy of analysis: how are sub-units treated?*

///

10.4.6. Software applied

A particular software has been developed for the analysis and calculation procedure. The language used is BASIC.

10.4.7. Hardware applied

Personal computer.

10.4.8. Availability of data (raw and aggregated data)

Data not yet available

10.4.9. Sub-units (Strata) available

///

10.4.10. Links to other information sources

Not known

10.5. RELIABILITY OF DATA

10.5.1. Check assessment

a) 2-3% of the field plots are visited a second time by check crews. The check crew is a well trained and experienced crew, mainly dealing with check assessments, but not regular field work. The field plots to be checked are randomly selected form the list of already assessed plots. All attributes are assessed a second time independently from the assessment of the regular field crews, i.e. no data recorded by the field crews are available for the check crews. The measurement procedures for locating the plot centre are repeated.

b) The check assessments are used for the following purposes:

- feed-back to field crews to improve the data quality
- results are used for the interpretation of inventory results
- if the field crews and the check crews differ, the results of the check crews instead of the field crews are used for further analysis

10.5.2. Error budgets

///

10.5.3. Procedures for consistency checks of data

See 2.5.3.

10.6. MODELS

10.6.1. Volume functions

a) outline of the model

The mean stem volume has been calculated in each single plot and for every single tree species and type. Volume functions have been derived using mean diameter and mean height (from stand height curve). The mean stem volume is derived from the INFI two-entry volume tables.

b) overview of prediction errors

See 2.6.1 b)

c) data material for derivation of model

See 2.6.1 c)

d) methods applied to validate the model

See 2.6.1 d)

10.6.2. Assortments

Not assessed.

10.6.3. Growth components

a) outline of the model

The mean annual increment has been calculated in each single plot and for every single tree species and type.

b) overview over prediction errors

///

c) data material for the derivation

///

d) methods applied to validate the model

///

10.6.4. Potential yield

Not assessed.

10.6.5. Forest functions

///

10.6.6. Other models applied

Not assessed.

10.7. INVENTORY REPORTS

10.7.1. List of published reports and media for dissemination of inventory results

Inventory	Year of publication	Citation	Language	Dissemination
LRFI	1988	Indagine conoscitiva sui boschi del Parco Naturale Regionale dei Monti Simbruini, 1988: SAF. Two volumes. 1. Relazione e allegati. 121 p. 2. Elaborati al calcolatore.	Italian	printed

10.7.2. List of contents of latest report and update level

Reference of latest report:
SAF (Società Agricola e Forestale), 1988. Indagine conoscitiva sui boschi del Parco Naturale dei Monti Simbruini.
1. Relazione e Allegati
2. Elaborati al calcolatore

Update level of inventory data used in this report:
Field data and questionnaire: 1987
Aerial photography: 1985-86

Content:

1. Relazione e Allegati

Chapter	Title	Number of pages
1	Introduction	2
2	Purpose, aim and methods of data assessment	22
3	Geographic, vegetation and ecological frame	6
4	Sampling frame and stratification	10
5	Data analysis on maps: forest units	12
6	Field assessment outcomes	31
7	Accessibility	9
8	Summary of site and soil figures	20
9	Forest management guidelines	7
10	Major findings	2

2. Elaborati al calcolatore

Chapter	Title	Number of pages
1	Introduction	
2	Preliminary notes	
3	General and summary tables	
4	Trees attributes summary tables	
5	Graphics	

10.7.3. Users of the results

Not directly assessed. Local administrations, agencies and bodies. Professionals.

10.8. FUTURE DEVELOPMENT AND IMPROVEMENT PLANS

///

10.9. MISCELLANEOUS

///

11. VALLE D'AOSTA REGIONAL FOREST INVENTORY

(INVENTARIO DELLE RISORSE FORESTALI E DEL TERRITORIO DELLA REGIONE AUTONOMA VALLE D'AOSTA)

11.1. NOMENCLATURE

11.1.1. List of attributes directly assessed

A) Geographic regions

Attribute	Data source	Object	Measurement unit
Mountain Community (*Comunità Montana)*	maps	sampling unit	categorical
Municipality *(Comune)*	maps	sampling unit	categorical

B) Ownership

Attribute	Data source	Object	Measurement unit
Ownership *(Proprietà)*	map, forest management plan maps	sampling unit	categorical

C) Wood production

Attribute	Data source	Object	Measurement unit
Diameter at breast height *(Diametro a petto d'uomo)*	field assessment	tree	cm
Total height *(Altezza dendrometrica)*	field assessment	tree	m
Tree type *(Origine della pianta)*	field assessment	tree	categorical
Radial increment *(Incremento diametrico)*	field assessment	tree (increment core)	mm

D) Site and soil

Attribute	Data source	Object	Measurement unit
Elevation *(Quota)*	map	sampling unit	m
Aspect *(Esposizione)*	field assessment	sampling unit	categorical
Slope *(Pendenza)*	field assessment	sampling unit	categorical

E) Forest structure

Attribute	Data source	Object	Measurement unit
Forest type *(Tipo di popolamento)*	field assessment, aerial photography	sampling unit	categorical
Silvicultural system *(Assetto evolutivo-colturale)*	field assessment, aerial photography	sampling unit	categorical
Stand development stage *(Stadio di sviluppo)*	field assessment	sampling unit	categorical
Crown canopy *(Copertura chiome)*	field assessment	sampling unit	percent
Age *(Età)*	field assessment,	tree (increment core)	integer
Tree species *(Rilievo specie)*	field assessment	sampling unit	categorical
Tending of stands: type *(Cure colturali: tipo)*	field assessment	sampling unit	categorical
Tending of stands: priority *(Cure colturali: priorità)*	field assessment	sampling unit	categorical

F) Regeneration

Attribute	Data source	Object	Measurement unit
Number of young trees *(Numero delle piantine)*	field assessment	sampling unit	integer

G) Forest condition

Attribute	Data Source	Object	Measurement unit
Number of dead trees *(Numero piante morte)*	field assessment	tree	integer
Predominant damage *(Danno prevalente)*	field assessment	tree	categorical
Damage intensity *(Intensità del danno)*	field assessment	tree	categorical
Defoliation and discolouration *(Defogliazione e ingiallimento)*	field assessment	tree	categorical
Grazing damage *(Danni da pascolamento)*	field assessment	tree	categorical

H) Accessibility and harvesting

Attribute	Data source	Object	Measurement unit
Accessibility *(Accessibilità)*	field assessment, map	sampling unit	categorical
Hauling system *(Sistema di esbosco)*	field assessment, map	sampling unit	categorical

J) Non-wood goods and services

Attribute	Data source	Object	Measurement unit
Forest function *(Destinazione prevalente)*	field assessment	stand	categorical

K) Miscellaneous

Attribute	Data source	Object	Measurement unit
Forest compartment number *(Numero di particella)*	Regional Forest Department	sampling unit	ordinal
Plot number *(Coordinate UTM dell'area di saggio)*	map	sampling unit	ordinal

11.1.2. List of derived attributes

C) Wood production

Attribute	Measurement unit	Input attributes
Basal area per hectare *(Area basimetrica per ettaro)*	m^2	d.b.h., number of stems, tree species
Top tree height *(Altezza dominante)*	m	d.b.h., total height, tree species
Number of stems per ha *(N°di piante per ettaro)*	integer	no. of stems, tree species and type
Growing stock per hectare *(Provvigione media per ettaro)*	m^3	no. of stems/ha, d.b.h., h, tree species

11.1.3. Measurement rules for measurable attributes

C) Wood production

Diameter at breast height

a) measured in 1.3 m height above ground, on slopes measured from uphill side. One reading with calliper pointing to plot centre. Border trees are alternatively measured.
b) minimum dbh: 7.5 cm
c) cm, recorded every single cm
d) rounded off and up to closest cm
e) calliper
f) field assessment

Total height

a) length of tree from ground level to top of tree. Measured in each plot on the dominant tree and on the closest tree to the plot centre.
b) no threshold
c) m
d) not available
e) Suunto clisimeter
f) field assessment

Radial increment

a) Measured in each plot on the dominant tree and on the closest tree to the plot centre.
b) data not available
c) data not available
d) data not available
e) ocular estimate on cores extracted with Pressler increment borer
f) field assessment

D) Site and soil

Elevation

a) data not available
b) no threshold
c) m
d) data not available
e) ocular estimate
f) 1:20000 topographic map, field assessment

E) Forest structure

Canopy cover

a) Crown covered area out of the total area of the sample plot
b) data not available
c) data not available
d) no rounding rules
e) ocular estimate
f) field assessment

Age

a) measured on increment cores counting increment rings. Measured in each plot on the dominant tree and on the closest tree to the plot centre.
b) no threshold
c) years
d) no rounding rules
e) ocular estimate on cores extracted with Pressler increment borer
f) field assessment

F) Regeneration

Number of young trees

a) assessed as the number of young trees with d.b.h. below measurement threshold
b) no threshold
c) integer
d) no rounding rules
e) ocular estimate
f) field assessment

G) Forest condition

Number of dead trees

a) assessed as the number of dead trees in the sample plot
b) no threshold
c) integer
d) no rounding rules
e) ocular estimate
f) field assessment

11.1.4. Definitions for attributes on nominal or ordinal scale

A) Geographic regions

Mountain Community
a) administrative unit joining several closed municipalities in the mountain area aimed at planning and managing mountain regions
b) eight Communities
c) map

Municipality
a) administrative unit
b) data not available
c) map

B) Ownership

Ownership
a) owner of the forest area under concern
b) 1) private ownership
2) public ownership
c) map

C) Wood production

Tree type
a) tree origin
b) shoot or seed tree
c) field assessment

D) Site and soil

Aspect
a) direction of exposure of the slope
b) nine categories
c) field assessment
d) remarks not available

Slope
a) along the ruling gradient line
b) data not available
c) field assessment
d) no remarks

E) Forest structure

Forest type
a) the forest types are defined on the basis of the main tree species
b) ten types
c) field assessment
d) no remarks

Silvicultural system
a) form of silvicultural management
b) six categories
c) field assessment
d) no remarks

Stand development stage
a) current stand development stage assessed on the basis of tree age and rotation period
b) six classes for even-aged high forest
four classes for coppice
two classes for uneven-aged or irregular high forest
c) field assessment
d) no remarks

Tending of stands: type
a) suggested silvicultural operation
b) 12 categories
c) field assessment
d) no remarks

Tending of stands: priority
a) urgency of the operation
b) three classes
c) field assessment
d) no remarks

G) Forest condition

Predominant damage
a) no definition
b) not available
c) field assessment
d) no remarks

Damage intensity
a) no definition
b) not available
c) field assessment
d) no remarks

Defoliation and discolouration
a) no definition
b) not available
c) field assessment
d) no remarks

Grazing damage
a) no definition
b) not available
c) field assessment
d) no remarks

H) Accessibility and harvesting

Accessibility
a) no definition
b) not available
c) field assessment
d) no remarks

Hauling system
a) no definition
b) not available
c) field assessment
d) no remarks

J) Non-wood goods and services

Forest function
a) main forest function
b) 1) wood production
2) protection
3) wood production/protection
4) nature conservation
5) recreation
6) natural development
c) field assessment
d) no remarks

K) Miscellaneous

Forest compartment number
a) number of the forest compartment as expressed in the current forest management plans
b) ordinal
c) regional forest department, maps
d) no remarks

Plot number
a) UTM coordinates of the sample plot
b) ordinal
c) km
d) no remarks

11.1.5. Forest area definition and definition of "Other Wooded Land"

Not available.

11.2. DATA SOURCES

C) aerial photography
- Regional Mapping Department
- 23*23 cm, colour photos,
- 1991
- the entire Region
- data not available
- data not available
- data not available

E) maps
- Regional Mapping Department;
- data not available
- 1:20000;
- analogue printed topographic map; analogue printed stock map; analogue printed cadastral maps; geologic map

11.3. ASSESSMENT TECHNIQUES

11.3.1. Sampling frame

The entire Region is covered by the assessment.

11.3.2. Sampling units

Field plot: fixed area (>200 m^2 <600 m^2) circular plot. 1648 temporary plots systematically distributed on a 500*500 m grid. The permanent plots are systematically distributed on a 1.5*1.5 km (any single plot represents 450 ha of forest area) which are permanently marked and azimuth and distance of tallied trees from plot centre recorded. In some Mountain Communities (one, seven and eight) field plots are alternatively assessed and each single plot represents 50 ha. Slope correction is done by increasing plot radii according to slope.

11.3.3. Sampling designs

The sampling design applied in the Valle d'Aosta RFI is characterised by a preliminary phase in which a forest/non-forest decision is made on the aerial photographs/map and land-use strata assigned to the photo plots. In the second phase forested plots are field assessed. The sampling intensity is different according to the relevance of the assessed areas. Nevertheless, plots are systematically distributed but on a different grid. The estimation of current values is based both on the permanent and temporary plots, the estimation of change will be based on the permanent plots only.

11.3.4. Techniques and methods for combination of data sources

Attributes for stratification are preliminary assessed on aerial photos and afterwards checked in field plots. Strata areas are calculated on the maps as plots represents respectively 25 ha, 50 ha (temporary plots) e 450 ha (permanent plots).

11.3.5. Sampling fraction

Data source and sampling unit	proportion of forested area covered by sample	represented mean area per sampling unit
field assessment, mean area sample plots	0.0024	25 ha

11.3.6. Temporal aspects

Inventory cycle	time period of data assessment	publication of results	time period between assessments	reference date
1st VdARFI	1991-94	not published	not quoted	not quoted

11.3.7. Data capturing techniques in the field

Data is recorded in the field by handheld computer (FW60) with immediate plausibility check.

11.4. DATA STORAGE AND ANALYSIS

11.4.1. Data storage and ownership

The data of the assessment are stored in a database system at the Forest Department of the Valle d'Aosta Regional Authority in Aosta.

11.4.2. Database system used

Data not available.

11.4.3. Data bank design

Data not available

11.4.4. Update level

Not quoted

11.4.5. Description of statistical procedures used to analyse data including procedures for sampling error estimation

Data not available

11.4.6. Software applied

Data not available

11.4.7. Hardware applied

Data not available

11.4.8. Availability of data (raw and aggregated data)

///

11.4.9. Sub-units (Strata) available

Results can be obtained for Mountain Communities and Municipalities.

11.4.10. Links to other information sources

///

11.5. RELIABILITY OF DATA

11.5.1. Check assessment

Data not available

11.5.2. Error budgets

Data not available

11.5.3. Procedures for consistency checks of data

Data not available

11.6. MODELS

11.6.1. Volume functions

a) outline of the model

Volume functions have been derived using d.b.h., and total height as input variables. Different volume tables for the Mountain Communities have been derived for dominant species, based on the height curves. The volume functions are applied for all trees.

b) overview of prediction errors

Not stated

c) data material for derivation of model

Not stated

d) methods applied to validate the model

Not stated

11.6.2. Assortments

Not assessed.

11.6.3. Growth components

Not available.

11.6.4. Potential yield

Not assessed.

11.6.5. Forest functions

Not assessed

11.6.6. Other models applied

////

11.7. INVENTORY REPORTS

11.7.1. List of published reports and media for dissemination of inventory results

Inventory	Year of publication	Citation	Language	Dissemination
VdARFI	not quoted	Inventario delle risorse forestali e del territorio regionale - Piano forestale delle proprieta private. Regione Autonoma Valle d'Aosta. Servizio Forestazione e Risorse naturali	Italian	printed

11.7.2. List of contents of latest report and update level

Reference of latest report:

Regione Autonoma Valle d'Aosta - Assessorato Agricoltura. Servizio Forestazione e Risorse naturali. Inventario delle risorse forestalli e del territorio regionale. Piano forestale delle proprietà private. Allegato n.1. - Metodologia adottata.

Update level of inventory data used in this report:
Field data: 1992-1994
Aerial photography: 1991

Content:

Chapter	Title	Number of pages
	Inventory frame	2
1	Regional territory land-use survey	6
2	Forest Inventory	18
3	Forest map and private forests management	1

11.7.3. Users of the results

Not directly assessed. Local administrations, agencies and bodies. Professionals.

11.8. FUTURE DEVELOPMENT AND IMPROVEMENT PLANS

///

12 ITALIAN NATIONAL INVENTORY OF FOREST DAMAGE
(INDAGINE SUL DEPERIMENTO FORESTALE - INDEFO)

12.1. NOMENCLATURE

12.1.1. List of attributes directly assessed

A) Geographic regions

Attribute	Data source	Object	Measurement unit
Region *(Regione)*	map	plot	categorical
Province *(Provincia)*	map	plot	categorical
Municipality *(Comune)*	map	plot	categorical

B) Ownership

Attribute	Data source	Object	Measurement unit
Ownership (*Proprietà)*	local information, cadastral	plot	categorical

D) Site and soil

Attribute	Data source	Object	Measurement unit
Elevation (*Altitudine*)	field (altimeter)	plot centre	10 m
Aspect (*Esposizione*)	field (compass)	plot	degree/categorical
Slope (*Pendenza*)	field (clisimeter)	plot	%/categorical
Site class fertility *(Fertilità)*	field	plot	categorical

E) Forest structure

Attribute	Data source	Object	Measurement unit
Wood type (*Tipo di bosco*)	field	plot	categorical
Type of stand (*Tipo di popolamento*	field	plot	categorical
Age of stand (*Età del soprassuolo*)	field	plot	year/categorical
Tree species *(Specie)*	field	single tree	categorical
Diameter at breast height (*Diametro a petto d'uomo*)	field	single tree	cm

G) Forest condition

Attribute	Data source	Object	Measurement unit
Tree damage: defoliation *(Defoliazione: intensità)*	field	single tree	categorical
Tree damage: decoloration *(Decolorazione: intensità)*	field	single tree	categorical
Tree damage: origin *(Causa dei danni)*	field	single tree	categorical

K) Miscellaneous

Attribute	Data source	Object	Measurement unit
IFNI plot code *(Codice punto IFNI)*	Forest Corps information	plot	categorical

12.1.2. List of derived attributes

G) Forest condition

Attribute	Measurement unit	Input attributes
Number of damaged trees *(Numero di alberi danneggiati)*	number	number of trees, tree species, tree damage types and classes
Percentage of damaged trees *(Percentuale di alberi danneggiati)*	%	number of trees, tree species, tree damage types and classes
Frequency of damage in main groups of species *(Ripartizione dei danni nei gruppi di specie principali)*	number/%	number of trees, tree species, tree damage types and classes

12.1.3 Measurements rules for measurable attributes

D) Site and Soil

Elevation

a) measured in plot centre
b) no threshold values
c) 10 m (31 categorical classes)
d) rounded to closest class limit
e) altimeter or reading of map
f) field assessment

Slope

a) The average slope of soil profile is measured on the maximum slope line through the plot centre, ranging two opposite stakes located uphill and downhill at 25 m distance from centre.
b) threshold values: 0% -
c) % (nine categorical classes)
d) rounded to closest 1% class limit
e) Suunto clinometer
f) field assessment

E) Forest structure

Diameter at breast height

a) Measured in 1.3 m height above ground, on slopes measured from uphill side; trees measured are not more than 30 and are selected by a centre-plot originated spiral
b) Minimum dbh: 12 cm
c) cm, recorded in 1 cm classes
d) rounded to the nearest (lower or upper) class limit
e) calliper
f) field assessment

Age of stands

a) This attribute is assessed only in even aged stands
b) no threshold
c) year (seven categorical classes)
d) no rounded
e) Pressler borer or ocular estimate
f) field assessment

12.1.4 Definitions for attributes on nominal or ordinal scale

A) Geographic regions

Region

a) The Region of the plot-centre area
b) 20 normal or autonomous Regions
c) local forest service and cadastral register
d) not relevant

Province

a) The Province of the plot-centre area
b) 95 Provinces
c) local forest service and cadastral register
d) not relevant

Municipality

a) The Municipality of the plot-centre area
b) all the Italian Municipality
c) local forest service and cadastral register
d) not relevant

IFNI plot code

a) The IFNI plot code of plot under concern
b) all the wooded IFNI point
c) local forest service
d) not relevant

B) Ownership

Ownership

a) The owner of the forest area under concern
b) three classes
c) local forest service and cadastral documents
d) not relevant

D) Site and soil

Site class fertility

a) Assessment of soil fertility
b) three classes
c) field assessment
d) not relevant

E) Forest structure

Site class fertility

a) Assessment of wood structure
b) six types of wood structure
c) field assessment
d) not relevant

Type of stand

a) Assessment made considering system and crop type
b) three types of stand
c) field assessment
d) assessment made only in wood classified high forest

Tree species

a) Scientific name (code) of selected list of 31 coniferous and 59 broadleaf tree species and two additional classes (other coniferous and other broadleaf species).
b) 90 tree species
c) field assessment
d) not relevant

G) Forest condition

Tree damage: defoliation

a) Synthetic assessment of tree defoliation intensity
b) four classes:
c) field assessment
d) not relevant

Tree damage: de-coloration

a) Synthetic assessment of tree de-coloration intensity
b) four classes:
c) field assessment
d) not relevant

Tree damage: origin

a) Assessed the main origin of damage for assessed trees
b) eight classes
c) field assessment
d) not relevant

12.1.5. Forest area definition and definition of "other wooded land"

These definitions are not relevant because the INDEFO survey is done only on wooded IFNI points.

12.2. DATA SOURCES

A) Field data

Type of stand: assessed on a visual area of approximate 0.5 ha.

12.3. ASSESSMENT TECHNIQUES

12.3.1. Sampling frame

The sampling frame of the INDEFO is given by the superimposition of the IFNI's grid: in fact, the entire country is covered by the assessment : the sample plots of INDEFO are on grid of 9*9 km and are assessed only the wooded one. The other land-use sample plots are excluded from the assessment.

12.3.2. Sampling units

Field assessment:

The plot are systematically distributed in the IFNI's grid in which measured trees are not more than 30 and are selected by a centre-plot originated spiral.

All plots are permanently monumented and documented (but marked hidden without giving an evident impact to the plot location).

12.3.3. Sampling designs

The sampling design applied in the INDEFO is one phase inventory with only a ground assessment of systematically distributed field plots: only the wooded IFNI plot are assessed.

12.3.4. Techniques and methods for combination of data sources

Since a one phase inventory is applied, every data are collected or derive from field assessment. Therefore no combination of data sources is applied.

12.3.5. Sampling fraction

Data are not area related and results are given as number of assessed trees.

12.3.6. Temporal aspects

Inventory cycle	time period of data assessment	publication of results	time period between assessment	reference date
INDEFO	1984-1995	yearly	1 year	summer of each year

12.3.7. Data capturing techniques in the field

Data are recorded on paper forms filled.

12.4. DATA STORAGE AND ANALYSIS

12.4.1. Data storage and ownership

The data of the assessment are stored in an Oracle data-base system in SIAN (*Sistema Informativo Agricolo Nazionale*) by the Direzione Generale delle Risorse Forestali, Montane e Idriche of Ministero delle Risorse Agricole, Forestali e Alimentari, Via Carducci 5, Roma. Dr. Fornaroli, Tel. +39-6-4665-7221, should be considered as contact person for information.

12.4.2. Database system used

The data are stored in a relational ORACLE database.

12.4.3. Data bank design

//

12.4.4. Update level

The reference date is each summer of the assessed year. The data assessed are used as they are without any specific updating.

12.4.5. Description of statistical procedures used to analyse data including procedures for sampling error estimation

Not available

12.4.6. Software applied

//

12.4.7. Hardware applied

//

12.4.8. Availability of data (raw and aggregated data)

The INDEFO data are stored in a relational database in PC. Requests of raw data and derived data by outside persons have to be addressed to the secretary's office of SIAN. Data requested by public institutions are normally available without particular costs if the purpose for which the data will be used is specified. In any case a regulation of data-concession is going to be made.

12.4.9. Sub-units (strata) available

Results are obtained for entire Italy.

12.4.10. Links to other information sources

//

12.5. RELIABILITY OF DATA

12.5.1. Check assessment

Internal

12.5.2. Error budgets

Not available

12.5.3. Procedures for consistency checks of data

Not available

12.6. MODELS

There are not models applied

12.7. INVENTORY REPORTS

12.7.1. List of published reports and media for dissemination of inventory results

Year of publication	Citation	Language	Dissemination
1984	Ministry of Agriculture - National Forestry Corp: *Indagine sul deperimento forestale - INDEFO I - anno 1984: Sintesi dei risultati*	Italian	printed
1985	Ministry of Agriculture - National Forestry Corp: *Indagine sul deperimento forestale - INDEFO II - anno 1985: Sintesi dei risultati*	Italian	printed
1986	Ministry of Agriculture - National Forestry Corp: *Indagine sul deperimento forestale - INDEFO III - anno 1986: Sintesi dei risultati*	Italian	printed
1987	Ministry of Agriculture - National Forestry Corp: *Indagine sul deperimento forestale - INDEFO IV - anno 1987: Sintesi dei risultati*	Italian	printed
1988	Ministry of Agriculture - National Forestry Corp: *Indagine sul deperimento forestale - INDEFO V - anno 1988: Sintesi dei risultati*	Italian	printed
1989	Ministry of Agriculture - National Forestry Corp: *Indagine sul deperimento forestale - INDEFO VI - anno 1989: Sintesi dei risultati*	Italian	printed
1990	Ministry of Agriculture - National Forestry Corp: *Indagine sul deperimento forestale - INDEFO VII - anno 1990: Sintesi dei risultati*	Italian	printed
1991	Ministry of Agriculture - National Forestry Corp: *Indagine sul deperimento forestale - INDEFO VIII - anno 1991: Sintesi dei risultati*	Italian	printed
1992	Ministry of Agriculture - National Forestry Corp: *Indagine sul deperimento forestale - INDEFO IX - anno 1992: Sintesi dei risultati*	Italian	printed

1993	Ministry of Agriculture - National Forestry Corp: *Indagine sul deperimento forestale - INDEFO X - anno 1993: Sintesi dei risultati*	Italian	printed
1994	Ministry of Agriculture - National Forestry Corp: *Indagine sul deperimento forestale - INDEFO XI - anno 1994: Sintesi dei risultati*	Italian	printed
1995	Ministry of Agriculture - National Forestry Corp: *Indagine sul deperimento forestale - INDEFO XII - anno 1995: Sintesi dei risultati*	Italian	printed

12.7.2. List of contents of report and update level

Reference of latest report:
Ministry of Agriculture - National Forestry Corp: Indagine sul deperimento forestale - INDEFO XII - anno 1995: Sintesi dei risultati. p. 27

Update level of inventory data used in this report:
Field data: 1995

Content:

Chapter	Title	Number of pages
1	Foreword	2
2	Trees health status at national level	6
3	Trees health status related to site attributes	8
4	Damage origin at national level	11

12.7.3. Users of the results

//

12.8. FUTURE DEVELOPMENT AND IMPROVEMENT PLANS

The methodology is going to change in 1996, reducing the number of sample plots.

12.9. MISCELLANEOUS

//

COUNTRY REPORT FOR LIECHTENSTEIN

Michael Köhl
Swiss Federal Institute for Forest, Snow and Landscape Research
Switzerland

with close cooperation of:
Felix Näscher
Norman Nigsch
Office for Forests, Nature and Landscape
Liechtenstein

Summary

1. Overview

The total area of Liechtenstein comprises 16,000 ha, of which 7,372 ha (46 %) are covered by forests. The main tree species are spruce, fir, pine and beech. The first National Forest Inventory (NFI) of Liechtenstein was conducted in 1986, the results were presented in 1988. The second inventory cycle will start in 1998.

The political mandate for Liechtenstein's NFI is with the Office for forest, nature and landscape *(Amt für Wald, Natur und Landschaft)*, Vaduz.

Since 1984 the condition of forests is monitored on the national level. The assessments are done according to the ICP regulations. In addition to the NFI, other inventories mainly concerning forest functions and natural protection are carried out.

2.1 Nomenclature

More than 90 attributes are directly assessed in field surveys, by aerial photo- interpretation, by enquiries or are taken from maps. Roughly 20 attributes are used for the identification of trees, plots, etc., roughly 90 attributes provide information which is used for the analysis or as input for derivations. The list of derived attributes contains at the moment little more than 25 attributes, which are mostly related to timber volume, assortments and accessibility.

The forest area decision is done in the field and follows the Swiss definition. The forest area definition was made up of the two attributes width of the forested patch and its crown closure. A functional relationship between these two attributes was set up for the forest/non-forest decision. The minimum crown closure is between 100 and 20 % cover, the minimum width is between 25 and 50m. Except for afforestations, young growth, mountain alder and mountain pine, the minimum top height of a forested patch has to be more than 3m to be classified as forest.

2.2 Data sources

The NFI of Liechtenstein utilises the following data sources: field assessments, enquiries and maps in printed format.

2.3 Assessment technigues

The sampling frame of Liechtenstein's National Forest Inventory is given by the forest area definition. The entire country is covered by the assessment.

The sampling design applied in Liechtenstein's NFI is a one-phase systematic sample. The sampling unit is a concentric circular plot. 362 field plots were assessed. The data was captured on tally sheets and later digitised. The period of the assessment covers one year (1986).

2.4 Data storage and analysis

The data of Liechtenstein's NFI are stored at the Swiss Federal Institute for Forest, Snow and Landscape Research, Birmensdorf, Switzerland. They were analysed with specially developed FORTRAN routines, which were also used to analyse the first Swiss NFI. The statistical analysis is done by a simple random sampling approach. The data is available with restrictions. Thematic and geographic sub-units can be formed with pleasure.

2.5 Reliability of data

Intensive training and check assessments are key elements to ensure the reliability of data. 15% of the field plots were visited a second time by check groups. The results of check assessments, however, were not used to correct data.

2.6 Inventory reports

The first inventory report was issued 1988. It is an internal report that is not published. The report contains five chapters, the content focuses mainly on the productive function of forests. The main user of the report is the Office for Forests, Nature and Landscape.

2.7 Future development and improvement plans

The second cycle of Liechtenstein's NFI will probably start in 1998. The methodology will follow the second Swiss NFI.

1. OVERVIEW

The total area of Liechtenstein comprises 16,000 ha, of which 7,372 ha (46%) are forested land. Out of the 7,372 ha of forested land, 6,866 ha are stocked with forests, 506 ha are rocky areas, boulders or erosion areas but are considered to be part of the forested land. 92% of the forests are publicly owned, 8% are privately owned. The average size of a stand (smallest management unit) is roughly 1 ha. The population of Liechtenstein amounts to 30,000 inhabitants, which results in slightly more than 4 inhabitants per ha of forested land (see Table 1).

Table 1. Statistics of Liechtenstein.

Forest area [ha]	Total area [ha]	Forest area [%]	Volume [m^3ha^{-1}]	Privately owned forest [%]	Inhabitants per ha forested land
7372	16000	46	387	8	4.07

source: forest functions mapping

The main tree species are spruce (*Picea abies* (L.) Karst.), fir (*Abies alba* Mill.), pine (*Pinus sylvestris* L.) and beech (*Fagus sylvatica* L.), which comprise 86 percent of all trees in Liechtenstein. The exact tree species proportions can be seen in Figure 1.

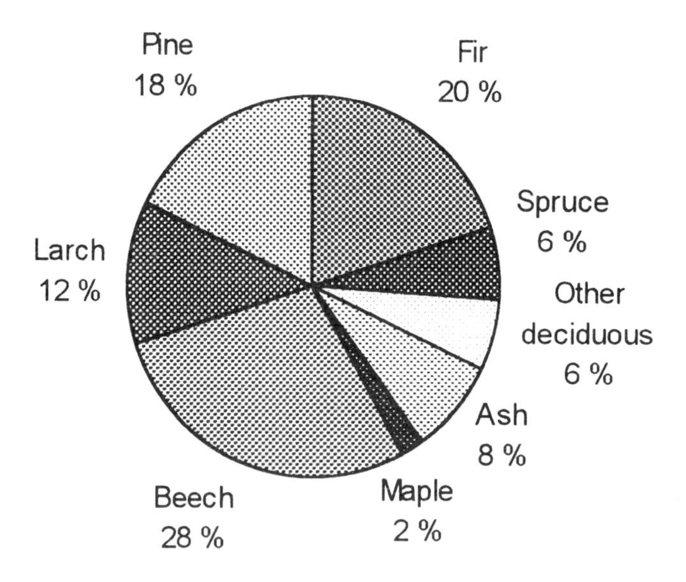

Figure 1. Tree species proportions (percent of standing volume) in Liechtenstein.

The average volume per hectare is 333 m^3ha^{-1} for entire Liechtenstein. At the moment no sample based results on increment are available. All results presented in this section have been analysed by the Swiss Federal Institute for Forest, Snow and Landscape Research, if not stated otherwise.

The NFI of Liechtenstein was conducted in close connection with the first Swiss NFI. The inventory method and the nomenclature were taken over from the Swiss NFI, the data analysis was carried out by the Swiss Federal Institute for Forest, Snow and Landscape Research in Birmensdorf, Switzerland. Liechtenstein will join the Swiss efforts to carry out a second NFI in 1998. Thus, the first and only completed inventory cycle is the first NFI of Liechtenstein.

Besides the NFI several other assessments are carried out in Liechtenstein.

1) The forest functions are mapped (*Waldfunktionenkartierung*) with a total area coverage. The data are stored in a GIS and used for forest management planning and nature conservation purposes. The mapping of forest functions started in the early 1980s.
2) Since 1992 damages by wildlife are monitored with 120 fenced plots (*Verbisskontrollzäune*). The plots are re-measured every four (lower altitudes) to six (higher altitudes) years.
3) An inventory of areas in which nature and landscape protection has priority has been conducted in 1990 and 1991 (*Naturvorrangflächen*). The inventory of those areas was done with total area coverage. Four inventories have been drawn up: (a) revised inventory of biotops in open areas *(revidiertes Biotopinventar im offenen Grünlandbereich)*, (b) natural protection areas in forested areas *(Naturschutzinventar im Waldareal)*, (c) inventory of protected landscapes *(Landschaftsschutzinventar)* and (d) inventory of natural monuments *(Inventar der Naturdenkmäler)*.
4) Mapping of plant societies *(pflanzensoziologische Kartierung)* in a scale of 1:5000 has been completed.

Forest condition is monitored in Liechtenstein since 1984 by three different programs.

1) On 15 to 20 permanent observation plots *(Dauerbeobachtungsflächen)*, each of which is stocked with about 100 trees, crown transparency and discoloration is assessed according to the ICP guidelines. The size of the plots ranges from 0.5 to 1 ha. Only spruce and fir from the upper layer are assessed. Some of the initially selected plots had to be replaced due to natural damages such as storm damage. No utilisation takes place on these plots. The natural losses of fir are considerably higher than the losses of spruce resulting in a shift of the tree species distribution on those permanent observation plots.
2) In 1986 and 1992 crown transparency and discoloration was assessed on 362 sample plots. Those are the same plots which were assessed for Liechtenstein's NFI. The assessment was done according to the ICP regulations.
3) Since 1984 forest condition is monitored by the interpretation of CIR aerial photographs. Each year special flights are conducted. Until 1986 the entire forested area of Liechtenstein was covered by aerial photographs (CIR, 1:9,000). In 1986 the total coverage was abandoned. After a transition period in which five locations were assessed a sector of Liechtenstein is annually covered by CIR aerial photographs (1:3,000) and interpreted.

1.2 OTHER IMPORTANT FOREST STATISTICS

1.2.1 Other forest data and statistics on the national level

Topic and spatial structure	Period	Responsibility	Data source	Availability
Statistical Yearbook (*Statistisches Jahrbuch*) (forest statistics, figures on tree species distribution, afforestations, timber utilisation, wildlife etc.)	annually	Office for forest, nature and landscape (forestry figures)	Accounting of forest enterprises, annual reports of foresters	available without restrictions
Operational accounting of forest enterprises (*Forstliche Betriebsabrechnung*)	annually	Office for forest, nature and landscape	Accounting, work reports, project reports	available without restrictions

1.2.2 Delivery of the statistics to UN and Community institutions

The statistics and information presented under 1.2.1. is available at the national and international level. Special investigations and analyses are done by various institutions upon request. Thus it is difficult to assess how and in which international institutions information on Liechtenstein forestry is utilised. As Liechtenstein is not a member of the EU, the delivery to community institutions is rather limited.

1.2.2.1 Responsibilities for international assessments.

a) FAO/ECE for the 1990 Forest Resource Assessment (FRA 1990)

Dr. Felix Näscher, Office for forest, nature and landscape (Amt für Wald, Natur und Landschaft), St. Florinsgasse 3, FL-9490 Vaduz, Tel: +41-75-2366400, Fax: +41-75-2366411

b) EUROSTAT

Data presented in the statistical yearbook are provided.

c) OECD (Environment Compendium)

Does not apply

d) to any other intergovernmental enquiries

Statistics in the scope of the Helsinki process in the working party "Protection of forests".

1.2.2.2 Data compilation.

No specific data compilation is conducted. Available figures are provided.

2 NATIONAL FOREST INVENTORY OF LIECHTENSTEIN

2.1 NOMENCLATURE

2.1.1 List of attributes directly assessed

A) Geographic region

Attribute	Data source	Object	Measurement unit
Number of topographic map 1:25000 *(Landeskartennummer 1:25000)*	map	field plot	Number
Number of topographic map 1:50000 *(Landeskartennummer 1:50000)*	map	field plot	Number
Number of topographic map 1:100000 *(Landeskartenummer 1:100000)*	map	field plot	Number
Forest district *(Forstkreis)*	map	field plot	Number
Community *(Gemeindenummer)*	map	field plot	Number
Forest/non-forest decision *(Wald/Nichtwald Entscheid)*	field assessment	field plot	Code

B) Ownership

Attribute	Data source	Object	Measurement unit
Ownership *(Eigentum)*	enquiry	field plot	Code
Superfizies/different owner of land and trees *(Superfizies)*	enquiry	field plot	Code

C) Wood production

Attribute	Data source	Object	Measurement unit
Diameter at breast height *(Brusthoehendurchmesser)*	field assessment	tree	cm
Upper stem diameter in 7m height (*Durchmesser in 7m Hoehe)*	field assessment	tree	cm
Tree height *(Baumhoehe)*	field assessment	tree	m
Shape of tree crown *(Kronenform)*	field assessment	tree	Code
Crown class *(Kronenklasse)*	field assessment	tree	Code
Crown length *(Kronenlaenge)*	field assessment	tree	Code
Kind of assortment *(Traemel/Langholz)*	enquiry	field plot	Code
Layer to which sample tree belongs *(Schicht)*	field assessment	tree	Code
Remarks on special features of sample tree *(Bemerkung)*	field assessment	tree	Code

D) Site and soil

Attribute	Data source	Object	Measurement unit
Azimuth of aspect *(Azimut der Exposition)*	field assessment	field plot	gon
Aspect *(Exposition)*	field assessment	photo plot	Code
Can aspect be determined? *(Exposition bestimmbar)*	field assessment	field plot	Code
Relief *(Relief)*	field assessment	field plot	Code
Relief *(Relief)*	field assessment	photo plot	Code
Soil characteristic depth *(Bodeneigenschaft Gruendigkeit)*	map	field plot	Code
Soil characteristic nutrient content *(Bodeneigenschaft Naehrstoffspeicher)*	map	field plot	Code
Soil characteristic skeleton content *(Bodeneigenschaft Skelettgehalt)*	map	field plot	Code

Soil characteristic soil moisture *(Bodeneigenschaft Vernaessung)*	map	field plot	Code
Soil characteristic water storage capacity *(Bodeneigenschaft Wasserspeicher-faehigkeit).*	map	field plot	Code
Soil characteristic risk of rock/soil slides *(Rutschungs-/Sackungsgefahr)*	map	field plot	Code

E) Forest structure

Attribute	Data source	Object	Measurement unit
Method for age determination *(Altersbestimmungsmethode)*	field assessment	tree	Code
Number of year rings *(Anzahl Jahringe)*	field assessment	tree	Number
Age of stand *(Bestandesalter)*	field assessment	field plot	Years
Tree species (including shrubs) *(Baumart)*	field assessment	tree	Code
Stand edge *(Bestandesgrenze)*	field assessment	field plot	Code
Stage of forest development *(Entwicklungsstufe)*	field assessment	field plot	Code
Mixture proportion (coniferous/ broadleaf) *(Mischungsgrad)*	field assessment	field plot	Code
Crown closure *(Schlussgrad)*	field assessment	field plot	Code
Closure *(Deckungsgrad)*	field assessment	field plot	Code
Social position *(Soziale Stellung)*	field assessment	tree	Code
Stand structure (vertical layers) *(Bestandesstruktur)*	field assessment	field plot	Code
Planting in advance *(Vorbau)*	field assessment	field plot	Code
Origin and management type of forest *(Waldform)*	field assessment	field plot	Code
Forest type *(Waldtyp)*	field assessment	field plot	Code

F) Regeneration

Attribute	Data source	Object	Measurement unit
Class of young growth (<1.3m, >1.3m) *(Jungwaldgruppe)*	field assessment	regeneration plot	Code
Class (height/d.b.h.) of regeneration *(Jungwaldklasse)*	field assessment	regeneration plot	Code
Tree and bush species on regeneration plot *(Pflanzenart)*	field assessment	tree/ seedling	Code
Species count on regeneration plot *(Pflanzenzählung)*	field assessment	regeneration plot	Number
Closure of regeneration (>0.1m height, <12 cm d.b.h.) *(Verjuengungs-Deckungsgrad)*	field assessment	field plot	Percent
Closure of regeneration on regeneration plot *(Gesamtdeckungsgrad)*	field assessment	regeneration plot	Percent
Protection of regeneration *(Schutz)*	field assessment	regeneration plot	Code
Type of regeneration *(Verjungungsart)*	field assessment	regeneration plot	Code
Condition of regeneration *(Gesundheitszustand)*	field assessment	regeneration plot	Code

G) Forest condition

Attribute	Data source	Object	Measurement unit
Stand stability *(Bestandesstabilitaet)*	field assessment	field plot	Code
Fire *(Brandspuren)*	field assessment	field plot	Code
Erosion caused by water *(Erosion durch Wasser)*	field assessment	field plot	Code
Slides *(Rutschungen)*	field assessment	field plot	Code
Tree damages *(Schaeden)*	field assessment	tree	Code
Location of damage *(Schadenort)*	field assessment	tree	Code

Size and type of damage *(Schadenbild-Groesse)*	field assessment	tree	Code
Cause of damage *(Schadenursache)*	field assessment	tree	Code
Damage caused by snow *(Schneespuren)*	field assessment	field plot	Code
Damage caused by rockfall *(Steinschlag)*	field assessment	field plot	Code
Pasture *(Beweidung)*	field assessment	field plot	Code
Intensity of grazing *(Beweidungsintensitaet)*	field assessment	field plot	Code
Heavy utilisation and disturbances *(Ueberbelastung und Stoerungen)*	field assessment	field plot	Code

H) Accessibility, harvesting and management

Attribute	Data source	Object	Measurement unit
Distance for timber extraction *(Rueckedistanz)*	enquiry	field plot	m
Management operation *(Nutzungsart)*	enquiry	field plot	Code
Utilisation class *(Nutzungskategorie)*	field assessment	field plot	Code
Kind of next silvicultural treatment *(Eingriffsart)*	enquiry	field plot	Code
Need for management operation *(Eingriffsdringlichkeit)*	field assessment	field plot	Code
Tools for timber harvest *(Art der Baumernte)*	enquiry	field plot	Code
Carrying out of timber harvest *(Ausfuehrung der Holzernte)*	enquiry	field plot	Code
Constraints for timber harvest *(Einschraenkungen fuer die Holzhauerei)*	field assessment	field plot	Code
Number of years since last cut *(Anzahl Jahre seit letzter Nutzung)*	enquiry	field plot	Years

Timber extraction method *(Rueckemittel)*	enquiry	field plot	Code
Kind of management plan *(Planungsgrundlage)*	enquiry	field plot	Code
Year when last management plan was created *(Jahrzahl der aktuellen Planungsgrundlagen)*	enquiry	field plot	Year
Direction of timber transport *(Rueckerichtung)*	enquiry	field plot	Code
Constraints for extraction method *(Einschraenkung fuer die (Rueckemittelwahl)*	field assessment	field plot	Code
Length of cable line *(Laenge der Seillinie)*	field assessment	field plot	m
Place to which timber is skidded after cut *(Rueckeziel)*	enquiry	field plot	Code
Accessibility *(Zugaenglichkeit)*	field assessment	field plot	Code
Proportion of unregulated fellings *(Anteil Zwangsnutzung)*	enquiry	field plot	Percent
Cause of unregulated fellings *(Ursache der Zwangsnutzung)*	enquiry	field plot	Code

I) Attributes describing forest ecosystems

Attribute	Data source	Object	Measurement unit
Dead trees *(Duerrstaender)*	field assessment	field plot	Code
Forest edge present *(Waldrand)*	field assessment	field plot	Code

J) Non-wood goods and services (NWGS), Forest functions other than production

Attribute	Data source	Object	Measurement unit
Type of gaps *(Luecken)*	field assessment	field plot	Code
Special situation (national park etc.) *(Besondere Verhaeltnisse)*	field assessment	map	Code

K) Miscellaneous

Attribute	Data source	Object	Measurement unit
Azimuth of single tree *(Azimut)*	field assessment	tree	gon
Date and time of assessment *(Aufnahmedatum und Zeit der Messung)*	field assessment	-	Date
Tree identification number *(Baumnummer)*	field assessment	tree	Number
Date of enquiry *(Datum der Umfrage)*	enquiry	-	Date
Date of field assessment *(Feldaufnahmedatum)*	field assessment	-	Date
Declination of compass *(Deklination)*	field assessment	-	gon
Distance of tree from plot centre *(Distanz)*	field assessment	tree	m
Identification of field group *(Aufnahmegruppennummer)*	field assessment	field plot	Number
Inventory identification *(Inventurnummer)*	field assessment	tree	Number
Plot radius of 500 m^2 field plot (horizontal: 12.62m) *(Probeflaechenradius gross)*	field assessment	field plot	m
Plot radius of 200 m^2 field plot (horizontal: 7.98m) *(Probeflaechenradius klein)*	field assessment	field plot	m
Number of stereo pair *(Nummer des Luftbildes (Landestop.)*	list	photo-plot	Number
Year of flight *(Flugjahr)*	list	photo-plot	Year
Number of flight line *(Nummer der Fluglinie)*	list	photo-plot	Number

2.1.2 List of derived attributes

Many attributes are not directly assessed but are derived using directly assessed attributes as input variables. This list contains all attributes, which are used for the presentation of the inventory results. Excluded are those attributes, which are used only in the analysis process and are not referred later on, e.g. dummy variables, weights, index values in databases, etc.

A) Geographic region

does not apply

B) Ownership

Attribute	Measurement unit	Input attributes
Ownership public/ private *Eigentumskategorien oeffentlich/ privat*	Code	ownership

C) Wood production

Attribute	Measurement unit	Input attributes
Proportion of basal area *(Basalflaechenanteil der Baeume mit BHD>50)*	percent	d.b.h., plot expansion factor
Single tree volume (*Einzelbaumvolumen*)	m^3	d.b.h., diameter in 7m height, total tree height, tree species
Type of assortment *(SORTART)*	Code	
Class of Assortment *(SORTKLD)*	Code	
Estimated volume of assortments without bark *(SORTVD)*	m^3	d.b.h., estimated d7, estimated tree height
Tariff volume of bole over bark *(TV)*	m^3	d.b.h., h_{dom}, site index, altitude
Estimated bole volume without bark *(VoRD)*	m^3	d.h.b., estimated d7, estimated tree height
Basal area increment *(BAI)*	cm^2	d.b.h., site index, estimated age of stand
Volume increment *(VI)*	m^3	d.b.h., site index, estimated age of stand
Estimated volume of branches (diameter ≥7cm) *(VAST)*	dm^3	d.b.h., estimated d7, estimated tree height, altitude, crown length
Estimated volume of twigs (diameter <7cm) *(VZWEIG)*	dm^3	d.b.h., estimated d7, estimated tree height, altitude, crown length

D) Site and soil

Attribute	Measurement unit	Input attributes
Acidity according to classification by Keller *(Aciditaet nach Keller)*	Code	4 geographic regions defined according to forest plant societies by Keller, geo-technical units on geo-technical map
Derived aspect *(Abgeleitete Exposition)*	Code	aspect
Site Index according to Keller *(GWL nach Keller)*	Code	4 geographic regions defined according to forest plant societies by Keller, soil acidity, geology (Trias, Lias, Limestone), aspect, relief, altitude
Site index for beech according to Keller *(Oberhoehenbonitaet fuer Buche nach Keller)*	m	site index, tree species
Site index for spruce according to Keller *(Oberhoehenbonitaet fuer Fichte nach Keller)*	m	site index, tree species
Site index for pine according to Keller *(Oberhoehenbonitaet fuer Kiefer nach Keller)*	m	site index, tree species
Site index for larch according to Keller *(Oberhoehenbonitaet fuer Laerche nach Keller)*	m	site index, tree species
Site index for fir according to Keller *(Oberhoehenbonitaet fuer Tanne nach Keller)*	m	site index, tree species

E) Forest structure

Does not apply

F) Regeneration

Does not apply

G) Forest condition

Does not apply

H) Accessibility, harvesting and management

Attribute	Measurement Unit	Input attributes
Accessibility for transport vehicles *Bodeneigenschaft Befahrbarkeit*	Code	soil type taken from geo-technical map
Period of utilisation *Abgeleiteter Nutzungszeitraum*	Code	if no utilisation assessed in 2nd NFI, period between assessments and period of utilisation as assessed in 1st NFI

I) Attributes describing forest ecosystem

Does not apply

J) Non-wood goods and services (NWGS)

Does not apply

K) Miscellaneous

Attribute	Measurement unit	Input attributes
Bifurcation *(Zwiesel)*	Code	remarks
Tree expansion factor for trees with d.b.h.≥16cm *(Repraesentierte Stammzahl mit BHD ≥ 16)*	number	d.b.h., plot expansion factor
Tree expansion factor for trees with d.b.h.≥40cm *(Repraesentierte Stammzahl mit BHD ≥ 40)*	number	d.b.h., plot expansion factor
Plot expansion factor *(repräsentierte Stammzahl)*	number	d.b.h., horizontal area of plot, slope, forest edge coordinates

2.1.3 Measurement rules for measurable attributes

The measurement rules for measurable attributes applied in Liechtenstein's NFI were taken over from the first Swiss NFI. A description can be found in the country report for Switzerland.

2.1.4 Definitions for attributes on nominal or ordinal scale

The definitions for attributes on nominal or ordinal scales applied in Liechtenstein's NFI were taken over from the first Swiss NFI. A description can be found in the country report for Switzerland.

2.1.5 Forest area definition and definition of "other wooded land"

The definitions for forest area and other wooded land applied in Liechtenstein's NFI were taken over from the first Swiss NFI. A description can be found in the country report for Switzerland. In contrast to the Swiss NFI the forest/non-forest decision was not made in aerial photographs but on the ground.

2.2 DATA SOURCES

In this section the data sources utilised in the forest resource assessments will be specified.

A) field data

The data in the field are assessed by sample plots. The sample plot design is described in detail in Chapter 2.3. No additional field data are assessed.

B) questionnaire

The questionnaire is designed by the national forest inventory department. For each forested plot a questionnaire has to be filled out. The field crews visit local foresters and fill out the questionnaire with their help. Thus non-responses do not occur. A copy of the questionnaire can be found in the field manual (Stierlin et al, 1993)

C) aerial photography

No aerial photography was applied.

D) spaceborne or airborne digital remote sensing

No spaceborne or airborne digital remote sensing techniques have been applied.

E) maps

The following maps are used:

- Topographic map (printed, 1:25,000), provided by Swiss Federal Office for Topography (*Bundesamt für Landestopographie*), Wabern, map sheets updated and printed between 1984 and 1992.

- Map of forest (*Waldpläne*), 1:5000, as used by local forest service.
- Forest management plans (*Wirtschaftspläne*), hand-drawn maps showing the stages of forest development

F) other geo-referenced data

Does not apply

2.3 ASSESSMENT TECHNIQUES

2.3.1 Sampling frame

The sampling frame of Liechtenstein's National Forest Inventory is given by the forest area definition. The entire country is covered by the assessment.

2.3.2 Sampling units

Field plot

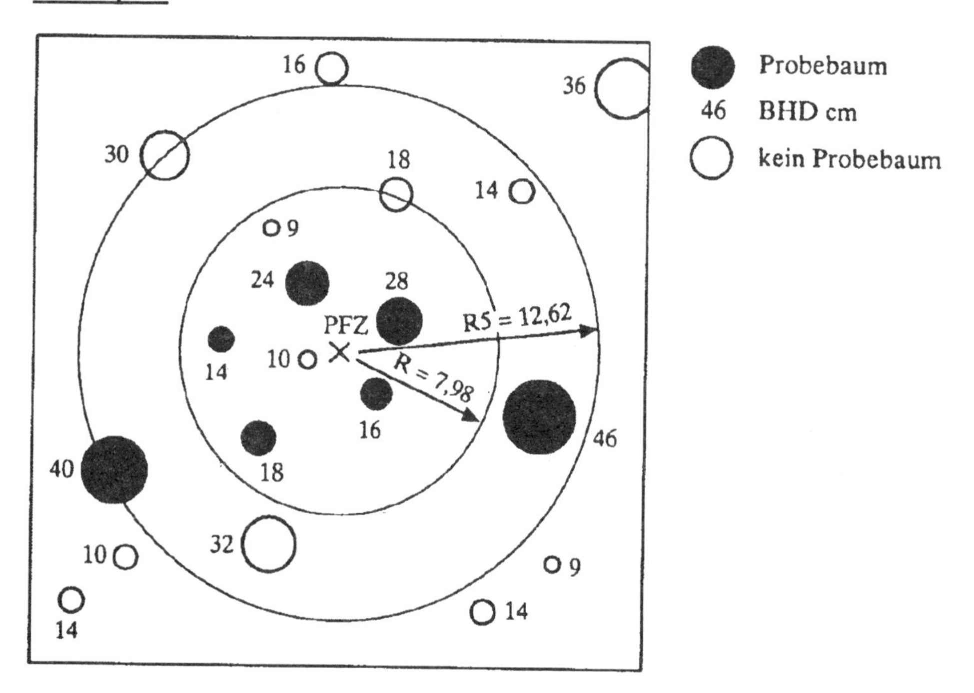

Figure 2. Field plot of the Liechtenstein's NFI (Probebaum=sample tree, BHD=d.b.h., kein Probebaum= no sample tree, PFZ= plot centre).

Concentric fixed area, circular plot. Smaller circle of 200 m² for trees with d.b.h. ≥ 12cm, larger circle of 500 m² for trees with d.b.h. >35cm. The plot-radii are corrected for slope. Interpretation area for area related data of 50*50m. All plots are permanently monumented and azimuth and distance of tallied trees from plot centre are recorded.

Regeneration plots

In the plot centre a circle with 3m radius is used for the assessment of the regeneration. For plants between 4 and 12 cm d.b.h. and plants smaller than 30 cm the number of plants on the regeneration plot is recorded. For plants with height between 0.3 and 1.3 m and plants with d.b.h. below 4 cm 30 plants are counted (starting from north and counting clockwise) and the azimuth of the 30th plant is recorded.

Aerial photo-plot

does not apply

2.3.3 Sampling designs

The sampling design applied in Liechtenstein's NFI is a one-phase systematic sample.

2.3.4 Techniques and methods for combination of data sources

Does not apply

2.3.5 Sampling fraction

Table 2. Sampling fraction per data source

Data source and sampling unit	Proportion of forested area covered by sample	Represented mean area per sampling unit
field assessment, permanent fixed area sample plots	0.0025	12.5 ha

2.3.6 Temporal aspects

Table 3. Inventory cycles

Inventory cycle	Time period of data assessment	Publication of results	Time period between assessments	Reference date
1st NFI	1986	1988		none

2.3.7 Data capturing techniques in the field

In Liechtenstein's NFI data were recorded on tally sheets and edited by hand into the computer.

2.4 DATA STORAGE AND ANALYSIS

2.4.1 Data storage and ownership

The data of the assessment are stored in a datafile system at the Swiss Federal Institute for Forest, Snow and Landscape Research (WSL) in Birmensdorf, Switzerland. The owner of the data is the Office for Forest, Nature and Landscape, Vaduz (Dr. Felix Näscher, Office for Forest, Nature and Landscape *(Amt für Wald, Natur und Landschaft)*, St. Florinsgasse 3, FL-9490 Vaduz, Tel: +41-75-2366400, Fax: +41-75-2366411. The submission of is possible upon request. Any publication of special analyses based on NFI-data has to be approved by the Office for Forest, Nature and Landscape.

2.4.2 Data base system used

The data were stored on a mainframe computer in a file system. No database system was used.

2.4.3 Data bank design (if applicable)

Does not apply.

2.4.4 Update level

For each record the date, when the data have been assessed, is available. The data are not updated to a common point in time and thus reflect the situation at the time of assessment. However, due to the short assessment period periodocity is not a problem.

2.4.5 Description of statistical procedures used to analyse data Including Procedures for sampling error estimation

a) area estimation

The forest area and forest area proportion is assessed on field plots. A systematic grid is laid over the entire country and for each point a forest/non-forest decision is made. The estimation of the proportion of forested area is done according to Cochran (1977):

$p = a/n$

$v(p) = s_p^2 \ pq/n$

$s_p = \sqrt[2]{v(p)}$

where:

p = proportion of forest land

q = 1-p = proportion of non-forested land

v(p) = variance of p

s_p = standard error of p

n = total number of points on the point grid

a = number of forested points on the point grid

The total forested area A_w is estimated by multiplying the total area of the entire country or sub-units, A, by the proportion of forested land, p.

$$A_W = n_W a = n_W \frac{A}{n} = Ap$$

with variance $v(A_w)$ and standard errors $s(A_w)$

$$v(A_w) = A^2 s_p^2$$

$$s(A_w) = \sqrt[2]{v(A_w)}$$

b) aggregation of tree and plot data

The aggregation of single tree data is done by weighting each single tree attribute Y_{ij} by

$w_{ij} = A/ a_{ij} = 1ha/a_{id}$

where ij stands for tree i on plot j.

As concentric sample plots are used the weighting factors (plot expansion factors), w_{ij}, have to be different for the 0.05 ha plot and the 0.02 ha plot to reflect the correct selection probability of a single tree. For plots lying entirely in a forested area the weights are constant.

$w_{ij} = w_{0.05} = A/a = 1ha/ 0.05\ ha = 20$, for trees standing on the 0.05 ha plot i

$w_{ij} = w_{0.02} = A/a = 1ha/ 0.02\ ha = 50$, for trees standing on the 0.02 ha plot i

As trees can be assigned to the two concentric plots regarding their d.b.h., the plot expansion factor can be given in relation to the d.b.h. of single trees.

$w_{ij} = w_{0.05} = A/a = 1ha/ 0.05\ ha = 20$, for trees with d.b.h. > 35cm

$w_{ij} = w_{0.02} = A/a = 1ha/ 0.02\ ha = 50$, for trees with 12cm d.b.h. 35 cm

The plot expansion factor has to be adjusted for plots not lying entirely in the forested area (plots at the forest edge). The procedure is treated below under e)'sampling at the forest edge'.

Once the single tree attributes have been related to unit area by multiplication with the plot expansion factor, the total values for plot i, Y_i, can be calculated by summing the individual single tree attributes.

$$Y_i = \Sigma Y_{ij} w_{ij}$$

c) estimation of total values

Total values and their variances were analysed by applying the general Simple Random Sampling estimators.

Total values, $\hat{Y}$ and their variances v($\hat{Y}$) are calculated as follows:

$$\hat{Y} = \hat{A}\hat{\bar{Y}} = \hat{A}\frac{\sum Y_i}{n}$$

$$v(\hat{Y}) = \hat{A}^2 v(\hat{\bar{Y}}) = \hat{A}^2 \frac{\sum_{i=1}^{n} \left(Y_i - \hat{\bar{Y}}\right)^2}{n}$$

where

$\hat{A}$ = estimated forested area

Y_i = observation on plot i

n = number of plots

The variance of the area estimation was not taken into account when calculating total values.

d) estimation of ratios

Ratios are obtained by calculating the ratio of means or total values, $\hat{R}$.

$$\hat{R} = \frac{\hat{\bar{Y}}}{\hat{\bar{X}}} = \frac{\hat{Y}}{\hat{X}}$$

$$v\left(\hat{R}\right) = \hat{R}^2 \left\{ \frac{v\left(\hat{\bar{X}}\right)}{\hat{\bar{X}}^2} + \frac{v\left(\hat{\bar{Y}}\right)}{\hat{\bar{Y}}^2} - 2\frac{S_{YX}}{n\hat{\bar{X}}\hat{\bar{Y}}} \right\}$$

where:

$$\hat{\bar{Y}} = \sum_{i=1}^{n} \frac{Y_i}{n}$$

$$\hat{Y} = \sum_{i=1}^{n} Y_i$$

$$\hat{\bar{X}} = \sum_{i=1}^{n} \frac{X_i}{n}$$

$$\hat{X} = \sum_{i=1}^{n} X_i$$

s_{YX} = covariance term

n = number of observations

e) sampling at the forest edge

As mentioned under b) 'aggregation of tree and plot data', the plot expansion factor has to be adjusted for plots that do not fall entirely into forests, but extend partially into non forested areas. The selection probability of a single tree, p_i, has to be taken into account for the estimation process. In Liechtenstein's NFI a tree concentric method is applied, by adjusting the plot expansion factor, w_{ij}, for each tree tallied on a plot at the forest margin.

f) estimation of growth and growth components (mortality, cut, in-growth)

As Liechtenstein's NFI was carried out only once, growth and growth components have not been estimated.

g) allocation of stand and area related data to single sample plots/sample points.

The class of categorical, area related attributes, in which the plot centre is located, is assigned to the entire plot. The only exception occurs for plots at the forest margin, where the exact plot area located within the forest is reported.

h) Hierarchy of analysis: how are sub-units treated?

Tree values are aggregated on the plot level. Plot related data (aggregated tree data and area related data) are used as input (observation) for the Simple Random Sampling estimators and ratio estimators. No regional results are presented.

2.4.6 Software applied

The analysis software of the first Swiss NFI was applied. It was a special development under FORTRAN 77, running on a mainframe computer.

2.4.7 Hardware applied

The data were stored in a file system and the analysis program ran on a CDC 6500 an a Cyber 174 under the operation system EMOS.

2.4.8 Availability of data (raw and aggregated data)

All raw data and derived attributes are stored in a file system. The ownership is with the Liechtenstein Forest Service.

2.4.9 Sub-units (strata) available

Results are available for entire Liechtenstein.

2.4.10 Links to other information sources

does not apply

2.5 RELIABILITY OF DATA

2.5.1 Check assessments

a) the way how check assessments are organised and carried out

Field plots

About 53 of the 362 of field plots are visited a second time by check crews. The field plots to be checked are randomly selected form the list of already assessed plots. All attributes are assessed a second time independently from the assessment of the regular field crews, i.e., no data recorded by the field crews are available for the check crews. The measurement procedures for locating the plot centre are not repeated. The field crews do not know in advance, which plots will be checked. The check crews visit the plots after the field crews have assessed them. The organisation and the analysis of the check assessments is done at the main office.

Questionnaire

No check assessments are conducted.

Analysis of check assessments

All data assessed by both field crews and check crews are analysed. For measured variables the mean difference deviation between field and individual check crews and its standard deviation is calculated and visually represented in graphs. Attributes on nominal or ordinal scale are presented in contingency tables with the classes for field crews and check crews as rows and columns, respectively.

b) Use of the results of check assessments in the inventory system

The check assessments are used for the following purposes:

- feed-back to field crews to improve the data quality
- interpretation of inventory results
- the results are not used to set up error budgets and to quantify the total error
- the results will not be used to correct any raw data or results.

2.5.2 Error budgets

No error budgets have been constructed for Liechtenstein. However, the main findings for the Swiss NFI apply.

2.5.3 Procedures for consistency checks of data

Data assessed in the field

The data were filled in tally sheets and digitised at the WSL. Before the data were released for analysis they were checked for consistency and manually corrected.

Questionnaire

No checks.

2.6 MODELS

See Swiss NFI. No models to estimate growth components were applied. Special volume tariffs were derived for Liechtenstein.

2.7 INVENTORY REPORTS

2.7.1 List of published reports and media for dissemination of inventory results

Table 4: Inventory reports

Inventory	Year of publication	Citation	Language	Dissemination
First NFI of Liechtenstein	1988	Xeroxed, unpublished report	German	internal report, copied

2.7.2 List of contents of latest report and update level

Reference of latest report:

Does not apply as the internal report was not printed and published.

Update level of inventory data used in this report:

Field data and questionnaire: 1986

Content:

The content of the inventory report is presented in Table 5 based on the German publication. For the French version the page numbers are slightly different.

Table 5. Content of inventory report

Chapter	Title	Number of pages
1	Introduction	4
2	Methods of data assessment and data analysis	3
3.1	Number of sampling units and forest area	4
3.2	Forest sites	11
3.3	Forest structure	6
3.4	Key statistics	8
3.5	Growth and yield	10
3.6	Forest condition	16
3.7	Regeneration	13
3.8	Accessibility and harvesting	9
3.9	Management and utilisation	11

2.7.3 Users of the results

- Office for Forest, Nature and Landscape
- Use for public relations activities

2.8 FUTURE DEVELOPMENT AND IMPROVEMENT PLANS

2.8.1 Next inventory period

A second NFI is planed for 1998. The assessment method will follow the second Swiss NFI. The stratification might not be made by the use of aerial photography but by maps.

2.8.2 Expected or planned changes

2.8.2.1 Nomenclature

Additional attributes focusing on the assessment of non-wood goods and services, especially protective function, forest diversity issues and nature protection, will be included.

2.8.2.2 Data sources

The data sources of the first and second NFI will be maintained. In addition other geo-referenced data will be analysed by means of a GIS. The forest functions will be transferred from the forest functions mapping, which is available in GIS-format. It is not yet decided if aerial photography will be used.

2.8.2.3 Assessment techniques

The methods of the second Swiss NFI will be modified and applied.

2.8.2.4 Data storage and analysis

The data will be stored in a data-base system both at the Office for Forest, Nature and Landscape in Vaduz and the Swiss Federal Institute for Forest, Snow and Landscape Research, Birmensdorf. The analysis software of the second NFI will be applied.

2.8.2.5 Reliability of data

Check assessments will be carried out to guarantee the data quality of the field assessments.

2.8.2.6 Models

Additional model for the quantification of biodiversity, biomass, naturalness, carbon cycle, habitat classification and others will be included

2.8.2.7 Inventory reports

Not yet decided.

2.8.2.8 Other forestry data

The integration of other national data on forestry and the NFI is planned.

2.9 MISCELLANEOUS

The second NFI of Liechtenstein will be conducted in close co-operation with the Swiss NFI.